Lecture Notes in Artificial Intelligence 2275

Subseries of Lecture Notes in Computer Science
Edited by J. G. Carbonell and J. Siekmann

Lecture Notes in Computer Science
Edited by G. Goos, J. Hartmanis and J. van Leeuwen

Nikhil R. Pal Michio Sugeno (Eds.)

Advances in Soft Computing – AFSS 2002

2002 AFSS International Conference on Fuzzy Systems
Calcutta, India, February 3-6, 2002
Proceedings

Series Editors

Jaime G. Carbonell,Carnegie Mellon University, Pittsburgh, PA, USA
Jörg Siekmann, University of Saarland, Saarbrücken, Germany

Volume Editors

Nikhil R. Pal
Electronics and Communication Sciences Unit
Indian Statistical Institute
203 B. T. Road, Calcutta, 700108 India
E-mail: nikhil@isical.ac.in

Michio Sugeno
Brain Science Institute, RIKEN
2-1 Hirosawa, Wako, Japan
E-mail: msgn@brain.riken.go.jp

Cataloging-in-Publication Data applied for

Die Deutsche Bibliothek - CIP-Einheitsaufnahme

Advances in soft computing : proceedings / AFSS 2002, 2002 AFSS International Conference on Fuzzy Systems, Calcutta, India, February 3 - 6, 2002. Nikhil R. Pal ; Michio Sugeno (ed.). - Berlin ; Heidelberg ; New York ; Barcelona ; Hong Kong ; London ; Milan ; Paris ; Tokyo : Springer, 2002
(Lecture notes in computer science ; Vol. 2275 : Lecture notes in artificial intelligence)
ISBN 3-540-43150-0

CR Subject Classification (1998): I.2, I.5.1

ISSN 0302-9743
ISBN 3-540-43150-0 Springer-Verlag Berlin Heidelberg New York

Springer-Verlag Berlin Heidelberg New York
a member of BertelsmannSpringer Science+Business Media GmbH

http://www.springer.de

Printed in Germany

Typesetting: Camera-ready by author, data conversion by Olgun Computergrafik
Printed on acid-free paper SPIN 10846199 06/3142 5 4 3 2 1 0

Preface

It is our great pleasure to welcome you all to the *2002 AFSS International Conference on Fuzzy Systems (AFSS 2002)* to be held in Calcutta, the great *City of Joy*. AFSS 2002 is the fifth conference in the series initiated by the Asian Fuzzy Systems Society (AFSS). AFSS 2002 is jointly being organized by the Indian Statistical Institute (ISI) and Jadavpur University (JU). Like previous conferences in this series, we are sure, AFSS 2002 will provide a forum for fruitful interaction and exchange of ideas between the participants from all over the globe. The present conference covers all major facets of soft computing such as fuzzy logic, neural networks, genetic algorithms including both theories and applications. We hope this meeting will be enjoyable academically and otherwise.

We are thankful to the members of the International Program Committee and the Area Chairs for extending their support in various forms to make a strong technical program. Each submitted paper was reviewed by at least three referees, and in some cases the revised versions were again checked by the referees. As a result of this tough screening process we could select only about 50% of the submitted papers. We again express our sincere thanks to all referees for doing a great job. We are happy to note that 19 different countries from all over the globe are represented by the authors, thereby making it a truly international conference. We are proud to have a list of distinguished speakers including Profs. Z. Pawlak, J. Bezdek, D. Dubois, and T. Yamakawa.

We are thankful to the Asian Fuzzy Systems Society and its members, and in particular, to Prof. M. Mukaidono and Prof. Z. Bien, who have extended their cooperation in many forms in spite of their busy schedule. We are grateful to the co-sponsoring societies including IFSA, ISFUMIP (India), SOFT (Japan), CFSAT (Taiwan), FMFSAC (China), the World Federation of Soft Computing, and other international affiliates of AFSS.

We are grateful to Prof. S.B. Rao, former Director, ISI and Prof. S.C. Som, former Vice-Chancellor, JU, for their active help in initiating this conference. Thanks are also due to Prof. A.N. Basu, Vice-Chancellor, JU and Prof. K.B. Sinha, Director, ISI, who have taken special interest on many occasions to help the organizer in many ways and continuously supported us in making this conference a reality. Thanks are due to the Finance Chair, Prof. R. Bandyopadhyay, and the Tutorial Chair, Prof. M.K. Chakraborty for organizing an excellent tutorial program.

We would like to express our sincere thanks to the members of the Organizing Committee for their whole hearted support. Special mention must be made of the organizing Co-chairs, Prof. D. Patranabis and Prof. J. Das, and the organizing coordinators, Dr. R.K. Mudi and Dr. S. Raha and our colleague Dr. Srimanta Pal for their initiative, cooperation, and leading roles in organizing the conference. We would also like to express our special thanks to Prof. Bimal Roy for his help and support. We gratefully acknowledge the help of Prof. S. Sanyal, Prof. K. Ray,

Dr. D.P. Mandal, and Ms. T. Pal. The staff members of the Electronics and Communication Sciences Unit of ISI have done a great job and we express our thanks to them. We are also grateful to the Computer and Statistical Services Center, ISI, for its continuous support. Things will remain incomplete unless we mention two names, Mr. D. Chakraborty and Mr. P. Mohanta without whose help, it would have been impossible for us to make this conference a successful one. We must have missed many other colleagues and friends who have helped us in many ways. We express our thanks to them also.

We gratefully acknowledge the financial support provided by different organizations as listed below. Without their support it would have been impossible to hold this conference on this scale.

Last, but surely not the least, we express our sincere thanks to Mr. Alfred Hofmann of Springer-Verlag for his excellent support in bringing out these proceedings on time.

December 2001

Michio Sugeno
Nikhil R. Pal

Funding Organizations

- All India Council of Technical Education
- The Council for Scientific and Industrial Research, India
- Zen & Art, USA
- Reserve Bank of India
- Avesta Computer Services Limited, USA
- MCC PTA India Corporation Private Limited
- Dept. of Higher Education, Govt. of West Bengal, India
- Dept. of Science & Technology, India
- Indian Space Research Organization
- Defence Research & Development Organization, India
- Indian Statistical Institute
- Jadavpur University

Message from the Organizing Co-chairs

With great pleasure we extend a very warm welcome to all delegates and participants at the International Conference on Fuzzy Systems, AFSS 2002, and wish an excellent and fruitful program for everyone. This conference is the first of its kind to be held in Calcutta, the heart of the culture and economy of eastern India. We hope this conference provides an excellent opportunity for the academic fraternity and industry personnel in the related field to interact on the state of the art in fuzzy logic and other soft computing technologies. We are pleased to bring out the proceedings of the AFSS 2002 containing 73 papers including 5 plenary talks and a few invited ones. We are thankful to the Program Chair, Prof. M. Sugeno, for doing an excellent job in selecting quality papers.

Soft Computing has acquired a huge dimension in recent years percolating through to almost all strata of life. Starting from navigational systems to health care, identification problem to control of domestic appliances, process control to load despatch – in fact, all known areas of social and techno-economic growth have been aptly supported by the technique in which fuzzy logic has been the predominant tool / factor. It is, therefore, quite appropriate that such an international symposium on soft computing be held in this part of the continent and be repeated as frequently as possible.

The symposium has received support from the experts in the area from all over the world and participation of them actively and in person has been of great significance considering the worldwide disturbance presently prevailing.

The organizing committee extends its thanks to all those who have been instrumental in making the symposium a success. Our special gratitude is due to the 'chief patrons' Prof. K. B. Sinha, Director, Indian Statistical Institute and Prof. A.N. Basu, Vice-chancellor, Jadavpur University. The committee is also grateful to the sponsors who have supported the conference financially and otherwise.

We hope the proceedings will make good reading for those interested in the relevant field of research and remain a valuable asset.

December 2001 D. Patranabis and J. Das

Referees

S. Abe
V.D. Andrea
M. Banerjee
R. Banerjee
J. Basak
S. Bhakat
U. Bhattacharya
Z. Bien
R. Biswas
P.J.C. Branco
M.K. Chakraborty
D. Chakraborty
U. Chakraborty
B. Chanda
B.N. Chatterjee
Y.Y. Chen
K. Chintalapudi
S.B. Cho
B.B. Choudhuri
P.P. Chowdhury
S. Chowdhury
V. Cross
P. Das
Phulendu Das
R. Dave
R. De
K. Deb
B.L. Deekshatalu
P. Diamond
L. Ding
D. Dubois
G. Fogel
T. Furuhashi
P. Gader
T. Gedeon
V.D. Gesu
A. Ghosh
S. Ghosh(De)
M. Grabisch
L. Hall
Y. Hata
R. Hathaway
Y. Hayashi
H. Hellendoorn
R. Hemasinha
K. Hirota
S. Ionita
H. Ishibuchi
L.C. Jain
J. Kacpryzk
N. Kasabov
O. Kaynak
J. Keller
A. Koenig
M. Koeppen
L. Kozcy
R. Krishnapuram
Satish Kumar
Senthil Kumar
H.V. Kumbhojkar
A. Laha
C.S.G. Lee
C.T. Lin
D.S. Malik
A.K. Mandal
D.P. Mandal
D.D. Majumder
M. Mashinchi
M. Mitra
S. Mitra
R.K. Mudi
M. Mukaidono
D.P. Mukherjee
Y. Nakano
S. Nanda
B.B. Pal
N.R. Pal
Srimanta Pal
T. Pal
S.K. Parui
A.V. Patel
D. Patranabis
Z. Pawlak
W. Pedrycz
I. Perfilieva
S. Raha
B. Raj
A. Ramer
K. Ray
S. Ray
A. Rosenfeld
B.K. Roy
T. Runkler
S.K. Samanta
P. Santiprahob
D. Sarkar
M.K. Sen
G. Sengupta
S. Sinha
M. Smith
A.K. Srivastava
P.N.Suganthan
M. Sugeno
S. Sural
D. Tibrewala
Y.V. Venkatesh
L. Wang
T. Yamakawa
H. Yan
Y.Y. Yao
B.Yegnanarayana
H. Zimmermann

Conference Organization

Chief Patrons:	K.B. Sinha (I.S.I., Calcutta, India) A.N. Basu (Jadavpur University, India)
General Chair:	Nikhil R. Pal (I.S.I., Calcutta, India)
Program Chair:	M. Sugeno (Reiken Brain Res. Inst., Japan)
Advisory Committee Co-chairs:	M. Mukaidono (Meiji University, Japan) D. Dutta Majumder (I.S.I., Calcutta, India)
Organizing Committee Co-chairs:	D. Patranabis (Jadavpur University, India) J. Das (I.S.I., Calcutta, India)
Organizing Coordinators:	R. K. Mudi (Jadavpur University, India) S. Raha (Vishwabharati University, India)
Program Coordinator:	D. P. Mandal (I.S.I., Calcutta, India)
International Coordination Committee Co-chairs:	Z. Bien (KAIST, Korea) S. Sanyal (Jadavpur University, India)
Tutorials Chair:	M.K. Chakraborty (I.S.I., Calcutta, India)
Finance Chair:	R. Bandyopadhyay (Jadavpur Univ., India)

Program Area Chairs

M.K. Chakraborty(India)
C.T. Lin(Taiwan)
W. Pedrycz(Canada)
R. Krishnapuram(India)
K. Hirota(Japan)
K. Roy(India)
N. Kasabov(New Zealand)
J. Keller(USA)
B.B. Chaudhuri(India)
L. Ding(Singapore)
J. Kacprzyk(Poland)
T. Yamakawa(Japan)
B.L. Deekshatulu(India)
T. Furuhashi(Japan)

Advisory Committee Members

S. Amari (Japan)
B.N. Chatterjee(India)
V. Di Gesu(Italy)
C. Lucas(Iran)
Mao-kang Luo(China)
R.A. Mashelkar(India)
H.T. Nguyen(USA)
A.N. Poo(Singapore)
V.S. Ramamurthy(India)
S.B. Rao(India)
A. Rosenfeld(USA)
E. Ruspini(USA)

P.Santiprahob(Thailand)
B.P. Sinha(India)
P. Smets(Belgium)
H. Szu(USA)
I. B. Turksen(Canada)
C. Wu(China)
T. Yamakawa(Japan)
S. Yasunobu(Japan)
T. Yokogawa(Japan)
M.M. Zahedi(Iran)
H.Zimmermann(Germany)

International Coordination Committee Members

L. Ding (Singapore)
J. Lee (Taiwan)
Y. M. Liu (China)
M. Mashinchi (Iran)
S. Miyamoto (Japan)
N. H. Phuong (Vietnam)
O. Kaynak (Turkey)
Y. Tsukamoto (Japan)

Program Committee Members

S. Abe(Japan)
S. Bagui(USA)
J. Basak(India)
J.C. Bezdek(USA)
Y.Y. Chen(Taiwan)
S.B. Cho(Korea)
V. Cross(USA)
R. Dave(USA)
K. Deb(India)
D. Dubois(France)
G. Fogel(USA)
T. Fukuda(Japan)
T. Gedeon(Australia)
M.M. Gupta(Canada)
M. Grabisch(France)
L. Hall(USA)
Y. Hata(Japan)
Y. Hayashi(Japan)
R. Hemasinha(USA)
H. Ishibuchi(Japan)
L. C. Jain(Australia)
A. Koenig(Germany)
M. Koeppen(Germany)
L. Kozcy(Hungary)
S. Kumar(India)
I. Perfilieva(Czech Rep.)
L. Kuncheva(UK)
S. Mitra(India)
S. Mitra(USA)
B. Raj(India)
S. Ray(Australia)
T. Runkler(Germany)
P.Santiprahob(Thailand)
M. Smith(Canada)
P.N.Suganthan(Singapore)
H. Tanaka(Japan)
Y.V. Venkatesh(India)
L. Wang(Singapore)
H. Yan(Australia)
B.Yegnanarayana(India)
J. Zurada(USA)

Organizing Committee Members

B. D. Acharya (India)
U. Bhattacharya (India)
S. N. Biswas (India)
B. Chanda (India)
P. Das (India)
C. Dasgupta (India)
D. Ghoshdastidar (India)
A. K. De (India)
R. De (India)
A. Dutta (India)
A. Ghosh (India)
A. Laha (India)
A. K. Majumder (India)
A. K. Mandal (India)
D. P. Mukherjee (India)
P. K. Nandy (India)
P. Pal (India)
S. Pal (India)
K. S. Ray (India)
K. Ray (India)
B. K. Roy (India)
S. K. Parui (India)

Table of Contents

Fuzzy Systems

Soft Computing – Theory and Applications

Neural Networks

Neuro-fuzzy Systems

Pattern Recognition

Image Processing

A New Perspective on Reasoning with Fuzzy Rules

D. Dubois[1], H. Prade[1], and L. Ughetto[2]

[1] IRIT – CNRS, Université Paul Sabatier, 31062 Toulouse Cedex 4, France
[2] IRIN, Université de Nantes, BP 92208, 44322 Nantes Cedex 3, France
{Didier.Dubois,Henri.Prade}@irit.fr, Laurent.Ughetto@irin.univ-nantes.fr

Abstract. Fuzzy rules are conditional pieces of knowledge which can either express constraints on the set of values which are left possible for a variable, given the values of other variables, or accumulate tuples of feasible values. The first type are implicative rules, while the second are based on conjunctions. Consequences of this view on inference and interpolation between sparse rules are presented.

1 Introduction

Fuzzy rule-based systems have been either used as a convenient tool for synthesizing control laws from data, or in a knowledge representation and reasoning perspective in Artificial Intelligence (AI) [3]. This paper focuses the second use. Indeed, fuzzy rules of the form "if X is A, then Y is B" (more generally, the condition part of a rule can be compound), where A and/or B are fuzzy sets, are often considered as a basic concept in fuzzy logic [20]. They have been used for knowledge representation, where implicative rules play an important place (e.g., [1]), as well as for data modeling, where conjunctive rules are preferred [16].

In AI, logic-based knowledge representation aims at delimiting a set of possible states of the world (the models of the propositions which constitute the knowledge base). Each interpretation satisfying the whole set of propositions in the base (representing the available knowledge) is then considered as possible, since it is not forbidden. Thus, the addition of new pieces of information will just reduce the set of possible states, since the set of models is reduced. The information is said to be complete if only one possible state for the represented world remains. A statement is true (resp. false) for sure if the set of its models contains (resp. rules out) the possible states of the world; in case of incomplete information, one can ignore if a statement is true or false.

With a different prospect, and a different tradition, databases generally use the *closed world assumption*. Only what is known as true is represented, and then this assumption allows for the deduction of what is regarded as false, by default: a statement is either true (present in the database) or considered as false, since it is not known as true. When new pieces of data are available, they are just added to the database, and the corresponding information, if considered as false until now, is then considered as true. There is no representation of ignorance, only the storage of accepted statements.

N.R. Pal and M. Sugeno (Eds.): AFSS 2002, LNAI 2275, pp. 1–11, 2002.

This basic distinction between positive and negative information is important, since (expert) knowledge is both made of restrictions or constraints on the possible values of tuples of variables (as in the AI view), often induced by general laws, and examples of values known for sure as being possible, generally obtained in the form of observations, or as reported facts (as in database practice). This distinction plays a central role for the representation as well as the handling of fuzzy information. The next section opposes implicative fuzzy rules which are of the constraint type, and conjunction-based rules which are of the data accumulation type. Section 3 discusses inference for the two types of rules, while section 4 deals with interpolation between sparse rules of the two types.

2 Different Types of Fuzzy Rules with Different Semantics

2.1 Implicative Rules

In possibility theory, the available information is represented by means of possibility distributions which rank-order the possible states of affairs in a given referential set or attribute domain. In fact, the main role of possibility distributions is to discard states of affairs inconsistent with the available knowledge. A piece of information "X is A_i", where X is a variable ranging on a domain U, and A_i is a subset of U (maybe fuzzy), means here "the (ill-known) value for X is for sure in A_i". It is represented by the constraint:

$$\forall u \in U,\ \pi_X(u) \leq \mu_{A_i}(u) \ , \tag{1}$$

where π_X is a possibility distribution restricting the values of X [19]. Several such pieces of information are naturally aggregated conjunctively into:

$$\forall u \in U,\ \pi_X(u) \leq \min_i\ \mu_{A_i}(u) \ . \tag{2}$$

Then, once all the constraints are taken into account, a minimal specificity principle is applied, which allocates to each value (or state of the world) the greatest possibility degree in agreement with the constraints. It leads to enforce the equality in (2).

Considering a knowledge base $\mathcal{K} = \{A_i \to B_i,\ i = 1, \ldots, n\}$, made of n parallel fuzzy rules (i.e., rules with the same input space U and the same output space V), each rule "if X is A_i, then Y is B_i" (denoted $A_i \to B_i$) is represented by a conditional possibility distribution $\pi^i_{Y|X} = \mu_{A_i \to B_i}$ (the membership function of $A_i \to B_i$), which is determined according to the semantics of the rule. X is the tuple of input variables (on which information can be obtained) and Y the tuple of output variables (about which we try to deduce information). According to (2), the possibility distribution $\pi^{\mathcal{K}}$ representing the base $\mathcal{K}$ is obtained as the (min-based) conjunction of the $\pi^i_{Y|X}$'s:

$$\pi^{\mathcal{K}} = \min_{i=1,\ldots,n}\ \pi^i_{Y|X} \ . \tag{3}$$

This equation shows that such rules are viewed as (fuzzy) constraints since the more rules, the more constraints, the smaller the number of values that satisfy them, and the smaller the levels of possibility. $\pi^{\mathcal{K}}$ is then an upper bound of possible values. Moreover, this conjunctive combination implies that some output values, which are possible according to some rules, can be forbidden by other ones. Then, the possibility degree $\pi^{\mathcal{K}}(u,v) = 0$ means that if $X = u$, then v is an impossible value for Y, i.e., (u,v) is an impossible pair of input/output values. By contrast, $\pi^{\mathcal{K}}(u,v) = 1$ denotes ignorance only. It means that for the input value $X = u$, no rule in $\mathcal{K}$ forbids the value v for the output variable Y. However, the addition of a new rule to $\mathcal{K}$ (expressing a new piece of knowledge) may lead to forbid this value. A possibility degree $\pi^{\mathcal{K}}(u,v) > 0$ means that the pair (u,v) is not known as totally impossible, with respect to the current knowledge.

According to the typology of fuzzy rules proposed in [13], there are two main kinds of implicative rules, whose prototypes are *certainty* and *gradual* rules.

Certainty rules are of the form "The more X is A, the more certainly Y lies in B", as in "The younger a man, the more certainly he is single". They correspond to the following conditional possibility distribution modeling the rule:

$$\forall(u,v),\ \pi_{Y|X}(v,u) \le \max(\mu_B(v), 1-\mu_A(u)) \ . \tag{4}$$

In (4), A and B are combined with Kleene-Dienes implication: $a \to b = \max(1-a, b)$. Given such a rule, and an observation A' for the input variable X, the (fuzzy) set B' denotes the induced restriction on the possible values for the output variable Y. For a precise input $A' = \{u_0\}$, the conclusion B' is given by $\forall v \in V,\ \mu_{B'}(v) \ge 1 - \mu_A(u_0)$, i.e., a uniform level of uncertainty $1-\mu_A(u_0)$ appears in B'. Then, "Y is B" is certain only to the degree $\mu_A(u_0)$, since values outside B are possible to the complementary degree. A similar behavior is obtained with the implication $a \to b = 1 - a \star (1-b)$, where $\star$ is the product instead of min. Clearly, certainty rules extend propositions with certainty levels by making the certainty level depend on the level of satisfaction of fuzzy properties.

Gradual rules are of the form "The more X is A, the more Y is B", as in "The redder the tomato, the riper it is". This statement corresponds to the constraint:

$$\forall u \in U,\ \mu_A(u) \star \pi_{Y|X}(v,u) \le \mu_B(v) \ , \tag{5}$$

where $\star$ is a conjunction operation. The greatest solution for $\pi_{Y|X}(v,u)$ in (5) (according to the minimal specificity principle which calls for the greatest permitted degrees of possibility) corresponds to the residuated implication:

$$\mu_{A \to B}(u,v) = \sup\{\beta \in [0,1],\ \mu_A(u) \star \beta \le \mu_B(v)\} \ . \tag{6}$$

When $\star$ is min, (6) corresponds to Gödel implication: $a \to b = 1$ if $a \le b$, and b if $a > b$. If only a crisp relation between X and Y is supposed to underlie the rule, it can be modeled by Rescher-Gaines implication: $a \to b = 1$ if $a \leq b$, and 0 if $a > b$.

For an input u_0, (6) provides an enlargement of the core of B, i.e., the less X satisfies A, the larger the set of values near the core of B which are completely possible for Y. This express a similarity-based tolerance: if the value of X is close to the core of A, then Y is close to the core of B. This is the basis for an interpolation mechanism when dealing with several overlapping rules ([7],[11]).

2.2 Conjunctive Rules

Observation-based information corresponds to converse inequalities w.r.t. the ones in Section 2.1. Let A_i be a subset of values testified as possible for X since all the values in A_i have been observed as possible for X by a source i (A_i may be a fuzzy set if values can be guaranteed possible to a degree). Then, the feasible values for X described by δ_X obey to the inequality:

$$\forall u \in U, \ \delta_X(u) \geq \mu_{A_i}(u) \ . \tag{7}$$

If several sources provide examples of possible values for X, all this information is aggregated disjunctively into:

$$\forall u \in U, \ \delta_X(u) \geq \max_i \ \mu_{A_i}(u) \ . \tag{8}$$

A principle of maximal specificity (converse of the one of minimal specifity used in the previous section), expressing that nothing can be guaranteed if it has not been observed, leads to limiting the set of feasible values for X to:

$$\forall u \in U, \ \delta_X(u) = \max_i \ \mu_{A_i}(u) \ . \tag{9}$$

In the fuzzy control tradition, rule-based systems are often made of conjunction-based rules, as Mamdani-rules for instance. These rules, denoted $A_i \wedge B_i$, summarize pieces of data, i.e., couples of jointly possible input/output values. Each rule is then represented by a joint "guaranteed possibility" distribution: $\delta^i_{X,Y} = \mu_{A_i \wedge B_i}$.

A first justification of this interpretation comes directly from the semantics of the conjunction. Moreover, given a precise input $A' = \{u_0\}$, and a conjunctive rule $A_i \wedge B_i$, if the rule does not apply (i.e., if $\mu_{A_i}(u_0) = 0$), then the inference mechanism leads to the conclusion $B' = \emptyset$, i.e., no output is guaranteed possible. This implies a disjunctive combination of the conjunctive rules, which appropriately corresponds to accumulation of data and leads to a set of values whose possibility/feasibility is guaranteed to some minimal degree. Then, the counterpart of (3) is:

$$\delta_{\mathcal{K}} = \max_{i=1,\dots,n} \delta^i_{X,Y} \ . \tag{10}$$

The distribution $\delta_{\mathcal{K}}$ is then a lower bound of possible values. Thus, a possibility degree $\delta_{\mathcal{K}}(u, v) = 1$ means that if $X = u$, then v is a totally possible value for Y. This is a guaranteed possibility degree. By contrast, $\delta_{\mathcal{K}}(u, v) = 0$ only means that if $X = u$, no rule can guarantee that v is a possible value for Y. By default, v is considered as not possible (since possibility cannot be guaranteed). A membership degree 0 to B' represents ignorance, while a degree 1 means a guaranteed possibility. Thus, a conclusion $B' = \emptyset$ should not be understood here as "all the output values are impossible", but as "no output value can be guaranteed".

As for implicative rules, there are two main kinds of conjunctive rules, called *possibility* and *antigradual* rules [13].

Possibility rules are of the form "the more X is A, the more possible Y lies in B", as in "the more cloudy the sky, the more possible it will rain soon". It corresponds to the following possibility distribution modeling the rule:

$$\forall(u, v) \in U \times V,\ \delta_{X,Y}(u, v) \geq \min(\mu_A(u), \mu_B(v)) \ . \tag{11}$$

These rules, modeled with the conjunction *min*, correspond to the ones introduced by Mamdani and Assilian in 1975 [15]. For an input value u_0 such that $\mu_A(u_0) = \alpha$, a possibility rule expresses that, when $\alpha = 1$, B is a set of possible values for Y (to different degrees if B is fuzzy). When $\alpha < 1$, values in B are still possible, but they are guaranteed possible only up to the degree α. To obtain B', the set B is then truncated from above. Finally, if $\alpha = 0$, the rule does not apply, and $B' = \emptyset$ as already said.

Antigradual rules, which are another type of conjunctive rule (see [13]), express that "the more X is A, the larger the set of guaranteed possible values for Y is, around the core of B", as in "the more experienced a manager, the wider the set of situations he can manage".

For an input $A' = \{u_0\}$, if $\mu_A(u_0) = \alpha < 1$, the values in B such that $\mu_B(v) < \alpha$, cannot be guaranteed. Such a rule expresses how values which are guaranteed possible can be extrapolated on a closeness basis.

3 Inference with Different Types of Rules

3.1 Inference with Implicative Rules

In this section, a set of implicative rules $\mathcal{K} = \{A_i \rightarrow B_i,\ i = 1, \ldots, n\}$ is considered. In order to compute the restriction induced on the values of Y, given a possibility distribution π'_X restricting the values of the input variable X, π'_X is then combined conjunctively with $\pi^{\mathcal{K}}$ and projected on V, the domain of Y (in agreement with (2)):

$$\pi_Y(v) = \sup_{u \in U} \min(\pi^{\mathcal{K}}(u, v), \pi'_X(u)) \ . \tag{12}$$

This combination-projection is known as *sup-min* composition (or Compositional Rule of Inference) and often denoted $\circ$. Then, given a set of rules $\mathcal{K}$ and an input set A', which means that the ill-known, real value for X lies in A', one can deduce the output B' given by:

$$B' = A' \circ \bigcap_{i=1}^{n} A_i \to B_i = A' \circ R^{\mathcal{K}} , \tag{13}$$

with $\mu_{R^{\mathcal{K}}} = \pi^{\mathcal{K}}$. The obtained fuzzy set B' is then an upper bound of the possible values for the output variable Y.

If, for a given precise input $A' = \{u_0\}$, the rule $A_i \to B_i$ does not apply, i.e., $\mu_{A_i}(u_0) = 0$, the *sup-min* composition yields the conclusion $B'_i = V$, the entire output space. This conclusion is in accordance with the conjunctive combination of the rules. Thus V plays the role of the neutral element for the aggregation operator. This is why implicative rules are combined conjunctively.

3.2 Inference with Conjunctive Rules

In this section, a set of conjunctive rules $\mathcal{K} = \{A_i \wedge B_i,\ i = 1, \ldots, n\}$ is considered. In this case, the computation of B' can no longer be achieved via the sup-min composition applied to the disjunctive aggregation of the rules, as it has sometimes been proposed, in order to extend fuzzy control systems to fuzzy inputs.

Indeed, what have to be computed, via an appropriate projection, are the conclusions which are guaranteed possible for Y when the value for X is constrained to be in A', i.e., the values for Y which are guaranteed possible whatever the value for X in A'. The sup-min composition applied to a too imprecise input, such that $A_i \cap A_{i+1} \subseteq A' \subseteq A_i \cup A_{i+1}$, leads to too large a conclusion $B_i \cup B_{i+1}$, since $A' \times (B_i \cup B_{i+1})$ contains values which are not guaranteed possiblen, as shown on Fig. 1. The expected conclusion, in terms of guaranteed possible values, is given for a non fuzzy input A' by:

$$\mu_{B'_\wedge}(v) = \inf_{u \in A'} \max_i \delta^i(u, v) . \tag{14}$$

When A' is fuzzy, this equation can be generalized by (see [12]):

$$\mu_{B'_\wedge}(v) = \inf_{u \in U} \left(\mu_{A'}(u) \to \max_i \delta^i(u, v) \right) , \tag{15}$$

where $\to$ is Gödel implication: $a \to b = 1$ if $a \leq b$ and b otherwise. It can be checked that for usual fuzzy partitions, if $A' = A_i$ in (15), then $B' = B_i$, a result that cannot be obtained using the sup-min composition (see [6]).

Generally speaking, the logical consequence mechanism works in an opposite way for $\pi^{\mathcal{K}_\to}$ and $\delta^{\mathcal{K}_\wedge}$ (see [8]). Indeed, a fuzzy set C is a consequence of the knowledge base $\mathcal{K}_\to$ iff $\pi^{\mathcal{K}_\to} \leq \mu_C$, since $\min(\pi^{\mathcal{K}_\to}, \mu_C) = \pi^{\mathcal{K}_\to}$. By contrast, C is a consequence of the base of examples $\mathcal{K}_\wedge$ iff $\delta^{\mathcal{K}_\wedge} \geq \mu_C$, since $\max(\delta^{\mathcal{K}_\wedge}, \mu_C) = \delta^{\mathcal{K}_\wedge}$. By the way, this is in agreement with the idea of defuzzification, which consists in choosing a value in the set $\{t,\ \delta^{\mathcal{K}_\wedge}(t) > 0\}$.

The focusing operation expressed by (14) and (15) (where A' is not combined disjunctively with the rest of the information corresponding to the rules), should not be confused with a Modus-Ponens-like counterpart for bases of examples,

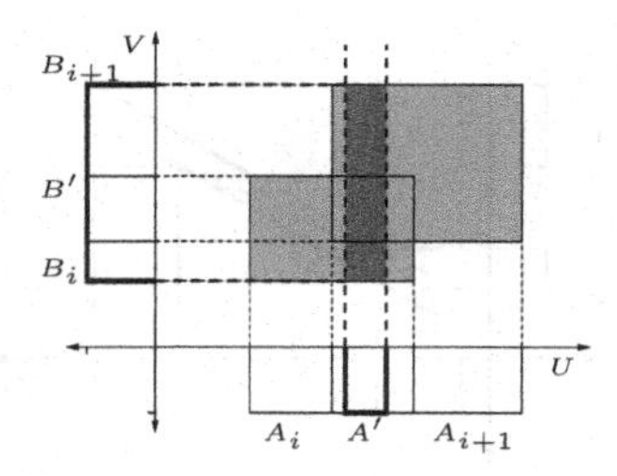

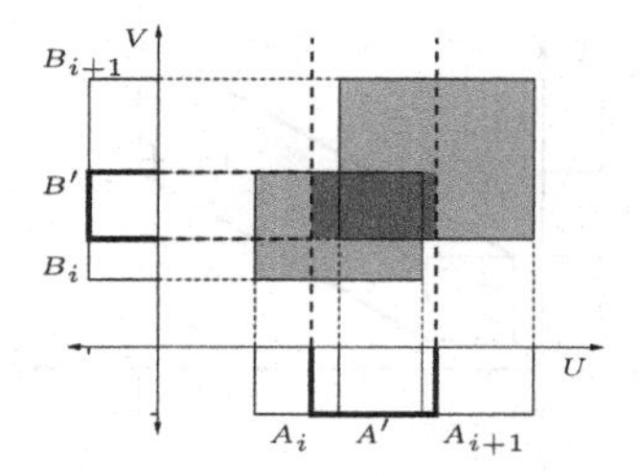

Fig. 1. B' gives the guaranteed possible values for Y, $\forall u \in A'$, for two different A'

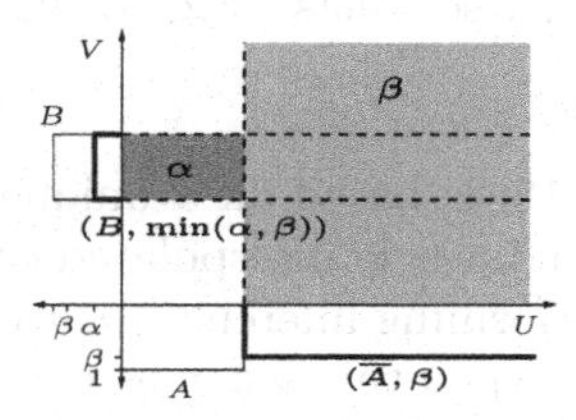

Fig. 2. Computation of guaranteed possible values

where the conjunctive rules and A' play the same role, unlike in (15). Indeed, let us consider the two premisses: i) the values in $\boldsymbol{A} \times \boldsymbol{B}$ are guaranteed possible at least at the degree α, i.e., $\left(\inf_{(u,v)\in A\times B}\ \delta(u,v)\right) \geq \alpha$, and ii) the values in $\overline{\boldsymbol{A}}\times V$ (where $\overline{A}$ is the complement of A on U) are guaranteed possible at least at the degree β, i.e., $\left(\inf_{(u,v)\in\overline{A}\times V}\ \delta(u,v)\right) \geq \beta$. Then one can deduce that the values for Y which lie in $\boldsymbol{B}$ are guaranteed possible at the degree $\min(\alpha,\beta)$, whatever $u \in U$, since $\inf_{(u,v)\in U\times B}\ \delta(u,v) \geq \min(\alpha,\beta)$, as it is clear on Fig. 2.

Formally, it corresponds to the following syntactic inference: $(A \times B,\, \alpha)$, $(\overline{A},\, \beta) \vdash (B,\, \min(\alpha,\beta))$ where the weights are interpreted in terms of the δ function.

4 Interpolation by Completion between Sparse Rules

When a rule base is made of incomplete knowledge, it may occur that for a given input A', no rule applies. Then, the obtained conclusion is either $B' = V$ with implicative rules, or $B' = \emptyset$ with conjunctive rules, both situations representing total ignorance, as expected. However, in such a case where the rules are sparse, it can be natural to interpolate between values given by (at least) two neighbouring rules, especially when a gradual and smooth variation of Y w.r.t. X is assumed.

The interpolation problem can be summarized as follows. Given a set of rules "if X is A_j, then Y is B_j", where the A_js are ordered on U, and an observation $A^\star$ such that $A_i \preceq A^\star \preceq A_{i+1}$ for a given i, the problem is to find an informative conclusion $B^\star$, such that $B_i \preceq B^\star \preceq B_{i+1}$ (in case of $B_i \preceq B_{i+1}$, assuming a non decreasing function).

Several interpolation mechanisms, based on differents principles have been proposed (e.g. [4] for some references). Among them, the completion-based ap-

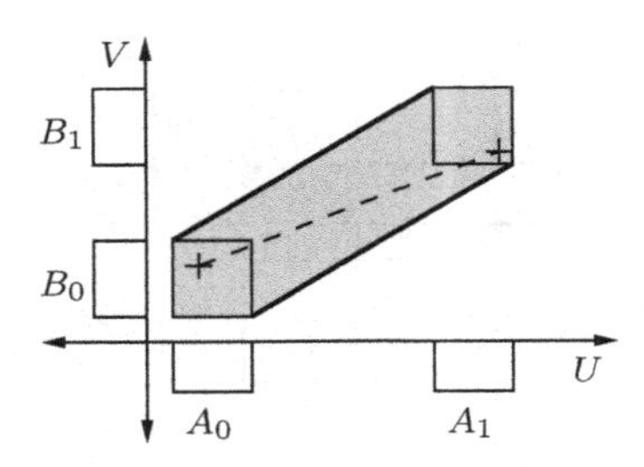

Fig. 3. Interpolation with imprecise points

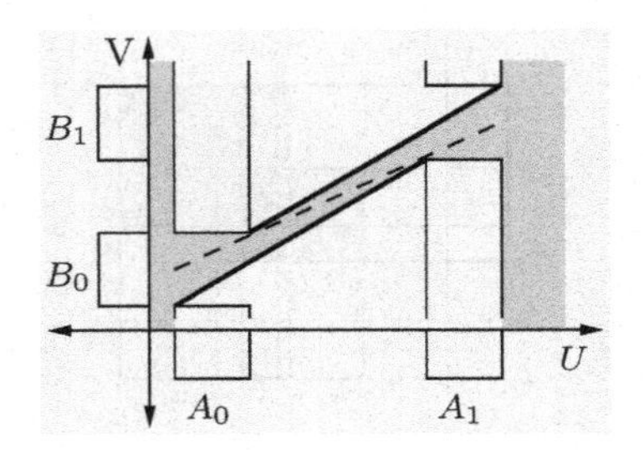

Fig. 4. Interp. between and on $A_0 \cup A_1$

proach to interpolation [17,18] is based on a unique principle which consists in completing the rule base with linearly interpolated rules. Then, interpolative reasoning simply consists in performing inference on the completed rule base. Based on this principle, several interpolative schemes can be proposed, depending on the semantics of the rules.

For the sake of simplicity, this section considers that rules have only one input (U is one-dimensional), and that the A_is and B_is are non-fuzzy closed intervals (see [17] for the extension to fuzzy sets and to multi-input rules). Moreover, as a convenient notation, the two considered sparse rules are denoted (A_0, B_0) and (A_1, B_1).

4.1 Rules as Data vs. Rules as Constraints

In the case of conjunction-based rules, the two rules between which the interpolation is performed represent points in $A_0 \times B_0$ and in $A_1 \times B_1$, which are guaranteed to be possible.

The envelop of all the straight lines between them, representing linear interpolation, leads to the relation Δ depicted on Fig. 3. Obviously, points $(u, v) \in \Delta$ can be s.t. $u \in A_i$ and $v \notin B_i$. This is incompatible with the semantics of implicative rules (of the form $A \rightarrow B$, where $\rightarrow$ is an implication), but is in agreement with the one of conjunctive rules (of the form $A \times B$, where $\times$ is the Cartesian product). Indeed, a conjunctive rule, modeled with a conjunction, means: "if X is (in) A, then Y *can* be (in) B (but may also be out of B)".

In the case of implicative rules, one rule $A \rightarrow B$ does not represent a point, but bounds on the range of Y, given that X is in A. Given a point (u, v) in the graph Δ of the function, if $u \in A$, then the rule means the constraint $v \in B$. This interpretation is in agreement with the semantics of implicative rules (classical rules since A and B are crisp intervals here).

Linear interpolation between two rules then consists in connecting the two bounded areas represented by the rules by means of a linear function achieving an interpolation between A_0 and A_1. Keeping the same linear model between and *on* A_0 and A_1 seems more natural. It leads to the interpolative scheme depicted on Fig. 4.

This scheme is more restrictive than the other. It can lead to incoherence cases, when it is impossible for a straight line to go through the grey area delimited by the constraints given by the rules.

4.2 Interpolation by Completion of the Set of Rules

The two linear interpolation schemes suggested in the previous section can be obtained by a unique principle which consists in completing the set of two rules $\mathcal{K}$, either by implicative or conjunctive-based rules. This interpolation principle simply consists in adding to $\mathcal{K}$ all the rules of the form $(A_\lambda\, B_\lambda)$, for $\lambda \in [0,1]$, where A_λ and B_λ are obtained by linear interpolation, by: $A_\lambda = \lambda \cdot A_1 + (1-\lambda) \cdot A_0$ and $B_\lambda = \lambda \cdot B_1 + (1-\lambda) \cdot B_0$. An infinite set of rules is obtained, namely $\mathcal{L} = \{(A_\lambda\, B_\lambda),\ \lambda \in [0,1]\}$. Now, it can be shown that the relations corresponding to $\mathcal{L}$ are the ones depicted Figs. 3 and 4.

Conjunctive rules, representing ill-known points correspond to the relation $A_i \times B_i$, where $\times$ is the Cartesian product. As these rules encode possible values, they are aggregated disjunctively (possible values add). Thus, the relation $R_\mathcal{L}$ corresponding to the set of rules $\mathcal{L}$ is $R_\mathcal{L} = \bigcup_{\lambda \in [0,1]} A_\lambda \times B_\lambda$, which correspond to the grey part of Fig. 3. It comes down to applying the extension principle to linear interpolation [14].

Implicative rules, representing parts of a fuzzy rules correspond to the relation $A \to B$ and are aggregated conjunctively. In this case, the relation $R_\mathcal{L}$ becomes $R_\mathcal{L} = \bigcap_{\lambda \in [0,1]} A_\lambda \to B_\lambda$, which corresponds to the grey part of Fig. 4. Thus, the proposed interpolation principle used with implicative rules corresponds to the case where the linear model has to be the same between A_0 and A_1 as on $A_0 \cup A_1$.

In the particular case where B_0 and B_1 are singletons, linear interpolation can be performed *between* A_0 and A_1 only (as soon as $B_0 \neq B_1$). The interpolation function is then a straight line. This means that the imprecision of the conclusions $B^\star$ only reflects the imprecision of the observations $A^\star$. Moreover, in agreement with classical deduction, and more generally with the Compositional Rule of Inference, if two inputs A' and A'' have a non-empty intersection, so have the two conclusions B' and B'', by contrast with some other methods, as the analogy-based one presented in [2,5].

The interpolation mechanism with a fuzzy input only consists in applying the appropriate inference method, as explained in section 3, to the set of rules $\mathcal{L}$ obtained from the two rules in $\mathcal{K}$, by construction of the relation $R_\mathcal{L}$. Then the fuzzy sets are trapezoidal, and the rules are either conjunctive, or Rescher-Gaines implicative, the computation can be done easily [18]. Note that if $A' = A_\lambda$, then $B' = B_\lambda$ is retrieved both in the case of implicative and of conjunctive rules, when applying the appropriate inference method.

4.3 Coherence of the Completed Set of Rules

In classical logic, two rules $A_0 \to B_0$ and $A_1 \to B_1$ are said to be *potentially* incoherent if they can be triggered by the same precise input value, i.e., if $A_0 \cap$

$A_1 \neq \emptyset$, and if for this value they lead to contradictory conclusions, i.e., if $B_0 \cap B_1 = \emptyset$ (see for instance [10]). The aggregation $B_0 \cap B_1$ indicates that the considered case is the one of implicative rules. Indeed, for the first interpolation scheme, obtained with conjunctive rules, the completed set of rules $\mathcal{L}$ *cannot* be incoherent since the corresponding relation is $R_{\mathcal{L}} = \bigcup_{\lambda \in [0,1]} A_\lambda \times B_\lambda$. As $B_0 \cup B_1$ cannot be empty, this case is always coherent.

By contrast, with implicative rules (i.e., all the other schemes), the rules are aggregated conjunctively by $R_{\mathcal{L}} = \bigcap_{\lambda \in [0,1]} A_\lambda \to B_\lambda$, and can be incoherent. However, the incoherence cases depends on properties on the A_is and B_is and can easily be detected.

Moreover, it can be shown that a case of incoherence occurs when the same straight line cannot be drawn between and on A_0 and A_1, satisfying the constraints enforced by the rules (see [17]). Then, the linear model should be applied on subsets of A_0 and A_1 only, or a non linear model should be considered as more accurate.

5 Conclusion

This paper has proposed a survey of different types of fuzzy rules, whose representation is based either on implications or on conjunctions. The two very different semantics justifying these two types of representation have been contrasted: constraints delimiting what is not impossible, or accumulation of feasible data. Theses two semantics leads to two specific and distinct inference modes. A general method by completion of the set of rules can be applied in both cases for interpolating between sparse rules. Besides, fuzzy rules and especially possibility rules have been shown recently to be useful in case-based prediction [9] for modeling the principle that similar inputs possibly have similar outputs; certainty rules are then used for interpreting the repertory of cases as constraints on the similarity relations.

References

1. A. Ayoun and M. Grabisch. Tracks real-time classification based on fuzzy rules. *International Journal of Intelligent Systems*, 12:865–876, 1997.
2. B. Bouchon-Meunier, J. Delechamp, C. Marsala, N. Mellouli, M. Rifqi, and L. Zerrouki. Analogy and fuzzy interpolation in case of sparse rules. In *Proc. of the EUROFUSE-SIC Joint Conference*, pages 132–136, 1999.
3. B. Bouchon-Meunier, D. Dubois, L. Godo, and H. Prade. Fuzzy sets and possibility theory in approximate and plausible reasoning. In J. Bezdek, D. Dubois, and H. Prade, editors, *Fuzzy sets in approximate reasoning and information systems*, The Handbooks of Fuzzy Sets, pages 15–190. Kluwer, Boston, 1999.
4. B. Bouchon-Meunier, D. Dubois, C. Marsala, H. Prade, and L. Ughetto. A comparative view of interpolation methods between sparse fuzzy rules. In *Proc. of the Joint 9th IFSA World Congress and 20th NAFIPS International Conference*, volume 5, pages 2499–2504, 2001.

5. B. Bouchon-Meunier, C. Marsala, and M. Rifqi. Interpolative reasoning based on graduality. In *Proc. of 9th Int. conf. on fuzzy systems (FUZZ-IEEE'2000)*, pages 483–487, 2000.
6. A. Di Nola, W. Pedrycz, and S. Sessa. An aspect of discrepancy in the implementation of modus ponens in the presence of fuzzy quantities. *International Journal of Approximate Reasoning*, 3:259–265, 1989.
7. D. Dubois, M. Grabisch, and H. Prade. Gradual rules and the approximation of control laws. In H.T. Nguyen, M. Sugeno, R. Tong, and R.R. Yager, editors, *Theoretical Aspects of Fuzzy Control*, pages 147–181. Wiley, 1994.
8. D. Dubois, P. Hajek, and H. Prade. Knowledge driven vs. data driven logics. *Journal of Logic, Language and Information*, 9:65–89, 2000.
9. D. Dubois, E. Hüllermeier, and H. Prade. Flexible control of case based prediction in the framework of possibility theory. In *Proc. of the 5th Eur. Work. on CBR (EWCBR'00)*, pages 61–73. LNCS 1898 – Springer Verlag, 2000.
10. D. Dubois, H. Prade, and L. Ughetto. Checking the coherence and redundancy of fuzzy knowledge bases. *IEEE Transactions on Fuzzy Systems*, 5(3):398–417, 1997.
11. D. Dubois, H. Prade, and L. Ughetto. Fuzzy logic, control engineering and artificial intelligence. In H.B. Verbruggen, H.-J. Zimmermann, and R. Babuska, editors, *Fuzzy Algorithms for Control Engineering*, pages 17–57. Kluwer Academic Publishers, 1999.
12. D. Dubois and H. Prade. Fuzzy rules in knowledge-based systems – modeling gradedness, uncertainty and preference. In R.R. Yager and L.A. Zadeh, editors, *An Introduction to Fuzzy Logic Applications in Intelligent Systems*, pages 45–68. Kluwer Academic Publishers, 1992.
13. D. Dubois and H. Prade. What are fuzzy rules and how to use them. *Fuzzy Sets and Systems*, 84(2):169–186, 1996.
14. D. Dubois and H. Prade. On fuzzy interpolation. *International Journal of General Systems*, 28:103–114, 1999.
15. E.H. Mamdani and S. Assilian. An experiment in linguistic synthesis with a fuzzy logic controller. *International Journal on Man-Machine Studies*, 7:1–13, 1975.
16. J. Mendel. Fuzzy logic systems for engineering: A tutorial. *Proc. IEEE – Special issue on fuzzy logic with engineering applications*, 83(3):345–377, 1995.
17. L. Ughetto, D. Dubois, and H. Prade. Fuzzy interpolation by convex completion of sparse rule bases. In *Proc. of 9th Int. conf. on fuzzy syst. (FUZZ-IEEE'2000)*, pages 465–470, 2000.
18. L. Ughetto, D. Dubois, and H. Prade. Interpolation lin aire par ajout de r gles dans une base incompl te. In *Proc. Rencontres Francophones sur la Logique Floue et ses Applications (LFA'00)*, pages 71–78. Cépaduès, Toulouse, 2000.
19. L.A. Zadeh. The concept of a linguistic variable and its application to approximate reasoning. *Information Sciences*, 1975. Part 1: 8:199-249; Part 2: 8:301-357; Part 3: 9:43-80.
20. L.A. Zadeh. The calculus of fuzzy if-then rules. *AI Expert*, 7(3):23–27, 1992.

On Interpretability of Fuzzy Models

T. Furuhashi

Dept. of Information Enegineering
Mie University
1515 Kamihama-cho, Tsu 514-8507, Japan
Tel.& Fax. +81-59-231-9456
furuhashi@pa.info.mie-u.ac.jp

Abstract. Interpretability is one of the indispensable features of fuzzy models. This paper discusses the interpretability of fuzzy models with/without prior knowledge about the target system. Without prior knowledge, conciseness of fuzzy models helps humans to interpret their input-output relationships. In the case where a human has the knowledge in advance, an interpretable model could be the one that explicitly explains his/her knowledge. Experimental results show that the concise model has the essential interpretable feature. The results also show that human's knowledge changes the most interpretable model from the most concise model.

1. Introduction

Problems of describing input-output relationships of unknown systems from data have attracted much attention in many fields. Fuzzy modeling is one of the effective tools for solving the problems. The distinguishing feature of fuzzy model is in that it is interpretable. However, there have been few reports on quantitative analysis of the interpretability of fuzzy models.

The interpretability of fuzzy models has been evaluated simply by the number of fuzzy rules, the number of membership functions [1], or the degree of freedom term of AIC (Akaike's Information Criterion) [2, 3]. Matsushita, Furuhashi, et al. [1] discussed hierarchical fuzzy modeling for identifying interpretable fuzzy models. However, automatically derived fuzzy models are not often linguistically interpretable, as recognized in [4 - 6]. The interpretability of fuzzy models also depends on other factors such as shape/allocation of membership functions, and more on interpreter's prior knowledge.

This paper studies the interpretability of fuzzy models by separating the cases where prior knowledge is available or not. Definition of conciseness of fuzzy models is introduced. In the case where prior knowledge is not available, input-output relationships of a concise fuzzy model are easy for a human to interpret. This paper introduces De Luca and Termini's fuzzy entropy [7] as a conciseness measure that evaluates the shape of membership function. This paper also presents a new measure derived on the analogy of relative entropy. This new measure is also a conciseness measure that evaluates the deviation of allocation of membership functions on the universe of discourse. A combination of these two measures is shown to be a good conciseness measure.

N.R. Pal and M. Sugeno (Eds.): AFSS 2002, LNAI 2275, pp. 12–19, 2002.

With *a priori* knowledge about the target system, a concise model is not always interpretable. In this case, an interpretable model could be the one that explicitly explains human's knowledge. Experimental results show that the obtained concise model has the essential interpretable feature. The results also show that human's knowledge changes the most interpretable model from the most concise model.

2. Interpretability and Conciseness

Interpretability of fuzzy models heavily depends on human's prior knowledge. If we have profound knowledge about the target system, an interpretable model could be the one that makes our knowledge explicit. Even though the fuzzy model had many parameters and the input-output relationships were highly non-linear, our knowledge helps us to interpret the relationships. Even a concise model is not interpretable, if it does not fit into our prior knowledge.

In the case where we have no *a priori* knowledge, a concise fuzzy model could be easy for a human to interpret. Assuming that three types of fuzzy models are given as shown in Fig.1. All the models in this figure are single-input single-output ones. Fig.1 (a) shows a case where crisp membership functions S, M, and B are allocated equidistantly on the universe of discourse in the antecedent. The output is depicted with granules O_B, O_M, and O_S. This model is interpreted as

If x is S then y is O_B
If x is M then y is O_M
If x is B then y is O_S.

The granules S, M, B, O_B, O_M, and O_S help us to grasp the input-output relationships. We interpret the models in Fig. 1 in the form of above rules with the granules. Thus the model in Fig. 1(a) with crisp and equidistantly allocated granules is most interpretable.

Fig.1(b) shows cases with Gaussian membership functions. Fig.1(c) is the case where triangular membership functions are allocated unevenly. Every model can be interpreted in terms of the above three rules. But it becomes more and more difficult to make correspondence between the models and the three rules. The model in Fig.1(c) is the most difficult to interpret among the three models in Fig.1.

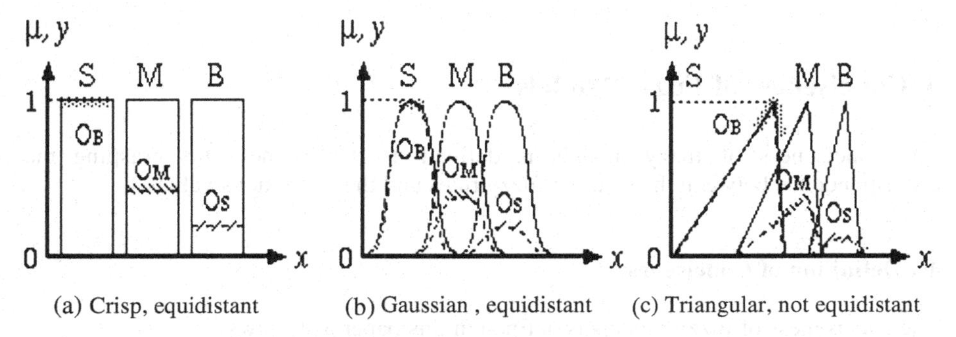

(a) Crisp, equidistant (b) Gaussian , equidistant (c) Triangular, not equidistant

Fig. 1. Membership functions and outputs as granules

Assume that we have the following knowledge about the target system: "the non-linearity of the system becomes stronger with larger x". In this case, the interpretability of the models in Fig. 1 changes drastically from the above observation. We may think that the model in Fig. 1(c) is the most interpretable, because this model fits our knowledge most.

Interpretability of fuzzy models depends on prior knowledge. For quantitative analysis of interpretability, this paper limits the discussions in the following sections to the cases where no prior knowledge is available.

3. Fuzzy Model

A single-input single-output fuzzy model with simplified fuzzy inference[8] is used in this paper.

The output y of a fuzzy model with the input x is given by

$$y = \sum_{i=1}^{N_m} \mu_i(x) c_i \tag{1}$$

where $\mu_i(x)$ and c_i ($i = 1,\ldots, N_m$) are grade of membership in the antecedent part and singleton in the consequent part, respectively. N_m is the number of membership functions.

The following conditions are used to make the discussion about the conciseness simple:

(a) For all $x \in X$, membership functions $\mu_i(x)$ ($i = 1,\ldots, N_m$) satisfy

$$\sum_{i=1}^{N_m} \mu_i(x) = 1 \tag{2}$$

(b) Two membership functions overlap where $1 > \mu_i(x) > 0$ ($i = 1,\ldots, N_m$).

(c) Each membership function $\mu_i(x)$ is similar with respect to the center point $x = a$, in the sense that

$$\mu_{li}(x) = \mu_{ri}(1 - \frac{1-a}{a} x) \tag{3}$$

where $\mu_{li}(x)$ and $\mu_{ri}(x)$ are left/right hand side membership functions, respectively.

(d) All the fuzzy sets are convex.

4. Conciseness of Fuzzy Models

The conciseness of fuzzy models is defined as the easiness for grasping the correspondence between the discrete fuzzy rules and the continuous values.

4.1 Definition of Conciseness

The conciseness of fuzzy models is defined in this paper as follows:

Definition 1 (Conciseness of Fuzzy Model)
Fuzzy model *A* is more concise than fuzzy model *B*, if the membership functions in *A* are more uniformly distributed across the universe of discourse than the membership functions in *B*, and the shapes of membership functions in *A* are less fuzzy than in *B*.

4.2 De Luca and Termini's Fuzzy Entropy

De Luca and Termini [7] defined fuzzy entropy of fuzzy set *A* as

$$d(A) = \int_{x_1}^{x_2} \{-\mu_A(x)\ln\mu_A(x) - \mu_A(1-x)\ln\mu_A(1-x)\} \tag{4}$$

where $\mu_A(x)$ is the membership function of fuzzy set *A*. If $\mu_A(x) = 0.5$ for all x on the support of *A*, then the fuzzy entropy of fuzzy set *A* is the maximum.

This fuzzy entropy can distinguish the shapes of membership functions, i.e. triangular, Gaussian, etc., and coincides with a part of the definition of conciseness.

With the conditions (a) and (b) in Section III, De Luca and Termini's entropy is simplified. Assuming that two membership functions $\mu_A(x)$ and $\mu_B(x)$ are overlapping and for all $x \in [x_1, x_2]$, $\mu_A(x) + \mu_B(x) = 1$, then

$$d(A) = \int_{x_1}^{x_2} \{-\mu_A(x)\ln\mu_A(x)\}. \tag{5}$$

4.3 Measure for Deviation of Membership Function

De Luca and Termini's entropy cannot distinguish similar membership functions shown in Fig.2. Two membership functions *A* and *C* are similar in the sense that the membership function *A* coincides with *C* by extending the horizontal axis *x* of the left hand side membership function μ_{lA} and shrinking that of the right hand side membership function μ_{rA}.

This paper defines a quantitative measure of deviation of a membership function from symmetry on the analogy of relative entropy. This is also a good candidate for the conciseness measure of fuzzy models. The membership function is assumed to satisfy the conditions (a) - (d) in Section III. This measure is defined by considering eq. (5).

Definition 2 (Measure for Deviation)
The measure for deviation of fuzzy set *A* from symmetry is given by

$$r(A) = \int_{x_1}^{x_2} \left(\mu_C \ln \frac{\mu_C}{\mu_A} \right) dx \tag{6}$$

where x_1 and x_2 are the left and right points of the support of fuzzy set *A*, respectively; $\mu_A(x)$ is the membership function of fuzzy set *A*; $\mu_C(x)$ is the symmetrical membership function of fuzzy set *C*, which has the same support as that of fuzzy set *A*.

Fig. 2 illustrates an example of fuzzy set *A* and *C*.

4.4 New Measure

One way of combining the two measures is summation. By summing the fuzzy entropy $d(A)$ in eq.(5) and the measure for deviation of a membership function $r(A)$ in eq.(6), a new measure $dr(A)$ is obtained.

$$dr(A) = d(A) + r(A) = -\int_{x_1}^{x_2} (\mu_C \ln \mu_A)\, dx. \qquad (7)$$

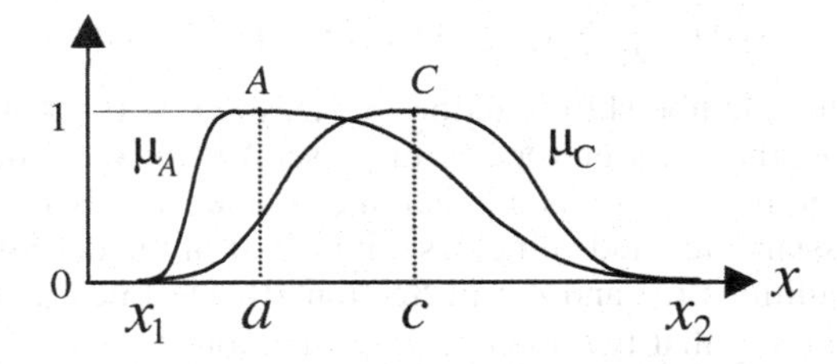

Fig. 2. Symmetrical/asymmetrical membership function

4.5 Average Measure

Average measure dr_{avr} is introduced to evaluate the shapes and allocations of N_m fuzzy sets A_i ($i = 1,..., N_m$) on the universe of discourse X on x-axis.

The average measure dr_{avr} is defined as

$$dr_{avr} = \frac{1}{N_m - 2} \sum_{i=2}^{N_m - 1} dr(A_i) \qquad (8)$$

where $dr(A)$ is the new measure in (7), which evaluates the shape and deviation of a membership function, N_m is the number of fuzzy sets A_i ($i = 1,\ldots, N_m$) on the universe of discourse X on x-axis.

5. Numerical Results

This section describes results of a numerical experiment to demonstrate the feasibility of the average fuzzy entropy for evaluation of the conciseness of fuzzy models. For simplicity, the following single-input single-output function is used as a target function. Fig. 3 depicts this function.

$$f(x) = \begin{cases} 1 - 2x & (0 \le x \le 1/2) \\ -4x^2 + 8x - 3 & (1/2 < x \le 1) \end{cases} \qquad (9)$$

Input-output pairs of data were generated using this function. The conditions (a)–(d) in Section III were imposed on the model. The accuracy of the obtained model was measured by mean squared error.

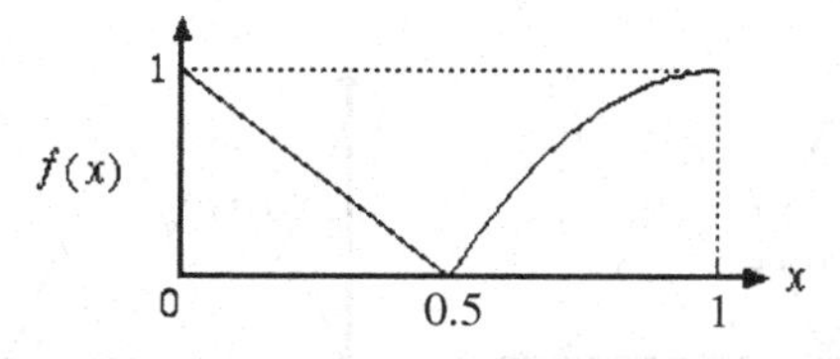

Fig. 3. Target function

To examine the relationships between the average measure dr_{avr} and the accuracy, 1000 fuzzy models were randomly generated and their values of average measure and accuracy were calculated. Fig. 4 shows the obtained results. Among them, the fuzzy models near the Pareto front are shown in Fig. 5.

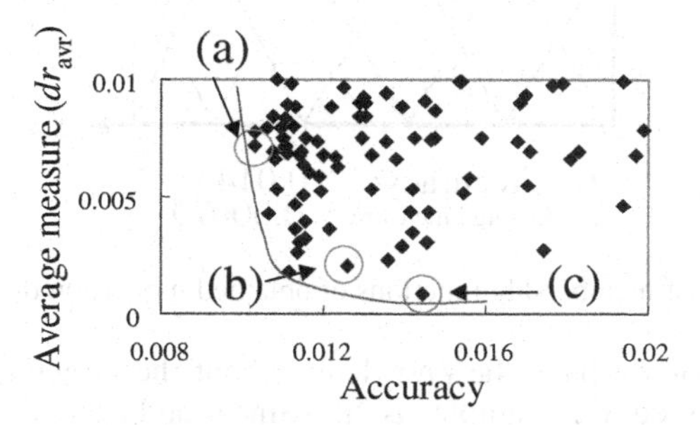

Fig. 4. Generated fuzzy models

In this paper, the shape of membership functions was fixed to triangular and the number of membership functions of a fuzzy model was set at 6. Each dot in Fig. 4 corresponds to a fuzzy model that has a unique allocation of membership functions. From Fig. 4, it is observed that the average measure and the accuracy are in conflict as indicated with the solid line.

Fig.5 (a)-(c) show the allocations of the membership functions of the fuzzy models (a)-(c), which were on the Pareto front in Fig.4 indicated by the circles. The less the average measure was, the more equidistant the allocation of membership functions was.

6. Discussion

Assume that we have no prior knowledge about the target system represented in eq.(9). From the collected input-output pairs of data, we have obtained the models in Fig. 5. The question here is which model is the most interpretable. The model in Fig. 5 (c) has equidistantly allocated membership functions. Although this model is less accurate, it is easier to get the rough idea about the input-output relationships from this model than from other models in Fig. 5. The average measure of the model in Fig. 5 (c) is the smallest, and this measure coincides with the observation of conciseness.

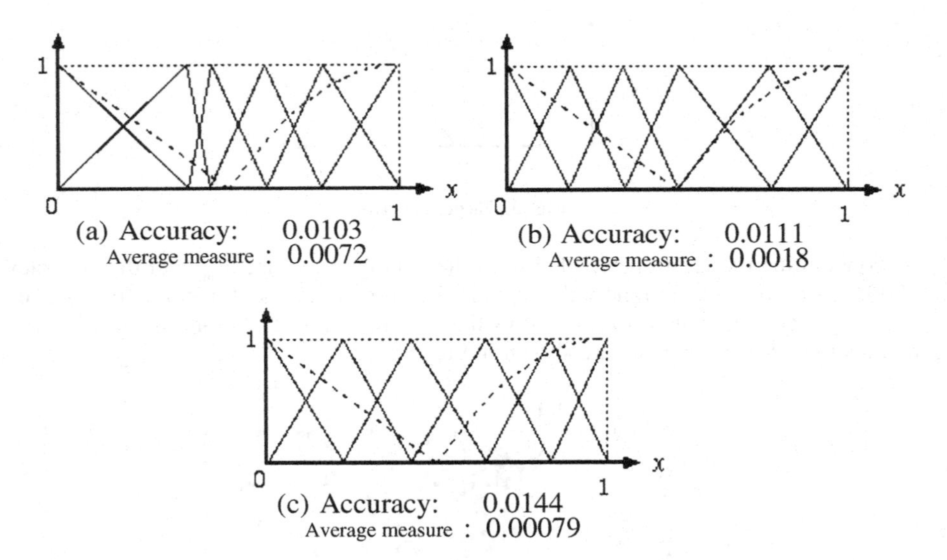

Fig. 5. Allocation of membership functions of obtained models on the Pareto front Fig. 4

Next, assume that we have the knowledge about the target system *a priori*. This knowledge is expressed, for example, as "it is linear and decreasing on the left half of the universe of discourse, and is sharply rising up from the central point with increasing x." In this case, we may think the model in Fig. 5 (a) is the most interpretable among the models in this figure, because this model fits our prior knowledge most. On the other hand, the model in Fig. 5 (c) is the least interpretable now. This case shows that the interpretability of models depends on our knowledge.

7. Conclusion

This paper discussed interpretability of fuzzy models by separating the cases with/without prior knowledge about target systems. Interpretability depends on the knowledge. For quantitative analysis of interpretability, this paper presented a new measure of conciseness of fuzzy models by focusing the cases where no prior knowledge was available. This paper defined the conciseness of fuzzy models, and quantified the conciseness by introducing De Luca and Termini's fuzzy entropy, and also defined a new measure of the deviation of a membership function from symmetry. By combining the new measure and De Luca and Termini's measure the new measure for the conciseness was derived. Based on the new measure, the average measure was defined to evaluate the shapes and the deviation of allocation of membership functions of fuzzy models. Experimental results showed that the new measure coincide with the observation of conciseness, and it was in conflict with the accuracy of fuzzy models. The results also showed that the least concise model was the most interpretable in the case where we had no prior knowledge. The interpretability was shown to be changed with *a priori* knowledge about the target system.

References

1. S. Matsushita and T. Furuhashi, et al., "Selection of Input Variables Using Genetic Algorithm for Hierarchical Fuzzy Modeling," Proc. of 1996 First Asia-Pacific Conference on Simulated Evolution and Learning, pp.106-113, 1996.
2. H. Akaike, "Information Theory and an Extension of the Maximum Likelihood Principle," 2nd International Symposium on Information Theory, pp.267-281, 1973.
3. H. Nomura, S. Araki, I. Hayashi and N. Wakami, "A Learning Method of Fuzzy Reasoning by Delta Rule," Proc. of Intelligent System Symposium, pp.25-30, 1992.
4. M. Setnes and R. Babuska and H.B.Verbruggen, "Rule-Based Modeling: Precision and Transparency," IEEE Trans. Syst., Man, Cybern., pt.C, Vol. 28, No. 1, pp.165-169, Feb. 1998.
5. J. Valente de Oliveira, "Semantic Constraints for Membership Function Optimization," IEEE Trans. Syst., Man, Cybern., pt.A, Vol. 23, No. 1, pp.128-138, Jan., 1999.
6. M. Setnes and H. Roubos, "GA-Fuzzy Modeling and Classification: Complexity and Performance," IEEE Trans. Fuzzy Syst., Vol. 8, No. 5, pp.509-522, Oct. 2000.
7. A. De Luca and S. Termini, "A Definition of a Nonprobabilistic Entropy in the Setting of Fuzzy Sets Theory," Information and Control, Vol. 20, pp.301-312, 1972.
8. M. Mizumoto, "Fuzzy Control Under Various Approximate Reasoning Methods," Proc. of Second IFSA Congress, pp.143-146, 1987.

Degree of Similarity in Fuzzy Partition

Rolly Intan and Masao Mukaidono

Department of Computer Science
Meiji University, Kawasaki-shi, Kanagawa-ken, Japan
{rolly,masao}@cs.meiji.ac.jp

Abstract. In this paper, we discuss preciseness of data in terms of obtaining degree of similarity in which fuzzy set can be used as an alternative to represent imprecise data. Degree of similarity between two imprecise data represented in two fuzzy sets is approximately determined by using fuzzy conditional probability relation. Moreover, degree of similarity relationship between fuzzy sets corresponding to fuzzy classes as results of fuzzy partition on a given finite set of data is examined. Related to a well known fuzzy partition, called *fuzzy pseudopartition* or *fuzzy c-partition* where c designates the number of fuzzy classes in the partition, we introduce *fuzzy symmetric c-partition* regarded as a special case of the fuzzy c-partition. In addition, we also introduce fuzzy covering as a generalization of fuzzy partition. Similarly, two fuzzy coverings, namely *fuzzy c-covering* and *fuzzy symmetric c-covering* are proposed corresponding to *fuzzy c-partition* and *fuzzy symmetric c-partition*, respectively.

1 Introduction

In the real word application, data are often imprecise in which degrees of preciseness of data are different. In this case, fuzzy set expression can be considered as an alternative to represent degree of preciseness of data. Degree of similarity between two imprecise data represented by fuzzy sets expression is approximately calculated by *fuzzy conditional probability relation.* In our previous paper [4,5], we proposed *weak fuzzy similarity relation* to be more realistic in representing similarity between two elements of data than fuzzy similarity relations [14] characterized by reflexive, symmetric, and (max-min)transitive properties. Naturally, degrees of similarity between elements of data in the real-world application are neither necessarily symmetric nor necessarily transitive as discussed in [11,10,5]. Here weak fuzzy similarity relation is regarded as a generalization of fuzzy similarity relation and fuzzy conditional probability relation can be considered as a concrete example of weak fuzzy similarity relation.

In this paper, based on fuzzy conditional probability relation, we examine and discuss similarity of fuzzy sets as results of fuzzy partition over a particular domain of data. A kind of fuzzy partition, called *fuzzy pseudopartition* or *fuzzy c-partition* where c designates the number of fuzzy classes in the partition, was introduced in [1,8]. Related to the fuzzy c-partition, we introduce *fuzzy symmetric c-partition* as a special case of fuzzy c-partition in which partition induced

N.R. Pal and M. Sugeno (Eds.): AFSS 2002, LNAI 2275, pp. 20–26, 2002.

by fuzzy symmetric c-partition will produce a symmetric similarity table. On the other hand, we consider fuzzy covering as a generalization of fuzzy partition by allowing fuzzy classes or fuzzy sets to take overlaps one to each other in dealing the element of data. In other words, a single element of data may totally (with the membership degree of 1) belong to more than one fuzzy class or fuzzy set. Similarly, *fuzzy c-covering* and *fuzzy symmetric c-covering* are introduced in fuzzy covering corresponding to fuzzy c-partition and fuzzy symmetric c-partition, respectively in fuzzy partition.

2 Preciseness and Similarity of Data

In the real world application, data are often imprecise in which degrees of preciseness might be different. For example two imprecise data *hot* and *about* 40°C have different degree of preciseness. Intuitively, *about* 40°C is more precise than *hot*. In general, preciseness of data is assigned from *total ignorance* to *crisp* where *total ignorance* and *crisp* express the most imprecise and the most precise of data, respectively. In this case, fuzzy sets expression can be used as an alternative to represent preciseness of data from *total ignorance* to *crisp*. The following definition shows how fuzzy sets can be used to represent imprecise data.

Definition 1. *Let D be an ordinary set of a given particular domain of data. An imprecise data, X over domain D regarded as a fuzzy set X on D is simply defined as a mapping from D to the closed interval [0,1] characterized by a membership function*

$$\mu_X : D \to [0,1], \tag{1}$$

where X is a label and μ_X is a membership function of the fuzzy set.

By Definition 1, *total ignorance* over D and *crisp* $d_i \in D$, where $D = \{d_1, d_2, ..., d_n\}$, are also simply defined as[5,6,7]:

$$\mu_{\text{TI}} = \{1/d_1, ..., 1/d_n\},$$
$$\mu_{d_i} = \{0/d_1, ..., 0/d_{i-1}, 1/d_i, 0/d_{i+1}, ..., 0/d_n\},$$

respectively, where i.e., $\mu_{\text{TI}}(d_1) = 1$ represented by $1/d_1$.
Here, TI is regarded as a fuzzy set representing universal set of domain D while crisp of d_i is a fuzzy sets that consists of a crisp element of data, d_i.

Example 1. Two imprecise data over domain **Temperature** in degree Celsius, *Warm*(W) and *Rather Hot*(RT), might be represented in the following membership functions of two fuzzy sets:

$$\mu_W = \{0.2/24, 0.5/26, 1/28, 1/30, 0.5/32, 0.2/34\},$$
$$\mu_{RH} = \{0.5/30, 1/32, 1/34, 0.5/36\}.$$

Based on their membership function, *Rather Hot* is more precise than *Warm*, because *Warm* takes wider space of area in **Temperature** than *Rather Hot*. Here, measures of preciseness is regarded as measures of specificity as proposed in [13]. Considering this reason, degree of similarity between two data is neither necessarily symmetric nor necessarily transitive as discussed in [11,10,4]. Formally, these properties belong to *weak fuzzy similarity relation* as a generalization of fuzzy similarity relation. For U is universal set of fuzzy sets defined on a given domain D, we have

Definition 2. *A* ***fuzzy similarity relation*** *[14] is a mapping,* $s : U \times U \rightarrow [0,1]$*, such that for* $X, Y, Z \in U$*,*

$$\begin{aligned}
&\text{(a) Reflexivity} : s(X,X) = 1,\\
&\text{(b) Symmetry} : s(X,Y) = s(Y,X),\\
&\text{(c) Max–min transitivity} : s(X,Z) \geq \max_{Y \in U} \min[s(X,Y), s(Y,Z)].
\end{aligned}$$

Definition 3. *A* ***weak fuzzy similarity relation*** *is a mapping,* $s : U \times U \rightarrow [0,1]$*, such that for* $X, Y, Z \in U$*,*

$$\begin{aligned}
&\text{(a) Reflexivity} : s(X,X) = 1,\\
&\text{(b) Conditional symmetry} : \text{if } s(X,Y) > 0 \text{ then } s(Y,X) > 0,\\
&\text{(c) Conditional transitivity} : \text{if } s(X,Y) \geq s(Y,X) > 0\\
&\qquad \text{and } s(Y,Z) \geq s(Z,Y) > 0 \text{ then } s(X,Z) \geq s(Z,X).
\end{aligned}$$

A fuzzy conditional probability relation is regarded as a concrete example of weak fuzzy similarity relations and formally defined as follows:

Definition 4. *Let* μ_X *and* μ_Y *be two membership function over a given domain* D *for two labels* X *and* Y *of a universe* U*. A fuzzy conditional probability relation is a mapping,* $R : U \times U \rightarrow [0,1]$*, defined by:*

$$R(X,Y) = \mathrm{P}(X \mid Y) = \frac{|X \cap Y|}{|Y|} = \frac{\sum_{d \in D} \min\{\mu_X(d), \mu_Y(d)\}}{\sum_{d \in D} \mu_Y(d)}, \tag{2}$$

where $R(X,Y)$ *means the degree* Y *supports* X *or the degree* Y *is similar to* X *and* $|Y| = \sum_{d \in D} \mu_Y(d)$ *is regarded as cardinality of* Y*.*

Fuzzy conditional probability in Definition 4 is principally the same as *fuzzy relative cardinality*(Dubois and Prade, 1982 [3]) as shown in the following equation:

$$I(F,G) = \frac{|F \cap G|}{|F|},$$

where $|F| = \sum_d \mu_F(d)$ and intersection is defined as minimum. Kosko [9] has pointed out the analogy between $I(F,G)$ and a conditional probability $\mathrm{P}(A|B)$ where B and F play the same role. In practical application, fuzzy conditional probability relations may be used as a basis of representing degree of similarity relationships between fuzzy sets in the universe U. In the definition of fuzzy conditional probability relations, the probability values may be estimated based on the semantic relationships between fuzzy sets by using the epistemological or subjective view of probability theory.

Example 2. Related to Example 1, degree of similarity between *Warm*(W) and *Rather Hot*(RH) is calculated by (2) as follows:

$$R(\mathrm{W},\mathrm{RH}) = \frac{\min(0.5, 0.5) + \min(0.2, 1)}{0.5 + 1 + 1 + 0.5} = \frac{0.7}{3.0},$$
$$R(\mathrm{RH},\mathrm{W}) = \frac{\min(0.5, 0.5) + \min(0.2, 1)}{0.2 + 0.5 + 1 + 1 + 0.5 + 0.2} = \frac{0.7}{3.4}.$$

Furthermore, some properties of (fuzzy) conditional probability relations are given in [5]: for $X, Y, Z \in U$,
(*r*0) $R(X,Y) = R(Y,X) = 1 \Longleftrightarrow X = Y$,
(*r*1) $[R(Y,X) = 1, R(X,Y) < 1] \Longleftrightarrow X \subset Y$,
(*r*2) $R(X,Y) = R(Y,X) > 0 \Longrightarrow |X| = |Y|$,
(*r*3) $R(X,Y) < R(Y,X) \Longrightarrow |X| < |Y|$,
(*r*4) $R(X,Y) > 0 \Longleftrightarrow R(Y,X) > 0$,
(*r*5) $[R(X,Y) \geq R(Y,X) > 0, R(Y,Z) \geq R(Z,Y) > 0] \Longrightarrow R(X,Z) \geq R(Z,X)$.

3 Fuzzy Partition and Degree of Similarity

A kind of fuzzy partition, called *fuzzy pseudopartition* or *fuzzy c-partition* proposed in [1,8], where c designates the number of fuzzy classes in the partition is defined as follows:

Definition 5. *Let* $D = \{d_1, ..., d_n\}$ *be a given domain of data. A fuzzy c-partition of* D *is a family of fuzzy subsets or fuzzy classes of* P*, denoted by* $P = \{P_1, P_2, ..., P_c\}$*, which satisfies*

$$\sum_{i=1}^{c} \mu_{P_i}(d_k) = 1 \quad \text{for all } k \in \mathbb{N}_n, \text{ and} \tag{3}$$

$$0 < \sum_{k=1}^{n} \mu_{P_i}(d_k) < n \quad \text{for all } i \in \mathbb{N}_c, \tag{4}$$

where c *is a positive integer and* $\mu_{P_i}(d_k) \in [0, 1]$.

(3) shows that every crisp element of data must totally belong to fuzzy classes in partition. All fuzzy classes in partition must entirely cover all crisp elements of data. On the other hand, (4) tells that empty class (the class or subset that has no any element of data) and universal class (the class which perfectly covers all elements of data) are meaningless, where universal class is considered as total ignorance in terms of preciseness of data (see Section 2).

Example 3. A fuzzy 3-partition of Education($\mathbb{E}$)={ES,JHS,SHS,BA,MS,PhD} might be arbitrarily given by:

$$\mu_{low\_edu} = \{1/ES, 0.8/JHS, 0.2/SHS\},$$
$$\mu_{mid\_edu} = \{0.2/JHS, 0.8/SHS, 0.4/BA\},$$
$$\mu_{hi\_edu} = \{0.6/BA, 1/MS, 1/PhD\}.$$

Similarity table concerning these three fuzzy classes calculated by (2) shows in Table 1 (Note: $R(low\_edu, mid\_edu) = 0.28$, but $R(mid\_edu, low\_edu) = 0.20$).

Table 1. Similarity $R(X,Y)$ of Partitions

$X \setminus Y$	low_edu	mid_edu	hi_edu
low_edu	1.00	0.28	0
mid_edu	0.20	1.00	0.15
hi_edu	0	0.28	1.00

Table 2. Similarity $R(X,Y)$ of Elements

$X \setminus Y$	ES	JHS	SHS	BA	MS	PhD
ES	1.0	0.69	0.15	0.0	0.0	0.0
JHS	0.8	1.0	0.36	0.27	0.0	0.0
SHS	0.2	0.41	1.0	0.56	0.0	0.0
BA	0.0	0.24	0.43	1.0	0.60	0.60
MS	0.0	0.0	0.0	0.44	1.0	1.0
PhD	0.0	0.0	0.0	0.44	1.0	1.0

On the other hand, every crisp element of data in education can also be represented as fuzzy sets in terms of their fuzzy classes. For instance,

$$\mu_{ES}(low\_edu) = \frac{\mu_{low\_edu}(ES)}{\sum_{X \in \mathbb{E}} \mu_{low\_edu}(X)} = \frac{1}{1 + 0.8 + 0.2} = 0.5.$$

We may consider $\mu_{ES}(low\_edu)$ as true value of preposition "if low_edu then ES" or true value of ES given low_edu. All results are given by:

$\mu_{ES} = \{0.5/low\_edu\}$, $\mu_{BA} = \{0.29/mid\_edu, 0.23/hi\_edu\}$,
$\mu_{JHS} = \{0.4/low\_edu, 0.14/mid\_edu\}$, $\mu_{MS} = \{0.385/hi\_edu\}$,
$\mu_{SHS} = \{0.1/low\_edu, 0.57/mid\_edu\}$, $\mu_{PhD} = \{0.385/hi\_edu\}$,

where the results are also satisfied (3) and (4).
Finally, similarity table concerning the elements of data in domain education is represented in Table 2.

Naturally, this may show an interesting concept that every element or object will have relation (similarity) to the others if they are involved in the same groups (classes). They will have stronger relationship (similarity) if they are involved in more the same groups (classes). On the other hand, an increasing number of elements in a class will reduce the degree of relationship (similarity) among the element involved in the class. Obviously Table 1 and 2 show that degrees of similarity between fuzzy classes as well as elements of data are not symmetric as a consequence of membership functions arbitrarily given in *low_edu, mid_edu,* and *hi_edu* where $|mid\_edu| < |low\_edu| < |hi\_edu|$, so that $|PhD| = |MS| < |ES| < |BA| < |JHS| < |SHS|$. However, it depends on the type of data and application in which we may provide the same cardinality of fuzzy classes in partition by applying the following equation (5) instead of (4) in Definition 3.

$$\sum_{k=1}^{n} \mu_{P_i}(d_k) = M, \quad for\ all\ i \in \mathbb{N}_c,\ where\ 0 < M < n. \tag{5}$$

Here, (5) is stronger than (4) or (4) is a generalization of (5). In order to distinguish this concept of partition from *fuzzy c-partition* as defined in Definition 3, the concept may be called ***fuzzy symmetric c-partition***, because it provides symmetric property in similarity table. Moreover, fuzzy symmetric c-partition may be considered as a basis of constructing *(fuzzy) proximity relation* [2,5,12]. Clearly, Example 3 is recalled in terms of fuzzy symmetric c-partition as follows.

Example 4. A fuzzy symmetric 3-partition of Education={ES,JHS,SHS,BA,MS, PhD} might be arbitrarily given by:

$$\mu_{low\_edu} = \{1/ES, 0.8/JHS, 0.2/SHS\},$$
$$\mu_{mid\_edu} = \{0.2/JHS, 0.8/SHS, 0.8/BA, 0.2/MS\},$$
$$\mu_{hi\_edu} = \{0.2/BA, 0.8/MS, 1/PhD\}.$$

Similarity table concerning these three fuzzy classes shows in Table 3.

Table 4. Similarity $R(X,Y)$ of Elements

$X \setminus Y$	ES	JHS	SHS	BA	MS	PhD
ES	1.0	0.8	0.2	0.0	0.0	0.0
JHS	0.8	1.0	0.4	0.2	0.2	0.0
SHS	0.2	0.4	1.0	0.8	0.2	0.0
BA	0.0	0.2	0.8	1.0	0.4	0.2
MS	0.0	0.2	0.2	0.4	1.0	0.8
PhD	0.0	0.0	0.0	0.2	0.8	1.0

Table 3. Similarity $R(X,Y)$ of Partitions

$X \setminus Y$	*low_edu*	*mid_edu*	*hi_edu*
low_edu	1.00	0.2	0
mid_edu	0.2	1.00	0.2
hi_edu	0	0.2	1.00

Similarly, elements of data in education can also be represented as fuzzy sets in terms of their partitions as follows:

$\mu_{ES} = \{0.5/low\_edu\}$, $\mu_{BA} = \{0.4/mid\_edu, 0.1/hi\_edu\}$,
$\mu_{JHS} = \{0.4/low\_edu, 0.1/mid\_edu\}$, $\mu_{MS} = \{0.1/mid\_edu, 0.4/hi\_edu\}$,
$\mu_{SHS} = \{0.1/low\_edu, 0.4/mid\_edu\}$, $\mu_{PhD} = \{0.5/hi\_edu\}$.

Clearly, they are also satisfied (3) and (5). Finally, similarity table concerning elements of data is represented in Table 4.

Again, considering the process of partition as shown in Definition 3, restriction in (3) shows that fuzzy classes in the partition seem to be 'disjoint' in dealing a single element of data. In other words, if equality in (3) is changed to be inequality as shown in the following equation (6), it allows the fuzzy classes to take overlaps one to each other in dealing the element of data.

$$\sum_{i=1}^{c} \mu_{P_i}(d_k) \geq 1 \quad for\ all\ k \in \mathbb{N}_n. \tag{6}$$

A single element of data may totally (with the degree of 1) belong to more than one fuzzy class. Here, the process must be called ***fuzzy covering*** instead of *fuzzy partition*, if we assume that fuzzy classes in partition must be disjoint in dealing a single element. Similarly, there are two fuzzy covering; ***fuzzy c-covering*** is restricted by (4) and (6) and ***fuzzy symmetric c-covering*** is restricted by (5) and (6). Fuzzy as well as crisp covering play important roles in generalization of rough set theory [4,10].

4 Conclusions

This paper discussed how fuzzy sets were used as an alternative to represent imprecise data. Degree of similarity relationship between two imprecise data represented in two fuzzy sets was approximately determined by using fuzzy conditional probability relation. Fuzzy partition as an alternative of constructing fuzzy granularity and mostly used in quantization and clustering data was examined in order to determine similarity of fuzzy classes as the result of partition as well as the elements of data because of their association in the classes. Every element, object, or whatever will have a relation (similarity) to the others because of their association in the same groups (classes). In this case, degree of similarity can be approximately calculated corresponding to their association in the same groups (classes) by using fuzzy conditional probability relation. Related to the symmetric property, we introduced *fuzzy symmetric c-partition* which was considered as a special case of fuzzy c-partition as proposed in [1,8]. Fuzzy symmetric c-partition is also regarded as a basis of constructing (fuzzy) proximity relation [2,5,12]. Moreover we also introduced fuzzy covering as a generalization of fuzzy partition. Similarly, there are two fuzzy covering; *fuzzy c-covering* and *fuzzy symmetric c-covering* corresponding to *fuzzy c-partition* and *fuzzy symmetric c-partition*, respectively.

References

1. Bezdek, J. C., *Pattern Recognition with Fuzzy Objective Function Algorithms*, (Plenum Press, New York, 1981).
2. Dubois, D., Prade, H., *Fuzzy Sets and Systems: Theory and Applications*, (Academic Press, New York, 1980).
3. Dubois, D., Prade, H., 'A Unifying View of Comparison Indices in a Fuzzy Set-Theoretic Framework', *Fuzzy Sets and Possibility Theory-Recent Developments*, Pergamon Press, (1982), pp. 1-13.
4. Intan, R., Mukaidono, M.,Yao, Y.Y., 'Generalization of Rough Sets with α-coverings of the Universe Induced by Conditional Probability Relations', *Proceedings of International Workshop on Rough Sets and Granular Computing*, (2001), pp.173-176.
5. Intan, R., Mukaidono, M., 'Conditional Probability Relations in Fuzzy Relational Database ', *Proceedings of RSCTC'00*, (2000), pp.213-222.
6. Intan, R., Mukaidono, M., 'Application of Conditional Probability in Constructing Fuzzy Functional Dependency(FFD)', *Proceedings of AFSS'00*, (2000), pp.271-276.
7. Intan, R., Mukaidono, M., 'Fuzzy Functional Dependency and Its Application to Approximate Querying', *Proceedings of IDEAS'00*, (2000), pp.47-54.
8. Klir, G.J., Yuan, B., *Fuzzy Sets and Fuzzy Logic: Theory and Applications*,(Prentice Hall, New Jersey, 1995).
9. Kosko, B., 'Fuzziness vs. Probability', *Int. J. of General Systems*, Vol. 17, (1990), pp. 211-240.
10. Slowinski, R., Vanderpooten, D., 'A Generalized Definition of Rough Approximations Based on Similarity', *IEEE Transactions on Knowledge and Data Engineering*, (2000), Vol.12, No.2, pp.331-336
11. Tversky, A., 'Features of Similarity', *Psychological Rev. 84(4)*, (1977), pp. 327-353.
12. Shenoi, S., Melton, A., 'Proximity Relations in The Fuzzy Relational Database Model', *Fuzzy Sets and Systems, 31*, (1989), pp. 285-296
13. Yager, R.R., 'Ordinal Measures of Specificity', *Int. J. General Systems, Vol.17*, (1990), pp. 57-72.
14. Zadeh, L.A., 'Similarity Relations and Fuzzy Orderings ', *Inform. Sci.3(2)*, (1970), pp. 177-200.

Fuzzy Set in Default Reasoning

Swapan Raha and Sanaul Hossain

Dept. of Mathematics, Visva-Bharati, Santiniketan 731235, India

Abstract. The purpose of this report is to draw attention to the use of fuzzy sets in default reasoning. An attempt has been made to represent and manipulate default consequences which are frequently encountered in reasoning about incompletely specified worlds, based on the theory of approximate reasoning. Here, the uncertainty associated with a typical individual sanctioned by default is represented by means of a composite vague certainty qualified statement. The vague certainty value assigned to a statement is based on only the consistency of the consequence of a default rule. Inferences are drawn using similarity based approximate reasoning.

1 Introduction

Many assertions about the real world express default properties of individuals or class of individuals. As for example, 'hard-workers are adults', 'intelligents prosper', 'teenagers are unmarried', 'snakes are highly-poisonous', 'birds fly faster than bees', 'Pathans are well-built' express default properties. Also it is interesting to see that they allow vague concepts in their expression. Reasoning by default corresponds to the process of deriving conclusions in the absence of total knowledge about the world. Basically, default inferences are in harmony with intuition — unsupported by confirmation.

Let us consider a typical instance of reasoning by default. Based on only supporting evidence and the absence of contradictory evidence, there may be times when conclusions like *Arpan is over 25 years old*, is believed to be drawn from the fact that *Arpan is a married person* as well as a frequently used conditional statement **A married person is over 25 years old**. At a latter stage, more information about Arpan suggests that he is *a student of undergraduate studies*. It is also well-known that **Undergraduate students are around twenty years old**. This would force us to revise our previous belief that *Arpan is over 25 years old.*

To cope with such form of reasoning through symbolic manipulation is a complex and difficult task. We may assert a fact when we may prove it and when there are no information available to the contrary. So in the absence of further information, we may derive information about Arpan's age from the fact that he is a married young and the general condition that **married persons are typically over 25 years old**. Of course, we may revise the same as and when some contradictory information is available. Likewise, the statement *Snakes are poisonous* means that for anything that is a snake, that thing is poisonous. Now

N.R. Pal and M. Sugeno (Eds.): AFSS 2002, LNAI 2275, pp. 27–33, 2002.

as we know that this statement is not absolutely true in the real world — a python is non-poisonous and it is not the only exception. But certainly it is an exception, in a list of few exceptions. Nevertheless, such a general remark about the characteristics of snakes is acceptable to many, if not to all. In these cases, the problem is to reject derivation of exceptional cases. A simple solution to this problem is to include the list of exceptions in the conditional statement. This is not a good suggestion because it would unnecessarily burden every default rule with an unmanageably large number of possible exceptions. Since our knowledge about the world is necessarily incomplete an explicit list of such exceptions is not even possible to make. Also our primary intention is with a belief in the consequence of a default rule and not with a list of possible exceptions. So we look for a mechanism, which would allow us to jump to conclusion, in the face of incomplete information and modification (revision) of such results, if necessary, at an appropriate time, in order to avoid the possibility of accepting the fact and its contradiction at the same time.

An attempt has been made to represent and manipulate incomplete and imprecise default knowledge, based on the theory of fuzzy sets. We find similarity is inherent in reasoning with vague default[8]. Therefore, we use similarity based approximate reasoning in decision making by default. Work in this direction is not numerous. In [10], Yager proposed to represent default knowledge through possibility-qualification and used the framework of Zadeh's concept of approximate reasoning for reasoning by default. Dubois and Prade [4] have considered a modified fuzzy set augmented, with a certainty factor for the representation of default values. Representation and manipulation are based on the theory of possibility distribution in an approximate reasoning framework.

In Yager's approach, a statement 'X is A is possible' is used to indicate a possibility-qualification. It should be mentioned here that, a possibility-qualification is a less restrictive data than the unqualified simple statement 'X is A'. The possibility-qualified statement induces a statement X is A^+, where A^+ is the set of all subsets of the power set of U, the universal set of the linguistic variable. A^+ is the set of all subsets of U which intersect A. Reasoning with default is performed in the framework of approximate reasoning based on the theory of possibility. Thus, from *if X is A is posssible then Y is B* and the data X is C, a consequence Y is D may be obtained as $\mu_D(v) = (1 - \text{Pos}[A \mid C]) \vee \mu_B(v)$.

Dubois and Prade have shown that,'the above treatment of default rules in an approximate reasoning framework fails to capture the uncertain nature of conclusions obtained via default rules' [4]. They proposed that, some uncertainty should be attached with each default conclusion, when relevant, in a rigorous manner. They have also observed that in Yager's approach, 'conclusions inferred from defaults and that inferred from rules without exceptions, have the same status, while clearly, they are not valid in the same sense: the latter are undisputable while the former may be interpreted as default values'. So, they proposed to modify the assignment for an attribute, with a special class of fuzzy sets augumented with a certainty factor, and deduced a technique for the propagation of uncertainty in default conclusions in the framework of the theory of possibility.

The authors in [6] represented a default value for an attribute X and an object u_0 by means of a special kind of fuzzy set A^λ, where

$$(\forall v)\mu_{A^\lambda}(v) = \max(\mu_A(v), 1 - \lambda), \lambda < 1. \quad (1)$$

The subset A gathers the a priori more plausible values among the possible ones; $1 - \lambda$ estimates to what extent it is possible that the value lies outside A, exceptional ones. Then reasoning with default rules is performed in the framework of conventional approximate reasoning. In [1], the authors propose to model default rules in the framework of possibility theory. They expressed that the exceptional situation is less possible than the normal state of affairs. However, human reasoning does not use only generic pieces of knowledge pervaded with exceptions; but also take advantage of independent assumptions. They made an attempt to describe the notion of independence in possibility theory [2] and applied it to default reasoning.

Motivated by the above works and in order to capture the uncertainty associated with a default value, in the present work, we represent a default rule as a fuzzily quantified conditional rule and a default value assigned to an attribute X corresponding to a typical individual u_0 from U is by means of a composite statement of the form $(X(u_0) = A)$ is λ_certain or simply, $(X(u_0) = A)$ is λ, where A and λ are linguistic labels and, therefore, may be vague [8]. As for example, *some intellectuals never prosper*, *most snakes are poisonous*, *many Pathans are well-built*, *few married persons are less than 25 years old*, *almost all birds fly faster than most bees*, certainly *Imran is well-built* and *Cobras are poisonous* or say, it is fairly certain that *Hornbills fly faster than honeybees* .

The vague certainty value assigned to a default conclusion is based on only the consistency of the consequence of a default rule. A '1' is assigned to a certain statement, a '0' to an uncertain statement and a positive proper fraction to a partial or graded certain statement. The possible certainty-value of a vague statement constitutes the continuum [0,1]. Thus, any linguistic certainty-value may be represented by a fuzzy set over the universe of discourse [0,1]. In the sequel, the statement 'X is A, is certain' may be taken as 'X is A' and the statement 'X is A is uncertain' may be taken as 'X is U', the universe of discourse of the linguistic variable.

Another important issue need to be considered in default reasoning is, a way of blocking undesirable transitivity which arises because of absence of an appropriate mechanism to incorporate explicit reference to the list of exceptional circumstances which would block its application. The certainty factor approach used by Dubois and Prade [4] helps to block such undesirable transitivity. Our certainty-qualification technique is also found to be useful in this regard.

2 Representation of Default Knowledge

Defaults are general rules subject to exceptions. Our knowledge about the exceptions is incomplete. Whereas a number of possible instances may be listed whose conclusions are, in some sense, reasonable in the light of what is known about the world. The set of exceptions, although not completely known, therefore, is fewer in number. Thus, we conclude that defaults are rules whose conclusions are almost always certain.

In the following, let us consider a few simple examples although, it is well known that, simple examples are always somewhat misleading. The actual test of any representation lies in whether it may cope with complexity. Nevertheless, these small and simple ones at least illustrate how the concept of certainty-qualification may be put to use in default reasoning.

Let us have a rule-base that contains within it the following rule, beside others:

Def 1: If a person has high blood pressure and is a heavy smoker then that person has a very high risk of heart attack.

Such a complex rule not only consists of several simple propositions connected with well known logical connectives but also admits vague concepts. In general, we have seen that, fuzzy logic provides a framework to represent such vague concepts in terms of fuzzy sets. In fuzzy logic, the rule may be interpreted as a fuzzy relation between the linguistic variables appearing in it(e.g., Pressure, Consumption, Risk). Explicitly, in a predicate like notation, we may represent the rule in a standard fuzzy formula as

If *Pressure(Blood)* = **high** and *Consumption(Smoke)* = **heavy** then *Risk(Heart Attack)* = **very high**.

The underlined terms may be conveniently represented using fuzzy sets and the rule may be represented by a fuzzy relation.

Next let us consider another rule:

Def 2: If the Horse is a colt and constitution healthy then, it can run fast.

In fuzzy logic, the above rule may be represented as

If *Age(Horse)* = **colt** and *Constitution(Horse)* = **healthy** then *Speed(Horse)* = **fast**.

Thus, a default rule may be represented by an expression of the form

$$A_1(X) \wedge A_2(X) \wedge \cdots \wedge A_m(X) \to B(X)$$

where $\{A_i(X) \mid i = 1, 2, \cdots, m\}, B(X)$ are all simple formulae as may be found in fuzzy logic. The concepts $\{A_i \mid i = 1, 2, \cdots, m\}, B$ are respectively called the prerequisites and consequent of the default and is interpreted as:

if X is A_1 and X is A_2 and $\cdots$ and X is A_m then infer that X is B.

Now, inference sanctioned by rules without exceptions are well established and hence acceptable to all. Whereas those sanctioned by default may be found to be contradictory with more knowledge pouring in. Since, we allow vague formulas in a default rule, therefore, anything inferred by default, in absence of proper justification, may or may not be true in graded term. This suggest a more interesting representation of a vague default value as $B(x_0)$ *is* λ_certain; where λ is a certainty value modifier and is usually vague, such as most, almost, fairly, absolutely, etc. In the theory of fuzzy set they may be represented as fuzzy subsets of the unit interval. This certainty qualification of a proposition actually estimates a degree of certainty that the value of the attribute B for the object u_0 does not correspond to an exception. This observation needs a matching and thereby demands the concept of similarity based framework for reasoning with vague default.

3 Reasoning with Vague Default

Let us now discuss the issues pertaining to reasoning with such possibly vague default. Actually, ordinary rules restrict the acceptable beliefs about the world, generated by the default rules. So, a typical proposition from the acceptable set of beliefs, as sanctioned by default in absence of justifications, may or may not hold with more information at a later time. Therefore, in absence of contradictory evidence we may assume that *Arnab being married is over 25 years of age* to be partially certain. That is why, in this work, we propose to take the beliefs sanctioned by defaults to be partially certain. This certainty-qualification is expressed through imprecise knowledge and is, therefore, represented by a fuzzy set over [0,1], the possible certainty values, rather than a precise number from the said universe. The problem then reduces to provide a mechanism to manipulate vague certainty.

Let us assume that the whole set [0,1] corresponds to the uncertain case. In order to provide a unified environment for the manipulation of vague concepts let us assume that, when X is A is certain we take it as simply X is A. From a partially certain proposition we may construct a valid proposition as and when necessary for manipulation. The statement X *is* A *is* λ_certain may be interpreted at the semantic level as 'it is partially certain that X is B' where, $\mu_B(u) = \mu_{\lambda\_\text{certain}}(\mu_A(u))$.

Next, we consider a typical inference pattern as:
Given a default rule *if* X *is* $A_1 \wedge X$ *is* $A_2 \wedge \cdots \wedge X$ *is* A_m then X *is* B
and a fact that X_0 *is* $A_1{}' \wedge X_0$ *is* $A_2{}' \wedge \cdots \wedge X_0$ *is* $A_m{}'$,
we expect a reasonable conclusion as X_0 *is* B' *is* λ_certain. In absence of any contradictory information about the $\{B_i \,:\, i = 1, 2, \cdots, m\}$ we take the conclusion to be partially certain, a fuzzy set defined over the unit interval [0,1] such as almost-certain, very-certain, certain, not very-certain, fairly-certain, etc, according to information available. For instance, we may take it as almost-certain, where

$$\mu(w) = \{\frac{1}{1+10(w-1)^2}\}^{1/2} \; ; \; w \in [0,1].$$

A default value for an attribute B and a particular object X_0 is represented by means of a certainty-qualified statement such as $B(X_0)$ *is* λ_i_certain. This partial certainty assignment although subjective must be meaningful and may act as an estimate of the consistency of the conclusion.

Let us represent the concepts A_i, A_i' as fuzzy subsets of U_i and B as a fuzzy subset of V. Then a consequence may be computed according to the following basic steps:

Algorithm : Default reasoning

Step 1. Compute $S(A_i, A_i')$ for $i = 1, 2, \cdots, m$ and set

$$s = \min\{S(A_1, A_1'), S(A_2, A_2'), \cdots, S(A_m, A_m')\}.$$

Step 2. Translate the default rule and compute $R(A_1, A_2, \cdots, A_m, B)$ using any suitable translating rule possibly, a T-norm operator.

Step 3. Modify R with s to obtain the modified conditional relation

$$R' = R(A_1, A_2, \cdots, A_m, B \mid A_1', A_2', \cdots, A_m')$$

according to $R' = 1 - (1 - R).s$, where $\mu_{R'}(u, v) = 1 - (1 - \mu_R(u, v)).s$.

Step 4. Use sup-projection operation on R' to obtain B' as $\mu_{B'(X_0)}(w) = \sup_{u \in U} \mu_{R'}(u, w)$.

For the computation of similarity value we may use the following definition.

Definition: Let $A = \sum_{u \in U} \{\mu_A(u)/u\}$ and $B = \sum_{u \in U} \{\mu_B(u)/u\}$ be two fuzzy sets defined over the same universe of discourse U. The similarity index of the pair $\{A, B\}$ is denoted by $S(A, B)$ and is defined by

$$S(A, B) = 1 - \left(\frac{1}{n} \sum_u \mid \mu_A(u) - \mu_B(u) \mid^q \right)^{1/q} \quad (2)$$

where n is the cardinality of the universe of discourse and q is the family parameter.

In the above scheme, when $A_i' = A_i$; $i = 1, 2, \cdots, m$; we may choose the translating rule in such a way that $B' = B$ holds. In any case, they are found to be close to each other, when A_i' and A_i are close i.e., s is close to one when $S(A, A')$ are so.

It is easy to see that this certainty value assignment to a default value obtained from the application of a default rule helps us to revise our belief about the world through manipulation of fuzzy sets only. In fact, if, at any instance of time, we find that $B'(X_0)$ is false, then the corresponding λ_certain becomes uncertain and as in the previous case, the conjunction of the certainty values becomes undefined and we reject all such consequences. Thus, the previous belief that $B'(X_0)$ is partially certain is rejected.

Next, let us consider an interesting case, where, $A'' \subseteq A' \subseteq A$ holds. Here, X_0 *is* A'' is more informative than X_0 *is* A'. Thus, from the above facts, we may make an inference X_0 *is* A'', the justifications being the same. It has already been discussed that in such a situation $S(A'', A) \leq S(A', A)$. Assume that $0 < S(A'', A) \leq S(A', A) \leq 1$. Then
$1 - (1 - r).S(A'', A) \geq 1 - (1 - r).S(A', A)$; $0 \leq r \leq 1$ and hence $C' \leq C''$. Thus more knowledge about the prerequisite does not always guarantee more knowledge about the inference, purely non-monotonic.

Let us now consider a simple rule of the form — $A \rightarrow B$ and a default rule $A \rightarrow_{def} B$ together with the values A' and A' is $\lambda\_certain$. Having a default rule labeled and a default conclusion identified we are now in a position to capture the following reasoning patterns.

- From $A \rightarrow B$ and A' is $\lambda\_certain$ we may conclude B' is $\lambda\_certain$.
- From $A \rightarrow_{def} B$ and A' is $\lambda\_certain$ we may conclude B' is $\mu\_certain$.
- From $A \rightarrow_{def} B$ and A' we may conclude B' is $\lambda\_certain$.

4 Conclusion

Obviously, this work represents a meagre beginning and suggests a few of the relevant issues involved in default reasoning. More research on the use of fuzzy set theory in the representation of default value is required to have a better understanding of the effect of the same on the cognitive processes involved in default reasoning. However, from this initial study we see that certainty-qualified vague statements and similarity based approximate reasoning are useful in reasoning by default. We hope to have established the fact that fuzzy set theory and similarity based approximate reasoning may be made popular because of the scope of its application in different wide and challenging fields of investigation and that fuzzy logic can also be used in non-monotonic reasoning.

References

1. Benferhat S., D.Dubois, H.Prade, 'Possibilistic independence and plausible reasoning', in Proc. FAPT'95, eds., G.Cooman, D.Ruan, E.E.Kerre, 1995, pp. 47-63.
2. Benferhat S., D.Dubois, H.Prade, 'Representing default rules in possibilistic logic', in Proc. of the 3rd international conf. on *Principles of Knowledge representation and reasoning(KR'92)*, eds. B.Nebel, C.Rich, W.Swartout, 1992, pp. 673-684.
3. Delgrande J.P., 'An approach to default reasoning based on a first order conditional logic: Revised report ', *Artificial Intelligence*, 1988, vol. 36, pp. 63-90.
4. Dubois D., H.Prade, 'Default reasoning and possibility theory', *Artificial Intelligence*, 1988, vol.35, pp. 243-257.
5. Poole D., 'A logical framework for default reasoning', *Artificial Intelligence*, 1988, vol.36, pp. 27-47.
6. Prade H., 'Reasoning with fuzzy default values', in Proceedings 15th IEEE International Symposium on Multiple-valued logic, Kingston, Ont. , 1985, pp. 191-197.
7. Raha S., 'Similarity based approximate reasoning', in methodologies for the conception, design and application of intelligent systems, Proc. IIZUKA'96, 4th Int. Conf. on Soft Computing, IIZUKA, Japan, 1996, pp. 414-417.
8. Raha S., K.S.Ray, 'Reasoning with vague default', *Fuzzy Sets and Systems*, 1997, vol. 91, no. 3, pp. 327-338.
9. Reiter, R., 'A logic for default reasoning', *Artificial Intelligence*, 1980, vol.13, pp. 81-132.
10. Yager R.R., 'Using approximate reasoning to represent default knowledge', *Artificial Intelligence*, 1987, vol.31, pp. 99-112.
11. Zadeh L.A., 'A theory of approximate reasoning', in J.E.Hayes, D.Mitchie, L.I.Mikulich, eds., Machine Intelligence, Vol.9, 1979, pp. 149-194, Wiley, New York.

Interpolation in Hierarchical Rule-Bases with Normal Conclusions

László T. Kóczy[1] and Leila Muresan[2]

[1] Dept. of Telecommunications & Telematics, Technical University of Budapest, Sztoczek u. 2, 1111 Budapest, Hungary
koczy@ttt-202.ttt.bme.hu

[2] Dept. of Autovehicles and Dept. of Telecom. & Telematics, Technical University of Budapest, Sztoczek u. 6, 1111 Budapest, Hungary
leila@jgi.bme.hu

Abstract. Combining fuzzy rule interpolation with the use of hierarchically structured fuzzy rule bases, as proposed by Sugeno leads to the reduction of the fuzzy algorithms' complexity. In this paper mainly the KH method and its versions are used for interpolation. One of the drawbacks of this method is that it often results in abnormal conclusions, so the hierarchical structures are impossible to use. This paper describes how this difficulty can be avoided by using a modified version of the KH method, the MACI algorithm.

1 Introduction

The classical fuzzy algorithms deal with dense rule bases where, for each dimension, the domains are fully covered by the antecedent fuzzy sets of the rule base, thus for every input there is at least one firing rule. The main problem with this approach is dealing with the high computational complexity that it implies.

If a fuzzy model contains k variables and maximum T linguistic (or other fuzzy) terms in each dimension, the number of necessary rules is of order $O(T^k)$. Decreasing T, or k, or both can decrease this expression. The first method leads to sparse rule bases and rule interpolation, first introduced by Kóczy and Hirota (see e.g. [4,5]). The second one, more effective, aims to reduce the dimension of the sub-rule bases (k's) by using meta-levels or hierarchical structures of fuzzy rule bases. The combination of the two was first attempted in [6]. An important obstacle in using both methods is the fact that the KH method does not always result in a (normal) fuzzy set. In order to avoid this situation, instead of the KH algorithm a new version of it is used, the MACI algorithm.

This research is supported by the Hungarian Ministry of Culture and Education (MKM) FKFP 0235/1997 and FKFP 0422/1997, and the National Science Research Fund (OTKA) T019671 and T030655, and the Australian Research Council, and the Australian Research Council.

N.R. Pal and M. Sugeno (Eds.): AFSS 2002, LNAI 2275, pp. 34–39, 2002.

In the next section we present the KH method and its application in hierarchical rule bases. Section 3 shows how abnormality of the conclusion and sub-conclusions can be avoided. We conclude the paper by showing the advantages of using the combination of the presented three techniques.

2 Fuzzy Interpolation and Hierarchical Fuzzy Rule Bases

In this section two complexity reduction techniques are presented: the fuzzy rule interpolation and the use of hierarchical rule bases.

In this paper we shall use the vector representation of fuzzy sets, which assigns to every fuzzy set a vector of its characteristic points. We will denote the representation of the piecewise linear fuzzy set A as the vector $\underline{a}=[a_{-m},\ldots,a_0,\ldots,a_n]$, where a_k ($k\in[-m,n]$) are the characteristic points of A and a_0 is the reference point of A having membership degree one. One can make a difference between the left flank $\underline{a}_L=[a_{-m},\ldots,a_0]$, and the right flank $\underline{a}_U=[a_0,\ldots,a_n]$ of A, respectively. If A is a CNF then, e.g. for the right flank, $a_i\geq a_j$, $i<j\in[0,n]$ should hold with monotone decreasing α levels. A partial ordering among CNF fuzzy sets is defined as: $A\prec B$ if $a_k\leq b_k$ ($k\in[-m,n]$).

2.1 Fuzzy Rule Interpolation

The basic idea of the fuzzy rule interpolation is formulated in the *Fundamental Equation of Rule Interpolation* (FERI): $D(A^*,A_1):D(A^*,A_2)=D(B^*,B_1):D(B^*,B_2)$.

In this equation A^* and B^* denote the observation and the corresponding conclusion, while $R_1=A_1\to B_2$, $R_2=A_2\to B_2$ are the rules to be interpolated, such that $A_1\prec A^*\prec A_2$ and $B_1\prec B_2$. If $D=\tilde{d}$ (the fuzzy distance family), linear interpolation between corresponding α-cuts is performed and the generated conclusion can be computed as below, (as it is first described in [2]):

$$b_k^*=\frac{\dfrac{b_{1k}}{d(a_{1k},a_k^*)}+\dfrac{b_{2k}}{d(a_{2k},a_k^*)}}{\dfrac{1}{d(a_{1k},a_k^*)}+\dfrac{1}{d(a_{2k},a_k^*)}} \tag{1}$$

where the first index (*1* or *2*) represents the number of the rule, while the second –*k* – the corresponding α-cut. From now on we shall consider $d(x, y)= |x - y|$, so that (1) becomes: ${}^{KH}b_k^*=(1-\lambda_k)b_{1k}+\lambda_k b_{2k}$, where $\lambda_k=(a_k^*-a_{1k})/(a_{2k}-a_{1k})$ (for the left and right side respectively). For a more general description see [4].

2.2 Hierarchical Rule Bases and Fuzzy Interpolation

The input space $X = X_1 \times X_2 \times ... \times X_m$ can be decomposed, so that some of its components, e. g. $Z_0 = X_1 \times X_2 \times ... \times X_p$ determine a subspace of X ($p<m$), so that in Z_0 a partition $\Pi = \{D_1, D_2, ..., D_n\}$ can be determined: $\bigcup_{i=1}^{n} D_i = Z_0$.

In each element of Π, i.e. D_i, a sub-rule base R_i can be constructed with local validity. In the worst case, each sub-rule base refers to exactly $X/Z_0 = X_{p+1} \times ... \times X_m$

The complexity of the whole rule base $O(T^m)$ is not decreased, as the size of R_0 is $O(T^p)$, and each R_i, $i > 0$, is of order $O(T^{m-p})$, $O(T^p) \times O(T^{m-p}) = O(T^m)$.

A way to decrease the complexity would be finding in each D_i a proper subset of $\{X_{p+1}, ..., X_m\}$, so that each R_i contains only less than $m - p$ input variables. The task of finding such a partition is often difficult, if not impossible, (sometimes such partition does not even exist).

There are cases when, locally, some variables unambiguously dominate the behaviour of the system, and consequently the omission of the other variables allows an acceptably accurate approximation. The bordering regions of the local domains might not be crisp or even worse, these domains overlap. For example, there is a region D_1, where the proper subspace Z_1 dominates, and another region D_2, where another proper subspace Z_2 is sufficient for the description of the system, however, in the region between D_1 and D_2 all variables in $[Z_1 \times Z_2]$ play a significant role ($[\cdot \times \cdot]$ denoting, the space that contains all variable that occur in either argument within the brackets). In this case, sparse fuzzy partitions can be used, so that in each element of the partition a proper subset of the remaining input state variables is identified as exclusively dominant. Such a sparse fuzzy partition can be described as follows: $\hat{\Pi} = \{D_1, D_2, ..., D_n\}$ and $\bigcup_{i=1}^{n} Core(D_i) \subset Z_0$ in the proper sense (fuzzy partition).

Even $\bigcup_{i=1}^{n} Supp(D_i) \subset Z_0$ is possible (sparse partition). If the fuzzy partition chosen is informative enough concerning the behaviour of the system, it is possible to interpolate its model among the elements of $\hat{\Pi}$, as we shall see below.

Each element D_i will determine a sub-rule base R_i referring to another subset of variables. The technical difficulty is how to combine the "sub-conclusions" B_i^* with the help of R_0 into the final conclusion.

Let us assume that the fuzzy partition has only two elements: $\hat{\Pi} = \{D_1, D_2\}$, and that $[Z_1 \times Z_2] = Z_1 \times Z_2$, i.e., Z_1 and Z_2 have no common component x_i. Consequently, $X = Z_0 \times Z_1 \times Z_2$. The rule base will have the following structure:

R_0 :	R_1 :	R_2 :
If z_0 **is** D_1 **then use** R_1	**If** z_1 **is** A_{11} **then y is** B_{11}	**If** z_2 **is** A_{21} **then y is** B_{21}
If z_0 **is** D_2 **then use** R_2	**If** z_1 **is** A_{12} **then y is** B_{12}	**If** z_2 **is** A_{22} **then y is** B_{22}

Let us assume that the observation on X is A^* and its projections are: $A_0^* = A^*/Z_0$, $A_1^* = A^*/Z_1$, $A_2^* = A^*/Z_2$. Using the Fundamental Equation, the two sub-conclusions, obtained from the two sub-rule bases R_1 and R_2 are:
${}^{KH}b_k^{1*} = (1-\lambda_k^1)b_{1k}^1 + \lambda_k^1 b_{2k}^1$, and ${}^{KH}b_k^{2*} = (1-\lambda_k^2)b_{1k}^2 + \lambda_k^2 b_{2k}^2$ respectively.
(The superscript shows the reference to the rule base R_1 and R_2.)
Finally, by substituting the sub-conclusions into the meta-rule base we get:

$$ {}^{KH}b_k^* = (1-\lambda_k^0)b_k^{1*} + \lambda_k^0 b_k^{2*} . \tag{2}$$

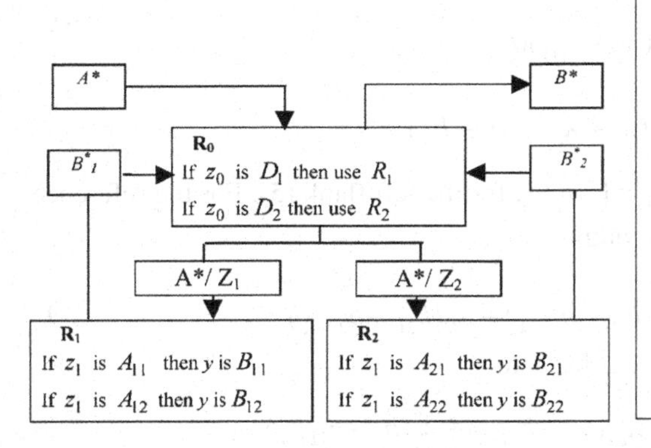

1. Determine the projection A_0^* of the observation A^* to the subspace of the fuzzy partition $\hat{\prod}$. Find the interpolating rules.
2. Determine λ_k^0.
3. For each R_i, determine A^*_i the projection of A^* to Z_i. Find the interpolating rules in each R_i.
4. Determine the sub-conclusions for each sub-rule base R_i.
5. Using the sub-conclusions from step 4, compute the final conclusion according to (2).

Fig. 1. Interpolation in a hierarchical model

3 Avoiding Abnormality of the Conclusion

It is easy to notice that the methods presented above can not be applied, unless the conclusions are also (normal) fuzzy sets. But the result of the KH interpolation sometimes is not a fuzzy set, so in this case hierarchical structures for rule bases cannot be used. In order to solve this problem we propose the use of the MACI method instead the KH interpolation method, such that the abnormality of the resulting sets is always avoided.

3.1 The MACI Algorithm

In order to avoid abnormality of the conclusion the MACI method was developed, which will be presented for the sake of simplicity, only for piecewise linear fuzzy sets that describe the antecedents and consequents of the fuzzy system. (We remark that the method can be applied also for arbitrary continuous CNF sets.)

In order to avoid abnormality the following inequality has to hold for the characteristic points of the conclusion:

$$b_i^* \geq b_j^* \forall i < j \in [-m, n] \tag{3}$$

The new method consist of three steps: choosing an appropriate coordinate system for the output space, computing the conclusion as it is done for the KH interpolation method, and finally this conclusion is transformed back into the original coordinate system. The condition (3) is assured by the choice of an appropriate transformation, which prevents the occurrence of abnormal conclusions. A detailed description of the method can be found in [7].

Using the following notations:

$$\lambda_k = \frac{a_k^* - a_{1k}}{a_{2k} - a_{1k}} \quad \text{and} \quad {}^{KH}b_k^* = (1-\lambda_k)b_{1k} + \lambda_k b_{2k}$$

(i.e. the value of the k coordinate is calculated according to the original KH approach), the conclusion of the MACI method is computed by the formulas:

$$b_k^* = {}^{KH}b_k^* + \sum_{i=0}^{k-1} (\lambda_i - \lambda_{i+1})(b_{2i} - b_{1i}), \tag{4}$$

$$b_k^* = {}^{KH}b_k^* + \sum_{i=k+1}^{0} (\lambda_i - \lambda_{i-1})(b_{2i} - b_{1i}), \tag{5}$$

$k \in [0, n]$ for the right flank (4), and $k \in [-m, 0]$, for the left flank (5). For the reference point the two equations give the same output.

Notice that from (4) results that

$$b_k^* - b_{k-1}^* = (1-\lambda_k)(b_{1k} - b_{1,k-1}) + \lambda_k (b_{2k} - b_{2,k-1}) \tag{6}$$

which applied recursively, for $k-1, \ldots, 2$ leads to

$$b_k^* = {}^{KH}b_0^* + \sum_{i=1}^{k} (1-\lambda_i)(b_{1i} - b_{1,i-1}) + \lambda_i (b_{2i} - b_{2,i-1})$$

Analogous relations hold for the left flank. From (5) it is obvious that $b_k^* - b_{k-1}^*$ is positive (it is a linear combination of two positive quantities), and thus the conclusion cannot be abnormal. It is also very easy to verify that if the observation coincides with one of the antecedents then the conclusion will be exactly the corresponding consequence.

Finally, we remark that multivariable antecedents can be handled analogously as the transformation described in this section affects only the consequent part. Common combined antecedent sets (and observation) can be calculated from the corresponding antecedents (observation) of each variable using Minkowski-type distance, where the weights are identical to 1.

3.2 Applications to Hierarchical Rule Bases

If we apply the previous method to the algorithm presented in section 2.2 from (2) and (6) it follows that:

$$b_k^* = {}^{KH}b_0^* + \sum_{i=1}^{k} S_i^0 \text{, where } S_i^0 = (1-\lambda_{ii}^0)(b_i^{1*} - b_{i-1}^{1*}) + \lambda_i^0 (b_i^{2*} - b_{i-1}^{2*})$$

Furthermore,

$$\begin{aligned} S_i^0 &= (1-\lambda_i^0)[(1-\lambda_i^1)(b_{1i}^1 - b_{1,i-1}^1) + \lambda_i^1 (b_{2i}^1 - b_{2,i-1}^1)] + \lambda_i^0 [(1-\lambda_i^2)(b_{1i}^2 - b_{1,i-1}^2) + \lambda_i^2 (b_{2i}^2 - b_{2,i-1}^2)] \\ &= (1-\lambda_i^0)[\underbrace{(1-\lambda_i^1)\Delta_{1,i}^1 + \lambda_i^1 \Delta_{2,i}^1}_{\leq \max(\Delta_{1,i}^1, \Delta_{2,i}^1)}] + \lambda_i^0 [\underbrace{(1-\lambda_i^2)\Delta_{1,i}^2 + \lambda_i^2 \Delta_{2,i}^2}_{\leq \max(\Delta_{1,i}^2, \Delta_{2,i}^2)}] \end{aligned}$$

Obviously, S_i^0 is not bigger than $\max\{\Delta_{1,i}^1, \Delta_{2,i}^1, \Delta_{1,i}^2, \Delta_{2,i}^2\}$, where $\Delta_{j,i}^k = (b_{ji}^k - b_{ji-1})$.

4 Conclusion

In this paper three techniques are combined, such that interpolation can be applied for the case of hierarchical rule bases, and abnormality of the conclusion is avoided. It was shown that it is possible to apply interpolation in fuzzy rule bases with meta-levels, even if the meta-rule base is sparse. The proposed method is similar to the approach of Mamdani, compared to the original CRI method of Zadeh, in the sense that every calculation is restricted to the projections of the observation, etc., reducing the computational complexity this way.

References

1. P. Baranyi, D. Tikk, Yeung Yam and L.T. Kóczy: "Investigation of a new alpha-cut Based Fuzzy Interpolation Method", *Tech. Rep., Dept. Of Mechanical and Automation Engineering,* The Chinese University of Hong Kong, 1999
2. S. Kawase and Q. Chen: "On fuzzy reasoning by Kóczy's Linear Rule Interpolation", *Tech.Rep,.* Teikyo Heisei University, Ichihara, Chiba, Japan, 1996, 9p.
3. L.T. Kóczy and K. Hirota: "Ordering, distance and closeness of fuzzy sets", *Fuzzy Sets and Systems* 59 1993, pp. 281-293
4. L.T. Kóczy and K. Hirota: "Approximate reasoning by linear rule interpolation and general approximation", *Internat. J. of Approximate Reasoning,* 9, 1993, pp. 197-225
5. L.T. Kóczy and K. Hirota: "Interpolative reasoning with insufficient evidence in sparse fuzzy rule bases", *Information Sciences* 71 1993, pp. 169-201
6. L.T. Kóczy and K. Hirota: "Interpolation in structured fuzzy rule bases", *FUZZ-IEEE'93,* San Francisco 1993, pp. 803-808
7. D.Tikk, P.Baranyi: "Comprehensive analysis of a new fuzzy rule interpolation method", *IEEE Trans. on Fuzzy Systems, 8 (3),* 2000, pp.281-296

The Dempster-Shafer Approach to Map-Building for an Autonomous Mobile Robot with Fuzzy Controller

Young-Chul Kim[1], Sung-Bae Cho[1], and Sang-Rok Oh[2]

[1] Dept. of Computer Science, Yonsei University, 134 Shinchon-dong, Sudaemoon-ku, Seoul, Korea
yckim@candy.yonsei.ac.kr, sbcho@csai.yonsei.ac.kr
[2] Bio-mimetic Control System Lab., KIST, 39-1 Hawolgok-dong, Seongbuk-ku, Seoul, Korea
sroh@amadeus.kist.re.kr

Abstract. This paper develops a sensor based navigation method that utilizes fuzzy logic and the Dempster-Shafer evidential theory for mobile robot in uncertain environment. The proposed navigator consists of two behaviors: Obstacle avoidance and goal seeking. To navigate reliably in the environment, we facilitate a map building process before the robot finds a goal position and create a robust fuzzy controller. In this work, the map is constructed on two-dimensional occupancy grids. The sensor values are fused into the map using the Dempster-Shafer inference rule. Whenever the robot moves, it catches new information about the environment and replaces the old map with new one. With that process the robot can wander and find the goal position. The usefulness of the proposed method is verified by a series of simulations.

1 Introduction

The basic feature of an autonomous mobile robot is the capacity to operate independently in unknown or partially known environments. To achieve this level of robustness, processes need to be developed to provide solutions for localization, globalization, map-building and obstacle avoidance. With a variety of sensor models, research is concentrated in two areas: the occupancy based map building [1] and beacon recognition and tracking [2]. The building of occupancy maps is well suited to path planning, navigation and obstacle avoidance because it explicitly models free space. However, occupancy maps are poor at localization. Also, beacon based methods have been successfully applied to the localization task. However, they fail to discern unknown types of obstacles or cluttered environments.

Path planning is one of the most vital tasks in navigation of autonomous mobile robots. The global path planning methods have been carried out in off-line manner in completely known environments. However, these methods are not suitable for navigation in complex and dynamically changing environments where unknown obstacles may be located on a priori planned path. Thus, the sensor-based local path planning methods carried out in on-line manner are required to perform the navigation of mo-

N.R. Pal and M. Sugeno (Eds.): AFSS 2002, LNAI 2275, pp. 40–46, 2002.

bile robots. Local path planning methods utilize the information provided by sensors such as ultrasonic sensor, vision, laser range finder, proximity sensor and bumper switch. Brooks [3] applied the force-field concept to the obstacle avoidance problem for mobile robots equipped with ultrasonic sensors whose readings are used to compute the repulsive forces. Borenstein and Koren proposed the virtual force field method for fast running mobile robots equipped with ultrasonic sensors and addressed the inherent limitations of this method [4].

In this paper we present a study on the map building using the Dempster-Shafer evidence theory and on the finding a goal position without collisions. We distinguish between obstacle avoidance and goal-seeking behavior. The proposed method enables the mobile robot to navigate through complex environment. Fig. 1 shows the proposed system in this paper.

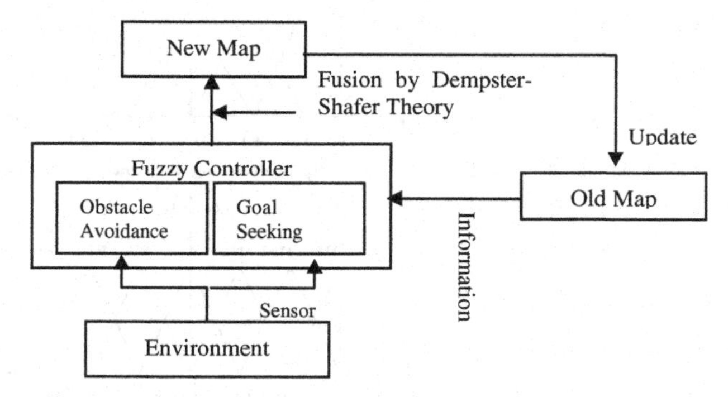

Fig. 1. The system proposed in this work

2 Fuzzy Controller for a Mobile Robot

Our work aims to design the navigator operating in uncertain environment without collision and to find goal position. The input variables of the fuzzy controller consist of the distance between each sensor suit and obstacle, $d_i, i = 1,2,\ldots,5$, obstacle's direction φ and heading difference ψ. The output variable of the fuzzy controller is the steering angle θ. The rule base of each behavior is constructed by fuzzy logic. We assume that the robot's moving distance is always the same at any time, therefore we should only consider robot's steering angle that is used to be defined robot's next position.

If the position of an obstacle is (x_{oi}, y_{oi}), then the coordinates of the obstacle are as follows :

$$x_{oi} = j * R * \cos(sT_i) + x_c \tag{1}$$

$$y_{oi} = j * R * \sin(sT_i) + y_c \tag{2}$$

where i denotes the number of sensor suit, j is the range of each sensor suit, R is the diameter of the robot, sT is the position of each sensor on the robot, and (x_c, y_c) is the coordinate of the robot's center position. Among the input variables, the distance between each sensor suit and obstacle d_i can be defined as follows:

$$d_i = \|X_{oi} - X_c\|, i = 1,2,...,5 \tag{3}$$

where $X_{oi}=(x_{oi,}y_{oi})$ is the coordinate of obstacle detected by sensor.

Fig. 2 shows the membership functions used in this work having triangular and trapezoidal shapes. Although there are many types of membership functions, triangular and trapezoidal shapes are used popularly because of their easiness to implement. Fig. 3 shows the rule base.

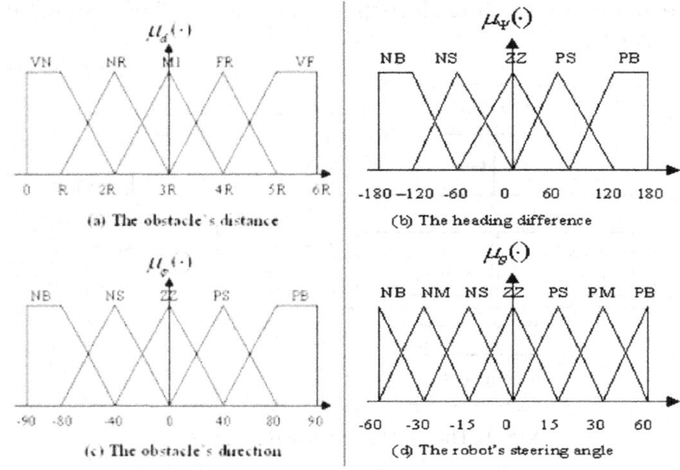

Fig. 2. The membership functions of the input/output variables

	VN	NR	MI	FR	VF
NB	NB	NB	NB	NB	NB
NS	NB	NB	NM	NM	NM
ZZ	ZZ	ZZ	NS	ZZ	ZZ
PS	PM	PM	PM	ZZ	PS
PB	PB	PB	PB	PB	PB

Fig. 3. The rule base used. X-axis means the linguistic value of the obstacle's distance. Y-axis does the obstacle's direction.

The input and output linguistic values have the following meanings:

VN : *very near* NR : *near* FR : *far*
NB : *negative big* NM : *negative medium*
NS : *negative small* ZZ : *zero*
PS : *positive small* PM : *positive medium*
PB : *positive big* MI : *Medium*

There are 5 obstacle's distance variables with their own rule bases. We use the same rule base in figure 3 to all the five distance variables. Once the union of the output fuzzy sets from each rule set is calculated and then we use a center of gravity defuzzification method to acquire crisp control actions, θ.

3 Map Building with Dempster-Shafer Theory

The Dempster-Shafer evidence theory is characterized by a frame of discernment (*FOD*), basic probability assignment (*Bpa*), belief (*Bel*), plausibility (*Pls*) functions and the Dempster's rule of combination [6]. The frame of discernment, denoted by Γ, is defined to be a finite set of labels representing mutually exclusive events. The basic probability assignment is the function, $m:\Psi \rightarrow [0,1]$, where Ψ is the set of all subsets of Γ. The function *m* can be interpreted as distributing probabilities to each of the labels in Ψ, with the following criteria satisfied

$$\sum_{A\subset\Psi} m(A)=1, \quad m(\phi)=0 \tag{4}$$

Thus, the label *A* is assigned a basic probability number *m*(*A*) describing the degree of belief that is committed to exactly *A*. However, the total evidence that is attributed to *A* is the sum of all probability numbers assigned to *A* and its subsets

$$Bel(A)=\sum_{\forall B:B\subseteq A} m(B) \tag{5}$$

The function $Bel:\Psi \rightarrow [0,1]$ is the quantity of evidence supporting the proposition *A* and has the following properties:

$$Bel(\phi)=0, \quad Bel(\Gamma)=1 \tag{6}$$

$$Bel(A)+Bel(\neg A)\leq 1 \tag{7}$$

$$Bel(A)\leq Bel(B), \quad if\ A\subset B \tag{8}$$

$$Bel(A\bigcap B)=\min(Bel(A),Bel(B)) \tag{9}$$

The plausibility of a proposition *A* can be thought of as the amount of evidence that does not support its negation $\neg A$. It is defined as $Pls:\Psi \rightarrow [0,1]$, with the following:

$$Pls(A)=1-Bel(\neg A)=1-\sum_{\forall B:B\not\subset A} m(B) \tag{10}$$

$$Pls(A)-Bel(A)\geq 0 \tag{11}$$

$$Pls(A \cup B) = \max(Pls(A), Pls(B)) \tag{12}$$

The state of each label (described by bpa) is updated by combining a new independent source of evidence using the Dempster's rule of combination

$$m_1 \oplus m_2(A) = \frac{\sum_{\forall B,C \subset \Psi : B \cap C = A} m_1(B)m_2(C)}{1 - \sum_{\forall B,C \subset \Psi : B \cap C = \phi} m_1(B)m_2(C)} \tag{13}$$

The *Bel* and *Pls* functions are often denoted as upper and lower probabilities. The amount of *Pls(A) – Bel(A)* is the additional undistributed evidence that is compatible with both hypotheses A and $\neg A$ being true.

To build an occupancy map of the environment, we firstly construct a grid representing the whole space. Every discrete region of the map (each cell) is discriminated by two states, *Empty* and *Full*. Thus, we define the frame of discernment, Γ, by the set $\Gamma = \{E, F\}$ where the E and F correspond to the possibilities that the cell is *Empty* or *Full*, respectively. The set of all subsets of Γ is the power set

$$\Lambda = 2^{\Gamma} = \{\phi, E, F, \{E, F\}\} \tag{14}$$

The state of each cell is described by assigning basic probability numbers to each label in Λ. However, we know that for each cell (i, j) in the grid

$$m_{i,j}(\phi) = 0 \tag{15}$$

$$\sum_{A \subset \Lambda} m_{i,j}(A) = m_{i,j}(\phi) + m_{i,j}(E) + m_{i,j}(F) + m_{i,j}(\{E, F\}) = 1 \tag{16}$$

Every cell in the map is at first initialized like (17) and (18).

$$m_{i,j}(E) = m_{i,j}(F) = 0 \tag{17}$$

$$m_{i,j}(\{E, F\}) = 1 \tag{18}$$

Then, as the mobile robot moves, scans of the environment are taken and fused into the map. If n cells exist on the arc, the basic probability assignment for the sensor arc is as follows :

$$m_{i,j}(F) = 1/n, \quad m_{i,j}(E) = 0 \qquad \forall cells(i, j) \in arc \tag{19}$$

$$m_{i,j}(F) = 0, \quad m_{i,j}(E) = 0 \qquad \forall cells(i, j) \notin arc \tag{20}$$

By adding subscripts S and M to the basic probability masses m, we can describe the basic probability assignments of both the sensor and the map as (21) through (23).

$$K = 1 - m_M(E)m_S(F) - m_M(F)m_S(E) \tag{21}$$

$$m_M \oplus m_S(E) = m_M(E)m_S(E) + m_M(E)m_S(\{E, F\}) + m_M(\{E, F\})m_S(E) / K \tag{22}$$

$$m_M \oplus m_S(F) = m_M(F)m_S(F) + m_M(F)m_S(\{E,F\}) + m_M(\{E,F\})m_S(F)/K \qquad (23)$$

4 Experimental Results

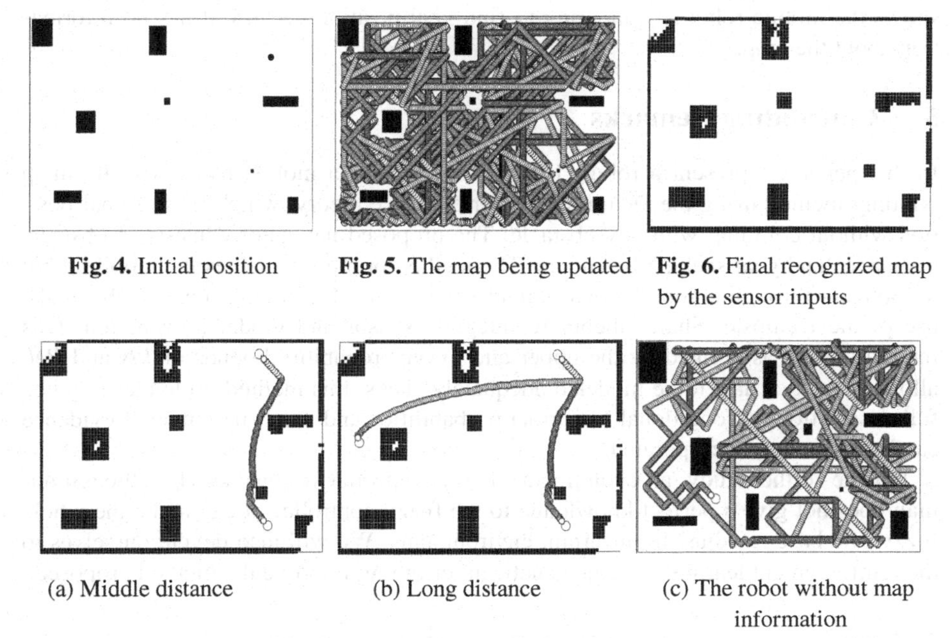

Fig. 4. Initial position **Fig. 5.** The map being updated **Fig. 6.** Final recognized map by the sensor inputs

(a) Middle distance (b) Long distance (c) The robot without map information

Fig. 7. The robot's goal finding trajectories

To show the usefulness of the proposed method, a series of simulations have been conducted using an arbitrarily constructed environment including several obstacles. Our experiment is divided into two steps: acquisition of the environment and finding goal position without collision. To succeed in our simulation tasks, it is the most important to complete the fuzzy controller that can be used to control the mobile robot. The measurement range of sensor is assumed to be six times larger than the radius of mobile robot. The number of fuzzy rules used in this experiment is only 25 rather than 75 that is the maximum number of the needed rules for this experiment. Because a lot of the fuzzy rules are observed to be the same or alike each other, we can get rid of them out of the whole fuzzy rule base. They would make the trajectory of the mobile robot smoother.

Fig. 4 shows the initial position of several obstacles and the robot. The small dot on the upper-right corner represents the robot's start position. As you can see, the size of one obstacle is much smaller than that of the robot. Fig. 5 shows the mobile robot is wandering in the environment and accepts the scan sensor input values from its sensor suits simultaneously. The sensor value is used to fuse with the old map information and generates the new map information. Fig. 6 shows the finally constructed map by the robot using the Dempster-Shafer evidence theory. It is almost the same with the initial map in terms of the position and size of the obstacles. White cells shown in some obstacles represent locations beyond the boundary of the robot's sensor. Fig. 7 (a) and (b) show the mobile robot's trajectories to the goal position without collision. The distances of goal position are changed from near to far. In all the experiments it is not difficult for the robot to find where the goal position is and obstacles are. Fig. 7 (c) shows the mobile robot is wandering to find goal position because it has no information about the map.

5 Concluding Remarks

In this paper we present a robust fuzzy controller for a mobile robot with the map building method using the Dempster-Shafer evidence theory, which finds a goal position without colliding with any obstacle. The proposed navigator consists of obstacle avoidance and goal-seeking behaviors. The method proposed enables the mobile robot to navigate through complex environment where a local minimum occurs. We make use of the Dempster-Shafer theory to integrate sensor and model information. It is interval based, as defined by the upper and lower probability bounds of *Pls* and *Bel*, allowing lack of data to be modeled adequately. Thus, this method no longer requires full description of conditional (or prior) probabilities and small incremental evidence can be adequately incorporated.

For the further study, an evolutionary fuzzy controller is considered. Although human does not give a perfect knowledge to the fuzzy controller, it can make the whole fuzzy rule base to adjust to uncertain environments. We will also devote ourselves to the reinforcement learning and the genetic algorithm to improve the method proposed.

References

1. A. Elfes, Sonar-based real world mapping and navigation, IEEE Transactions on Robotics and Automation, vol. RA3, pp. 249–265, June 1987.
2. B. Barshan and R. Kuc, Differentiating sonar reactions from corners and planes by employing an intelligent sensor, IEEE Transactions on Pattern Analysis and Machine Intelligence, vol. 12, pp. 560–569, 1990.
3. R.A. Brooks, A robust layered control system for a mobile robot, IEEE Transactions on Robotics and Automation, vol. RA-2, no. 1, pp. 14-23, March 1986.
4. J. Borenstein and Y. Koren, Real-time obstacle avoidance for fast mobile robot, IEEE Transactions on Systems, Man and Cybernetics, vol. 19, no. 5, pp. 1179-1187, 1989.
5. L.A. Zadeh, Fuzzy set, Information and Control, vol. 8, pp. 338-353, 1965.
6. G. Shafer, A Mathematical Theory of Evidence, Princeton, NJ: Prince-ton Univ. Press, 1976.
7. G. Oriolo, G. Ulivi and M. Vendittelli, Real-time map building and navigation for autonomous robots in unknown environments, IEEE Transactions on Systems, Man and Cybernetics, Part B, vol. 28, no. 3, pp. 316-333, June 1998.
8. J.A. Castellanos, J.M.M. Montiel, J. Neira, J.D. Tardos, The SPmap: A probabilistic framework for simultaneous localization and map building, IEEE Transactions on Robotics and Automation, vol. 15, no. 5, pp. 948-952, 1999.

The Fuzzy Model for Aircraft Landing Control

Silviu Ionita and Emil Sofron

University of Pitesti, Electronics Department,
Targul din Vale 1, Pitesti 0300, Romania
{ionis,sofron}@upit.ro

Abstract. This paper deals the issue of aircraft landing maneuvers from the fuzzy logic perspective. Generally this part of flight needs to be strongly assisted by human pilot. Our purpose is to evaluate the output parameters from a fuzzy system. We have developed a fuzzy control model for three kinematics parameters to do a suitable landing control. We have tried to demonstrate the potential of fuzzy rules based control systems for high precision maneuvers. Features, the approximate reasoning and precision could be simultaneously reached by empirical refining of the fuzzy model. All in all, the result of our work aims to develop the appropriate models for the complete automates of landing process and related tasks.

1. Introduction

The autonomous aircraft landing is an issue that implies three main aspects: the performance of equipment, the process models and the ethics. Generally, the landing is not a standard flight task as it could be thinking. We consider it a nonstandard flight stage because it has a very high sensitivity versus environment perturbation and to the psyhological factors. The other complementary problems are related of measurement precision of equipment and the reliability of the systems (hardware and software). The above-mentioned aspects reveal the complexity of the automatization of aircraft landing procedure.

One of the best ways to solve this problem is to approach the artificial intelligence modeling technology based on fuzzy logic [2]. The capability of fuzzy modeling for various problems of dynamical processes control has been already demonstrated in a number of applications [5], [9], [10].

The goal of this paper is to demonstrate the possibility to shynthetis the three essential parameters for aircraft landing control: the direction, the altitude and the speed. Our study is based on related researches on flight control issues [3], [7].

2. The Reference Model of Landing

Aircraft landing process enhanced several phases that define the so-called standard landing trajectory. The landing operation concerning two controlled maneuvers: first for guiding the aircraft in the horizontal plane, in order to align it onto the axe of the runway and the second, for aircraft guiding in the vertical plane in order to do the approaching of runway surface. In Fig. 1 is depicted the basic cinematic model of

N.R. Pal and M. Sugeno (Eds.): AFSS 2002, LNAI 2275, pp. 47–54, 2002.

landing. Basically, the automatic landing systems provide the information for instrument navigation along the standard trajectory. However, the decisions in aircraft command should take by human pilot. The very high precise and reliable controller could be able to do this task, but we think the human supervision on board is still required.

The basic analytical model of aircraft movement is described by differential equations as follow:

$$\begin{cases} dx/dt = V_a \cdot \cos \varphi_H \\ dy/dt = V_a \cdot \sin \varphi_H \\ dz/dt = V_a \cdot \cos \varphi_V \\ d\varphi_H = \theta_H \\ d\varphi_V = \theta_V \end{cases} \tag{2.1}$$

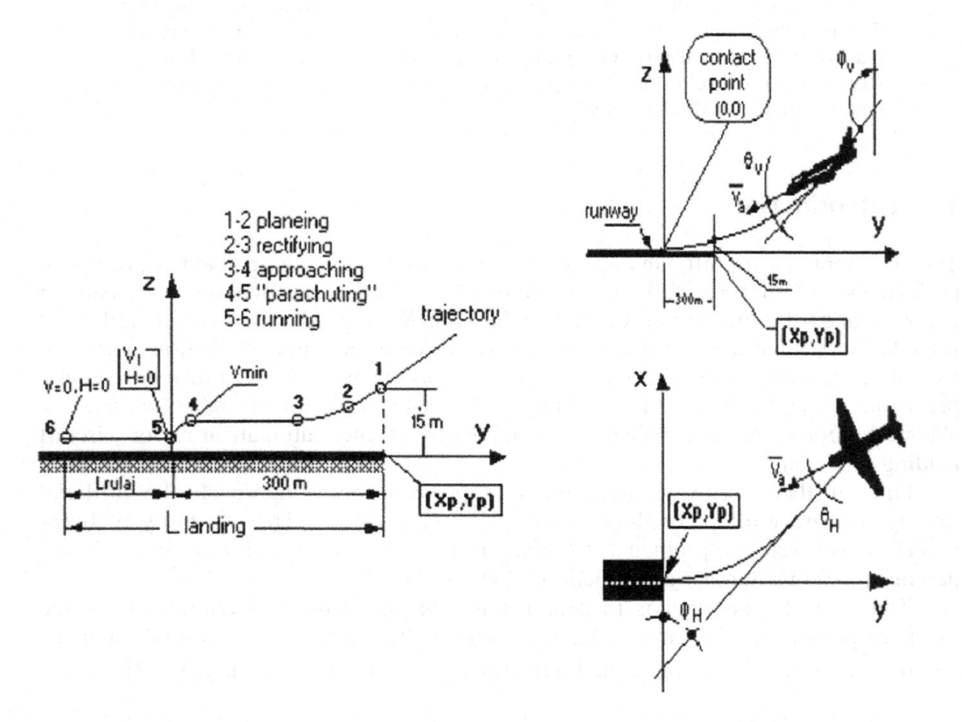

Fig. 1. The standard phases of the airplane landing and the cinematic basic schemata of landing process

3. The Aircraft Landing Control by Fuzzy Logic

The aircraft landing procedures admit a linguistic describing. This is practiced, for example, in case of guiding for landing in non-visibility conditions or in piloting learning. This approach permits to build a model for landing control based on the reasoning rules using the fuzzy logic. The process requires the control of the following parameters: the current altitude to runway surface (*H*), the current distance from the beginning of the runway (*D*), the aircraft's vertical speed (V_V) and aircraft flight speed (V_a). The goal of the control is formulated as follow: the aircraft should touch the runway (*H* becomes 0) at the conventional point of landing (usually imposed at a distance D^*=300m) with admitted vertical touch speed $V_V \approx$0.5m/s and the recommended landing speed $V_{landing}$, (V_a becomes $V_{landing}$).

Our fuzzy model treats the following tasks related to aircraft landing stages: the control of flight direction in order to correct approach from the airport, the control aircraft vertical movement and the control of flight speed.

We defined the formal relationships between the parameters, which govern the process as follows:

$$\theta_H = f_H(x, \varphi_H) \quad (3.1)$$

$$\theta_V = f_V(z, \varphi_V) \quad (3.2)$$

$$\Delta V_a = f_s(\mathrm{V_a}, y) \quad (3.3)$$

where the aircraft curent speed is a function of many parameters, for example: $V_a = f(T, m, y, \theta_V, \rho, K_a)$.

The symbols in the above formulae denote: *x, y, z*- the current coordinates of the aircraft, φ_H- the angle of flight direction, θ_H- the direction correction angle, φ_V- the angle of landing slope, θ_V- the slope correction angle, ΔV_a- the aircraft speed correction, *T*- the propulsor thrust, *m*- the aircraft mass, ρ- the athmosphere density and K_a- a constant.

The information about the state of the system is available from the dedicated board sensors and from the ground equipment. That means the system is currently affected by systematical errors and noise that is other reason to model this kind of task with non-deterministic approaching.

3.1 The Fuzzy Sets

The steps for developing the fuzzy model are briefly presented. In Tab. 1 are the fuzzy parameters and their assigned linguistic labels. They were convenient abbreviated. The sign rule for the angular parameters is the counterclockwise (the positive sign is to right turn direction). The fuzzy sets are defined on the universe of discourse of each those parameters. Tab. 2 contains the chosen domains of discourse of the considered parameters according with their practice domain. The membership functions (*m*) have been designed in three diagrams using the triangular and trapezoidal shapes, depicted in Fig. 2.

3.2 The Rules and Strategies for Control

The rules for control are *IF-THEN* assertions using the logic operator *AND*. We develop the sets of fuzzy rules as the strategies for control for each of the functional relationships (3.1-3.3), for example: *IF x='Left' AND* φ_H*='Big toward left' THEN*

θ_H= *'Positive small'*, and so on. Combining all linguistic values associated to each fuzzy variable that compose the rules (see Tab. 1), the proposed control strategies results in Tab. 3 and Tab. 4 as the Fuzzy Rules Bases (FRB). We mention that the direction control strategy (θ_H parameter) and the vertical control strategy (θ_V parameter) are identical. Both are contained in the same Tab. 3 and reflect the continuous approaching of the aircraft on runway with the minimum errors. We also notice that aircraft current speed is obviously a positive amount, which is recursively adjusted ($V_a \pm \Delta V$), in order to meet the landing goal. As the mater of fact the upper-right domain is practically used for speed control strategy (see Tab.4).

These samples come out from an empiric procedure based on trial-error process under supervision of the human expert (i.e. the pilot). As the matter of fact, we have designed certain FRBs using a more complex methodology as the off-line process, which involves the successive tests for different arrangements in FRBs cells as well as the current adjustments of the membership functions shape (i.e. their distribution). The technique of adaptive fuzzy rule base could be considered too [5].

Table 1. The linquistic labels

X, y, z	φ_H, φ_V	θ_H, θ_V, ΔV
L- left	**B**- big toward left	**b**- negativ big
l- left center	**U**- medium toward left	**m**- negativ medium
C- center	**V**- small toward left	**s**- negativ small
r- right center	**E**- vertical	**z**- zero
R- right	**v**- small toward right	**S**- pozitiv small
	u- medium toward right	**M**- pozitiv medium
	b- big toward right	**B**- pozitiv big

Table 2. The universes of discours of the parameters

$0 \le x \le 10{,}000$ m	$-90° \le \varphi_H \le 270°$	$V_{min} \le V_a \le V_{max}$
$0 \le y \le 10{,}000$ m	$-90° \le \varphi_V \le 270°$	(the limits are specific to current
$0 \le z \le 1{,}000$ m	$-45° \le \theta_H \le 45°$	aircraft speed features)
	$-45° \le \theta_V \le 45°$	$-30 \le \Delta V \le 30$ m/s

Table 3. FRB for control θ_H *and* θ_V

φ_H/φ_V \ x/z	L	l	C	r	R
B	S	M	M	B	B
U	s	S	M	B	B
V	m	s	S	M	B
E	m	m	z	M	M
v	b	m	s	s	M
u	b	b	m	s	S
b	b	b	m	m	s

Table 4. FRB for control ΔV

V_a \ y	L	l	C	r	R
B	b	b	b	m	s
M	b	b	m	s	z
S	b	b	s	z	S
C	z	z	z	S	M
s	z	z	z	z	z
m	z	z	B	B	B
b	z	z	B	B	B

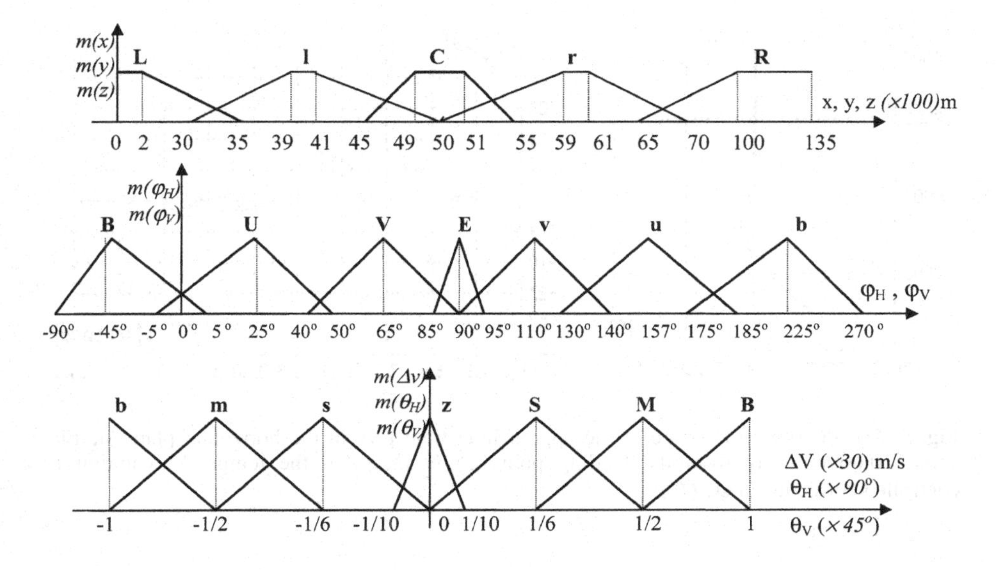

Fig. 2. The membership functions

We also could take in consideration the aspect of learning the rule base from examples (practice). This approach aims to develop the algorithms for goal-generated FRBs but it has not used in this case. A certain approach is pointed out in our related work [4]. It should also take in consideration the complementary intelligent solution with a high learning capability as neural networks [6], [8].

4. Results

Using a soft computing approach, in this section are presented a set of diagrams of the response of the model in simulation mood. The algorithm of the problem has been implemented in a high level procedural language. The application program provides both numeric and graphic outputs. The program allows a rapid testing off-line process in order to refine our fuzzy model. The relevant diagrams have been selected from a number of simulation cases, which reflect the realistic conditions. We notice a good control in direction, finishing with a perfect aircraft alignment to the axe of the runway (θ_H=0), as you see in Fig. 3. The vertical control is consistent with the purpose: to bring (and to keep) the plane at the constant low height over the runway until the moment of landing contact (see the Fig.4). We also remark in Fig.5 an appropriate speed control by currently adjusting of aircraft speed in order to realize the imposed amount right to the recommended runway contact point. (In this case we have chosen V_a=150m/s).

The results highlight the good qualitative behavior of our fuzzy rules based model. As the matter of fact, the final simulated goal is reached and the plane has landed in the imposed bounds of the parameters. The quantitative results encourage us but they still need some improvements that represent part of our current research.

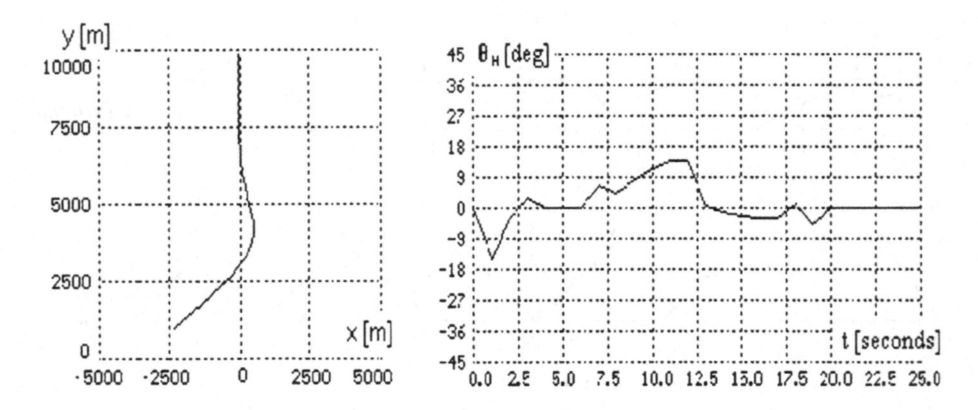

Fig. 3. *Left diagram*: the shape of the approaching trajectory in the horizontal plane starting about 10,000m away from the landing point. *Right diagram*: the temporal evolution of controlled correction angle (θ_H).

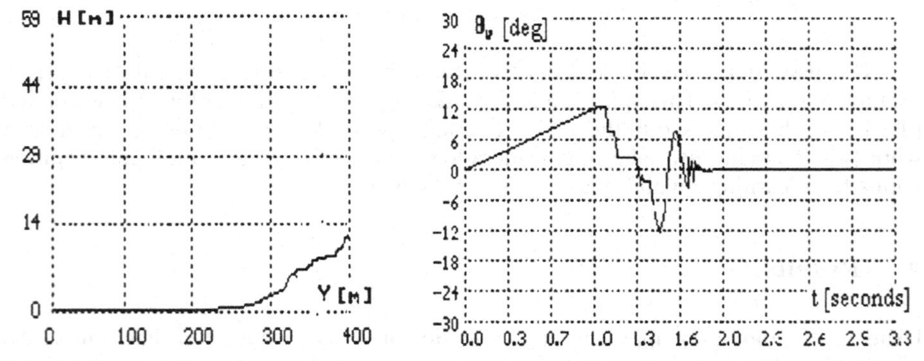

Fig. 4. The evolutions for vertical movement parameters: the trajectory profile on the latest 400m range to runway (*left diagram*); the temporal evolution of controlled correction angle θ_V (*right diagram*).

At the glance, we could estimate the precision of this model relative to the recommended values by ICAO (International Civil Aviation Organization) regarding the deviations from standard landing trajectory. The accepted values of position error at the touch point are ±5m in horizontal plane and ±*0.5*m on vertically [1]. The proposed model is capable to control the parameters under these limits.

5. Conclusions

Current developments of artificial intelligence enable an appropriate approach of high precision control tasks. This paper reveals some aspects of soft computing in fuzzy control engineering. Our particular goal was to demonstrate the potential of fuzzy rules based systems for high precision maneuvers required by aircraft landing. The proposed model reveals the functional aspect for realistic simulation data. On one hand the information about the current state of the landing system is available from the different sensors, which are affected by errors, and on the other hand the human operator himself could be source of uncertainty in the control loop. This fact is a strong reason to model this problem with non-deterministic approaching. Shortly, the robustness of control is the main quality of this kind of models.

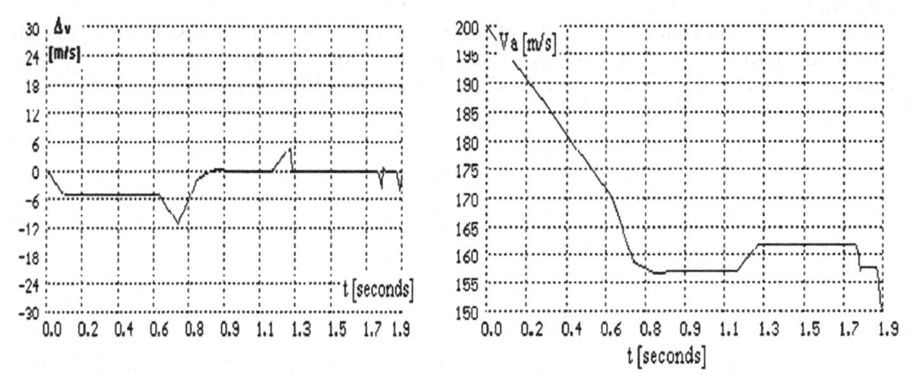

Fig. 5. Diagrams with information about aircraft speed: the speed corrections along the latest 300m range (over the runway) to touch point (*left diagram*); the approaching flight speed evolution (*right diagram*).

We demonstrate that using very simple functional relationships between the parameters of the process we have obtained the consistent responses. The model is very flexible working like an expert system for landing procedures. It has an intrinsic approximate reasoning capability that means it has the so-called *human consistency*. More over the proposed model could be refined as much as it is need in order to optimize its precision as well as the robustness.

Joining this model with other related methodologies like neural systems and evolutionary strategies it will become an optimal learning system.

Finally our goal is to implement a high tested model with fuzzy logic as an embedded controller for aircraft landing tasks. The proposed methodology could be extended to the other specific application as spacecrafts orbital coupling control and maneuvers control for aircraft flight refueling.

References

1. Aron, I., Lungu, R., Cismaru, C.: Aerospace Navigation Systems, (in Romanian). Scrisul Romanesc, Craiova, (1989)
2. Bezdek, J.: Fuzzy Models: What are they and Why. IEEE Transactions of Fuzzy Systems, Vol.1 1 (1993) 1-6
3. Ionita, S.: Contributions on Optimal Guiding of Aircraft's towards High Speed Targets, (in Romanian). Doctoral Thesis, Technical Military Academy, Bucharest, Romania, (1997)
4. Ionita, S.: The Fuzzy Rules Generating for a Guiding Aircraft Expert System, (in Romanian). The XXVII-th Conference, Technical Military Academy, Bucharest, Romania, 13-14 Nov. (1997) 78-85

5. Kosko, B.: Neural Networks and Fuzzy Systems. Prentice Hall, Englewood Cliffs (1992).
6. Rokhsaz, K., Steck, J.E.: Use of Neural Network in Control of High-Alpha Maneuvers. AIAA Journal of Guidance, Control, and Dynamics, Vol.16. 5. Sept-Oct. (1993) 934-939
7. Sofron, E., Bizon, N., Ionita, S., Raducu, R.: Fuzzy Control Systems - Modeling and Computer Aided Design, (in Romanian). ALL Educational Publishing House. Bucharest, Romania (1998)
8. Troudet, T., Garg, S., Merrill, W.: Neurocontrol Design and Analysis for a Multivariable Aircraft Control Problem. AIAA Journal of Guidance, Control, and Dynamics, Vol.16. 4. July-August (1993) 738-747
9. Yamakawa, T.: Stabilization of an Inverted Pendulum by a High-speed Fuzzy Logic Controller Hardware System. Fuzzy Sets and Systems, Vol.32. (1989) 161-180
10. Zimmermann, H.J.: Fuzzy Set Theory-and its Applications. Boston (1991)

Implementation of Nonlinear Fuzzy Models Using Microcontrollers

S. Himavathi and B. Umamaheswari

School of Electrical and Electronics Engineering,
Anna University, Chennai-600025, India

Abstract. This paper presents an algorithm for implementation of fuzzy systems using inexpensive general-purpose hardware. The two major difficulties in fuzzy system implementation using 16 bit fixed point micro processors/controllers is in handling floating point arithmetic and the determination of the floating-point membership value. Existing membership functions (mfs), do not satisfy simultaneously ease in optimization and low end hardware implementation, hence new membership function that satisfies the two contradicting requirements have been proposed by us [1]. An algorithm for hardware implementation of fuzzy systems using fixed-point micro-controllers is proposed. The worst-case execution time is computed as a function of the number of inputs and the number of rules. Optimized fuzzy model of a benchmark system has been coded and implemented using Intel 8XC196KC micro controller.

1 Introduction

Fuzzy modeling and Fuzzy control has emerged as the two major branches of Fuzzy system theory. From a mathematical point of view a fuzzy system is a universal function approximator [2]. The capability of fuzzy logic system to approximate any continuous function to any degree of accuracy has been reported [3],[4]. Irrespective of its application a fuzzy system relates a set of input data to an output data. The knowledge description in fuzzy modeling can be viewed as having two classes. The Class A as suggested by Takagi and Sugeno[6] and the Class B developed by Mamdami [6].

In this paper the zero order Class A type is used as it is simpler and more suitable for hardware implementation. The complexity of the Fuzzy Logic Systems(FLS) has made its implementation in low cost hardware unattractive. Intensive research is in progress to provide hardware support to fuzzy systems. Currently fuzzy systems are implemented using DSP chips[7] fuzzy ICs like WARP 2[8], and custom designed fuzzy ICs using VHDL [9],[10].

The major difficulty in implementing fuzzy systems is in handling the floating-point arithmetic and the determination of the membership value. Choices of membership functions(mfs) form an important aspect of fuzzy system design. The two most popularly used mfs are the triangular and Gaussian mfs. Triangular mfs are comparatively easier to implement than gaussian mfs, but offline design and optimization can only be achieved through random search techniques. This makes the

N.R. Pal and M. Sugeno (Eds.): AFSS 2002, LNAI 2275, pp. 55–61, 2002.

design procedure more random and time consuming. Gaussian mfs on the other hand can be optimized using powerful gradient optimization techniques, but is extremely difficult to implement in low-end hardware. Hence new mfs, satisfying the two contradictory requirements of systematic offline design and easy hardware implementation, have been suggested by us [1]. A general-purpose algorithm using new mfs for implementation of fuzzy systems using low-end hardware has been developed.

2 Fuzzy System Design

A Fuzzy Logic system (FLS) consists of four basic elements, the fuzzifier, the fuzzy rule base, inference engine and the defuzzifier. A number of design techniques with different properties are available in literature for each of the four basic elements. In this paper the following approach is used. The off line design package has been developed as a MATLAB m file.

(i) The premise parameters are obtained using c-means clustering.
(ii) The consequent parameters of the zero order Sugeno inference are obtained using least square estimate.
(iii) The model developed is optimized using Adaptive network based fuzzy inference system(ANFIS) [11].

3 New Membership Function [1]

The proposed membership function is negative powers of 2. The membership value $\mu(x)$ is defined by the expression

$$\mu(x) = 2^{-w} \qquad \text{where} \quad w = \text{abs}(c-x)/\sigma \tag{1}$$

where c and σ are the design parameters. This distribution looks more or less like a triangular membership plot with a long tail. The plot of the membership function is shown in Fig .1

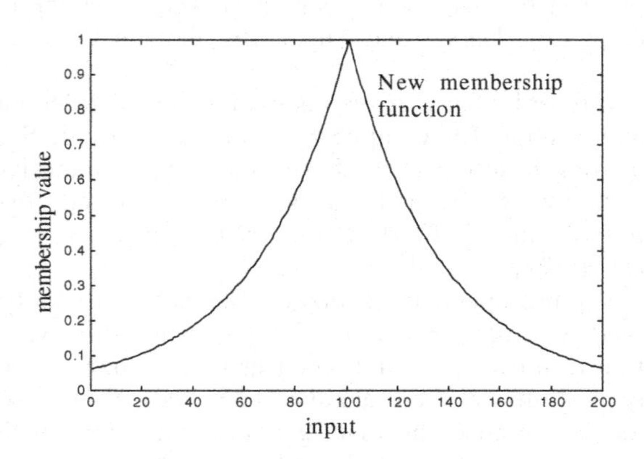

Fig. 1. Plot of new membership function

4 Implementation Using Micro Controllers/Processors

The step-by-step implementation of a Fuzzy logic system using fixed-point micro controllers is detailed in this section. Let x_i (i =1,2..,m) be the m system inputs. Let c be the number of rules. The premise design parameters are, the center(c_{ij}) and the variance (σ_{ij}) where (i=1,2,..,m) and (j =1,2,..,c).

Step 1: Input x_i, c_{ij} and σ_{ij} for all i and j where (i=1,2,..,m) and (j=1,2,..,c)

Step 2: Determine p_{ij} where

$$p_{ij} = (\text{abs}(c_{ij}-x_j)) \;/\; \sigma_{ij} \tag{2}$$

and $p_{ij=}$ $p1_{ij}+p2_{ij}$ where $p1_{ij}$ is a whole number is obtained as quotient using division instruction and $p2_{ij}$ the fractional part obtained using Algorithm 1

Step 3: Obtain w_j defined as $w_j = \sum_{i=1}^{m} p_{ij}$ (3)

Also

$$w_j = w1_j + w2_j \tag{4}$$

Step 4: Determine the weight of each rule $r_j = 2^{-w_j}$ using Algorithm 2

Step 5: Using Algorithm 3, obtain the normalized weight of each rule $\bar{r}_j$

$$\bar{r}_j = \frac{r_j}{\sum_{j=1}^{c} r_j} \tag{5}$$

Step 6: Input the consequent parameters z_j . The output y is now obtained using

$$y = \sum_{j=1}^{c} z_j \bar{r}_j \tag{6}$$

The algorithm can be easily implemented using any 16 bit fixed point micro processors/controllers which support fundamental arithmetic/logic operation. Three specific algorithms are developed to easily handle the fuzzy computation. Fuzzy systems are implemented using triangular mfs and the membership value is generally computed using look up table method. Fuzzy systems can also be implementation using special fuzzy ICS like WARP 2, ST52 Dualogic 8 bit micro controller and Motorola 68HC12 micro controller. All these ICs support triangular mfs. The disadvantage is the non availability and cost of hardware.

5 Proposed New Algorithms

To implement the fuzzy system three algorithms are proposed . They are referred to in section 4. The algorithm is presented as a MATLAB m file. The algorithm uses simple arithmetic operations. It is assumed that n is the word length of the processor/controller

5.1 Algorithm 1

The objective is to compute a/b when a < b . The result d is a n bit fraction

```
        d=0
        if   a >= 0.5                       (check MSB)
        a=a/2;                              (right shift a)
        b=b/2                               (right shift b)
    end
    for  i=1:n-1
           a=a*2;                           (left shift a )
          if  a>=b
              a=a-b;
              d=d+1;
          end
         d=d*2;                             (left shift c )
    end
```

5.2 Algorithm 2

To determine $r = 2^{-x}$ when $0<x<n$ and $x=w_1+w_2$ where w_1 is a n bit whole number and w_2 is a n bit fraction. Let the (1x n) vector $h = [\ (2)^{-2^{-1}}\ \ (2)^{-2^{-2}}\ \dots (2)^{-2^{-n}}\ ]$

```
  r=1;
    for i=1:n
      if w2>= 0.5             (check MSB)
         r=r *h(i);
      end
    w2=w2*2;                  (left shift w2)
    end
    for i=1: w1
        r=r/2;                (right shift m)
    end
```

5.3 Algorithm 3

The algorithm computes normalized rule weights nr. Let r be the rule strength vector rs be the sum of rule strength

```
    rs=0;
    for i=1:c
        rs=rs+r(i);
        if rs>=1
           rs=rs/2;                        (right shift rs)
           for i=1:c
                r(i)=r(i)/2;               (right shift r)
           end
        end
    end
    for i=1:c
        a=r(i);
        b=rs;
        d=0
        if   a >= 0.5                       (check MSB)
        a=a/2;                              (right shift a)
        b=b/2                               (right shift b)
        end
```

```
        for i=1:n-1
            a=a*2;                      (left shift a )
            if a>=b
               a=a-b;
               d=d+1;
            end
            d=d*2;                      (left shift c )
        end
      nr(i)=d;
end
```

6 Execution Time

The general purpose algorithm has been implemented and tested using Intel 8XC196KC micro controller. The worst case execution time (number of t states) for the program as a function of the number of inputs (m), and number of rules (c) is obtained as

$$\text{number of t states (nt)} = 50+681cm+1656c+24c^2 \quad (7)$$

For a given system, and a specified accuracy, the feasibility of hardware implementation, the optimum number of rules, and the corresponding sampling interval can be obtained using (7). The number of t states for different values of m and c is shown in Fig.2.

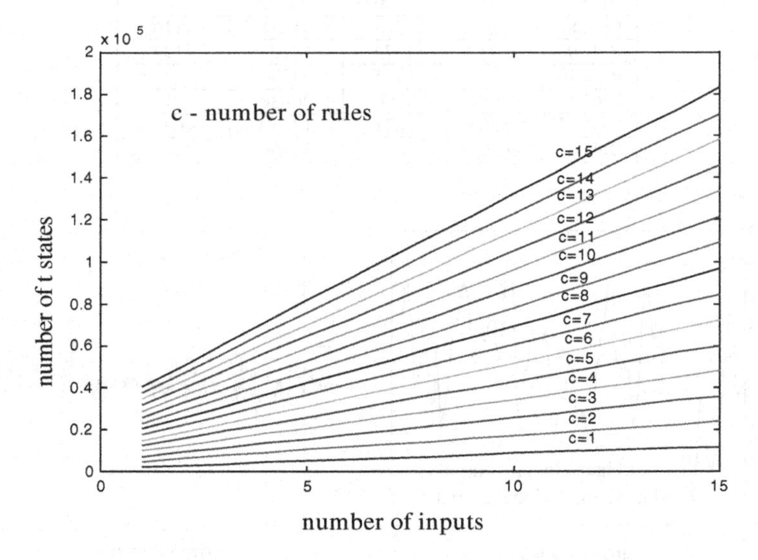

Fig. 2. Plot of maximum execution time

7 Numerical Example

The well known example of system identification given by Box and Jenkins[12],is modeled using the new membership functions. As the word length of the micro controller is 16 bits, the input and output values are scaled to the range $0 - 2^{16}$. Using the scaled data the premise and consequent parameters are obtained using the detailed modeling methodology given in Section 2.1 . The model output is scaled down to the actual range for computation of mse. The range of u is [– 4 to +4] and that of y is [40 to 72] is scaled to [$0 - 2^{16}$] ,the design parameters used are shown in Table I . The mse obtained after scaling down to the actual range is 0.1056. The actual output and model output are shown in Fig.3a. Since the mismatch is not clearly seen , the modeling error curve is shown in Fig.3b

Table 1. Premise parameters of Fuzzy model for micro controller implementation

Rules	y(k-1)		u(k-4)	
	center	variance	center	variance
Rule 1	23348	1922	39315	1897
Rule 2	29265	1479	32570	1069
Rule 3	34848	2778	24905	767
Rule 4	18255	2221	47618	5906
Rule 5	25496	1458	38101	3316
Rule 6	36307	1527	19282	2949
Rule 7	16172	4300	53327	3560
Rule 8	32152	3335	31641	4116
Rule 9	36732	1574	18707	4134
Rule 10	26129	1125	24336	489
Rule 11	22014	857	26630	5271
Rule 12	45302	2831	13459	4697
Rule 13	40308	2573	32769	7276

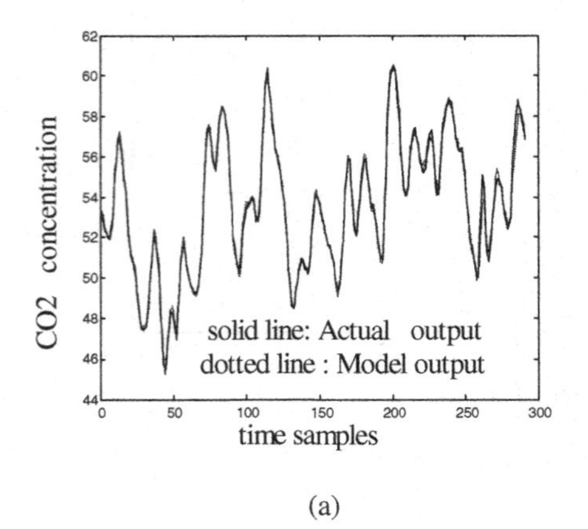

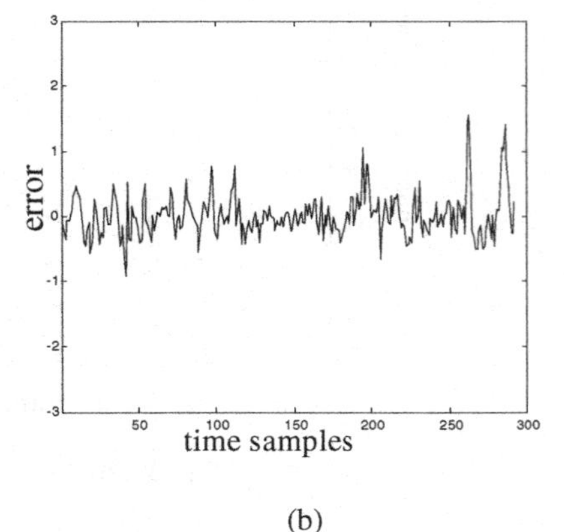

(a) (b)

Fig. 3. (a) Comparison of model outputs and system output, (b) Error curve

8 Conclusion

This paper presents implementation of fuzzy systems using inexpensive general-purpose hardware. New membership function that provides a systematic approach to design and allow simple hardware implementation is used. The algorithm for hardware implementation has been tested using Intel 8XC196KC micro controller. The maximum execution time has been computed as a function of the number of inputs and number of rules. A Fuzzy model of a Box and Jenkins furnace implemented using micro controller is presented. Given the inputs and design parameters the proposed algorithm helps to implement any fuzzy system namely fuzzy models, fuzzy predictors and fuzzy controllers using inexpensive general purpose hardware. The proposed algorithm will help realize portable hand held nonlinear models, and effective low cost controllers.

References

1. S. Himavathi and B. Umamaheswari: Design and implementation of Fuzzy systems sing new Membership functions". IEEE rans on Sys, Man., Cybern. Part B.
2. Ke Zeng, Nai. Y. Zhang, Wen-Li Xu: A comparative Study on Sufficient Conditions for Takagi-Sugeno Systems as Universal Approximator. IEEE Trans. Fuzzy Syst. vol.8. (2000) 773- 780.
3. Hao Ying: General SISO Tagaki-Sugeno Fuzzy systems with linear rule consequents are Universal Approximators. IEEE Trans. Fuzzy Syst., vol.6. (1998) 582-586.
4. Yongsheng. D. Haoing, Shihuang. S: Necessary conditions on Minimal System Configuration for General MISO Mamdami Fuzzy Systems as Universal Approximators". IEEE Trans. Sys, Man, Cybern. vol 30. (2000) 857-863.
5. T. Tagaki and M. Sugeno: Fuzzy Identification of systems and its applications to modeling and control. IEEE Trans. Sys. Man., Cybern, vol.5.(1985).
6. E. Mamdami: Advances in the linguistic synthesis of Fuzzy controllers, Int J. Man machine studies. vol.8. (1976) 669-678.
7. Donald. S. Reay, M. Mirkazemi-Moud, Tim. C. Green, Barry .W. Williams: Switched reluctance motor control via Fuzzy Adaptive systems. IEEE Control Syst. (1995) 8-14.
8. Ph. Cheynet, R. Velazeo, S. Rezgui, L. Peters et. al: Digital Fuzzy control: a robust Alternative Suitable for space applications, IEEE Trans. Nucl. Sci. vol 45. (1998) 2941-2947.
9. Giuseppe. A, Incenzo. C, Marco. R: VLSI Hardware Architecture for complex Fuzzy systems IEEE Trans. Fuzzy Syst., vol.7.(1999) 553-569.
10. A. Costa, A.D. Gloria, F. Giuditi and M. Olivieri: Fuzzy Logic Micro controller, IEEE Micro. (1997) 66-74.
11. Jyh-Shing, Roger Jang: ANFIS: Adaptive-Network Based Fuzzy Inference system. IEEE Trans. Sys, Man., Cybern, vol.23. (1993) 665-683.
12. George E.P. Box and Gwilym. M. Jenkins: Time Series Analysis-Forecasting and control. Holden day.(1976).

A Gain Adaptive Fuzzy Logic Controller

Rajani K. Mudi, Kalyan Majumdar, and Chanchal Dey

Department of Instrumentation & Electronics Engg.
Jadavpur University, Salt-lake Campus
Calcutta 700098, India.
ieeju@cal2.vsnl.net.in

Abstract. We propose a PI-type gain adaptive fuzzy logic controller (GA-FLC). Input scaling factors (SFs) of the GA-FLC for error (e) and change of error (Δe) are updated on-line, by a single non-linear parameter computed through a unipolar sigmoid function defined on the current process states (e and Δe). The performance of the proposed GA-FLC for step set-point change and load disturbance is compared with those of a conventional PI-type FLC (FPIC) and a Ziegler-Nichols tuned (ZN-tuned) non-fuzzy PI-controller (NFPIC). Results for various higher order processes show that GA-FLC provides much improved performance over both FPIC and ZN-tuned NFPIC.

1 Introduction

FLCs have been successfully implemented for a variety of complex and non-linear processes [1] and in many applications such controllers outperform their conventional counterparts. But unlike conventional controllers, standard tuning schemes for FLCs are not available till date. Proper choice of the tunable parameters of an FLC plays the key role for the successful design of such a controller. Among the various tunable parameters, tuning of SFs is assigned the highest priority due to their global effect on the controller performance. To ensure satisfactory performance for practical systems having non-linearity and dead-time, the FLC parameters need to be tuned on-line to adapt to changes in the process operating conditions. A lot of research activity on the on-line tuning of FLCs has been reported where either the SFs or the membership functions (MFs) are tuned to match the current plant characteristics [2-7]. With a view to achieving an improved overall performance, here we attempt to develop an on-line tuning scheme for updating the input SFs of a PI-type FLC (FPIC), based on the current process states thereby making it a gain adaptive FLC (GA-FLC). The effectiveness of the scheme is verified on various types of second order systems.

2 The Proposed GA-FLC

The block diagram of the proposed GA-FLC is shown in Fig. 1. We used triangular type MFs for e, Δe and Δu (incremental change in control output) as shown in Fig. 2. In a FPIC the values of the actual inputs e and Δe are mapped to the interval [-1, 1] by

N.R. Pal and M. Sugeno (Eds.): AFSS 2002, LNAI 2275, pp. 62–68, 2002.

the input SFs, G_e and G_{Δ_e} respectively. The defuzzified output Δu_N is translated into the actual output Δu by the output SF, G_u. In the proposed GA-FLC (Fig. 1) the effective input SFs G_e^{GA} and $G_{\Delta e}^{GA}$ for e and Δe are respectively defined as

$$G_e^{GA} = G_e(k_2\beta) \quad \text{and} \quad G_{\Delta e}^{GA} = G_{\Delta e}(1 + k_3\beta) \tag{1}$$

In the above equation, β is the non-linear gain tuning factor, which causes the required variations in the input SFs from their initial settings, G_e and G_{Δ_e}, in order to achieve the desired performance. k_2 and k_3 are positive constants which determine the extent to which the SFs are modified by β. The value of β at the kth sampling instant is determined by the following unipolar (+ve) sigmoid function:

$$\beta(k) = 1/\left(1 + e^{-\alpha(k)/k_1}\right) \tag{2}$$

$$\text{where} \quad \alpha(k) = \{e_N(k-1) \times \Delta e_N(k-1)\} \tag{3}$$

and k_1 is a positive constant. Figure 3 shows the variation of β with α for different values of k_1. Thus we see that the magnitude of k_1 affects two aspects of β : (i) It determines the rate at which β changes non-linearly about its mean value (0.5) for variations in the value of α about zero. Hence k_1 determines the sensitivity of G_e^{GA} and $G_{\Delta e}^{GA}$ for given e and Δe. (ii) It determines the range over which β varies as α changes over its entire domain [-1 1], which in turn determines the possible range of variations of G_e^{GA} and $G_{\Delta e}^{GA}$. Therefore by choosing a proper value of k_1 we can provide the required non-linear variation of G_e^{GA} and $G_{\Delta e}^{GA}$ in different operating points towards achieving the desired performance. The justification behind the choice of such a heuristic scheme will be explained in the next section.

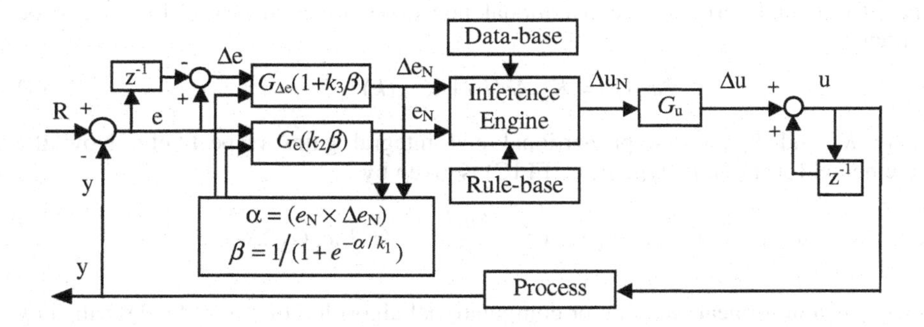

Fig. 1. Block diagram of the proposed GA-FLC

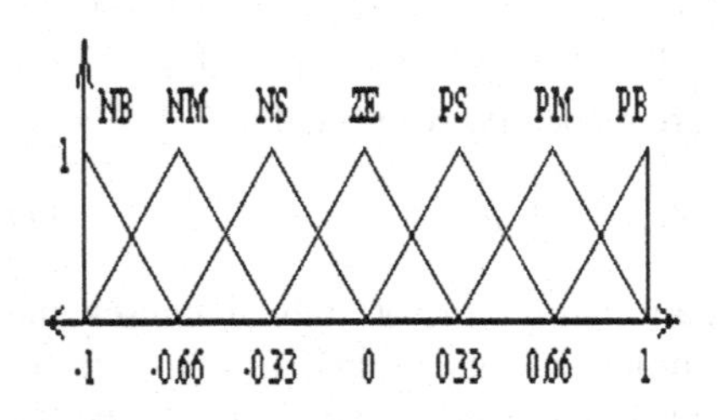

N = Negative, P = Positive, ZE = Zero
B = Big, M = Medium, S = Small

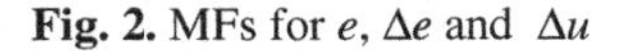

Fig. 2. MFs for e, Δe and Δu

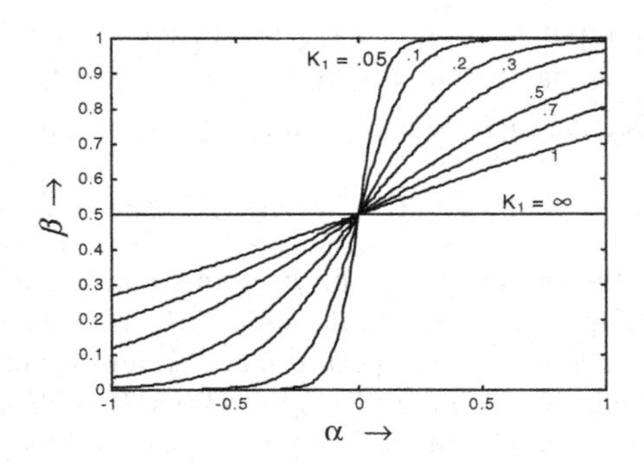

Fig. 3. Plot of α vs β for different values of k_1.

The incremental change in the controller output Δu (for FPIC) is determined by rules of the form: IF e is E and Δe is ΔE THEN Δu is ΔU. The rule base for computing Δu is shown in Fig. 4, which is a commonly used rule base designed in sliding mode principle [8]. The final controller output u is given by $u(k) = u(k-1) + \Delta u(k)$.

Δe/e	NB	NM	NS	ZE	PS	PM	PB
NB	NB	NB	NB	NM	NS	NS	ZE
NM	NB	NM	NM	NM	NS	ZE	PS
NS	NB	NM	NS	NS	ZE	PS	PM
ZE	NB	NM	NS	ZE	PS	PM	PB
PS	NM	NS	ZE	PS	PS	PM	PB
PM	NS	ZE	PS	PM	PM	PM	PB
PB	ZE	PS	PS	PM	PB	PB	PB

Fig. 4. Rule-base for computation of Δu

2.1 Tuning Strategy

The incremental form of a conventional non-fuzzy PI controller (NFPIC) can be written as

$$\Delta u_c(k) = K_p.\Delta e(k) + K_i.e(k) \tag{4}$$

where K_p and K_i are the proportional and integral gains respectively. Now the incremental form of a PI-type FLC (FPIC) is given by

$$\Delta u_N(k) = f(e_N, \Delta e_N) \quad \text{or} \quad \frac{\Delta u(k)}{G_u} = f(G_e.e, G_{\Delta e}.\Delta e), \tag{5}$$

where f is a non-linear function or computational algorithm of e and Δe, described by the rule-base of Fig. 4 and the associated inferencing scheme. Since a FPIC is found to be equivalent to a linear NFPIC [9], the Eqn. (5) may be approximated as

$$\Delta u(k) = G_u\{G_e.e(k) + G_{\Delta e}.\Delta e(k)\} \tag{6}$$

From Eqns. (4) and (6) we find that a FPIC is analogous to a NFPIC with proportional gain (K_{fp}) of $G_u G_{\Delta_e}$ and integral gain (K_{fi}) of $G_u G_e$. Note that, since the output SF, G_u is a common term for both K_{fp} and K_{fi}, we can adjust them by updating only G_u. Such a strategy is developed in [5] with the help of a very *complicated fuzzy rule-based scheme*. Therefore, we attempt to develop a *simple non-fuzzy tuning scheme* to modify G_e and G_{Δ_e} independently in order to provide appropriate K_{fp} and K_{fi} for achieving improved control performance. Observe that, in our scheme $K_{fp} = G_u G_{\Delta e}^{GA} = G_u G_{\Delta e}(1+k_3\beta)$ and $K_{fi} = G_u G_e^{GA} = G_u G_e(k_2\beta)$, therefore possible variations of integral gain (K_{fi}) is much more than that of proportional gain (K_{fp}). For example, in our simulation study we have considered $k_1 = 0.3$, $k_2 = 2$ and $k_3 = 0.3$ (these are empirical values). Therefore, maximum change of integral gain (K_{fi}) will be almost 100% whereas that of proportional gain (K_{fp}) will be only around 30% from their initial settings. Next we explain how such gain variations will try to improve the transient as well as steady state control performance.

i) When e is large and Δe is small or medium but they are of opposite sign (*e.g.*, the controlled variable is far from the set-point and is slowly moving towards it) then the proportional gain (K_{fp}) should be higher to speed up the response. But the integral gain (K_{fi}) should be small to prevent large accumulation of the controller output which may cause an excessive overshoot in future. In such situations, values of β will be considerably small since α is negative small or medium. Hence, our GA-FLC, will provide the desired smaller K_{fi} and larger K_{fp} compared to a FPIC towards achieving a better performance.

ii) To prevent large overshoot/undershoot and to ensure rapid convergence there should a considerable variation of the gains around the set-point. For example, when the controlled variable is close to the set-point and is rapidly moving towards it, the integral gain should be reduced to minimize the overshoot or undershoot. Note that, in this case, e is small but Δe is large and they are of opposite sign, which means α is negative so β becomes sufficiently small. On the contrary, when the controlled variable is close to the set-point but is moving away from it, then both the K_{fp} and K_{fi} should be sufficiently increased to prevent further deterioration of the process state. Observe that, in this situation, e is small and Δe is medium or large, but e and Δe have the same sign, therefore, α becomes positive resulting in a comparatively larger β. Thus from Eqns. (1), (2) and (3) we see that the proposed GA-FLC provides the desired variation of gains around the set-point.

iii) There should be a good regulation against load disturbances. This can be achieved by setting maximum allowable gains. After the occurrence of a large load disturbance, e is small but Δe becomes large and both are of the same sign (*i.e.*, α is positive, which means larger β). In such situations, GA-FLC will increase its both integral and proportional gains. Therefore, recovery of the controlled variable will be faster, and hence, performance due to load fluctuations will be better.

iv) To improve the steady state response both proportional and integral gains should be set to a moderately large value around the steady state, $e \approx \Delta e \approx 0$. The proposed GA-FLC meets these requirements. Since at $e \approx \Delta e \approx 0 \Rightarrow \beta \approx 0.5$, therefore, at or around steady state we have $K_{fi} = G_e \times k_2\beta \approx G_e$ and $K_{fp} = G_{\Delta e}(1+k_3\beta) \approx$

$1.15 \times G_{\Delta e}$ (as we considered $k_2 = 2$ and $k_3 = 0.3$). Which means the proportional gain is enhanced by 15% and the integral gain becomes considerably large. Thus an improved steady state performance will be achieved.

From the above discussions we find that our proposed on-line tuning scheme attempts to provide improved transient responses due to both step set-point change and load disturbance (i, ii, iii), as well as improved steady state performance (iv). Further note that, the scheme is completely *model independent*. To make it clear, Eqns.(1), (2) and (3) used for the adjustments of SFs are solely determined from the instantaneous process states and does not involve any process parameters.

3 Results

We present the simulation results for the following two processes with moderate to large dead-time (L) under step set-point change and load disturbance.

$$\frac{d^2y}{dt^2} + 0.5y\frac{dy}{dt} = u(t-L) \tag{7}$$

$$\frac{d^2y}{dt^2} + \frac{dy}{dt} = u(t-L) \tag{8}$$

The performance of the proposed GA-FLC is compared with those of a FPIC, and ZN-tuned PI controllers (ZN-PIC). Several performance measures such as peak overshoot (%OS), settling time (t_s), integral absolute error (IAE) and time integral absolute error (ITAE) have been evaluated. We have used Mamdani type inferencing and height method of defuzzification [10]. Figures 5 and 6 respectively show the responses of (7) and (8) for both step set-point change and load disturbance. In all figures *dashed* (- -) curves represent the responses due to the ZN-PIC, *dash-dot* (-·-) due to FPIC, and *solid* (—) curves due to the GA-FLC. Performance indices of these controllers for (7) and (8) are provided in the Table I. From the results (Figs. 5 and 6, and Table I) it is found that in each case GA-FLC shows remarkably improved overall performance over both FPIC and ZN-PIC. From Figs. 5b and 6b we see that when the dead-time (L) is increased the processes under ZN-PIC become unstable in each case. But GA-FLC maintains almost the same level of performance compared to FPIC without making the system unstable.

Table 1. Performance analysis of different controllers

		Performance indices for							
		Non-linear system in (7)				Integrating system in (8)			
L	**FLC/PI**	**%OS**	**t_s(s)**	**IAE**	**ITAE**	**%OS**	**t_s(s)**	**IAE**	**ITAE**
0.3	ZN-PIC	99.4	Barely stable			98.5	18.1	18.1	228.6
	FPIC	42.6	31.4	12.3	99.7	36.4	18.4	8.9	43.9
	GA-FLC	24.5	27.4	10.4	63.2	10.3	13.2	7.3	25.3
0.5	ZN-PIC	Unstable				Unstable			
	FPIC	54.0	51.8	16.3	196.1	52.6	24.7	12.1	93.5
	GA-FLC	36.4	40.6	12.4	112.9	23.7	17.8	9.0	51.1

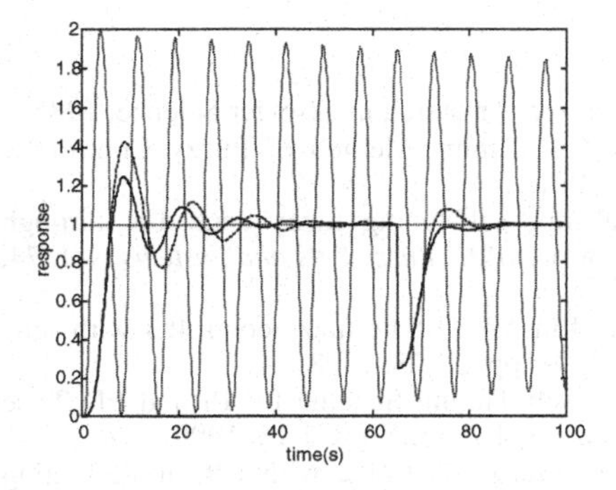

Fig. 5a. Responses of (7) with L = 0.3

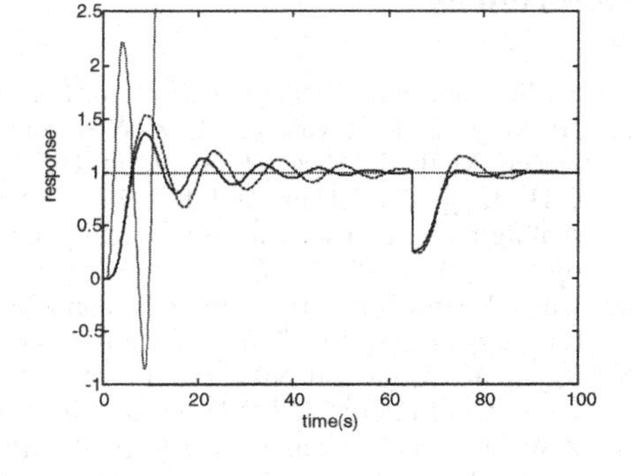

Fig. 5b. Responses of (7) with L = 0.5

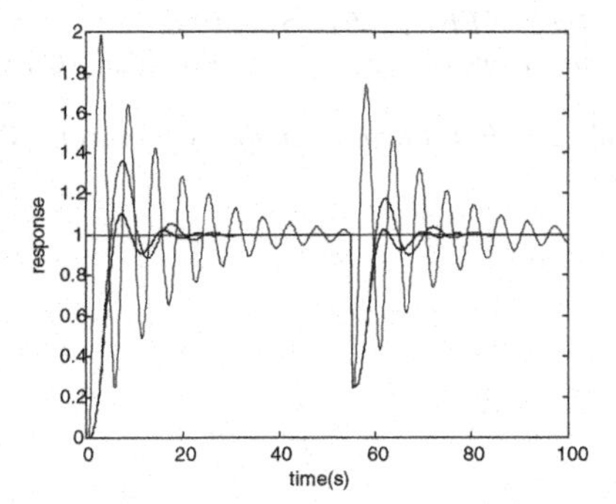

Fig. 6a. Responses of (8) with L = 0.3

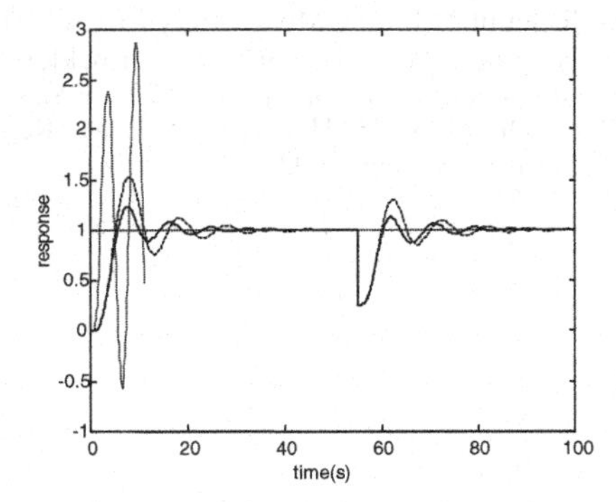

Fig. 6b. Responses of (8) with L = 0.5

4 Conclusion

In this investigation we proposed a simple model-free on-line tuning scheme for a conventional PI-type FLC (FPIC). Here equivalent proportional and integral gains of the FPIC are dynamically adjusted by modifying its input SFs thereby making it a gain adaptive FLC. For the adjustment of the input SFs we used a single parameter β (+ve sigmoid) which provides the required non-linear gain to achieve an improved control performance. The most important feature of the proposed scheme is that it is independent of the parameters of the process being controlled. Results showed that the proposed GA-FLC outperformed FPIC as well as ZN-PIC.

Acknowledgement

We are grateful to Prof. S. Bandyopadhyay, Dept. of Instrumentation & Electronics Engg., Jadavpur University and AICTE, Govt. of India, for partly supporting this work under grant No.-8017/RDII/TAP/TEG(954)/98-99.

References

1. M. Sugeno, *Industrial Applications of Fuzzy Control*, Amsterdam: Elsevier Science, 1985.
2. H. Nomura, I. Hayashi and N. Wakami, "A Self-Tuning Method of Fuzzy Control by Decent Method," *Proc. IFSA '91*, pp. 155-158, 1991.
3. C.H. Jung, C.H. Ham and K.I. Lee, "A real-time self-tuning fuzzy controller through scaling factor adjustment for the steam generator of NPP," *Fuzzy Sets and Systems*, vol. 74, pp. 53 - 60, 1995.
4. Hung-Yuan Chung, Bor-Chin Chen and Jin-Jye Lin, "A PI-type fuzzy controller with self-tuning scaling factors," *Fuzzy Sets and Syst.*, vol. 93, pp. 23-28, 1998.
5. Rajani K. Mudi and Nikhil R. Pal, "A Robust Self-Tuning Scheme for PI and PD Type Fuzzy Controllers," *IEEE Trans. on Fuzzy Systems*, vol.7, no. 1, pp. 2-16, 1999.
6. Z.W. Woo, H.Y. Chung and J.J. Lin, "A PID type fuzzy controller with self-tuning scaling factors," *Fuzzy Sets and Systems,* vol. 115, no. 2, pp. 321 - 326, 2000.
7. Rajani K. Mudi and Nikhil R. Pal, "A Self-Tuning Fuzzy PI Controller," *Fuzzy Sets and Systems,* vol. 115, no. 2, pp. 327 - 338, 2000.
8. R. Palm, " Sliding Mode Fuzzy Control," *Proc. Fuzz IEEE,* pp. 519-526, 1992.
9. H. Ying, W. Siler, and J. J. Buckley, "Fuzzy Control Theory: A Nonlinear Case," *Automatica*, vol. 26, pp. 513-520, 1990.
10. D. Dirankov, H. Hellendoorn & M. Reinfrank, *An Introduction to Fuzzy Control,* NY: Springer-Verlag, 1993.

A Traffic Light Controlling FLC Considering the Traffic Congestion

WanKyoo Choi[1], HongSang Yoon[1], Kyungsu Kim[2], IlYong Chung[3], and SungJoo Lee[3]

[1] Division of Computer, Electronics & Communiation Engineering, Kwangju University, 591-1 Jinwol-dong Nam-gu, Kwangju, Korea
{wkchoi,hsyoon}@kwangju.ac.kr

[2] Kwangju Institute of Science and Technology (K-JIST), 1 Oryong-dong Puk-gu, Kwangju, 500-712, Korea
arieskim@kjist.ac.kr

[3] School of Computer Engineering, Chosun University, Kwangju, 501-759, Korea
{iyc,sjlee}@mail.chosun.ac.kr

Abstract. The existing many fuzzy traffic controllers could not be applied to the traffic congestion situation because they have been developed and applied for a non-congestion traffic flow. In this paper, therefore, we propose a traffic light controlling FLC that is able to cope with traffic congestion appropriately. In order to consider such situation as missing the green signal because of spillback of upper crossroad, it uses as an input variable a degree of traffic congestion of upper roads, which vehicles on a crossroad are to proceed to. It uses the first-order Sugeno fuzzy model for modeling nonlinear traffic flow. In experiment, we compared and analyzed the fixed traffic signal controller and the proposed FLC. As a result of experiment, the proposed FLC showed more enhanced performance than the fixed traffic signal controller.

1 Introduction

Fuzzy logical controller (FLC) is able to use knowledge of experts and operators as the control rules and able to process the ambiguous information. For that reason, it has been applied to control of the complex nonlinear systems or to control of systems that have no a mathematical model[1].

There have been many researches[2, 6, 7, 8, 9] on the fuzzy traffic control that used such advantages as the linguistic description and the qualitative modeling of fuzzy logical controller. They used the number of entering vehicles at the green signal, the number of waiting vehicles during the red signal, the mean density of vehicles, the duration of the red signal and etc. as input variables, and used the duration of the green signal as output variable[2].

However, they could not be applied to the traffic congestion situation because they have been developed and applied for a non-congestion traffic flow. Under the traffic congestion situation, spillback on crossroads happens. Spillback on crossroads leads to over-saturation of roads. If spillback on upper crossroad happens, missing the green signal on lower crossroad is unavoidable[3]. Therefore,

N.R. Pal and M. Sugeno (Eds.): AFSS 2002, LNAI 2275, pp. 69–75, 2002.

the traffic controllers not considering the traffic congestion situation can't cope successfully with the traffic congestion and the change of traffic situation.

In this paper, therefore, we propose a traffic light controlling FLC that can cope with the traffic congestion appropriately. It uses as an input variable a degree of traffic congestion of upper roads, which vehicles on a crossroad are to proceed to. It is the first-order Sugeno fuzzy model of 3-inputs and 1-output, which is constructed from Mamdani-type FLC tuned by *the membership function modification algorithm*[4].

We compared and analyzed the fixed traffic signal controller and the proposed FLC by using the delay time and the proportion of passed vehicles to entered vehicles. As a result of comparison, the proposed controller showed more enhanced performance than the fixed traffic signal controller.

2 Membership Function Modification Algorithm

For the purpose of enhancing performance of the fuzzy logical controller(that is, FLC), the membershp function modification algorithm of fig.1 modifies the size and shape of the triangular membership functions of FLC by clustering the input/output data from its fuzzy inference system[4].

```
procedure MFM(FIS, fitness)
// FIS is the fuzzy inference system with the same size and shape.
// fitness is the threshold value.
// Generates the tunning data by the ramdom number generator.
inoutdata ← GetTuningData(FIS);
// Get the control knowledge from rules of FIS.
controlknowledge ← GetControlKnowledge(FIS);
newFIS = FIS;
while TRUE do
// Deletes elements from inoutdata which don't satisfy controlknowledge.
newinoutdata ← Delete(inoutdata, controlknowledge);
// Computes fitness of inoutdata by using the fitness measure.
newfitness ← Evaluate(inoutdata, newinoutdata);
if(newfitness <= fitness) then break; end
inoutdata ← newinoutdata;
// Modifies newFIS by clustering inoutdata
// using the KMeans clustering algorithm.
newFIS ← MakeNewFis(newFIS, inoutdata);
// Gets the output values by using the input values as input of newFIS.
NewOutputValues(inoutdata, newFIS);
end
FIS ← newFIS;
end MFM
```

Fig. 1. The membership functions modification algorithm

In this algorithm, the fitness measure is defined as follows:

$$fitness = |X_{incons}|/|X_{all}| \tag{1}$$

where $|X_{incons}|$ is the number of input/output data which does not satisfy the control knowledge, and $|X_{all}|$ is the cardinality of input/output data set.

3 Fuzzy Traffic Controller Considering the Traffic Congestion

The purpose of the intelligent traffic control using the fuzzy logical controller is to lighten the traffic congestion and to decrease the delay time on crossroads by finding out the traffic situation of the current crossroads and by changing the traffic signal appropriately.

In order to achieve this purpose, we design a traffic light controlling FLC based on the following assumptions.

1. Sensors can detect the congestion of corresponding roads.
2. The width of crossroad and the length of vehicles decide the duration of yellow signal.

Input variables of the designed FLC for controlling the traffic light are x_1, x_2 and x_3. x_1 indicates the duration of the red signal on a crossroad, and x_2 indicates the number of the waiting vehicles during the red signal, and x_3 indicates a degree of traffic congestion of upper roads, which vehicles on a crossroad are to proceed to. A degree of traffic congestion of upper roads is computed as follows:

$$DC = NV/((LR/LV) \times NL) \times 100 \tag{2}$$

where DC indicate a degree of traffic congestion, and NV means the number of vehicles traveling or stopping on road, and LR indicates the length of road, and LV indicates of the average length of vehicle, and NL indicate the number of lanes on road.

Output variable is y, which indicates the duration of the green signal on a crossroad. The first-order linear equation for decisions of FLC is as follows.

$$y = p_0 + p_1x_1 + p_2x_2 + p_3x_3 \tag{3}$$

Fig.2 shows the membership functions of input variables are x_1, x_2, x_3. Table 1 shows rules for conditionals of the proposed FLC.

4 Experiment and Results

We developed the traffic-flow simulator by using Matlab, and estimated the performance of the proposed fuzzy traffic controllers on a crossroad. In our simulation, situation of a crossroad was assumed as show in fig.3.

In situation of a crossroad as like fig.3, the length of vehicles was assumed as $4m$ and length of all roads was $50m$. We did not consider crosswalk and vehicles

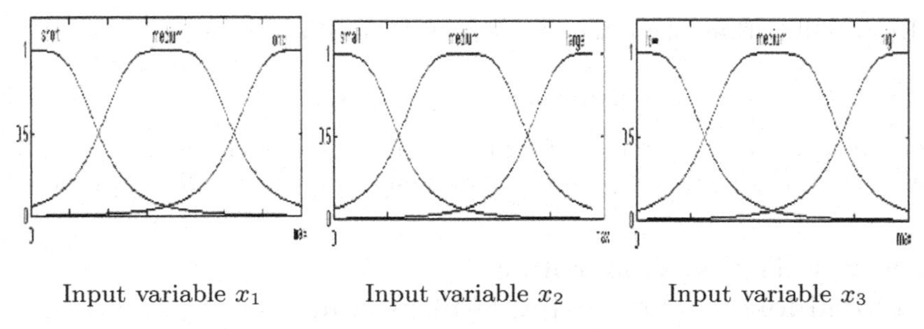

Input variable x_1 Input variable x_2 Input variable x_3

Fig. 2. The membership functions for the input variables of the designed FLC

Table 1. Rules for conditionals of the designed FLC

Rule	x_1	x_2	x_3	Rule	x_1	x_2	x_3	Rule	x_1	x_2	x_3
1	short	small	low	10	medium	small	low	19	long	small	low
2	short	small	medium	11	medium	small	medium	20	long	small	medium
3	short	small	high	12	medium	small	high	21	long	small	high
4	short	medium	low	13	medium	medium	low	22	long	medium	low
5	short	medium	medium	14	medium	medium	medium	23	long	medium	medium
6	short	medium	high	15	medium	medium	high	24	long	medium	high
7	short	large	low	16	medium	large	low	25	long	large	low
8	short	large	medium	17	medium	large	medium	26	long	large	medium
9	short	large	high	18	medium	large	high	27	long	large	high

of right-turn. Signal-cycle of the upper roads (that is, A′, B′, C′ and D′ in fig.3) was decided randomly in range of [50 80] and the duration of their green signal was decided by a degree of their complexity. The green signal was circulated based on signal for straight and left-turn (that is, A→B→C→D→A→...). We assigned 18 seconds for straight and left-turn and 2 seconds for yellow signal. We used $4\times(18+2)=80$ seconds as basic signal-cycle.

Therefore, ranges of input variables x_1, x_2 and x_3 of the proposed FLC are [0 70], [0 25] and [0 100] respectively. Table 2 shows the coefficients for decisions of the proposed FLC. Coefficients are retrieved from Mamdani-type FLC tuned by *the Membership Function Modification Algorithm.*

For simulation, velocity and occurrence of vehicles was assumed as five levels. The number of the occurred vehicles were 350±50, 600±50, 1100±50, 1450±50 and 1800±50 respectively. Velocities of vehicles were $10km/h$, $20km/h$, $30km/h$, $40km/h$ and $50km/h$ respectively.

During green signal, only if vehicles on the waiting roads(that is, A, B, C and D in fig.3) could enter upper roads, they entered a crossroad and proceeded to upper roads. Only if the occurred vehicles could enter current roads, they entered to current roads.

In situation of the same velocity of vehicles, the same controller and the same number of the occurred vehicles, we simulated 20 times respectively.

Service levels of a crossroad with traffic signal are decided by the delay time. Delay time indicates a loss of fuel and it is the measure of discomfort. Because

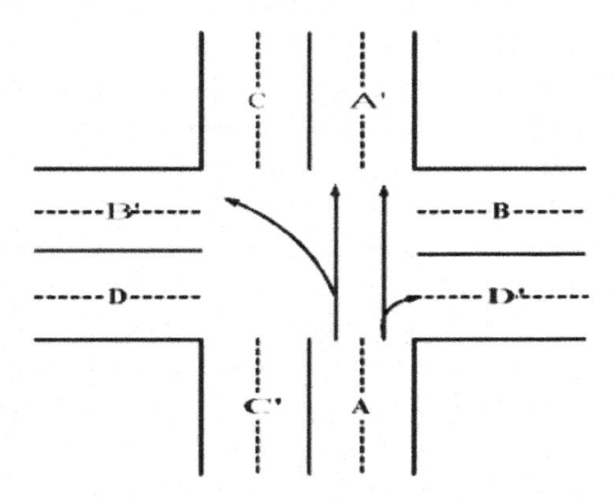

Fig. 3. Situation of crossroad

Table 2. Coefficients for decisions of the poposed FLC

Rule	p_0	p_1	p_2	p_3	Rule	p_0	p_1	p_2	p_3
1	-0.01950	0.38852	-0.05853	10.79292	15	0.03524	-0.00409	-0.32359	34.39597
2	0.02464	0.15028	-0.03351	10.14033	16	-0.17693	0.38142	-0.01547	17.21890
3	0.00153	0.07194	-0.53159	53.79191	17	0.05489	0.07655	-0.21064	20.02599
4	-0.14096	1.26741	-0.05632	2.69337	18	0.00825	0.07876	-0.37951	38.50978
5	-0.11028	0.08930	-0.15790	21.57437	19	-0.07724	0.04836	0.12491	8.15390
6	0.06031	-0.15257	-0.35749	39.84764	20	-0.10405	0.16311	0.02180	9.48079
7	-0.00232	0.19459	0.14001	12.11409	21	-0.04824	0.07414	-0.02004	9.27996
8	0.00693	0.28749	-0.13661	14.14908	22	-0.16664	0.17622	0.07081	11.86633
9	0.05418	-0.09648	-0.28683	34.20968	23	-0.11883	0.07644	0.01182	9.88934
10	-0.12793	0.17118	0.06010	7.82502	24	-0.01978	0.27917	0.05008	-3.42197
11	-0.11130	0.14118	0.02259	6.81630	25	-0.22981	0.29956	-0.06481	19.09516
12	-0.12890	-0.14107	0.04363	4.59792	26	-0.33633	0.08413	-0.03334	31.64426
13	-0.25188	0.68808	-0.06762	10.38073	27	-0.00371	-0.19136	-0.51709	57.49267
14	-0.28770	0.95889	-0.02953	9.54964					

delay time is not always decided according to a degree of saturation, it is difficult to decide service level based on a degree of saturation[5]. Therefore, we estimated performance of controllers by using not a degree of saturation but the delay time.

The number of passed vehicles, which is commonly used for estimating performance of fuzzy traffic controller, did not assumed the traffic congestion on a crossroad. Therefore, it is not used for estimating performance of the proposed controllers. Instead of it, we use a degree of passage(that is, DP), which means the proportion of passed vehicles to entered vehicles.

$$DP = Number\ of\ passed\ vehicles\ /\ Number\ of\ entered\ vehicles \qquad (4)$$

The traffic congestion of a crossroad results to missing the green signal, and decreases the number of passed vehicles on a crossroad, and increases a degree of traffic complexity on roads, and obstructs vehicles from entering roads.

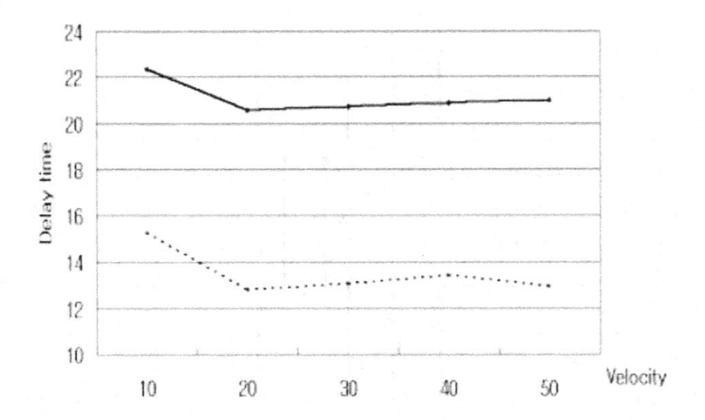

Fig. 4. Delay time according to velocity of vehicles

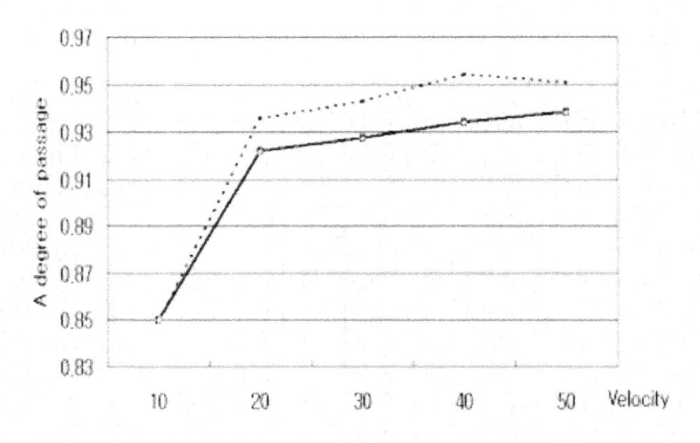

Fig. 5. DP according to occurrence of vehicles

Fig.4 and fig.5 show results of experiment. In figures, the dotted line indicates the fixed signal controller, and the solid line indicates the proposed FLC. In fig.4, the delay time of the proposed FLC is lower than the fixed traffic signal controller. In fig.5, DP of the proposed FLC are higher than the fixed traffic signal controller.

Table 3 shows the means of delay-time and DP. The proposed FLC decreases the delay time about 35%, and enhances a degree of passage.

5 Conclusion

In this paper, we proposed a traffic light controlling FLC that can cope with the traffic congestion well. In order to consider such situation as missing the green signal caused by the spillback of upper crossroad, we used as an input variable a degree of traffic congestion of upper roads, which vehicles on a crossroad are to proceed to.

Table 3. Mean values of the performace measures

Controller	Fixed traffic signal controller	Proposed FLC
Delay time	21.1068	13.5156
DP	0.9142	0.9265

In situation of the same velocity of vehicles, the same controller and the same number of the occurred vehicles, we simulated 20 times respectively. We evaluated performance of the proposed FLC and the fixed traffic signal controller, by using the delay time and the proportion of passed vehicles to entered vehicles.

As a result of experiment, the proposed controller showed more enhanced performance than the fixed traffic signal controller. That is, the proposed FLC decreased the delay time about 35% and enhanced a degree of passage.

Acknowledgement

This work was partially supported by Agricultural R&D Promotion Center (ARPC).

References

1. Y. T. Kim, Y, J. Lee, S. H. Lee, T.S. Chung, Z. N. Bien: A Survey on the Fuzzy Control Systems with Learning/Adaptation Capability. Journal of KFIS 5(1995) 11-35.
2. J. W. Kim: A Traffic Light Controlling FLC Adaptable to Various Traffic Volumes. Journal of KISS(B) 24(1977) 976-985.
3. K. H. Lee: Green-Split Coordination Strategy In Over-saturated Signal System. Journal of KTRS 11(1993) 87-103.
4. W. K. Choi, M. J. Chung: Performance Improvement of the FLC by Membership Function Modification Algorithm. The KIPS Transactions 8-B(2001) 123-129.
5. J. M. Won, J. S. Choi: Traffic Engineering, Parkyoungsa:Seoul(1993).
6. Pappis, C.P. and Mamdani, E.H: A Fuzzy Logic Controller for a traffic junction. IEEE Transactions on Systems, Man, and Cybernetics SMC-7(1977) 707-717.
7. Favilla, J, Machion, A., and Gomide, F.: Fuzzy Traffic Control: Adaptive Strategies. Proc. Of 2nd IEEE int'l Conf. On Fuzzy Systems(1993), 506-511.
8. Jamshidi, M., Kelsey, R., and Bissel, K.: traffic Fuzzy Control: Software and Hardware Implementations. Proc. Of the Fifth IFSA Word Congress(1993) 907-910.
9. Jongwan Kim: A Fuzzy logic Control Simulator for Adaptive Traffic Management. Proc. Of IEEE Int'l conf. On Fuzzy Systems(1997) 1519-1524.

Vehicle Routing, Scheduling and Dispatching System Based on HIMS Model

Kaoru Hirota[1], Fanyan Dong[1], Kewei Chen[2], Yasufumi Takama[1]

[1] Hirota Lab., G3 Bldg., Tokyo Institute of Technology,
4259 Nagatsuta-Cho, Midori-Ku, Yokohama City 226-8502, Japan
{ hirota, fydong, takama }@hrt.dis.titech.ac.jp
[2] J & F CO., LTD. Sumito-Komagata Building 3F,
Komagata 2-7-4, Taitouku, Tokyo 111-0043, Japan,
VZW03122@nifty.ne.jp

Abstract: A concept of the VRSDP/SD problem and its formularization are proposed in order to bridge the gap between conventional methods and complex situations in the real world. A HIMS model is also proposed for the VRSDP/SD problem. Experiments with two cases (full or partial working), are performed. The experimental results and the evaluations by experts from practice show that the HIMS model provides a feasible, fast, and efficient tool for the real world problem.

1. Introduction

Vehicle routing, scheduling, and dispatching problems have grown to be an important research field in the past 20 years. Many researchers [1,2] have been trying to answer fundamental research questions, utilizing SA [3,4], Tabu search [5,6], and GA [7,8]. Although applications [9,10] have also been developed, they are not sufficient because there are many gaps between conventional methods and complex situations in the real world. In this article, a concept of the VRSDP/SD problem (***Vehicle Routing & Scheduling Dispatching Problem with Single Depot***) is described to model those complex situations of real world in a synthetic manner. A HIMS model (*a calculation model with **HI**erarchical **M**ultiplex **S**tructure*) is also proposed as a solution for the practice VRSDP/SD problem.

A concept of the VRSDP/SD problem and its formulization based on fuzzy set theory are proposed in section 2. The structure of the HIMS model and its operation criteria are mentioned in section 3. Experimental results and evaluations are elaborated in section 4.

N.R. Pal and M. Sugeno (Eds.): AFSS 2002, LNAI 2275, pp. 76–84, 2002.

2. VRSDP/SD problem and Its Formulation

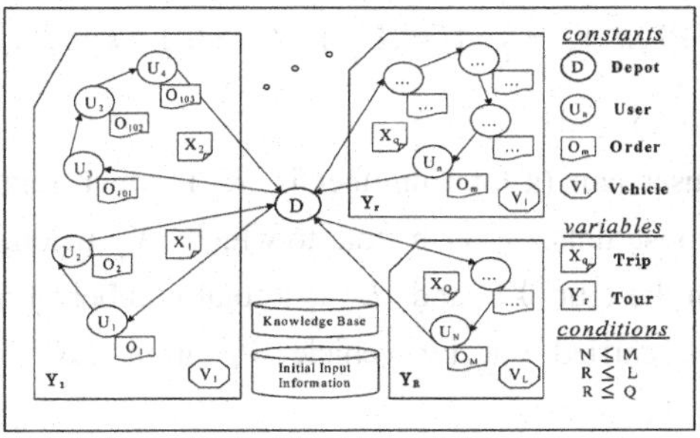

Item	Universal Set
D	D
U_n	$U = \{U_1, .., U_n, .., U_N\}$
O_m	$O = \{O_1, .., O_m, .., O_M\}$
V_l	$V = \{V_1, .., V_l, .., V_L\}$
X_q	$X = \{X_1, .., X_q, .., X_Q\}$
Y_r	$Y = \{Y_1, .., Y_r, .., Y_R\}$

Fig. 1 VRSDP/SD problem

The VRSDP/SD problem is introduced for daily delivery activities in the real world (Fig.1). There are many practical applications for the VRSDP/SD problem such as transporting food and drink product by truck and oil product delivery by tank lorry.

2.1. Description of the VRSDP/SD problem

A depot D (a delivery center which holds all goods in stock) has L vehicles $\{V_l\}$ for delivery, and there are several types of vehicles with different capacity. When M orders $\{O_m\}$ (needs for some goods in a certain time window) are obtained from N users $\{U_n\}$ (consumers of some orders and each user's parking space for a delivery vehicle, business hours, and so on are different from others), a delivery schedule must be made for all of these M orders by R vehicles of various types until the next day. The problems are how a plan can be made with optimum routing and efficient scheduling for all delivery jobs, and with suitable dispatching for vehicles of various types while satisfying some constraints.

2.2. Constants & Variables in VRSDP/SD Problem

$$D=([Bt_D, Et_D]) \quad (1) \qquad O_m=(C_m,[Bt_m,Et_m],U_n^m),\ C_m \le SU_n^m \quad (3)$$

$$U_n=([Bt_n,Et_n],SU_n) \quad (2) \qquad V_l=(SV_l,RWt_l,EWt_l) \quad (4)$$

Bt_D (Depot Beginning Time) and Et_D (Depot End Time) are integer values in minutes. The integer interval $[Bt_n, Et_n]$ is a time window of the business hours and SU_n is a max size of the vehicles that can enter in U_n. The C_m indicates the capacity of an order O_m by a user U_n^m and the integer interval $[Bt_m, Et_m]$ is a desired time window for delivery. SV_l is a max capacity that the vehicle can load, RWt_l is a restriction time until when V_l must work full, and EWt_l is a max extended time until when V_l can work more.

Variables of trip and tour represent a solution to the VRSDP/SD problem.

$$X_q = ([D, J^q_{m1}, J^q_{m2}, \ldots, J^q_{mK_q}, D], Y^q_r, k_q), \; J_m = (O_m, U^m_n), \; K_q \geq 1, \; 1 \leq k_q \leq K^q_r \quad (5)$$

$$Y_r = ([X^r_{q1}, X^r_{q2}, \ldots, X^r_{qK_r}], V^r_l); \; K_r \geq 1. \quad (6)$$

Here, J_m is a sub-job with order/user pair (a total number is K_q), which represents order information and related user's constraints, Y^q_r is the tour to which X_q belongs, k_q indicates an index number of X_q in the tour Y^q_r, and K^q_r is a total number of trips in the tour Y^q_r. Eq.(6) shows a tour carried out by vehicle V^r_l in a day. Where, $[X^r_{q1}, \ldots, X^r_{qK_r}]$ is a sequence of trips in Y_r.

2.3. Constraints for VRSDP/SD Problem

Some constraints must be satisfied in the VRSDP/SD problem.

(1). Vehicle Capacity

$$C_{cnst} = \sum_{q=1}^{Q} U^q_{cnst} = 0, \quad C^q_{cnst} = \begin{cases} 0 & if \; \sum_{j=1}^{K_q} C^q_{mj} \leq SV^q_l \\ 1 & otherwise \end{cases} \quad (7)$$

Eq.7 represent the constraint to vehicles that the total capacity of orders in a trip cannot be over the capacity of the corresponding vehicle.

(2). User Condition

$$U_{cnst} = \sum_{q=1}^{Q} U^q_{cnst} = 0, \quad U^q_{cnst} = \begin{cases} 0 & if \; SV^q_l \leq SU^{mj}_n \; for \; ^\forall j = 1, \ldots, K_q \\ 1 & otherwise \end{cases} \quad (8)$$

Eq.8 means a vehicle size must be equal to or less than the max size that can enter in all users' parking spaces in a trip.

(3). Depot Condition

$$D_{cnst} = \sum_{q=1}^{Q} D^q_{cnst} = 0, \quad D^q_{cnst} = \begin{cases} 0 & if \; Bt_D \leq At^q_D \leq Et_D \\ 1 & otherwise \end{cases} \quad (9)$$

Eq.9 indicate that all loading works must be done while the depot is open. Where, At^q_D is the start time that the trip X_q at the depot.

2.4. Evaluation for VRSDP/SD problem

(1). Total Working Time

$$T_{cost} = \sum_{q=1}^{Q} T^q_{cost} \Rightarrow \min, \qquad T^q_{cost} = T^q_{Running} + T^q_{Loading} + T^q_{Unloading} \quad (10)$$

(2). Average Loading Capacity

$$C_{cost} = (\sum_{q=1}^{Q} C^q_{cost}) / Q \Rightarrow \max, \qquad C^q_{cost} = (\sum_{j=1}^{K_q} C^q_{mj}) \Big/ SV^q_l \quad (\in [0,1]) \quad (11)$$

Here, C^q_{mi} is a capacity of an order O^q_{mi} in X_q, and SV^q_l is the capacity of the vehicle carrying out the tour Y^q_r.

(3). Working Balance

$$B_{cost} = \sum_{r=1}^{R} | B^{r}_{cost} - B^{mean}_{cost} | \Big/ R \Rightarrow \min, \ B^{q}_{cost} = \sum_{i=1}^{K_r} T^{qi}_{cost}, \ B^{mean}_{cost} = \sum_{r=1}^{R} B^{r}_{cost} \Big/ R \qquad (12)$$

The working balance of each vehicle is considered, which is important for equality in a labor condition among all drivers. B^{r}_{cost} is a working time for a tour Y_r and T^{qi}_{cost} is a working time of a trip X^{r}_{qi} and is defined by Eq.10.

(4). Working Capacity

$$V_{cost} = L - R \quad \Rightarrow \max \qquad (13)$$

To decrease the transport cost, a suitable (less) number of vehicles would like to be used according to the number of orders.

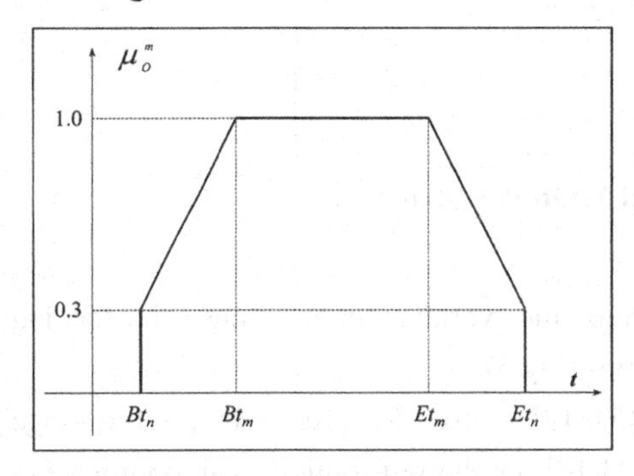

Fig 2. Fuzzy function of Service 1

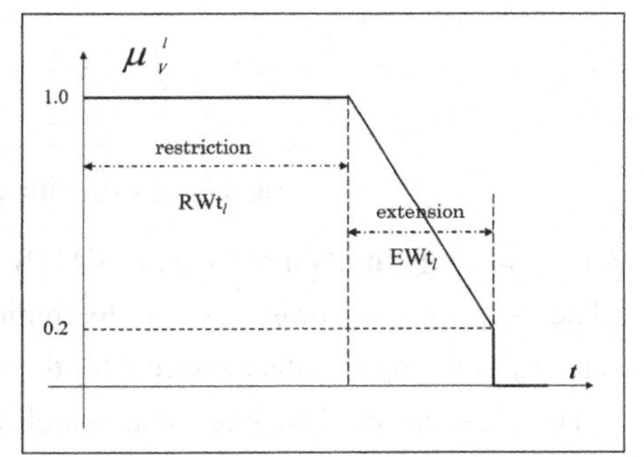

Fig 3. Fuzzy function of Service 2

(5). Service 1 (User Time Service)

$$S_T = \sum_{q=1}^{Q} S^{q}_{T} / Q \Rightarrow \max, \quad S^{q}_{T} = \sum_{j=1}^{K_q} \mu^{m}_{O} \Big/ K_q \qquad (14)$$

Where μ^{m}_{O} is assumed as the satisfaction degree of the user U^{m}_{n} about the delivered time of O_m. Service 1 (Fig.2) is the time service for users (Eq.14).

(6). Service 2 (Driver Working Service)

Service 2 (Fig.3) is the constraints about vehicle working (Eq.15).

$$S_V = \sum_{l=1}^{L} \mu^{l}_{V} \Big/ L \quad \Rightarrow \max \qquad (15)$$

3. HIMS Calculation Model

The VRSDP/SD problem has been studied and discovered that there are several levels in which one of the main elements (orders, trips, tours) is very active. This fact is utilized to make a calculation model of ***HIerarchical Multiplex Structure*** for the VRSDP/SD problem that is called the HIMS model (Fig.4).

There are three levels in the HIMS model: Atomic level is an active area of system energy (e.g. running cost). Molecular level is a reflection area of system properties (e.g.

the base jobs). Individual level is a forming area of a system architecture (a state with best planning), in which the objective character can be adjusted exactly by means of inference from knowledge base.

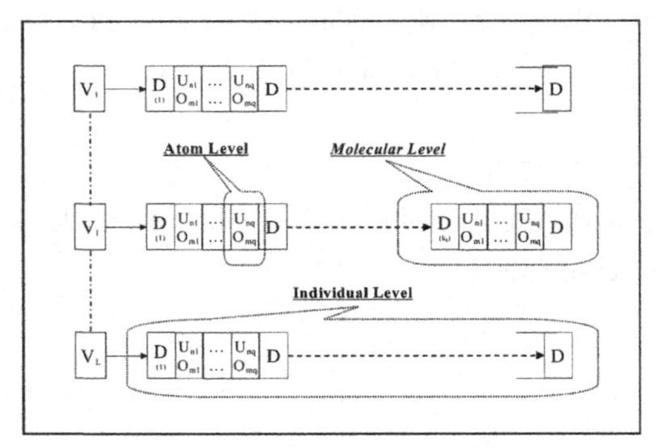

Fig 4. Hierarchical Multiplex Structure

3.1 Strategy in Atomic Level of HIMS

The purpose in Atomic level is to minimize the working time, to raise the loading capacity, and improve the Service 2 for the trips (Fig.5).

The heuristic method like Tabu search [5,6,13] is used for order Move / Exchange operations, and SA (Simulated Annealing) [11,12] for shortest routing as an optimization algorithm in Atomic level.

$\exists X_a = ([D, J_{a1}, .., J_{ai}, .., J_{aI}, D], ..),\ 1 \le a \le Q;$

$\exists X_b = ([D, J_{b1}, .., J_{bj}, .., J_{bJ}, D], ..),\ 1 \le b \le Q;$

1. Order Move Operation

$X_{a'} = ([D, J_{a1}, .., J_{ai}, J_{bj}, J_{ai+1}, .., J_{aI}, D], ..),$

$X_{b'} = ([D, J_{b1}, .., J_{bj-1}, J_{bj+1}, .., J_{bJ}, D], ..),$

2. Orders Exchange Operation

$X_{a'} = ([D, J_{a1}, .., J_{ai-1}, J_{bj}, J_{ai+1}, .., J_{aI}, D], ..),$

$X_{b'} = ([D, J_{b1}, .., J_{bj-1}, J_{ai}, J_{bj+1}, .., J_{bJ}, D], ..),$

To Satisfy:

I. $TSP(X_{a'}),\ TSP(X_{b'}),$

II. $T^{a'}_{cost} + T^{b'}_{cost} \le T^{a}_{cost} + T^{b}_{cost}$

Fig. 5 Criteria in Atomic Level

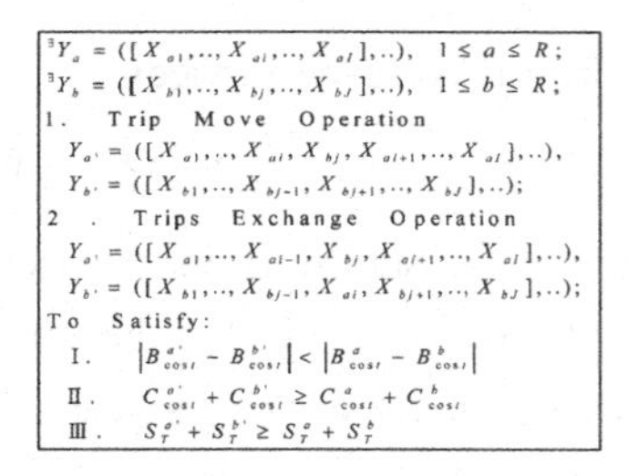
$\exists Y_a = ([X_{a1}, .., X_{ai}, .., X_{aI}], ..),\ 1 \le a \le R;$

$\exists Y_b = ([X_{b1}, .., X_{bj}, .., X_{bJ}], ..),\ 1 \le b \le R;$

1. Trip Move Operation

$Y_{a'} = ([X_{a1}, .., X_{ai}, X_{bj}, X_{ai+1}, .., X_{aI}], ..),$

$Y_{b'} = ([X_{b1}, .., X_{bj-1}, X_{bj+1}, .., X_{bJ}], ..);$

2 . Trips Exchange Operation

$Y_{a'} = ([X_{a1}, .., X_{ai-1}, X_{bj}, X_{ai+1}, .., X_{aI}], ..),$

$Y_{b'} = ([X_{b1}, .., X_{bj-1}, X_{ai}, X_{bj+1}, .., X_{bJ}], ..);$

To Satisfy:

I. $|B^{a'}_{cost} - B^{b'}_{cost}| < |B^{a}_{cost} - B^{b}_{cost}|$

II. $C^{a'}_{cost} + C^{b'}_{cost} \ge C^{a}_{cost} + C^{b}_{cost}$

III. $S^{a'}_{T} + S^{b'}_{T} \ge S^{a}_{T} + S^{b}_{T}$

Fig. 6 Criteria in Molecular Level

3.2 Strategy in Molecular Level

In Molecular level, the aim is to raise loading capacity, to adjust working balance, and to improve service 1 for all tours (Fig.6). Tabu search is also used for trip Move / Exchange operations similarly in Atomic level. Furthermore, a global search is also used for improving Service 1 because the number of trips is not so many in a tour.

3.3 Fuzzy Inference in Individual Level

The aims of operation in this level are to improve working balance and to assign appropriate number of vehicles. Fig.7 is a fuzzy membership function for trip balance operation to different vehicle types. The datum lines of trip balance, μ_L (lower boundary) and μ_U (upper boundary), are given. μ_T (Trip Balance's current value) can be calculated and is used to decide whether to adjust trip balance or not. Fig.8 is also a fuzzy membership function for vehicle balance operation in the same type. The datum lines of vehicle balance, λ_L (lower boundary) and λ_U (upper boundary), are also given. λ_V (Vehicle Balance's current value) can be obtained and is used to decide whether to adjust vehicle balance or not. By trips ± and vehicles ± operations, not only the relations among trips and tours can be adjusted, but also the vehicles with different type can be dispatched appropriately.

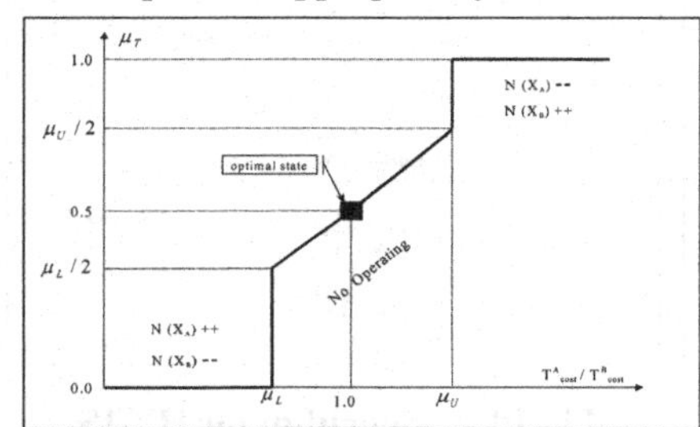

Fig 7. Trip Balance Operation **Fig 8. Vehicle Balance Operation**

3.4 Evaluation and Selection

(1). Evaluation Criteria of Objective State

$$g_1(X^k) = 1 - T_{\cos t}^{[k]} / T_{\cos t}^{[0]} \quad \in [0,1] \qquad (16)$$

$$g_2(X^k) = C_{\cos t}^{[k]} \quad \in [0,1] \qquad (17)$$

$$g_3(Y^k) = 1 - B_{\cos t}^{[k]} / B_{\cos t}^{[0]} \quad \in [0,1] \qquad (18)$$

$$g_4(Y^k) = 1 - 1 / (1 + V_{\cos t}^{[k]}) \quad \in [0,1] \qquad (19)$$

$$g_5(X^k) = S_T^{[k]} \quad \in [0,1] \qquad (20)$$

$$g_6(Y^k) = S_V^{[k]} \quad \in [0,1] \qquad (21)$$

Here the superscript [0] indicates an initial value of each objective state and the superscript [k] represents a state value of kth generation.

$$g(X^k, Y^k) = \sum_{i=1}^{6} \rho_i \, g_i(X^k | Y^k); \qquad \rho_i \in R^+ \qquad (22)$$

Eq.22 describes a general evaluation for all objective items, where ρ_i is a weighted coefficient which represents an importance of each objective item.

(2). Selection Criteria of Local Optimum State

Eq.23 defines a fitness of the HIMS model, which can be used to capture a local optimum state. Eq.24 indicates how to make a decision to select the better state of X and Y, where $a_1, .., a_6$ are the lowest goal value for each objective item.

$$eval(k) = \frac{g(X^k, Y^k) - g(X^0, Y^0)}{g(X^*, Y^*) - g(X^0, Y^0)} \tag{23}$$

$$g(X^*, Y^*) = \begin{cases} g(X^k, Y^k) & if \;\; eval(k) > 1 \\ g(X^*, Y^*) & \text{otherwise} \end{cases} \tag{24}$$

$$s.t. \; g_1 \geq a_1, g_2 \geq a_2, g_3 \geq a_3, g_4 \geq a_4, g_5 \geq a_5, g_6 \geq a_6$$

3.5 The Implementation of HIMS Model

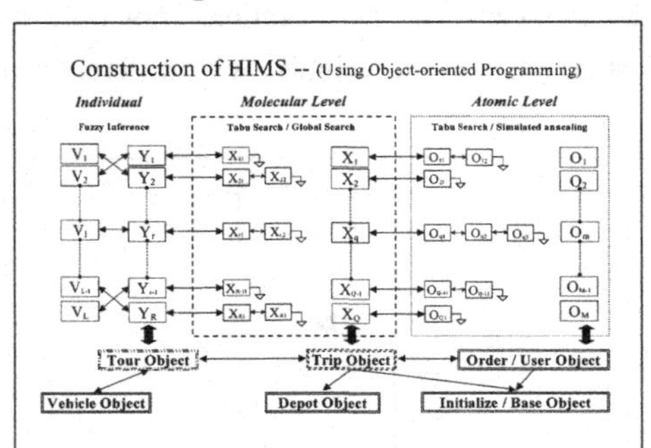

Fig 9. HIMS Data Structure

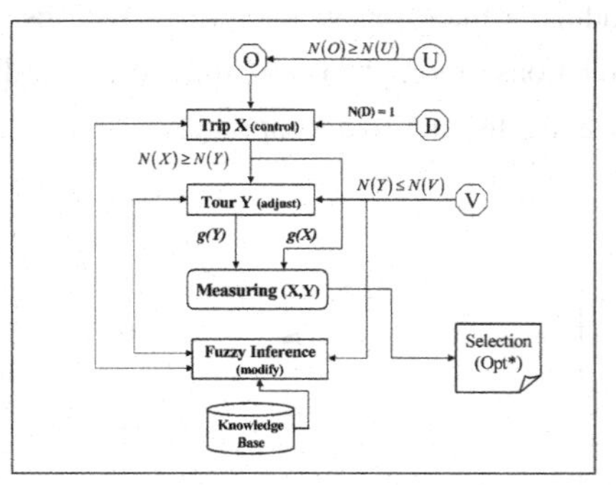

Fig 10. Algorithm for HIMS

A data structure for the HIMS model (Fig.9), which is suitable for object-oriented programming, is proposed. An optimization algorithm (Fig.10) with multiplex heuristic method is also described.

4. Experiments and Discussion

The component of the HIMS model is applied to the test data set used in an actual oil company, where 39 gasoline stations in Izu peninsula area (Japan) is serviced from an oil center by a transport company with 10 tank lorries. The transportation networks are also known by a digital map (GIS), from which the cost information can be calculated by Dijkstra algorithm.

Table 1. Comparison between HIMS and Experts

Operator		T_{cost}	C_{cost}	B_{cost}	V_{cost}	S_T	S_V
1	HIMS	5250	95.57	28.40	0	98.74	86.26
	Experts	*5381*	*87.32*	*43.20*	*0*	*83.26*	*81.63*
2	HIMS	5095	93.91	14.69	3	96.88	93.25
	Experts	*5273*	*85.62*	*58.34*	*1*	*85.68*	*77.26*

The two experiments are mainly for transport companies in some common cases (full working when the needs is too higher, or partly working when a periodical inspection, repairs, etc. are considered). Table 1 shows the results obtained by HIMS model and experts in the two experiments.

Table 2. Comparison with Algorithms

Algorithm	Process	Data Type	Data Structure	Optimal Means
HIMS	Synthesis	Fuzzy [0,1]	Object-oriented	Meta + Fuzzy
Conventional	Individual	Crisp {0,1}	Linear	Meta or IP
Expert	Individual	Crisp {0,1}	--------	Experience

Table 2 shows the difference among the HIMS and others.

Table 3. Comparison with Applications

Application	Parameters	Functions	Balance	Time	Flexibility
HIMS	Few	Many	Good	Fast	Good
Conventional	Many	Few	Bad	Slow	Bad

Table 3 shows the advantages of HIMS model as a software component in the practical system applications.

5. Conclusion

A concept of the VRSDP/SD problem in the real world, including both the *Combinatorial Optimization Problem* (COP) and the *Constraint Satisfaction Problem* (CSP), has been introduced and is formalized in terms of fuzzy set theory. A calculation model with hierarchical multiplex structure, called the HIMS model, and its operation strategies based on heuristics, optimization, and fuzzy inference, have also been proposed. The HIMS model is constructed as a software component using object-oriented programming, and the corresponding optimization algorithm is represented through meta-programming and fuzzy inference. Experiments with two cases (full or partial working) are performed, which show that the HIMS model is a feasible, fast, and efficient tool for the application in the real world VRSDP/SD problem ([14] for detail).

As an extension of the VRSDP/SD problem, a new problem of VRSDP/MD (***Vehicle Routing & Scheduling Dispatching Problem with Multiple Depot***) will be defined and a corresponding HIMS$^+$ model will be also verified as part of future work. The HIMS model gives the foundation to some other problems such as problems involving both COP and CSP.

Reference

1. M.O. Ball, T.L. Magnanti, C.L. Monma, and G.L. Nemhauser: Network Routing, vol.8, Elsevier Science, Amsterdam, 1995.
2. G. Laporte and I.H. Osman: Routing problems: A bibliography, Ann. Oper. Res., vol.61, pp.227-262, 1995.
3. A.V. Breedam: Improvements heuristics for the vehicle routing problem based on simulated annealing, Eur. J. Oper. Res., vol.86, pp.480-490, 1995.
4. A.S. Alfa, S.S. Heragu, and M. Chen: A 3-opt based simulated annealing algorithm for the vehicle routing problem, Comp. Ind. Eng., vol.21, p.635, 1991.
5. E. Taillard and P. Badeau: A tabu search heuristic for the vehicle routing problem with soft time windows, Transportation Science, vol.31, pp.170-186, May 1997.
6. C. Duhamel, J.Y. Potvin, and J.M. Rousseau: A tabu search heuristic for the vehicle routing problem with backhauls and time windows, Transp. Science, vol.31, pp.49-59, 1997.
7. F. Leclerc and J.Y. Potvin: Genetic algorithms for vehicle dispatching, Int. Trans. Opl. Res., vol.4, no.5/6, pp.391-400, 1997.
8. R. Cheng, and M. Gen: Vehicle routing problem with fuzzy due-time using genetic algorithms, Journal of Fuzzy Theory and Systems, vol.7, no.5, pp.1050-1061, Japan, 1995.
9. L.D. Bodin: Twenty years of routing and scheduling, Oper. Res., vol.38, pp.571-579, 1990.
10. P.K. Bagchi and B.N. Nag: Dynamic vehicle scheduling: An expert systems approach, Int. J. Physical Dist. & Log. Manag., vol.21, no.2, pp.10-18, 1991.
11. G. Laporte: The traveling salesman problem: An overwiew of exact and approximate algorithms, Eur. J. Oper. Res., vol.59, pp.231-247, 1992.
12. L. Ingber: Simulated Annealing: Practice versus Theory, Mathl. Comput. Modelling, Vol.18, No.11, pp.29-57, 1993.
13. F. Glover, E. Taillard and D.D. Werra: A user's guide to tabu search, Annals of Operations Research, Vol.41, pp.3-28, 1993.
14. K. Chen, Y. Takama, K. Hirota: A Calculation Model with Hierarchical Multiplex Structure for Vehicle Routing, Scheduling, and Dispatching Problem With Single Depot, Journal of Japan Society for Fuzzy Theory and System, vol.13, no.2, pp.187-198, 2001.

Similarity between Fuzzy Multi-objective Control and Eligibility

Hwan-Chun Myung and Z. Zenn Bien

Dept. of Elec. Eng. & Comp. Sci. KAIST
373-1 Yusung-ku, Kusong-dong, Taejon, 305-701, Korea
mhc@ctrsys.kaist.ac.kr, zbien@ee.kaist.ac.kr

Abstract A fuzzy multi-objective control problem has been handled in many different ways such as neural network and reinforcement learning etc. Among them, reinforcement learning solves a fuzzy multi-objective control problem without any priori knowledge about an environment. In this paper, a new method of reinforcement learning for a fuzzy multi-objective control problem is proposed in consideration of newly defined objective TD(λ), where TD stands for a temporal difference. The proposed method reformulates a fuzzy multi-objective control problem into a problem similar to a reinforcement learning problem under non-Markov environment, where objective eligibility is considered for handling multi-rewards, similarly as TD(λ) is applied to a reinforcement learning problem under a non-Markov environment.

1 Introduction

A multi-objective problem is considered a difficult task in a field of decision making, and the same problem is very often encountered in a control area, too. For some simple examples such as a cart-pole balancing system and a crane system, the desired position and angle, tolerable angle range, energy consumption, and performance of each state can be multi-objectives that should be satisfied at the same time. Specifically, our concern is to handle those kinds of systems under unknown dynamics with a reinforcement learning scheme. From multi-objective problem's point of view, a reinforcement learning problem is changed into multi-reward handling one. A.Blumel [1] and S.Yoshizawa [2] have proposed some heuristic ways to solve such a problem through an evolutionary algorithm, without any analytical result on convergence. Meanwhile, C.T.Lin [1] has presented a neuro-fuzzy combiner to analytically handle such multi-rewards and showed its convergence in spite of the limited number of objectives.

In the paper, we show a new reinforcement learning method applied to a multi-objective control problem. The main idea comes from a multi-reward handling-like process of a eligibility method [3]. A newly defined objective TD(λ) enables eligibility of multi-rewards to be considered on an objective scale, which makes the previous research result on the convergence of eligibility applicable.

N.R. Pal and M. Sugeno (Eds.): AFSS 2002, LNAI 2275, pp. 85–90, 2002.

2 Problem Formulation

In general, a reinforcement learning problem can be described with a fuzzy state value function for a policy π

$$V^{\pi}(s)=E_{\pi}\{R_t \mid s_t=s\}=E_{\pi}\{\sum_{k=0}^{\infty}\gamma^k r_{t+k+1} \mid s_t=s\}, \tag{1}$$

where s is a state, $0\le\gamma\le1$, r_{t+k+1} is a reward at $t+k+1$, and E_{π} stands for an expected value under the given policy π. The above equation is changed into a Bellman equation for V^{π}

$$V^{\pi}(s)=\sum_{a}\pi(s,a)\sum_{s'}P_{ss'}^{a}[R_{ss'}^{a}+\gamma V^{\pi}(s')], \tag{2}$$

where ‘a ’ is an action, $P_{ss'}^{a}$ is a transition probability from s to s', and $R_{ss'}^{a}$ is an expected reward for the transition. And, given by

$$\pi\ge\pi' \text{ if } V^{\pi}(s)\ge V^{\pi'}(s) \text{ for all } s\in S, \tag{3}$$

an optimal fuzzy state-value function is defined as

$$V^{*}(s)=\max_{\pi}V^{\pi}(s). \tag{4}$$

As a result, a reinforcement learning problem means policy evaluation and policy improvement to find the optimal fuzzy state value function. Similarly, a multi-objective problem can be regarded as an extended version of the above definitions. It is noted that a multi-objective problem corresponds to a multi-reward handling problem from reinforcement learning's point of view. Here, assume that r_t^i is the reward of an *ith* objective. Then, at the first step, make all the objectives in order, considering their relative importance, such as

Less objective number $\propto$ Importance

$$O_1>O_2>\cdots>O_i>\cdots>O_M, \tag{5}$$

where O_i is an important degree of an *ith* objective and M is a number of objectives. Similarly as in eq. (2), define an objective fuzzy state-value function as

$$\begin{aligned}\Lambda^{\pi_o}(s^1)&=E_{\pi_o}\{L_t \mid s_t^1=s^1\}\\&=E_{\pi_o}\{\sum_{i=1}^{M}\sum_{k=0}^{\infty}v_{ik}\gamma^k r_{t+k+1}^{i} \mid s_t^1=s^1\},\end{aligned} \tag{6}$$

in which π_o is a policy of reinforcement learning for a multi-objective problem, and s^i means a state related to *ith* objective. In case of $M=1$ and $v_1=1$, we have

$$\Lambda^{\pi_o}(s^1) = V^{\pi}(s). \tag{7}$$

Here, a problem is to find an optimal objective fuzzy state-value function $\Lambda^*(s^1)$.

3 Eligibility of Reinforcement Learning: Review [3]

Consider temporal difference learning as follows

$$V(s_{t+1}) \leftarrow V(s_t) + \alpha[r_{t+1} + \gamma V(s_{t+1}) - V(s_t)], \tag{8}$$

where $0 \le \alpha \le 1$. Here, to develop a general version of temporal difference learning, define n-step prediction

$$\begin{aligned} R_t^{(1)} &= r_{t+1} + \gamma V_t(s_{t+1}) : \text{1-step prediction,} \\ &\vdots \\ R_t^{(n)} &= r_{t+1} + \gamma r_{t+2} + \cdots + \gamma^{n-1} r_{t+n} + \gamma^n V_t(s_{t+n}) : \text{n-step prediction,} \\ &\vdots \\ R_t &= r_{t+1} + \gamma r_{t+2} + \cdots + \gamma^{T-t-1} r_T,\ T : \text{a final time step} \end{aligned} \tag{9}$$

The 1-step prediction can be replaced with the n-step prediction in eq. (8) to generalize a temporal difference learning method. Moreover, combination of all the predictions, called TD(λ) [3], can be used to get faster state-value expectation such as

$$R_t^{\lambda} = (1-\lambda)\sum_{n=1}^{\infty} \lambda^{n-1} R_t^{(n)} = (1-\lambda)\sum_{n=1}^{T-t-1} \lambda^{n-1} R_t^{(n)} + \lambda^{T-t-1} R_t, \tag{10}$$

where $0 \le \lambda \le 1$. In a sense, the combined predictions can be regarded as handling multi-rewards, which gives a motivation to a new method for multi-objective problem.

Assuming Markov property

$$\begin{aligned} &\Pr\{s_{t+1} = s', r_{t+1} = r \mid s_t, a_t, s_{t-1}, a_{t-1}, \cdots, s_0, a_0\} \\ &= \Pr\{s_{t+1} = s', r_{t+1} = r \mid s_t, a_t\}, \end{aligned} \tag{11}$$

an eligibility trace is introduced as a different view point about a TD(λ) method and, in particular, is preferred because of its conceptual and computational simplicity. The eligibility trace is denoted by

$$e_t(s) = \begin{cases} \gamma\lambda e_{t-1}(s) & \text{if} \quad s \ne s_t \\ \gamma\lambda e_{t-1}(s) + 1 & \text{if} \quad s = s_t \end{cases}, \tag{12}$$

in which $e_t(s) \in R^+$. On each step, the eligibility traces for all states decay by $\gamma\lambda$, and the eligibility trace for the one state visited on the step is incremented by 1. This kind of trace is also called an accumulating trace because it accumulates each time a state is visited, then fades away gradually when the state is not visited. With the introduced eligibility trace, define the new temporal difference error

$$\Delta V_t(s) = \beta \delta_t e_t(s) , \tag{13}$$

where

$$\delta_t = r_{t+1} + \gamma V(s_{t+1}) - V(s_t) . \tag{14}$$

The historical characteristic such as accumulating and fading is useful to overcome the Markov property of an environment which is normally assumed in a reinforcement learning problem. In general, the two different views about TD(λ) method are called forward and backward view, respectively, due to their non-causal and causal property. Despite their different perspective, however, the mathematical equivalence of both the methods in an episode is proved [3].

4 Multi-reward Handling Method

Define a *ith* objective prediction as follows:

$$\begin{aligned} L_t^1 &= r_{t+1}^1 + \gamma(1-\lambda)V_t^1(s_{t+1}^1) , \\ &\vdots \\ L_t^i &= r_{t+1}^i + \gamma(1-\lambda)V_t^i(s_{t+1}^i) , \\ &\vdots \\ L_t^M &= r_{t+1}^M + \gamma(1-\lambda)V_t^M(s_{t+1}^M) . \end{aligned} \tag{15}$$

With all the objective predictions, consider the combination with parameters λ and γ

$$\begin{aligned} L_t &= \sum_{i=1}^{M} (\gamma\lambda)^{i-1} \gamma L_t^i \\ &= \sum_{i=1}^{M} w_i L_t^i = \sum_{i=1}^{M} \sum_{k=0}^{\infty} v_{ik} \gamma^k r_{t+k+1}^i , \end{aligned} \tag{16}$$

where $w_i = (\gamma\lambda)^{i-1}\gamma$, $v_{i0} = w_i$, and $v_{ik} = (1-\lambda)w_i$ for $k \neq 0$. It is noted that the combination of objective predictions plays the same role on an ordered objective scale as a TD(λ) method does on a time-scale. Meanwhile, in order to consider a physical view of the objective TD(λ) method, define an Objective Markov Property(OMP)

$$\begin{aligned} &\Pr\{s_{t+1}^1 = s^1, r_{t+1}^1 = r \mid s_t^1, a_t^1, s_t^2, a_t^2, \cdots, s_t^M, a_t^M\} \\ &= \Pr\{s_{t+1}^1 = s^1, r_{t+1}^1 = r \mid s_t^1, a_t^1\} . \end{aligned} \tag{17}$$

With eq. (17), an OMP means that the fuzzy state-value function for the first objective is independent of the other fuzzy state-value functions in terms of probability. In order to consider a non-OMP case, an updating rule for handling multi-rewards is given by

$$
\begin{aligned}
\Delta\Lambda(s_t^1) &= \sum_{k=1}^{M} (\gamma\lambda)^{k-1}\gamma\delta_k \\
&= (\gamma\lambda)^0 \gamma[r_{t+1}^1 + \gamma V_t(s_{t+1}^1) - \frac{1}{\gamma}V_t(s_t^1)] \\
&\quad + (\gamma\lambda)^1 \gamma[r_{t+1}^2 + \gamma V_t(s_{t+1}^2) - V_t(s_{t+1}^1)] \\
&\quad \vdots \\
&\quad + (\gamma\lambda)^{M-1} \gamma[r_{t+1}^M + \gamma V_t(s_{t+1}^M) - V_t(s_{t+1}^{M-1})] \\
&= -V_t(s_t^1) \\
&\quad + (\gamma\lambda)^0 \gamma[r_{t+1}^1 + \gamma V_t(s_{t+1}^1) - \gamma\lambda V_t(s_{t+1}^1)] \\
&\quad \vdots \\
&\quad + (\gamma\lambda)^{M-1} \gamma[r_{t+1}^M + \gamma V_t(s_{t+1}^M) - \gamma\lambda V_t(s_{t+1}^{M-1})] \\
&\approx -V_t(s_t^1) + (1-\lambda)\lambda^1[r_{t+1}^1 + \gamma V_t(s_{t+1}^1)] \\
&\quad + (1-\lambda)\lambda^2[r_{t+1}^1 + \gamma r_{t+1}^2 + \gamma^2 V_t(s_{t+1}^2)] \\
&\quad \vdots \\
&\quad + (1-\lambda)\lambda^M [r_{t+1}^1 + \cdots + \gamma^{M-1} r_{t+1}^M + \gamma^M V_t(s_{t+1}^M)],
\end{aligned}
\tag{18}
$$

where $0 < \lambda < 1$,

$$
\delta_k = \begin{cases} r_{t+1}^1 + \gamma V_t(s_{t+1}^1) - \frac{1}{\gamma}V_t(s_t^1) & \text{for } k = 1 \\ r_{t+1}^k + \gamma V_t(s_{t+1}^k) - V_t(s_{t+1}^{k-1}) & \text{for } k \neq 1 \end{cases}, \tag{19}
$$

and

$$
\Delta V_t(s_t^i) = r_{t+1}^i + \gamma V_t(s_{t+1}^i) - V_t(s_t^i) \text{ for } i = 2, \cdots, M . \tag{20}
$$

The updating rule in (18) is obtained based upon the concept of eligibility, where the definition of eq. (19) causes the proposed updating rule is in a form of eligibility shown in the last step of eq. (18). If $M = \infty$, the last step in eq. (18) exactly corresponds to the proposed updating rule. As a result, such a relation results in similarity between multi-objective problem and eligibility.

5 Concluding Remarks

The new method for handling multi-rewards is proposed based upon the concept of eligibility in reinforcement learning. Since the proposed multi-reward handling method almost corresponds to TD(λ) on an objective scale, it is advantageous that the previous research result on convergence of TD(λ) can be used to prove its convergence. For our further study, the speed of convergence is required to be improved, analytically and generally.

References

1. A.Blumel and B.A.White, "Multi-objective Optimization of Fuzzy Logic Scheduled Controllers for Missile Autopilot Design," *IFSA World Congress*, Vancouver, 2001.
2. S. Yoshizawa et al., " Robust Control Configured Design Method for Systems with Multi-objective Sepcifications," *IFSA World Congress,* Vancouver, 2001.
3. R.S.Sutton and A.G.Barto, *Reinforcement Learning*, MIT Press, 1998.
4. G.J.Klir and T.A.Folger, *Fuzzy Sets, Uncertainty, and Information,* Prentice-Hall, 1992

A Fuzzy Goal Programming Approach for Solving Bilevel Programming Problems

Bhola Nath Moitra and Bijay Baran Pal

Dept. of Mathematics, University of Kalyani, Kalyani - 741235, INDIA
bbpal18@hotmail.com

Abstract. This paper presents a fuzzy goal programming procedure for solving linear bilevel programming problems. The concept of tolerance membership functions for measuring the degree of satisfactions of the objectives of the decision makers at both the levels and the degree of optimality of vector of decision variables controlled by upper-level decision maker are defined first in the model formulation of the problem. Then a linear programming model by using distance function to minimize the group regret of degree of satisfactions of both the decision makers is developed. In the decision process, the linear programming model is transformed into an equivalent fuzzy goal programming model to achieve the highest degree (unity) of each of the defined membership function goals to the extent possible by minimizing their deviational variables and thereby obtaining the most satisfactory solution for both the decision makers. To demonstrate the approach, a numerical example is solved and compared the solution with the solutions of other two fuzzy programming approaches [11,12] studied previously.

1 Introduction

Bilevel programming problem (BLPP) is a special case of a multilevel programming problem (MLPP) of a large hierarchical decision system. In a BLPP, two decision makers (DMs) are located at two different hierarchical levels, each independently controlling one set of decision variables and with different and perhaps conflicting objectives.

In the hierarchical decision process, the lower-level DM (the follower) executes his/her decision powers, after the decisions of the upper-level DM (the leader). Although the leader independently optimizes its own benefits, the decision may be affected by the reaction of the follower. As a consequence, decision deadlock arises frequently and the problem of distribution of proper decision power is encountered in most of the practical decision situations.

The concept of BLPP was first introduced by Candler and Townsley [4] in 1982. Thereafter, various versions of bilevel programming (BLP) have been presented in [2,3]. During 1980's several solution approaches have been developed for solving BLPPs as well as MLPPs in general from the view points of their potential applications to hierarchical decentralized decision systems as economic systems, warfare, network designs, and is specially applicable for conflict resolution. Most of the classical approaches developed so far are based on the vertex enumeration method

N.R. Pal and M. Sugeno (Eds.): AFSS 2002, LNAI 2275, pp. 91–98, 2002.

[4] and transformation approach [3] which are actually the extensions of Stackelberg strategy for solving two-person non-zero and non-cooperative games. They are computationally very inefficient and highly complex due to the lack of defining explicit relationship between the leader and follower. Also, the use of these approaches leads to a paradox that the followers decision power dominates the leader. To overcome this situation, multiobjective solution techniques with post optimality analysis on the objective values to obtain efficient solutions have been suggested by Wen & Hsu [14]. In their method, three efficient compromise solutions, threat-point, ideal-point and ideal-threat-point dependent solutions depending on the DMs preference have been provided. But, there is a possibility of obtaining undesirable solution because of the inconsistency of the objectives and decision variables in a highly conflict hierarchical decision situation.

Now, in a hierarchical decision making context, it has been realized that each DM should have a motivation to cooperate with other, and a minimum level of satisfaction of the DM at a lower-level must be considered for overall benefit of the organization. The use of the concept of membership function of fuzzy set theory to BLPPs as well as to general MLPPs for satisfactory decisions was first introduced by Lai [6] in 1996. Thereafter, Lai's satisfactory solution concept was extended by Shih et al. [11] and a supervised search procedure with the use of max-min operator of Bellman and Zadeh [1] was proposed. Recently, Shih & Lee [12] have used the compensatory γ-operator within the framework of the same model to make a reasonable balance of decision powers of the DMs. The basic concept of these fuzzy programming (FP) approaches is the same as implies the follower optimizes his/her objective function, taking a goal or preference of the leader into consideration. In the decision process, considering the membership functions of the fuzzy goals for the decision variables of the leader, the follower solves a FP problem with a constraint on a overall satisfactory degree of the leader. If the proposed solution is not satisfactory to the leader, the solution search is continued by redefining the elicited membership functions until a satisfactory solution is reached.

The main difficulty arises with the FP approach of Shih et al. is that there is possibility of rejecting the solution again and again by the leader and re-evaluation of the problem is repeatedly needed to reach the satisfactory decision, where the objectives of the DMs are over conflicting. Even inconsistency between the fuzzy goals of the objectives and the decision variables may arise. This makes the solution process a lengthy one.

To overcome the above undesirable situation, fuzzy goal programming (FGP) technique with the use of the concept of distance function for minimizing the group regret to achieving the highest degree (unity) of satisfaction to the extent possible for both leader and follower is proposed. A numerical example is solved to illustrate the proposed approach, and the solution is compared with the Shih et als' satisfactory solutions.

2 Problem Formulation

Let the BLPP in a decision situation is such that each of the DMs at both levels, leader and follower, takes overall satisfactory balance between both the levels into consideration and tries to maximize his/her own objective function, paying serious attention to the preferences of the other. Then, let f_1 and f_2, respectively, be the objective functions of the leader and follower, and let the vectors of decision variables $\mathbf{x}_1$ and $\mathbf{x}_2$ are under the control of leader and follower respectively. Such a BLPP can be stated as:

$$\underset{\mathbf{x}_1}{\text{Max}} f_1(\mathbf{x}_1, \mathbf{x}_2) = \mathbf{c}_{11}\mathbf{x}_1 + \mathbf{c}_{12}\mathbf{x}_2 \tag{1}$$

where, for given $\mathbf{x}_1$, $\mathbf{x}_2$ solves

$$\underset{\mathbf{x}_2}{\text{Max}} f_2(\mathbf{x}_1, \mathbf{x}_2) = \mathbf{c}_{21}\mathbf{x}_1 + \mathbf{c}_{22}\mathbf{x}_2 \tag{2}$$

$$\text{subject to } \mathbf{A}_1\mathbf{x}_1 + \mathbf{A}_2\mathbf{x}_2 \leq \boldsymbol{b} \tag{3}$$
$$\mathbf{x}_1 \geq \mathbf{0}, \mathbf{x}_2 \geq \mathbf{0}$$

where $\mathbf{c}_{11}$, $\mathbf{c}_{12}$, $\mathbf{c}_{21}$, $\mathbf{c}_{22}$ and $\boldsymbol{b}$ are constant vectors and A_1 and A_2 are constant matrices. The functions f_1 and f_2 are assumed to be linear and bounded.

2.1 FP Approach to BLPP

To formulate the FP model of the BLPP under consideration, the objective functions f_1 and f_2 and the decision vector $\mathbf{x}_1$ of the leader are required to be transformed into the fuzzy goals by means of assigning an aspiration level to each of them. Then, they are characterized by their membership functions by defining the tolerance limits for achievement of the aspired levels of the goals.

Construction of Membership Function. Since the leader and follower both are interested of maximizing their own objective functions over the same feasible region defined by the system constraints (3), the optimal solutions of both of them calculated in isolation would be the aspiration levels of their associated fuzzy goals.

Let $(\mathbf{x}_1^U, \mathbf{x}_2^U; f_1^U)$ and $(\mathbf{x}_1^L, \mathbf{x}_2^L; f_2^L)$ be the optimal solutions of leader and follower, respectively, when calculated in isolation. Then the fuzzy goals of leader and follower appear as:

$$f_1 \gtrsim f_1^U, \quad \mathbf{x}_1 \gtrsim \mathbf{x}_1^U, \quad f_2 \gtrsim f_2^L.$$

It may be noted that the two solutions $(\mathbf{x}_1^U, \mathbf{x}_2^U)$ and $(\mathbf{x}_1^L, \mathbf{x}_2^L)$ are usually different because the objectives of leader and follower are conflicting in nature. So it can reasonably be assumed that the value f_1^L $[= f_1(\mathbf{x}_1^L, \mathbf{x}_2^L)] < f_1^U$ and all beyond of it is absolutely unacceptable to the leader. As such, f_1^L can be considered as the (lower) tolerance limit of the fuzzy objective goal of the leader. Similarly, f_2^U $[= f_2(\mathbf{x}_1^U, \mathbf{x}_2^U)] < f_2^L$ can be considered as the (lower) tolerance limit of the fuzzy objective goal of the follower.

Again, it is to be noted that any decision higher than $\mathbf{x}_1^U$ is absolutely acceptable to the leader. But, to make a satisfactory balance of decisions at both the levels, the leader would have to give a possible relaxation of the decision $\mathbf{x}_1^U$ and that depends on the decision context. Let $\mathbf{x}_1^m$ $(\mathbf{x}_1^L < \mathbf{x}_1^m < \mathbf{x}_1^U)$ be the tolerance limit of the fuzzy decision goal of the leader.

The membership functions of the defined fuzzy goals successively appear as follows:

$$\mu_{f_1}(f_1(\mathbf{x}_1, \mathbf{x}_2)) = \begin{cases} 1 & \text{if } f_1(\mathbf{x}_1, \mathbf{x}_2) > f_1^U \\ \dfrac{f_1(\mathbf{x}_1, \mathbf{x}_2) - f_1^L}{f_1^U - f_1^L} & \text{if } f_1^L \le f_1(\mathbf{x}_1, \mathbf{x}_2) \le f_1^U \\ 0 & \text{if } f_1(\mathbf{x}_1, \mathbf{x}_2) < f_1^L \end{cases} \quad (4)$$

$$\mu_{\mathbf{x}_1}(\mathbf{x}_1) = \begin{cases} 1 & \text{if } \mathbf{x}_1 > \mathbf{x}_1^U \\ \dfrac{\mathbf{x}_1 - \mathbf{x}_1^m}{\mathbf{x}_1^U - \mathbf{x}_1^m} & \text{if } \mathbf{x}_1^m \le \mathbf{x}_1 \le \mathbf{x}_1^U \\ 0 & \text{if } \mathbf{x}_1 < \mathbf{x}_1^m \end{cases} \quad (5)$$

$$\mu_{f_2}(f_2(\mathbf{x}_1, \mathbf{x}_2)) = \begin{cases} 1 & \text{if } f_2(\mathbf{x}_1, \mathbf{x}_2) > f_2^L \\ \dfrac{f_2(\mathbf{x}_1, \mathbf{x}_2) - f_2^U}{f_2^L - f_2^U} & \text{if } f_2^U \le f_2(\mathbf{x}_1, \mathbf{x}_2) \le f_2^L \\ 0 & \text{if } f_2(\mathbf{x}_1, \mathbf{x}_2) < f_2^U \end{cases} \quad (6)$$

Now, in any decision situation, achievement of a membership function to its highest degree (unity) means decision is absolutely satisfactory for the fuzzy goal associated with it. In this decision context, each of the DMs' aim is to achieve each his/her membership function values to the highest degree to the extent possible. But, in actual practice, achievement of all the membership function values to the highest degree is not possible. So, decision policy for minimizing the regrets of the leader and follower is to be taken into consideration. In such a case, the concept of distance function can be used to make a satisfactory balance of decision powers between them.

2.2 Use of Distance Function

The notion of distance function has been widely used to several multiobjective decision making (MODM) problems to arrive at the compromise decision.

The distance function can be presented as [15]:

$$d_p(u(\mathbf{x})) = [\sum_{i=1}^{n} (u_i^* - u_i(\mathbf{x}))^p]^{1/p}, \; p \ge 1$$

where $d_p(u(\mathbf{x}))$ indicates the distance between the utopia point u^* and the actual utilities resulting from the decision $\mathbf{x}$. The parameter p plays the role of measuring decisions in different decision making situations.

In this decision situation, u_i^* represents the ideal point consisting of the highest degree of each of the membership function values and $u_i(\mathbf{x})$ correspond to the membership function values. Now, when p=1, $d_1(u(\mathbf{x}))$ indicates sum of individual regret, which may be interpreted as the group regret.

The problem here is that of making decision in such a way by which the group regret can be minimized.

The problem can be formulated as follows:

$$\text{Minimize } d_1 = [1 - \mu_{f_1}(f_1(\mathbf{x}_1, \mathbf{x}_2))] + [I_1 - \mu_{\mathbf{x}_1}(\mathbf{x}_1)] + [1 - \mu_{f_2}(f_2(\mathbf{x}_1, \mathbf{x}_2))]$$

$$\begin{aligned} \text{subject to} \quad & \mu_{f_1}(f_1(\mathbf{x}_1, \mathbf{x}_2)) \le 1 \\ & \mu_{\mathbf{x}_1}(\mathbf{x}_1) \le I_2 \\ & \mu_{f_2}(f_2(\mathbf{x}_1, \mathbf{x}_2)) \le 1 \\ & \mathbf{A}_1 \mathbf{x}_1 + \mathbf{A}_2 \mathbf{x}_2 \le \boldsymbol{b} \\ & \mathbf{x}_1 \ge \mathbf{0}, \mathbf{x}_2 \ge \mathbf{0} \end{aligned} \qquad (7)$$

where I_1 and I_2 are respectively the row and column vectors with all elements equal to 1 and the dimension of each depends on $\mu_{\mathbf{x}_1}(\mathbf{x}_1)$ or $\mathbf{x}_1$.

The problem (7) is equivalent to the following problem:

$$\text{Maximize } d_1' = \mu_{f_1}(f_1(\mathbf{x}_1, \mathbf{x}_2)) + \mu_{\mathbf{x}_1}(\mathbf{x}_1) + \mu_{f_2}(f_2(\mathbf{x}_1, \mathbf{x}_2))$$

$$\begin{aligned} \text{subject to} \quad & \mu_{f_1}(f_1(\mathbf{x}_1, \mathbf{x}_2)) \le 1 \\ & \mu_{\mathbf{x}_1}(\mathbf{x}_1) \le I_2 \\ & \mu_{f_2}(f_2(\mathbf{x}_1, \mathbf{x}_2)) \le 1 \\ & \mathbf{A}_1 \mathbf{x}_1 + \mathbf{A}_2 \mathbf{x}_2 \le \boldsymbol{b} \\ & \mathbf{x}_1 \ge \mathbf{0}, \mathbf{x}_2 \ge \mathbf{0} \end{aligned} \qquad (8)$$

The problem (8) is similar to the additive fuzzy linear programming (FLP) model proposed by Tiwari et al. [13]. However, to take the most satisfactory solution, the FGP technique, as a robust and most flexible technique for solving MODM problems, is used to solve the problem (8).

3 FGP Solution Approach

FGP is an extension of the conventional goal programming (GP) introduced by Charnes and Cooper [5] in 1961. As a robust tool for MODM problems, GP has been studied extensively in [7] for the last 35 years. In the recent past, FGP in the form of classical GP has been introduced by Mohamed [8] and further studied in [9,10].

Now, in problem (8), it is to be observed that maximization of d_1' means achievement of each of the membership goals to 1 to the extent possible subject to the given system constraints. Therefore, under the framework of *minsum* GP, the equivalent FGP model of problem (8), can be explicitly formulated as [8]:

Find $(\mathbf{x}_1, \mathbf{x}_2)$ so as to:

$$\text{Minimize } Z = w_1^- d_1^- + w_2^- d_2^- + w_3^- d_3^-$$

$$\text{subject to} \quad [f_1(\mathbf{x}_1, \mathbf{x}_2) - f_1^L] / [f_1^U - f_1^L] + d_1^- - d_1^+ = 1$$
$$[\mathbf{x}_1 - \mathbf{x}_1^m] / [\mathbf{x}_1^U - \mathbf{x}_1^m] + d_2^- - d_2^+ = I_2$$
$$[f_2(\mathbf{x}_1, \mathbf{x}_2) - f_2^U] / [f_2^L - f_2^U] + d_3^- - d_3^+ = 1$$
$$\mathbf{A}_1 \mathbf{x}_1 + \mathbf{A}_2 \mathbf{x}_2 \le \boldsymbol{b}$$
$$\mathbf{x}_1, \mathbf{x}_2 \ge \mathbf{0}$$
$$d_i^-, d_i^+ \ge 0 \text{ with } d_i^-.d_i^+ = 0, \ i=1, 2, 3 \qquad (9)$$

where d_i^- (≥ 0) and d_i^+ (≥ 0) represent the under- and over-deviational variables, respectively, from the aspired levels of the goals and Z represents the fuzzy achievement function consisting of the weighted under-deviational variables. The numerical weights w_i^- (≥ 0), associated with d_i^-, i = 1, 2, 3, represent the relative importance of achieving the aspired levels of the respective fuzzy goals subject to the given set of constraints.

To assess the relative importance of the fuzzy goals properly, the weighting scheme suggested by Mohamed [8] can be used to assign the values to w_i^- (i = 1, 2, 3). In the present formulation, the values of w_i^- are determined as:

$$w_1^- = 1 / (f_1^U - f_1^L), \qquad w_2^- = 1 / (\mathbf{x}_1^U - \mathbf{x}_1^m), \qquad w_3^- = 1 / (f_2^L - f_2^U) \qquad (10)$$

The FGP model (9) provides the most satisfactory decision by achieving the aspired levels of the membership goals to the extent possible in the decision environment.

The solution procedure is straightforward and illustrated via the following example presented by Shih et al. [11].

4 An Illustrative Example

A BLPP for satisfactory decision to make balance between the export trade and profit.

$\underset{x_1}{\text{Max}}\, f_1(x_1, x_2) = 2x_1 - x_2$ (balancing export trade – **leader's problem**)

and, for given x_1, x_2 solves

$\underset{x_2}{\text{Max}}\, f_2(x_1, x_2) = x_1 + 2x_2$ (profit – **follower's problem**)

$$\text{subject to} \quad 3x_1 - 5x_2 \le 15$$
$$3x_1 - x_2 \le 12$$
$$3x_1 + x_2 \ge 27$$
$$3x_1 + 4x_2 \le 45$$
$$x_1 + 3x_2 \le 30$$
$$x_1 \ge 0, x_2 \ge 0 \qquad (11)$$

The individual optimal solutions of the leader and the follower are $(x_1^U, x_2^U) = (7.5, 1.5)$ with $f_1^U = 13.5$ and $(x_1^L, x_2^L) = (3, 9)$ with $f_2^L = 21$. Then it is found that $f_1^L = -3$ which has no meaning and can not be allowed . So, $f_1^L = 0$ is considered here, which serves as a (lower) tolerance limit of the leader's fuzzy objective goal. Also, f_2^U =

10.5 is obtained and it acts as a (lower) tolerance limit of follower's fuzzy objective goal. Again, let the leader feel that the value of the control variable can be relaxed up to 4.5 but not beyond of it in the decision making situation. So $x_1^m = 4.5$, $x_1^L < 4.5 < x_1^U$, is considered, which acts as a tolerance limit of the decision x_1.

Based on the above numerical values, the membership functions $u_{f_1}(f_1(x_1, x_2))$, $\mu_{x_1}(x_1)$ and $\mu_{f_2}(f_2(x_1, x_2))$ can now be obtained by (4), (5) and (6).

Then following the proposed procedure the resulting executable model, the FGP model (9), is obtained as:

Find (x_1, x_2) so as to:

$$\text{Minimize } Z = 2/27\, d_1^- + 1/3\, d_2^- + 2/21 d_3^-$$

$$\text{subject to } \quad (2x_1 - x_2) / 13.5 + d_1^- - d_1^+ = 1$$

$$(x_1 - 4.5)/3 + d_2^- - d_2^+ = 1$$

$$(x_1 + 2x_2 - 10.5)/10.5 + d_3^- - d_3^+ = 1$$

and to the given system constraints in (11),

$$x_1, x_2, d_i^-, d_i^+ \geq 0 \text{ with } d_i^-, d_i^+ = 0, i = 1, 2, 3.$$

The obtained solution is $(x_1, x_2) = (8.6, 4.8)$ with $f_1 = 12.4$ and $f_2 = 18.2$. The resulting membership values are $\mu_{f_1}(f_1(x_1,x_2))=0.91851$, $\mu_{x_1}(x_1) = 1$ and $\mu_{f_2}(f_2(x_1, x_2)) = 0.733$.

The solution achieved here is the most satisfactory solution to both the leader and the follower.

Note: The solution of the problem obtained by Shih et al. [11] using max-min fuzzy operator is $(x_1, x_2) = (7.26, 5.23)$ and $(f_1, f_2) = (9.29, 17.72)$ and that obtained by Shih & Lee [12] using fuzzy compensatory γ–operator is $(x_1, x_2) = (7.26, 5.23)$ and $(f_1, f_2) = (9.29, 17.72)$ with same membership values $\mu_{f_1}(f_1(x_1, x_2)) = 0.69$, $\mu_{x_1}(x_1) = 0.95$ and $\mu_{f_2}(f_2(x_1, x_2)) = 0.69$.

A comparison shows that a better solution is achieved here in terms of satisfactions of both the leader and the follower.

5 Conclusion

This paper reveals how the concept of distance function can be used for modeling a BLPP through FP and solving it efficiently using FGP technique for satisfactory distribution of decision powers between the leader and follower in a decision making environment. In the proposed approach, the computational load is not involved with the upper bound restrictions on the membership functions because of the consideration of the highest degree of satisfaction as an aspiration level of each of the membership function goals. Further, the question of computational complexity (as involved in the previous FP approaches) of searching the higher degree of satisfaction for the leader by redefining the membership functions with different degrees of achievement (not the highest degree) does not arise here.

The proposed method can also be extended for solving bilevel decentralized decision problems without involving any computational complexity. In future studies, the proposed approach can be extended to solve multiobjective (more than one objective at a level) MLPPs.

Finally, it is hoped that the concept of solving BLPP presented here can contribute to future studies in the other fields of MODM problems.

Acknowledgement

The authors are grateful to the reviewers for their helpful suggestions on the paper.

References

1. Bellman, R.E. and Zadeh, L.A.: Decision-making in a fuzzy environment. Management Science 17 (1970) B141-B164
2. Bialas, W.F. and Karwan, M.H.: On two-level optimization. IEEE Transactions on Automatic Control 27 (1982) 211-214
3. Bialas, W.F. and Karwan, M.H.: Two-level linear programming. Management Science 30 (1984) 1004-1020
4. Candler, W. and Townsley, R.: A linear two-level programming problem. Computer & Operations Research 9 (1982) 59-76
5. Charnes, A. and Cooper, W.W.: Management Models of Industrial Applications of Linear Programming (Appendix B), Vol-I, John Wiley & Sons, New York (1961)
6. Lai, Y.J.: Hierarchical optimization. A satisfactory solution. Fuzzy Sets and Systems 77 (1996) 321-335
7. Lin, W.T.: A survey of goal programming applications. Omega 8 (1980) 115-117
8. Mohamed, R.H.: The relationship between goal programming and fuzzy programming. Fuzzy Sets and Systems 89 (1997) 215-222
9. Pal B.B. and Moitra, B.N.: A goal programming procedure for multiple objective fuzzy linear fractional programming problem. In: J. C. Misra (eds.), Applicable Mathematics Its Perspectices and Challenges, Narosa Pub., New Delhi (2000) 347-362
10. Pal, B.B. and Moitra, B.N.: A goal programming procedure for solving problems with multiple fuzzy goals using dynamic programming. (to be published), European Journal of Operational Research, 2001
11. Shih, H.-S., Lai, Y.-J and Lee, E.S.: Fuzzy approach for multi-level programming problems. Computer & Operations Research 23 (1996) 73-91
12. Shih, H.-S and Lee, S.: Compensatory fuzzy multiple level decision making. Fuzzy Sets and Systems 114 (2000) 71-87
13. Tiwari, R.N., Dharmar, S. and Rao, J.R.: Fuzzy goal programming — An additive model. Fuzzy Sets and Systems 24(1987) 27-34
14. Wen, U.P. and Hsu, S. -T.: Efficient solutions for the linear bilevel programming problem. European Journal of Operational Research 62 (1991) 354-362
15. Yu, P.L.: A class of solutions for group decision problems. Management Science 19 (1973) 936-946

Material Handling Equipment Selection by Fuzzy Multi-criteria Decision Making Methods

S.K. Deb, B. Bhattacharyya, and S.K. Sorkhel

Department of Production Engineering,
Jadavpur University,
Calcutta-32,
West Bengal, India
sudip_d2000@yahoo.com

Abstract: Very little attention has been paid for selecting accurate material handling equipment (MHE) from a set of alternatives under fuzzy environment. One of the real difficulties in developing and using models is due to the natural vagueness associated with the inputs to the models. In this paper, Fuzzy multicriteria decision-making methodology is used to aggregate the rating attitudes of decision makers (DMs) and trade-off various selection criteria to find ranking values of the fuzzy suitability indices. A hypothetical problem is designed and coded in Turbo C language to obtain the final ranking of the alternatives automatically.

1 Introduction

MHE selection problem deals with the appropriate equipment selection to allow greater productivity and flexibility. Owing to the unstructured nature of the problem, many researchers have proposed various approaches, which have not been very successful to deal with the qualitative factors associated with the problem. In common practice MHE is selected with the help of arbitrary guidelines formulated from the knowledge and experience of the plant managers. Regardless of the type of data, there is an element of vagueness or fuzziness in it. Traditional selection methods are based on quantitative analysis of flow and cost [1,2]. Recently, a hybrid methodology has been applied to select minimum cost MHE for each move of a heavy manufacturing system by integrating a knowledge base with the optimization part in the process of facilities layout design [3]. Fuzzy set theory has been recently applied in various areas of production management. This theory has been applied successfully to develop facility layout design heuristic employing fuzzy closeness rating of the different activities [4]. A linguistic pattern approach is employed considering production link, environmental link and distance link for developing a block layout depending on adjacency of two departments [5]. Fuzzy decision making system has got its potential application to determine facility selection routine for their placement in open continual plane under multiple criteria approach consisting of information link, equipment link, supervision link and safety link to develop manufacturing facility layout [6,7]. Most of the models and algorithms available in the literature are suitable for handling exact measure and crisp evaluation [8]. One of the real difficulties of

N.R. Pal and M. Sugeno (Eds.): AFSS 2002, LNAI 2275, pp. 99–105, 2002.

selecting appropriate MHE under manufacturing environment is the presence of inexact and vague data and yet to work in a mathematically strict and vigorous way. In the real life, to evaluate MHE selection suitability, measures for the subjective criteria, e.g. level of supervision, working condition, environmental condition and safety condition etc may not be preciously defined for the decision-makers. Hence, the precision-based evaluation may not be practical. Besides, the evaluating data of the MHE selection suitability under different criteria as well as the weight of the criteria are often expressed in linguistic terms, e.g. very low, medium, high etc. Therefore, the present research work follows to integrate various linguistic assessments and weights to evaluate equipment interrelation suitability and determine the best alternative selection order by using fuzzy multi-criteria decision-making methodology.

2 Fuzzy Set and Basic Operations

The fuzzy set theory was introduced by Zadeh [9] to deal with problems in which the absence of preciously defined criteria is involved. Formally, if X={x} is a set of objects, then the fuzzy set A on X is defined by its membership function $f_A(x)$ which assignees to each element x ε X a real number in the interval [0, 1] which represents the grade of membership of x in A. Thus, A can be written as: $A=\{(f_A(x)/x)|x \in X\};X\rightarrow[0, 1]$. Trapezoidal fuzzy number, as given in the above equation can be denoted by $(\alpha,\beta,\gamma,\delta)$ and its membership function is shown in figure 1.

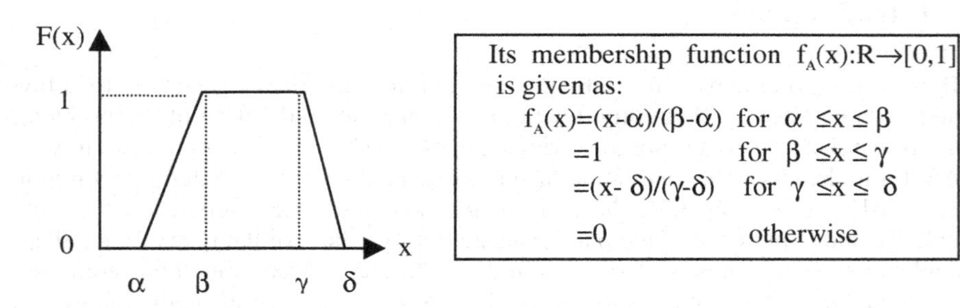

Fig. 1. Membership function of a trapezoidal fuzzy number

With this notation and by the extension principle proposed by Zadeh (1965), the extended algebraic operations like addition and multiplication on trapezoidal fuzzy numbers can be expressed as:

- Addition: $(\alpha_1,\beta_1,\gamma_1,\delta_1)\oplus(\alpha_2,\beta_2,\gamma_2,\delta_2)=(\alpha_1+\alpha_2, \beta_1+\beta_2, \gamma_1+\gamma_2, \delta_1+\delta_2)$
- Multiplication: $(\alpha_1,\beta_1,\gamma_1,\delta_1)\otimes(\alpha_2,\beta_2,\gamma_2,\delta_2)\cong(\alpha_1\alpha_2, \beta_1\beta_2, \gamma_1\gamma_2, \delta_1\delta_2)$

Trapezoidal fuzzy numbers are easy to use and interpret. For example, 'approximately equal to 24' can be represented by (20, 24, 24, 30). The concept of linguistic variable is very useful in dealing with situations, which are too complex or too ill defined to be reasonably described in conventional quantitative expressions. A linguistic variable is a variable whose values are words in natural or artificial language. The approximate reasoning of fuzzy set theory can represent the Linguistic value. For example, the linguistic variable of weight may be considered as {VL, L, M, H, VH} where VL=Very Low, L=Low, M=Medium, H=High and VH=Very High.

3 Problem Formulation and Procedure

In this paper, MHE selection under a fuzzy environment is proposed. The subjective (qualitative) as well as objective (quantitative) criteria associated with the problem are assigned by the DMs within approximated information value available to determine the ranking of the alternatives.

3.1 Model Generation and Selection Criteria

The concept of hierarchical structure analysis with two distinct levels is used in this paper. The first level is to evaluate the fuzzy importance of the decision criteria (e.g. availability of safety, supervision, working condition, fixed and operating cost etc). The second level is to assign to various alternatives under each decision criteria. A group of 'm' DMs is assumed to employ rating sets to evaluate preference information. The DMs assess the suitability of 'n' alternatives under each criterion. Let R_{ijk} be the rating assigned to alternative (i) by DM (j) for criteria (k). Similarly W_{kj} be the weight given to criteria (k) by DM (j). Thus the committee has to first aggregate the ratings R_{ijk} for each alternative versus each criterion to form the rating R_{ik}. Each aggregated R_{ik} for i=1,n; k=1,p; can further be weighted by a weight W_k according to the relative importance of the criteria. The fuzzy suitability index F_i of each alternative can be obtained by aggregating R_{ik} and W_k for all selection criteria to form a suitability vector. Finally, applying maximizing and minimizing fuzzy number the corresponding ranking values of the fuzzy suitability indices are obtained to decide about the best selection based on highest-ranking value.

3.2 Preference Rating System

The preference rating adopted in the present problem are fuzzy members and linguistic variable values. The DM employs linguistic weighting set W={VL, L, M, H, VH} to evaluate the importance of the criteria through a designed rating scale in the range of [0,1] whose membership functions of the linguistic values are shown as:
VL(0,0,0,0.3); L(0,0.3,0.3,0.5); M(0.2,0.5,0.5,0.8); H(0.5,0.7,0.7,1); VH(0.7,1,1,1).
The DMs also employ a linguistic rating set R={VP, P, F, G, VG} to evaluate the suitability of various alternatives versus each subjective criteria. The membership function of linguistic ratings are represented as:
VP(0,0,0,0.2);P(0,0,0.2,0.4); F(0,0.2,0.4,0.6); G(0.2,0.4,0.6,0.8);VG(0.4,0.6,0.8,1)
In order to ensure compatibility between fuzzy cost values of the objective criterion and linguistic ratings of subjective criteria; fuzzy cost values must be converted to

dimensionless indices. The alternative with the minimum cost value should have the maximum rating. The rating of alternative (i) for objective criterion can be written as:
$RF_i=\{F_i \otimes [F_1^{-1} \oplus F_2^{-1} \oplus F_3^{-1} \oplus .... \oplus F_n^{-1}]\}^{-1}$ ....(1)
Where F_i is the value of cost for alternative (i). Thus, $RFi = R_{ip}$.

3.3 Aggregation of Fuzzy Assessment

The mean operator is most commonly used to aggregate the DMs fuzzy assessments. Let $R_{ijk}= (r_{ijk}^1, r_{ijk}^2, r_{ijk}^3, r_{ijk}^4)$, for i=1,n; j=1,m; k=1,p. be the linguistic rating assigned to alternative (i) by DM (j) for criterion (k) and let $W_{kj}=(w_{kj}^1, w_{kj}^2, w_{kj}^3, w_{kj}^4)$, for k=1,p; j=1,m; be the linguistic weight given to subjective criteria (1,2,...,p-1) and objective criterion (p) by DM (j). The average linguistic rating and weight are written as:

$R_{ik}= (1/n) \otimes [R_{ik1} \oplus R_{ik2} \oplus .... \oplus R_{ikm}]$ for i=1,n; k=1,p-1.
$=(r_{ip}^1, r_{ip}^2, r_{ip}^3, r_{ip}^4)$ for i=1.n; k=p.
and $W_k=(1/m) \otimes [w_{k1} \oplus w_{k2} \oplus .... \oplus w_{km}]$ for k=1,p.
then $R_{ik}=(r_{ik}^1, r_{ik}^2, r_{ik}^3, r_{ik}^4)$ ....(2)
and $W_k=(w_k^1, w_k^2, w_k^3, w_k^4)$ ....(3)

R_{ik} and W_k are further aggregated by averaging the corresponding products over all the criteria. The fuzzy suitability index of the i th alternative can be obtained by standard arithmetic method written as:
$F_i=(i/p) \otimes [(R_{i1} \otimes W_1) \oplus (R_{i2} \otimes W_2) \oplus .... \oplus (R_{ip} \otimes W_p)]$ ....(4)
which provides a trapezoidal fuzzy number $F_i= (\alpha_i, \beta_i, \gamma_i, \delta_i)$.

3.4 Ranking Values of Alternatives

The ranking values of the alternatives are determined by using Chen's method [10] of maximizing set (M) and minimizing set (N) as given below:
$M=\{(x, f_M(x)) \mid x \varepsilon R\}$ with membership function values as

$f_M(x)=[(x-x_1)/(x_2-x_1)]^k$ for $x_2 \geq x \geq x_1$
$=0$ otherwise.

The membership function for minimizing set is given as:

$F_N(x) =[(x-x_2)/(x_1-x_2)]^k$ for $x_2 \geq x \geq x_1$
$=0$ otherwise.

Where k>0, x_1=inf D, x_2=sup D, $D=U_{i=1,n}D_i$, $D_i=\{x|f_{Fi}(x)>0\}$
The value of k depends on DMs preference. The ranking value of fuzzy suitability index can be obtained by the ranking value of trapezoidal fuzzy number $F_i=(\alpha,\beta,\gamma,\delta)$ with the help of equation,

$V(F_i)=[(\delta_i-x_1)/((x_2-x_1)-(\gamma_i-\delta_i))+1-(x_2-\alpha_i)/((x_2-x_1)+(\beta_i-\alpha_i))]/2$ ....(5)

The ranking values of fuzzy suitability indices of n alternatives are determined. Based on the ranking values, the DMs can easily make the best MHE selection for the alternative having highest-ranking value.

Table 1. DMs assessment for alternative Vs criteria and weight Vs criteria

Criteria DM	Supervision DM1	DM2	Safety DM1	DM2	Environment DM1	DM2	Cost ($*$10^5$) DM1	DM2
Weight	M	VH	H	VH	VH	H	VH	H
MHE1	F	G	P	F	G	VG	(18,22,28,32)	
MHE2	G	F	F	G	F	P	(22,26,32,38)	
MHE3	P	G	G	VG	P	F	(22,25,28,33)	
MHE4	VG	G	F	F	P	G	(16,20,28,33)	

(VL=Very Low, M=Medium, H=High, VP=Very Poor, P=Poor, F=Fair, G=Good)

4 Example

In this section, a hypothetical MHE selection problem is designed to demonstrate the application of the procedure. The problem statement is given below:
No. of MHE=4, no. of DMs=2 and no. of subjective criteria=3 (supervision, safety and environment) and one objective criterion as operating cost. The DMs assessment table for alternatives versus criteria and weight assigned to four criteria are shown in table 1. The aggregated weighing (W_k) of the DMs is obtained by using equation (3)

W_1=(0.385, 0.675, 0.675, 0.865); W2=(0.515, 0.750, 0.750, 0.950);
W_3=(0.624, 0.850, 0.850, 0.956); W_4=(0.655, 0.925, 0.925, 1.000).

Table 2. Average fuzzy subjective and objective ratings

K_I	Supervision	Safety	Environment	Cost
E1	(.23,.38,.58,.74)	(.17,.36,.44,.65)	(.51,.69,.78,.92)	(.22,.25,.32,.36)
E2	(.52,.71,.76,.93)	(.22,.38,.45,.65)	(.34,.54,.72,.92)	(.24,.29,.41,.54)
E3	(.22,.43,.62,.82)	(.28,.48,.68,.89)	(.32,.48,.62,.84)	(.26,.32,.43,.56)
E4	(.26,.38,.47,.64)	(.24,.36,.47,.69)	(.20,.32,.46,.62)	(.23,.27,.36,.48)

By using equation (2) with reference to the DMs assessment of alternatives versus each criteria as shown in table 1, the average evaluation of alternatives for subjective criteria (R_{ik} for k=1,p-1) are computed. Similarly by using equation (1) the fuzzy objective rating (R_{ik} for k=p) is obtained and they are presented in table 2. Fuzzy suitability indices are obtained after aggregating R_{ik} and W_k by averaging the corresponding products over all the criteria by using equation (4). The ranking values of the fuzzy suitability indices are obtained by using equation (5) and presented in table 3. The ranking order of fuzzy suitability for the four alternatives is MHE_2, MHE_3, MHE_4 and MHE_1. Therefore, the best selection is MHE 2.

Table 3. Table 3. Fuzzy suitability indices and ranking values

MHEs	Fuzzy suitability indices	Ranking values
E1	(0.170, 0.322,0.463,0.620)	0.386
E2	(0.182,0.344,0.480,0.673)	0.446
E3	(0.164,0.375,0.462,0.648)	0.431
E4	(0.156,0.362,0.485,0.612)	0.427

5 Effectiveness of the Proposed Methodology

The most of the existing conventional approaches are based on quantitative factors, failing to recognize many important qualitative factors such as flexibility, environmental hazards and safety issues etc. involved with the selection of material handling equipment under manufacturing environment. The conventional methods of MHE selection are based on the concept of total investment and operating cost (quantitative) along with some arbitrary decision rules framed by the plant managers from their knowledge and experience [3]. The quantitative approaches like discounted cash flow and net present value methods cannot consider the intangible factors involved in the selection process. The Analytical Hierarchy Process (AHP) of Satty for multi-criteria decision making using pair wise comparison theory is also based on crisp evaluation, i.e., the measuring values of criteria must be defined precisely and numerically [11]. The selection order obtained by applying Satty's AHP is in total agreement with the results presented following the proposed methodology. Moreover, the procedure takes less computational time than AHP. In real life, the measure of subjective criteria (qualitative) is very difficult to define preciously and numerically for the decision-makers. Further, the evaluation data under subjective criteria are commonly expressed in linguistic terms. As is evident by the discussion above that the proposed methodology is very effective in conveying the imprecise or vague nature of the linguistic assessments for selecting MHE.

The framework of the proposed methodology is based on fuzzy linguistic assessment under multiple decision makers. The approach enhances the quality of decision by eliminating the inconsistencies in selection policy through the aggregation process [12, 13]. One of the real difficulties faced by the authors to compare the effectiveness of the proposed methodology with the existing ones and present experimental results, is due to non-availability of relevant data regarding cost coefficients of supervision, safety and environmental effect. The present work simply demonstrates the potential applicability of fuzzy set theory and offers a systematic guidance to the decision makers in selecting material handling equipment under fuzzy environment. Research is being carried out by the authors in this direction as a second phase of the present work to establish some relevant data that are compatible to both existing and proposed methodology so as to provide comparative statements in terms of monetary value.

6 Discussions and Conclusion

In this paper, a decision procedure is proposed to solve the MHE selection under fuzzy environment. The conventional approaches are less sensitive in making effective decision when the assessments of alternatives versus criteria and the importance weights are given in linguistic terms. The proposed methodology considers both objective and subjective factors in such a manner that the viewpoints of total decision-making body can be expressed without any constraints. Thus by conducting fuzzy linguistic assessments and fuzzy objective assessments, the decision makers can have the final ranking of the alternatives automatically.

References

1. Apple, J.M., Material handling system design, 1978, John Wiley and sons.
2. Tomkins, J.A. and White, J.A., Facilities planning, 1984, New York, Wiley.
3. Deb, S.K., Bhattacharyya, B., Sorkhel, S.K., Management of machine layout and material handling system selection using hybrid approach, 1st International conference on logistic and supply chain management, ILSCM-2001, PSG Tech, India, pp. 50-55.
4. Karwowski, W. and Evans, G.W., 1987, A layout design heuristic employing theory of fuzzy set, Int. J. of Production Research, vol. 25, pp. 1431-1450.
5. Raoot, A.D. and Rakshit, A., 1993, A linguistic pattern approach for multiple criteria facility layout design, Int. J. of Production Research, vol. 31, pp. 203-222.
6. Deb, S.K., Bhattacharyya, B., Sorkhel, S.K., Fuzzy decision making system for facility layout under manufacturing environment, Proceeding of National symposium on manufacturing Engineering for 21st century, I.I.T., Kanpur, 2001, pp 69-72.
7. Deb, S.K., Bhattacharyya, B., Sorkhel, S.K., Computer aided facility layout design using fuzzy logic under manufacturing environment, Proceedings NCCIDM-2001, Coimbatore, pp-17-22.
8. Spohere, G.A. and Kmak, T.R., 1984, Qualitative analysis used in evaluating alternative plant location, Industrial Engineering, August,pp.52-56.
9. Chen, S.H., 1985, Ranking fuzzy numbers with maximizing and minimizing set, Fuzzy sets and system, vol. 17, pp. 113-129.
10. Zadeh, L.A., 1975, The concept of linguistic variable and its application to approximate reasoning, Information Science, 8, 199-249.
11. Saaty, T.L., The Analytic Hierarchy Process, McGraw-Hill, New York, 1988.
12. Deb, S.K., Bhattacharyya, B., Sorkhel, S.K.,, Fuzzy multi-criteria decision making methodology for facilities layout planning, Int. conference on Information Technology, CIT-2001, NIST (accepted for publication in December 2001).
13. Deb, S.K., Bhattacharyya, B., Sorkhel, S.K.,, Fuzzy ranking methods of facility layout alternatives under manufacturing environment, International conference of Mechanical Engineering, ICME-2001, BUET, Dhaka (Accepted for publication in December 2001).

The Rough Set View on Bayes' Theorem

Zdzisław Pawlak

University of Information Technology and Management
ul. Newelska 6, 01–447 Warsaw, Poland
zpw@ii.pw.edu.pl

MOTTO:
"It is a capital mistake to theorise before one has data"
Sherlock Holms
In: A Scandal in Bohemia

Abstract. Rough set theory offers new perspective on Bayes' theorem. The look on Bayes' theorem offered by rough set theory reveals that any data set (decision table) satisfies total probability theorem and Bayes' theorem. These properties can be used directly to draw conclusions from objective data without referring to subjective prior knowledge and its revision if new evidence is available.
Thus the rough set view on Bayes' theorem is rather objective in contrast to subjective "classical" interpretation of the theorem .

1 Introduction

In his paper [2] Bayes considered the following problem: "*Given* the number of times in which an unknown event has happened and failed: *required* the chance that the probability of its happening in a single trial lies somewhere between any two degrees of probability that can be named."

In fact "... it was Laplace (1774 – 1886) – apparently unaware of Bayes' work – who stated the theorem in its general (discrete) from" [3].

Currently Bayes' theorem is the basic of statistical interference.

"The result of the Bayesian data analysis process is the posterior distribution that represents a revision of the prior distribution on the light of the evidence provided by the data" [5].

Bayes' based inference methodology rised many controversy and criticism. For example,

"Opinion as to the values of Bayes' theorem as a basic for statistical inference has swung between acceptance and rejection since its publication on 1763" [4].

"The technical results at the heart of the essay is what we now know as *Bayes' theorem*. However, from a purely formal perspective there is no obvious reason why this essentially trivial probability result should continue to excite interest" [3].

Rough set theory offers new insight into Bayes' theorem. The look on Bayes' theorem offered by rough set theory is completely different to that used in the

N.R. Pal and M. Sugeno (Eds.): AFSS 2002, LNAI 2275, pp. 106–116, 2002.

Bayesian data analysis philosophy. It does not refer either to prior or posterior probabilities, inherently associated with Bayesian reasoning, but it reveals some probabilistic structure of the data being analyzed. It states that any data set (decision table) satisfies total probability theorem and Bayes' theorem. This property can be used directly to draw conclusions from the data without referring to prior knowledge and its revision if new evidence is available. Thus in the presented approach the only source of knowledge is the data and there is no need to assume that there is any prior knowledge besides the data.

Moreover, the rough set approach to Bayes' theorem shows close relationship between logic of implications and probability, which was first observed by Łukasiewicz [6] and also independly studied by Adams [1] and others. Bayes' theorem in this context can be used to "invert" implications, i.e., to give reasons for decisions. This is a very important feature of utmost importance to data mining and decision analysis, for it extends the class of problem which can be considered in these domains.

Besides, we propose a new form of Bayes' theorem where basic role plays strength of decision rules (implications) derived from the data. The strength of decision rules is computed from the data or it can be also a subjective assessment. This formulation gives new look on Bayesian method of inference and also simplifies essentially computations.

It is also interesting to note a relationship between Bayes' theorem and flow graphs.

Let us also observe that the rough set view on Bayes' theorem is rather objective in contrast to subjective "classical" interpretation.

2 Information Systems and Approximation of Sets

In this section we define basic concepts of rough set theory: information system and approximation of sets. Rudiments of rough set theory can be found in [7, 10].

An information system is a data table, whose columns are labeled by attributes, rows are labeled by objects of interest and entries of the table are attribute values.

Formally, by an *information system* we will understand a pair $S = (U, A)$, where U and A, are finite, nonempty sets called the *universe*, and the set of *attributes*, respectively. With every attribute $a \in A$ we associate a set V_a, of its *values*, called the *domain* of a. Any subset B of A determines a binary relation $I(B)$ on U, which will be called an *indiscernibility relation*, and defined as follows: $(x, y) \in I(B)$ if and only if $a(x) = a(y)$ for every $a \in A$, where $a(x)$ denotes the value of attribute a for element x. Obviously $I(B)$ is an equivalence relation. The family of all equivalence classes of $I(B)$, i.e., a partition determined by B, will be denoted by $U/I(B)$, or simply by U/B; an equivalence class of $I(B)$, i.e., block of the partition U/B, containing x will be denoted by $B(x)$.

If (x, y) belongs to $I(B)$ we will say that x and y are *B-indiscernible* (*indiscernible with respect to B*). Equivalence classes of the relation $I(B)$ (or blocks of the partition U/B) are referred to as *B-elementary sets* or *B-granules*.

If we distinguish in an information system two disjoint classes of attributes, called *condition* and *decision attributes*, respectively, then the system will be called a *decision table* and will be denoted by $S = (U, C, D)$, where C and D are disjoint sets of condition and decision attributes, respectively.

Thus the decision table determines decisions which must be taken, when some conditions are satisfied. In other words each row of the of the decision table specifies a decision rule which determines decisions in terms of conditions.

Observe, that elements of the universe are in the case of decision tables simply labels of decision rules.

Suppose we are given an information system $S = (U, A)$, $X \subseteq U$, and $B \subseteq A$. Our task is to describe the set X in terms of attribute values from B. To this end we define two operations assigning to every $X \subseteq U$ two sets $B_* (X)$ and $B^* (X)$ called the *B-lower* and the *B-upper approximation* of X, respectively, and defined as follows:

$$B_* (X) = \bigcup_{x \in U} \{B (x) : B (x) \subseteq X\},$$

$$B^* (X) = \bigcup_{x \in U} \{B (x) : B (x) \cap X \neq \emptyset\}.$$

Hence, the B-lower approximation of a set is the union of all B-granules that are included in the set, whereas the *B-upper* approximation of a set is the union of all B-granules that have a nonempty intersection with the set. The set

$$BN_B (X) = B^* (X) - B_* (X)$$

will be referred to as the *B-boundary region* of X.

If the boundary region of X is the empty set, i.e., $BN_B (X) = \emptyset$, then X is *crisp* (*exact*) with respect to B; in the opposite case, i.e., if $BN_B(X) \neq \emptyset$, X is referred to as *rough* (*inexact*) with respect to B.

3 Rough Membership

Rough sets can be also defined employing instead of approximations rough membership function [9], which is defined as follows:

$$\mu_X^B : U \rightarrow [0, 1]$$

and

$$\mu_X^B (x) = \frac{|B (x) \cap X|}{|B (x)|},$$

where $X \subseteq U$ and $B \subseteq A$ and $|X|$ denotes the cardinality of X.

The function measures the degree that x belongs to X in view of information about x expressed by the set of attributes B.

The rough membership function has the following properties:

1. $\mu_X^B(x) = 1$ *iff* $x \in B_*(X)$
2. $\mu_X^B(x) = 0$ *iff* $x \in U - B^*(X)$
3. $0 < \mu_X^B(x) < 1$ *iff* $x \in BN_B(X)$
4. $\mu_{U-X}^B(x) = 1 - \mu_X^B(x)$ for any $x \in U$
5. $\mu_{X \cup Y}^B(x) \geq max\left(\mu_X^B(x),\ \mu_Y^B(x)\right)$ for any $x \in U$
6. $\mu_{X \cap Y}^B(x) \leq min\left(\mu_X^B(x),\ \mu_Y^B(x)\right)$ for any $x \in U$

Compare these properties to those of fuzzy membership. Obviously rough membership is a generalization of fuzzy membership.

The rough membership function, can be used to define approximations and the boundary region of a set, as shown below:

$$B_*(X) = \{x \in U : \mu_X^B(x) = 1\},$$

$$B^*(X) = \{x \in U : \mu_X^B(x) > 0\},$$

$$BN_B(X) = \{x \in U : 0 < \mu_X^B(x) < 1\}.$$

4 Information Systems and Decision Rules

Every decision table describes decisions determined, when some conditions are satisfied. In other words each row of the decision table specifies a decision rule which determines decisions in terms of conditions.

Let us describe decision rules more exactly.

Let $S = (U, C, D)$ be a decision table. Every $x \in U$ determines a sequence $c_1(x), \ldots, c_n(x), d_1(x), \ldots, d_m(x)$ where $\{c_1, \ldots, c_n\} = C$ and $\{d_1, \ldots, d_m\} = D$.

The sequence will be called a *decision rule induced by* x (in S) and denoted by $c_1(x), \ldots, c_n(x) \to d_1(x), \ldots, d_m(x)$ or in short $C \to_x D$.

The number $supp_x(C, D) = |C(x) \cap D(x)|$ will be called a *support* of the decision rule $C \to_x D$ and the number

$$\sigma_x(C, D) = \frac{supp_x(C, D)}{|U|},$$

will be referred to as the *strength* of the decision rule $C \to_x D$. With every decision rule $C \to_x D$ we associate the *certainty factor* of the decision rule, denoted $cer_x(C, D)$ and defined as follows:

$$cer_x(C, D) = \frac{|C(x) \cap D(x)|}{|C(x)|} = \frac{supp_x(C, D)}{|C(x)|} = \frac{\sigma_x(C, D)}{\pi(C(x))},$$

where $\pi(C(x)) = \frac{|C(x)|}{|U|}$.

The certainty factor may be interpreted as a conditional probability that y belongs to $D(x)$ given y belongs to $C(x)$, symbolically $\pi_x(D|C)$.

If $cer_x(C, D) = 1$, then $C \to_x D$ will be called a *certain decision* rule in S; if $0 < cer_x(C, D) < 1$ the decision rule will be referred to as an *uncertain decision rule* in S.

Besides, we will also use a *coverage factor* of the decision rule, denoted $cov_x(C, D)$ defined as

$$cov_x(C, D) = \frac{|C(x) \cap D(x)|}{|D(x)|} = \frac{supp_x(C, D)}{|D(x)|} = \\ = \frac{\sigma_x(C, D)}{\pi(D(x))},$$

where $\pi(D(x)) = \frac{|D(x)|}{|U|}$.

Similarly

$$cov_x(C, D) = \pi_x(C|D).$$

If $C \rightarrow_x D$ is a decision rule then $D \rightarrow_x C$ will be called an *inverse decision rule*. The inverse decision rules can be used to give *explanations* (*reasons*) for decisions.

Let us observe that

$$cer_x(C, D) = \mu_{D(x)}^C(x) \text{ and } cov_x(C, D) = \mu_{C(x)}^D(x).$$

That means that the certainty factor expresses the degree of membership of x to the decision class $D(x)$, given C, whereas the coverage factor expresses the degree of membership of x to condition class $C(x)$, given D.

Decision rules are often represented in a form of "*if* ... *then* ..." implications. Thus any decision table can be transformed in a set of "*if* ... *then* ..." rules, called a *decision algorithm*.

Generation of minimal decision algorithms from decision tables is a complex task and we will not discuss this issue here. The interested reader is advised to consult the references.

5 Probabilistic Properties of Decision Tables

Decision tables have important probabilistic properties which are discussed next.

Let $C \rightarrow_x D$ be a decision rule in S and let $\Gamma = C(x)$ and let $\Delta = D(x)$. Then the following properties are valid:

$$\sum_{y \in \Gamma} cer_y(C, D) = 1 \tag{1}$$

$$\sum_{y \in \Delta} cov_y(C, D) = 1 \tag{2}$$

$$\pi(D(x)) = \sum_{y \in \Gamma} cer_y(C, D) \cdot \pi(C(y)) = \\ = \sum_{y \in \Gamma} \sigma_y(C, D) \tag{3}$$

$$\pi(C(x)) = \sum_{y \in \Delta} cov_y(C,D) \cdot \pi(D(y)) = \sum_{y \in \Delta} \sigma_y(C,D) \tag{4}$$

$$cer_x(C,D) = \frac{cov_x(C,D) \cdot \pi(D(x))}{\sum_{y \in \Delta} cov_y(C,D) \cdot \pi(D(y))} = \frac{\sigma_x(C,D)}{\pi(C(x))} \tag{5}$$

$$cov_x(C,D) = \frac{cer_x(C,D) \cdot \pi(C(x))}{\sum_{y \in \Gamma} cer_y(C,D) \cdot \pi(C(y))} = \frac{\sigma_x(C,D)}{\pi(D(x))} \tag{6}$$

That is, any decision table, satisfies (1),...,(6). Observe that (3) and (4) refer to the well known *total probability theorem*, whereas (5) and (6) refer to *Bayes' theorem*.

Thus in order to compute the certainty and coverage factors of decision rules according to formula (5) and (6) it is enough to know the strength (support) of all decision rules only. The strength of decision rules can be computed from data or can be a subjective assessment.

6 Decision Tables and Flow Graphs

With every decision table we associate a *flow graph* , i.e., a directed, connected, acyclic graph defined as follows: to every decision rule $C \to_x D$ we assign a *directed branch* x connecting the *input node* $C(x)$ and the *output node* $D(x)$. Strength of the decision rule represents a *throughflow* of the corresponding branch. The throughflow of the graph is governed by formulas (1),...,(6).

Formulas (1) and (2) say that an outflow of an input node or an output node is equal to their inflows. Formula (3) states that the outflow of the output node amounts to the sum of its inflows, whereas formula (4) says that the sum of outflows of the input node equals to its inflow. Finally, formulas (5) and (6) reveal how throughflow in the flow graph is distributed between its inputs and outputs.

7 Illustrative Examples

Now we will illustrate the ideas considered in the previous sections by simple tutorial examples. These examples intend to show clearly the difference between "classical" Bayesian approach and that proposed by the rough set philosophy.

Example 1. This example will clearly show the different role of Bayes' theorem in classical statistical inference and that in rough set based data analysis.

Let us consider the data table shown in Table 1.

Table 1. Data table

	T^+	T^-
D	95	5
$\overline{D}$	1998	97902

In Table 1 the number of patients belonging to the corresponding classes is given. Thus we start from the original data (not probabilities) representing outcome of the test.

Now from Table 1 we create a decision table and compute strength of decision rules. The results are shown in Table 2.

Table 2. Decision table

fact	*D*	*T*	*support*	*strength*
1	+	+	95	0.00095
2	−	+	1998	0.01998
3	+	−	5	0.00005
4	−	−	97902	0.97902

In Table 2 D is the condition attribute, wheras T is the decision attribute. The decision table is meant to represent a "cause–effect" relation between the disease and result of the test. That is, we expect that the disease causes positive test result and lack of the disease results in negative test result.

The decision algorithm is given below:

1') *if* (*disease, yes*) *then* (*test, positive*)
2') *if* (*disease, no*) *then* (*test, positive*)
3') *if* (*disease, yes*) *then* (*test, negative*)
4') *if* (*disease, no*) *then* (*test, negative*)

The certainty and coverage factors of the decision rules for the above decision algorithm are given is Table 3.

The decision algorithm and the certainty factors lead to the following conclusions:

- 95% persons suffering from the disease have positive test results
- 2% healthy persons have positive test results
- 5% persons suffering from the disease have negative test result
- 98% healthy persons have negative test result

Table 3. Certainty and coverage

rule	*strength*	*certainty*	*coverage*
1	0.00095	0.95	0.04500
2	0.01998	0.02	0.95500
3	0.00005	0.05	0.00005
4	0.97902	0.98	0.99995

That is to say that if a person has the disease most probably the test result will be positive and if a person is healthy the test result will be most probably negative. In other words, in view of the data there is a causal relationship between the disease and the test result.

The inverse decision algorithm is the following:

1) *if* (*test, positive*) *then* (*disease, yes*)
2) *if* (*test, positive*) *then* (*disease, no*)
3) *if* (*test, negative*) *then* (*disease, yes*)
4) *if* (*test, negative*) *then* (*disease, no*)

From the coverage factors we can conclude the following:

- 4.5% persons with positive test result are suffering from the disease
- 95.5% persons with positive test result are not suffering from the disease
- 0.005% persons with negative test results are suffering from the disease
- 99.995% persons with negative test results are not suffering from the disease

That means that if the test result is positive it does not necessarily indicate the disease but negative test results most probably (almost for certain) does indicate lack of the disease.

That is to say that the negative test result almost exactly identifies healthy patients.

For the remaining rules the accuracy is much smaller and consequently test results are not indicating the presence or absence of the disease. □

Example 2. Let us now consider a little more sophisticated example, shown in Table 4.

Attributes *disease, age* and *sex* are condition attributes, wheras *test* is the decision attribute.

The strength, certainty and coverage factors for decision table are shown in Table 5.

Below a decision algorithm associated with Table 5 is presented.

1) *if* (*disease, yes*) *and* (*age, old*) *then* (*test,* +)
2) *if* (*disease, yes*) *and* (*age, middle*) *then* (*test,* +)
3) *if* (*disease, no*) *then* (*test,* −)
4) *if* (*disease, yes*) *and* (*age, old*) *then* (*test,* −)

Table 4. Decision table

fact	*disease*	*age*	*sex*	*test*	*support*
1	*yes*	*old*	*man*	+	400
2	*yes*	*middle*	*woman*	+	80
3	*no*	*old*	*man*	−	100
4	*yes*	*old*	*man*	−	40
5	*no*	*young*	*woman*	−	220
6	*yes*	*middle*	*woman*	−	60

Table 5. Certainty and coverage

fact	*strength*	*certainty*	*coverage*
1	0.44	0.92	0.83
2	0.09	0.56	0.17
3	0.11	1.00	0.23
4	0.04	0.08	0.09
5	0.24	1.00	0.51
6	0.07	0.44	0.15

5) *if* (*disease, yes*) *and* (*age, middle*) *then* (*test,* −)

The flow graph for the decision algorithm is presented in Fig. 1.

The certainty and coverage factors for the above algorithm are given in Table 6

Table 6. Certainty and coverage factors

rule	*strength*	*certainty*	*coverage*
1	0.44	0.92	0.83
2	0.09	0.56	0.17
3	0.36	1.00	0.76
4	0.04	0.08	0.09
5	0.07	0.44	0.15

The certainty factors of the decision rules lead the following conclusions:

- 92% ill and old patients have positive test result
- 56% ill and middle age patients have positive test result
- all healthy patients have negative test result
- 8% ill and old patients have negative test result
- 44% ill and old patients have negative test result

In other words:

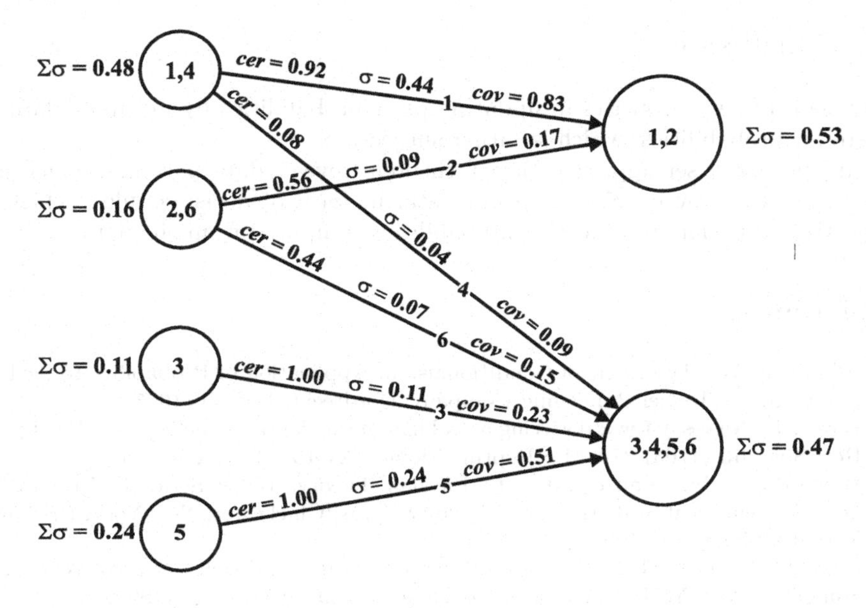

Fig. 1. Flow graf

- ill and old patients most probably have positive test result (probability = 0.92)
- ill and middle age patients most probably have positive test result (probability = 0.56)
- healthy patients have certainly negative test result (probability = 1.00)

Now let us examine the inverse decision algorithm, which is given below:

1') *if* (*test*, +) *then* (*disease*, *yes*) *and* (*age*, *old*)
2') *if* (*test*, +) *then* (*disease*, *yes*) *and* (*age*, *middle*)
3') *if* (*test*, −) *then* (*disease*, *no*)
4') *if* (*test*, −) *then* (*disease*, *yes*) *and* (*age*, *old*)
5') *if* (*test*, −) *then* (*disease*, *yes*) *and* (*age*, *middle*)

Employing the inverse decision algorithm and the coverage factor we get the following explanation of test results:

- reasons for positive test results are most probably disease and old age (probability = 0.83)
- reason for negative test result is most probably lack of the disease (probability = 0.76) □

If is clearly seen from examples 1 and 2 the difference between Bayesian data analysis and the rough set approach. In the Bayesian inference the data is used to update prior knowledge (probability) into a posterior probability, whereas rough sets are used to understand what the data are telling us.

8 Conclusion

Bayesian inference consists in updating prior probabilities by means of data to posterior probabilities, which is rather subjective.

In the rough set approach Bayes' theorem reveals data patterns, which are used next to draw conclusions from data, in form of decision rules, which is objective and refers to objective probabilities computed from the data.

References

1. Adams, E. W.: The Logic of Conditionals, an Application of Probability to Deductive Logic. D. Reidel Publishing Company, Dordrecht, Boston (1975)
2. Bayes, T.: An essay toward solving a problem in the doctrine of chances. Phil. Trans. Roy. Soc., **53** (1763) 370–418; Reprint Biometrika **45** (1958) 296–315
3. Bernardo, J. M., Smith,A. F. M.: Baysian Theory, Wiley Series in Probability and Mathematical Statistics. John Wiley & Sons, Chichester, New York, Brisbane, Toronto, Singapore (1994)
4. Box, G.E.P., Tiao, G.C.: Bayesian Inference in Statistical Analysis. John Wiley and Sons, Inc., New York, Chichester, Brisbane, Toronto, Singapore (1992)
5. Berthold, M., Hand, D.J.: Intelligent Data Analysis, An Introduction. Springer-Verlag, Berlin , Heidelberg, New York (1999)
6. Łukasiewicz, J.: Die logishen Grundlagen der Wahrscheinilchkeitsrechnung. Kraków, 1913. In: L. Borkowski (ed.), Jan Łukasiewicz – Selected Works, North Holland Publishing Company, Amsterdam, London, Polish Scientific Publishers, Warsaw (1970)
7. Pawlak, Z.: Rough Sets – Theoretical Aspect of Reasoning about Data. Kluwer Academic Publishers, Boston Dordrech, London (1991)
8. Pawlak, Z.: New look on Bayes' theorem – the rough set outlook. Proceeding of International Workshop on Rough Set Theory and Granular Computing (RSTGC-2001), Matsue, Shimane, Japan, May 20-22, S. Hirano, M. Inuiguchi and S. Tsumoto (eds.), Bull. of Int. Rough Set Society vol. **5** no. **1/2** 2001 1–8
9. Z. Pawlak, A. Skowron, Rough membership functions, advances in the Dempster-Shafer theory of evidence. R, Yager, M. Fedrizzi, J. Kacprzyk (eds.), John Wiley & Sons, Inc., New York (1994) 251–271

Verbalizing Computers
– A Way to Everyday Language Computing –

Michio Sugeno

Brain Science Institute, RIKEN
2-1 Hirosawa, Wako, Japan

Abstract. This paper describes Everyday Language Computing Project that we are currently promoting. An idea of the project is to solve the so-called 'information divide' in IT revolution. To this aim, we propose to verbalize computers like the brain to deal with the meanings of information. Verbalizing computers will be achieved by the semiotic base as a stored-intelligence together with text understanding/generation, language communication protocol, and language-based applications.

1 Introduction

In recent years we have being used computers more often for information processing rather than for numerical computation. In a sense, computation has changed its meaning from calculating numbers with computers to processing information with computers. With this tendency, more people, not only experts but also ordinary people, are being involved in computation. To use computers, we have to know 'computer language' and even 'computer architecture'. It is, however, not the case when we use telephones and televisions. We speak ordinary language in everyday life but computers do not. This simple fact enlarges 'information divide'.

In order to solve this problem, the author and his research group are now engaged in research on everyday language computing [1,2]. It aims to develop computing systems that anybody can access and use with his/her own everyday language. In short we verbalize computers just as we are doing in thinking process, where we verbalize our brains. For example, we verbalize interface, communication in computer network and software/hardware applications. A key technology for it, we use an idea of semiotic base, and make understanding/generation of meanings.

Our project is, on the other hand, concerned with 'Creating the Brain' which is one of the research areas of brain science. That is, the human brain consists of two parts: hardware and software. By software what we mean is language. As far as the human brain is concerned, it is characterized by language rather than neural network. In this context, we might be able to artificially create the brain in a sense of realizing a hardware which accepts human language.

The next section discusses objectives and some research issues of the authors research group at Laboratory for Language-Based Intelligent Systems, the Brain Science Institute.

N.R. Pal and M. Sugeno (Eds.): AFSS 2002, LNAI 2275, pp. 117–120, 2002.

2 Everyday Language Computing

2.1 Objectives

The methodology of our research in the field of Creating the Brain is characterized by fusing linguistics and engineering together. We use Systemic Functional Linguistic Theory [4] initiated by M.A.K. Halliday and information technology together with systems theory.

As it is said that the human brain has evolved in connection with language, one of the outstanding functions of the human brain is its capability of handling highly complex language. The language system as social semiotic is a system of meanings built up through human social activities. Human explicit intelligence is supported by language.

The essential difference between the human species and all other species is the fact that human species has created software, called language. Human society and culture are nothing but what has been weaved by this software.

From this point of view, our research aims at realizing in the sense of engineering brain-style intelligent systems in which the human language system is embedded. According to Halliday, the human language system is the most complex system in the cosmos. No doubt the language system is the supreme masterpiece created by the brain.

Most existential parts of the language system are stored outside of the brain, in society, but its central parts concerning processes of meaning understanding and meaning generation exist inside of the brain as its higher order functions.

Based on Systemic Functional Linguistic Theory, we apply systems theory to reveal the structure of the language system as meaning potential, develops an algorithm of dealing with meanings through language, and implements these into computer hardware with the aid of information technology. In realizing a process of computing meanings on computers, we estimate the complexity of a part of the brain corresponding to that of the language system and contribute to analyze the architecture of the brain related to t he higher order functions.

2.2 Some Research Issues

Our research on Everyday Language Computing is aimed at realizing a computing environment in which all information processing on computers can be performed by natural language as a meta-language. In other words, we aim to make a paradigm shift from number-based computing to language-based computing. To this aim, we make the conceptual design of computing systems storing language-based intelligence.

Semiotic Base

Semiotic base plays a key role as language-based intelligence stored in computers for everyday language computing. It is a collection of meaning resources and knowledge to deal with meanings of social semiotic symbols, especially language.

We describe language system with special reference to its use in context from the systemic functional linguistic point of view, and compile the results together with knowledge as meaning resources into database. These data are stratified into three layers constructing the semiotic base: context base, meaning base and lexicogrammar base [5].

The meaning base and lexicogrammar base are structurized according to systemic functional linguistics, where the former represents semantic features of language (text) and the latter represents lexicogrammatical features of language (text). The context base describes context of culture and context of situation. Currently we are only dealing with situation ; the context base is a set of situation types. Each situation type is characterized by field (what is taking place), tenor(who are acting) and mode (what channels, spoken or written, are used).

In a sense, the semiotic base is a knowledge base embedded in language system, where a piece of knowledge with verbal expression is characterized by context, meaning and lexicogrammar.

Text Understanding/Generation by Semiotic Base

The semiotic base is a framework for understanding and generating social semiotic symbols. Restricting semiotic symbols to language, we study algorithms of text understanding and text generation using a semiotic base.

In the case of understanding, we first analyze a given text to find its lexicogrammatical features. Then through the features, we guess a situation type in which the text is located. We finally understand the meaning of the text based on its contextual and lexicogrammatical features. In the case of generation, given contextual and semantic features concerning what we want to express, we find its lexicogrammatical features and generate a text.

Language-Based Computing System

A language computing system is a computing environment to realize everyday language computing, on conventional computer platforms. It includes the following issues.

(a) language communication protocol [3]
Language communication protocol is a fundamental concept together with the semiotic base in everyday language computing. It is a protocol to transmit meanings on computer networks, delivering the resources in the semiotic base through language.

(b) language-based applications
Language-based applications are all applications of software and hardware used with computers that can be driven by natural language. We design the architecture of language application interface in order to manipulate and manage conventional computer applications through language.
We also design an environment for language-based programming, where we consider algorithms for automatic generation of simple executable computer programs.

(c) language-based interface
A language-based interface is an intelligent interface with a virtual 3D client's office and a client's secretary with a user model for a client to use computers through his/her everyday language.

(d) language-based agent system
A language-based agent system is a system consisting of agents who collect, transmit and manage information according to a client's linguistic instructions.

(e) language operation system
A language operating system is to manage a language-based computing environment and consists of a client operating system and a network operating system. These operating systems include language communication protocol and language-based file management as common functions. In particular, a client operating system has a function of language-based interface and a network operating system has a function of language-based process management.

3 Conclusions

This paper has described a project of Everyday Language Computing which enables us to use computers with our everyday language without any specific knowledge about computing. The idea will be realized by verbalizing computers where the semiotic base, the language communication protocol, and language-based applications will play key roles.

References

1. Sugeno, M.: Toward intelligent computing, in: Y.Yam and K.S.Lenng (ed.), Future Directions of Fuzzy Theory and Systems, World Scientific (1995) 10-18
2. Sugeno, M.: Language-Based Computing Environment for Internet Communication between the Brain and Society, a plenary talk, Joint 9th IFSA World Congress and 20th NAFIPS International Conference, Vancouver (2001)
3. Kobayashi, I. et al.: Toward a Computational Environment for Everyday Language Communication, Proc. of Joint 9th IFSA World Congress and 20th NAFIPS International Conference, Vancouver (2001) 663-668
4. Halliday, M.A.K.: An Introduction to Functional Grammar, 2nd edn. Edward Arnold (1994)
5. Halliday M.A.K. and Matthiessen C.M.I.M.: Construing Experience through Meaning -A Language-Based Approach to Cognition, Cassell (1999)

Soft Computing Based Emotion/Intention Reading for Service Robot

Z. Zenn Bien, Jung-Bae Kim, Dae-Jin Kim, Jeong-Su Han, and Jun-Hyeong Do

Div. of EE, Dept. of EECS, KAIST 373-1 Guseong-Dong, Yuseong-Gu, Daejeon 305-701, KOREA
zbien@ee.kaist.ac.kr,
{jbkim,djkim,pyodori,jhdo}@ctrsys.kaist.ac.kr

Abstract. Due to its tolerance to imprecision, uncertainty and partial truth, the soft computing technique deals well with human related signals such as voice, gesture, facial expression, bio-signal, etc. In this paper, we propose architecture of soft computing based recognition for a class of biosign. Especially, the problem of inferring emotion and intention reading from such recognition is considered with applications for service robots that interact with human. It is shown that proposed architecture renders good performance in a few experimental systems for rehabilitation.

1 Introduction

The service robots are mainly designed to serve humans directly or indirectly by helping or replacing humans in the works that usually require human flexibility under unstructured, possibly varying environments and sometimes intense-interactions. They may immensely differ from the industrial robots that repeat only those works predefined in a structured workspace.

The service robots take various forms and functions. For examples, they include housekeeping home robots, entertainment robots, rehabilitation robots for the disabled, intelligent robot house, etc. For these service robots, an important basic technology which needs a special attention is "human friendly interface" including voice recognition, gesture recognition, object recognition, user's intention reading, etc. This technique focuses on human-machine interaction because the service robots receive direct human command or cooperate with human.

To recognize *biosigns* such as voice, gesture, facial expression and bio-signals, we need an intelligent recognition method that is tolerant of imprecision, uncertainty and partial truth of biosign. Here, bio-signals include ECG (Electrocardiogram: heart signal), EMG (Electromyogram: muscle signal), EEG (Electroencephalogram: brain signal), etc. The soft computing method, which differs from the conventional hard computing paradigm, is known to have those characteristics and potential to solve many real-world problems [8]. The soft computing techniques contain fuzzy logic, neural network, probabilistic reasoning, evolutionary algorithms, chaos theory, belief networks, learning theory, etc [9].

N.R. Pal and M. Sugeno (Eds.): AFSS 2002, LNAI 2275, pp. 121–128, 2002.

This paper proposes an architecture of a soft computing-based recognition system that deals with human mind level signals, especially the user's emotion and intention. Reading human emotion and/or intention is a challenging research topic, because there are apparently no appropriate sensors/measurement devices for many biosign-related variables, and, even if possible, subjectiveness of the result may have an ill effect.

The rest of the paper is organized as follows: Section 2 introduces emotion and intention and how to perceive them. Section 3 introduces more details about signal flow in man-machine interaction systems. In Section 4, we suggest an architecture of soft computing-based recognition system. We deal with application systems in Section 5. This paper concludes in Section 6.

2 Preliminary

2.1 Emotion/Intention

The word 'emotion' is used very often in our daily lives. According to [5], it is very difficult to answer the question such as 'What is the emotion?' because of its wide usage and subjective characterization [6]. However, we use the term 'emotion' to express our natural feeling of happiness, joy, sadness, surprise, anger, greeting, love, hate and so on. In this paper, the word 'emotion' is also used to represent such feelings as well as mood and affection [7].

Intention is an act or instance of determining mentally some action or result. It is a direct representation of the user's purpose, whereas emotion is an indirect one. For example, "bringing the cup to the user's mouth" is a good example of direct representation of the user's purpose, and we may relate it with an intention of the user. On the other hand, a negative reaction such as "shutting the user's mouth when the robot serves" may be interpreted as an emotional state to express that the user does not want to eat anything, which may be interpreted as a kind of indirect representation of the user's purpose, and we may relate it with emotion of the user.

From a psychological point of view, there have been many attempts to understand "how a human can recognize emotions/intentions of the other humans". Mehrabian proposes an emotion-space model called "PAD Emotional State Model" [14]. It consists of three nearly independent dimensions that are used to describe and measure emotional states: *Pleasure-displeasure, Arousal-nonarousal, and Dominance-submissiveness.* "Pleasure-displeasure" distinguishes the positive-negative affective quality of emotional states, while "arousal-nonarousal" refers to a combination of physical activity and mental alertness. And "dominance-submissiveness" is defined in terms of control versus lack of control. Visual stimuli-based approach by Ekman et al. is also very popular. They proposed that many emotions or intentions in human's face may be recognized by combination of various facial muscular actions, so called "AU (Action Unit)" [15]. Dellaert et al. attempted to find elements that can affect emotions from speech signals [16].

On the basis of these psychological approaches, many researchers have been also trying to recognize human emotions (or intentions) for engineering purpose. An emotional agent proposed by Breazeal can recognize emotions of human beings based on PAD emotional state model [17]. This agent can recognize and represent many emotions based on PAD emotional model with mechanical structures. Vision-based approaches based on Ekman's theory show promising results [4]. With soft computing techniques, machine can effectively recognize emotions of human beings based on images of facial expression. Nicholson made an attempt to recognize emotions from speech signals using artificial neural networks [18].

2.2 Soft Computing Tool Box

Soft computing techniques are convenient tools to solve many real world problems. It is known to exploit the tolerance for uncertainty and imprecision to achieve tractability, robustness, and low solution cost [9]. Key methodologies include the fuzzy logic theory (FL), neural networks (NN), evolutionary computation (EC), and the rough set theory (RS). Complementary combination of these methodologies may exhibit a higher computing power that "parallels the remarkable ability of the human mind to reason and learn in an environment of uncertainty and imprecision [8]."

Two concepts play a key role within FL [9]. One is the concept of *linguistic variable* and the other is the *fuzzy if-then rules*. FL mimics the remarkable ability of the human mind to summarize data and focus on decision-relevant information.

NN is a massively parallel computing system made up of simple processing units, called *neurons*, which has a natural propensity for storing experiential knowledge and making it available for use in decision making. Nonlinearity of neuron, input-output mapping, adaptivity, and fault tolerance are useful properties of NN [19].

EC can be described as a two-step iterative process, consisting of *random variation* followed by *selection*. In the real world, EC offers considerable advantages such as adaptability to changing situations, generation of good enough solutions quickly, and so on [20].

By applying RS into a data set that is incomplete, imprecise, and vague, we can extract knowledge in a form of a minimal set of rules [21]. RS provides many advantages including efficient algorithms for finding hidden patterns in data, data reduction, methods for evaluating significance of data, etc.

To summarize, FL, NN, EC and RS can be appropriate tools for rule induction leaning, optimization and rule reduction, respectively.

3 Signal Flow in Man-Machine Interaction System

Fig. 1 shows a model which we propose to describe signal flow from human's *mind level* to machine's *action decision making module*. Emotion and intention in mind level induce various biosigns through many human's physical organs such as face, hand, muscle, brain and vocal cord in the *body level*. These biosigns include biosignals, gesture, facial expression, voice, eye gaze, etc.

The machine senses biosigns using various sensors in *acquisition module* and recognizes emotion and (or) intention in the *emotion/intention reading module* [Fig. 1]. Finally, the machine's actions are made between human and service robots.

To deal with the biosign, which has imprecision, uncertainty and partial truth, soft computing tool box is used in emotion/intention reading module and *action decision making module*. The detailed part from the acquisition module to emotion/intention module is dealt in Section 4. As the man shows some biosign to the machine and the machine recognizes the biosign and produces some actions to the man, it makes the man-machine interaction.

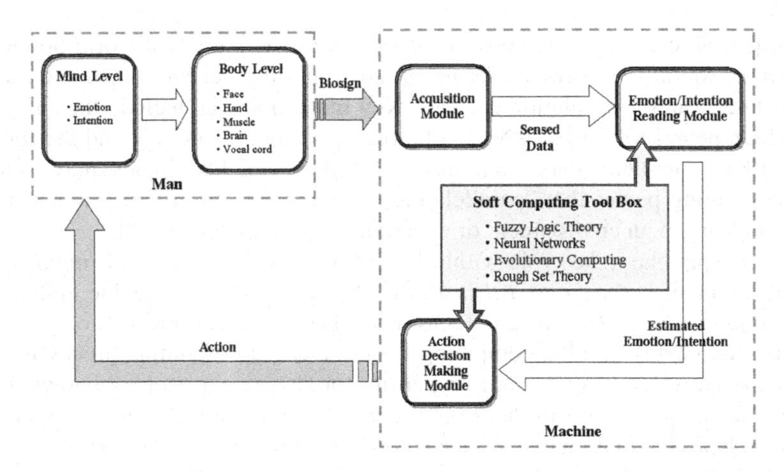

Fig. 1. Soft computing-based emotion/intention reading procedure from human mind level to action decision making level

4 An Architecture of Soft Computing-Based Recognition System

As in cases of human, the partner's intention or emotion can be inferred not only from language but also from behavior. Typically, inferred intentions or emotions are vague and not necessarily expressible, but they play a key role for conservative decision making as in the case of design in consideration of safety or for smooth cooperation for comfort. A human being also tries to read the other party's intention or emotion subjectively. Thus, any classical probability or statistics may not be appropriate to express one's intention or emotion in a mathematical way [1]. Hence, we need appropriate methods, such as soft computing techniques, to deal with these types of vague and uncertain knowledge.

We propose a soft computing-based recognition system for the biosign as shown in Fig. 2. It is a modified figure of the fundamental step of digital image processing [10]. The input of the architecture is biosign and the output is the recognized intention, emotion, information and exogenous event.

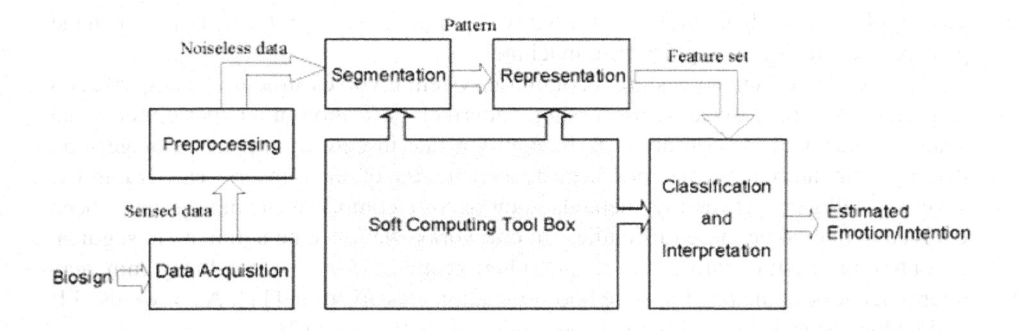

Fig. 2. An Architecture of soft computing-based recognition system

The starting block of the system is "data acquisition", that is, acquiring biosigns. The sensors for acquisition could be microphone, camera, glove device, motion capture device, EMG signal detector, etc. After the biosign is obtained, the next step deals with preprocessing. The preprocessing block typically deals with enhancing the signal and removing noise. The next stage deals with segmentation. It means partitioning a biosign into constituent signals. In general, it contains two segmentation parts: spatial segmentation and temporal segmentation. The former means selecting the meaningful signal from a signal mixed with background signal, and the latter means selecting isolated signal from a continuous signal.

The output of the segmentation stage needs to be converted into a form suitable for computer processing. This involves representation of raw data. It contains the feature extraction process. The last stage of Fig. 2 involves classification and interpretation. Classification is the process that assigns a label to an object based on the information provided by its features. Interpretation involves assigning meaning to an ensemble of classified objects.

To deal with biosign, we need prior knowledge in the processing modules in Fig. 2. We implement it with soft computing technique. As we mentioned, FL, RS, EC and NN may be appropriate method for rule induction, rule reduction, optimization and learning respectively. So, we apply FL and NN to the segmentation stage, FL and RS to the representation stage, and FL, NN and EC to the classification and interpretation stage. As auxiliary methods, state automata and Hidden Markov Model are used for segmentation and classification stage.

5 Application Systems

5.1 Gesture Recognition System [11][12]

To overcome inconveniences of human-machine communication tools such as keyboards and mouse, the hand gesture method has been developed to accommodate a variety of commands naturally and directly. In spite of its usefulness, however, hand gesture is difficult to recognize by a machine.

Construction of a hand gesture recognition system involves structural categorization of gesture, real-time dynamic processing, pattern classification in a hyper dimensional space, coping with deterioration on recognition rate in case of expansion of gesture, dealing with ambiguity and nonlinearity constraints of the sensors, etc. Naturally, several intelligent processing methods such as soft computing technique have been evolved to overcome these difficulties. In our works, we use state automata to segment a continuous gesture into a set of individual gestures. And we use fuzzy min max neural network in the hand posture and orientation classification [11]. Also, we use FL and Hidden Markov Model in the hand motion classification [12].

5.2 Facial Emotional Expression Recognition System [4]

In general, the problem of recognizing emotion from a face is known to be very complex and difficult because individuality may come in expressing and observing emotions. It is interesting to note, however, that human beings can successfully understand facial expressions in a seemingly easy way. In our work, various soft computing techniques are used effectively for recognizing a positive expression of happiness [5]. This work has adopted NN, FS, and RS theory. To handle the recognition system by employing a traditional FL framework, a novel concept termed as "fuzzy observer" is proposed to indirectly estimate a linguistic variable from conventionally measured data.

As an application, we have been developing an intention reading system for intelligent visual servoing for wheelchair-mounted rehabilitation robot system, KARES II [2]. Among many principal tasks of KARES II, we focus on the task of "bringing the cup to the user's mouth". During this task, we use the mouth region to infer the user's intention such as a will to drink or not. According to the extracted user's intention, the robot may approach to the user's face or get away from it. A fuzzy rule base is constructed using linguistic variables such as 'mouth openness', 'positive intention', and 'negative intention'. Here, the fuzzy observer is also effectively used for extracting user's intention during service [3].

5.3 Bio-Signal Recognition System [13]

The EMG control is well known from the operation of some prosthesis with small DOF. Its application to the user's high level of movement paralysis is limited because the useful signals often interfere with the EMG signals from another muscle groups. The soft computing technique allows effective extraction of informative signal features in cases of high interference between the useful EMG signals and another muscle EMG signals.

To read the user's movement intentions effectively, we have proposed the minimal feature set extraction algorithm [14] based on the fuzzy c-means algorithm (FCM), and RS. We can obtain the intervals of each feature by FCM to make condition rules, and then apply the rough set theory to extract a minimally sufficient set of rules for classification. After extracting numerous rules for classification and reduction done by RS, we can find the best feature set by measuring the separability of each feature in each rules. By use of fuzzy min-max neural network (FMMNN) as a pattern recognizer with the extracted minimal feature sets, we can classify the eight primitive arm motions with high classification rates [14].

5.4 Service Robot System with Emotion Monitoring Capability [7]

To help human mentally and emotionally, a service robot system is designed to understand the user's emotion and react depending on the monitored information.

In our works, we build an intelligent robot agent for emotion-invoking action and emotion monitoring to combine the user's emotion and emotional model of the agent. For emotion monitoring the robot is to observe the user's behavior pattern that may be caused by some changes in the surroundings or by some initial robot action, to understand the user's emotional condition and then act by ingratiating itself with the user. For learning, the robot agent gets a feedback from the user's response so that it can behave properly, depending on any situation for the user's sake. Most important problem is to establish a mapping concept from the user's behavior pattern to the user's emotional state and from the emotional state to the robot's action ingratiating the user. Since each mapping rule depends on the personality of the user, it is difficult to determine universal affective properties in the user's behavior pattern and robot's action. By the proposed NN structure, it is described how the robot would understand the user's emotional condition and how it shows its reaction, depending on the user's emotional state as a service robot [7].

6 Concluding Remark

Soft computing techniques can deal with many real-world problems effectively. Among many possible applications of soft computing techniques, human-machine interface or interaction procedures for the service robots are found to be very suitable because of its capability to deal with uncertainty and ambiguity. In this paper, we have proposed a novel scheme for emotion/intention reading based on various soft computing techniques. And four successful applications are given as examples based on the proposed scheme.

Acknowledgement

This work was supported by the Human-friendly Welfare Robot System Engineering Research Center, KAIST.

References

1. Y. Inagaki, et al., "Behavior-based intention inference for intelligent robots cooperating with human," Proc. of Int'l Conf. of the 4th FUZZ-IEEE, vol. 3, pp. 1695–1700, 1995
2. Z. Bien, and W.-K. Song, "Novel wheelchair-based robotic arm with visual servoing capability for human-robot interaction", Proc. of Workshop on Service Automation and Robotics, CIDAM, pp. 5-17, June 19-20, 2000.
3. D.-J. Kim, W.-K. Song, and Z. Bien, "Effective Intention Reading in Rehabilitation Robitcs", In the Proceeding of HWRS2001 Workshop, KAIST, Korea, Jan. 15-16, 2001.
4. Gyu-tae Park and Zeungnam Bien, "Neural network-based fuzzy observer with application to facial analysis" , Pattern Recognition Letters, vol. 21, pp. 93-105, February, 2000.
5. Yong-Min Kim, "Emotion"(in Korean), Han-Gil co., 1999.
6. Eun-Kyung Cho et al., "Cognitive Psychology", Hak-Ji co., 1999.
7. J.-H. Do, K.-H. Park, et al., "A Development of Emotional Interactive Robot," Proc. of 32nd ISR2001, pp.544-549, Seoul, Korea, 2001.
8. Jang. J.S., et al., Neuro-fuzzy and soft computing, Prentice-Hall, 1997.
9. Zadeh L.A., Soft computing and fuzzy logic, IEEE software, vol. 11, pp. 48–56, 1994.
10. Rafael C. Gonzalez and Richard E. Woods, Digital Image Processing, Addison-Wesley Publishing Co. 1993.
11. J.-S. Kim, W. Jang and Z. Bien, "A Dynamic Gesture Recognition System for the Korean Sign Language (KSL)", IEEE Trans. on Systems, Man, and Cybernetics, Vol. 26, No. 2, pp. 354-359, 1996.
12. J.-B. Kim, K.-H. Park, W.-C. Bang, J.-S. Kim and Z. Bien, "Continuous Korean Sign Language Recognition using Automata based Gesture Segmentation and Hidden Markov Model", ICCAS2001 submitted, 2001.
13. J.-S. Han, Z. Bien, et al., "New EMG Pattern Recognition based on Soft Computing Techniques and Its Application to Control of a Rehabilitation Robotic Arm," Proc. of 6th IIZUKA2000, pp.890-897, Iizuka, Japan, Oct. 1-4, 2000.
14. A. Mehrabian, "Basic dimensions for a general psychological theory: Implications for personality, social, environmental, and developmental studies," Oelgeschlager, Gunn & Hain, Cambridge, MA, 1980.
15. P. Ekman and W.V. Friesen, The Facial Action Coding System, Consulting Psychologist Press, Inc., San Francisco, CA, 1978.
16. F. Dellaert, T. Polzin and A. Waibel, "Recognition emotion in speech," Spoken Language, 1996.
17. C. Breazeal, "Robot in Society: Friend or Appliance?," Agents'99 workshop on emotion-based agent architectures, pp. 18-26, 1999.
18. J. Nicholson, K. Takahashi and R. Nakatsu, "Emotion recognition in speech using neural networks," Proceedings of ICONIP'99, vol. 2, pp. 495-501, 1999.
19. S. Haykin, Neural Networks, Prentice-Hall, 2nd edition, 1999.
20. D.B. Fogel, "What is Evolutionary Computation?", IEEE Spec. pp. 26-32, Feb., 2000.
21. Z. Pawlak, "Why Rough Sets?", Proc. Of the fifth IEEE Inter. conf. on. Fuzzy Systems, vol. 2, pp. 738-743, 1996.

Generalization of Rough Membership Function Based on α-Coverings of the Universe

Rolly Intan and Masao Mukaidono

Department of Computer Science,
Meiji University, Kawasaki-shi, Kanagawa-ken, Japan
{rolly,masao}@cs.meiji.ac.jp

Abstract. In 1982, Pawlak proposed the concept of *rough sets* with practical purpose of representing indiscernibility of elements. Even it is easy to analyze, the rough set theory built on a partition induced by equivalence relation may not provide a realistic view of relationships between elements in the real-world application. Here, coverings of, or non-equivalence relations on, the universe can be considered to represent a more realistic model instead of partition in which a generalized model of rough sets was proposed. In this paper, based on α-coverings of the universe, a generalized concept of rough membership functions is proposed and defined into three values, minimum, maximum and average. Their properties are examined.

1 Introduction

Rough set theory generalizes classical set theory by allowing an alternative to formulate sets with imprecise boundaries. A rough sets is basically an approximate representation of a given crisp set in terms of two subsets derived from a crisp partition defined on the universal set involved [5]. The two subsets are called a lower and an upper approximation. In a partition, an element belongs to one equivalence class and two distinct equivalence classes are disjoint. Rough membership functions of an element in the presence of a subset $A \subseteq U$ is defined as the following [7]:

$$\mu_A(x) = \frac{|[x]_R \cap A|}{|[x]_R|},$$

where $|.|$ denotes the cardinality of a set, and $[x]_R$ denotes the equivalence class in U/R that contains $x \in U$.

However, as pointed out in [8], even it is easy to analyze, the rough set theory built on a partition induced by equivalence relations may not provide a realistic view of relationships between elements in the real-world application. Instead a covering of the universe might be considered as an alternative to provide a more realistic model of rough sets. A covering of the universe, $C = \{C_1, ..., C_n\}$, is a family of subset of non-empty universe U such that $U = \bigcup\{C_i | i = 1, ..., n\}$. Two distinct sets in C may have a non-empty overlap. An arbitrary element x of U may belong to more than one set in C. The sets in C may describe *different types* or various degrees of similarity between elements of U.

N.R. Pal and M. Sugeno (Eds.): AFSS 2002, LNAI 2275, pp. 129–136, 2002.

In [1], *conditional probability relations* was proposed and regarded to be more realistic in representing relationships between two elements of data than equivalence relations. Naturally, relationship between elements of data in the real-world application is not necessarily symmetric as characterized by conditional probability relation. Moreover, conditional probability relations can be considered as a generalization of equivalence relations. In [2], a generalized concept of rough approximations was introduced based on α-coverings of the universe induced by conditional probability relation. Corresponding to the generalization of rough sets, objective of this paper is to extend and generalize the concept of rough membership function by α-coverings of the universe dealing with conditional probability relations in constructing the coverings. Moreover generalized rough membership functions are defined into three values, minimum, maximum and average. Their properties are examined.

2 Conditional Probability Relations

The concept of *conditional probability relations* was introduced in the context of fuzzy relational databases [1]. It may be considered as a concrete example of *weak fuzzy similarity relation*, which in turn is a special type of fuzzy binary relation.

Definition 1. *A* ***fuzzy similarity relation*** *is a mapping,* $s : U \times U \to [0, 1]$, *such that for* $x, y, z \in U$,

$$\begin{aligned}
&\text{(a) Reflexivity}: s(x, x) = 1,\\
&\text{(b) Symmetry}: s(x, y) = s(y, x),\\
&\text{(c) Max–min transitivity}: s(x, z) \geq \max_{y \in U} \min[s(x, y), s(y, z)].
\end{aligned}$$

Definition 2. *A* ***weak fuzzy similarity relation*** *is a mapping,* $s : U \times U \to [0, 1]$, *such that for* $x, y, z \in U$,

$$\begin{aligned}
&\text{(a) Reflexivity}: s(x, x) = 1,\\
&\text{(b) Conditional symmetry}: \text{if } s(x, y) > 0 \text{ then } s(y, x) > 0,\\
&\text{(c) Conditional transitivity}: \text{if } s(x, y) \geq s(y, x) > 0 \text{ and}\\
&\qquad s(y, z) \geq s(z, y) > 0 \text{ then } s(x, z) \geq s(z, x).
\end{aligned}$$

Definition 3. *A* ***conditional probability relation*** *is a mapping,* $R : U \times U \to [0, 1]$, *such that for* $x, y \in U$,

$$R(x, y) = \mathrm{P}(x \mid y) = \mathrm{P}(y \to x) = \frac{|x \cap y|}{|y|},$$

where $R(x, y)$ *means the degree* y *supports* x *or the degree* y *is similar to* x.

By definition, a fuzzy similarity relation is regarded as a special case (or type) of weak fuzzy similarity relation, and a conditional probability relation is an example of weak fuzzy similarity relations. The conditional probability relations may be used as a basis of representing degree of similarity relationships between elements in the universe U. In the definition of conditional probability relations, the probability values may be estimated based on the semantic relationships between elements by using the epistemological or subjective view of probability theory. When elements in U are represented by sets of features or attributes as in the case of binary information tables, we have a simple procedure for estimating the conditional probability relation as shown in Definition 3, where $|\cdot|$ denotes the cardinality of a set.

The notion of binary information tables can be easily generalized to fuzzy information tables by allowing a number in the unit interval $[0, 1]$ for each cell of the table. The number is the degree to which an element has a particular attribute. Each element is represented as a fuzzy set of attributes. The degree of similarity two elements can be calculated by a conditional probability relation on fuzzy sets [1,2]. In this case, $|x| = \sum_{a \in At} \mu_x(a)$, where μ_x is membership function of x over a set of attribute At, and intersection is defined by minimum.

Definition 4. *Let μ_x and μ_y be two fuzzy sets over a set of attribute At for two elements x and y of a universe of elements U. A fuzzy conditional probability relation is defined by:*

$$R(x, y) = \frac{\sum_{a \in At} \min\{\mu_x(a), \mu_y(a)\}}{\sum_{a \in At} \mu_y(a)}.$$

It can be easily verified that R satisfies properties of a weak fuzzy similarity relation. Additional properties of similarity as defined by conditional probability relations can be found in [1].

3 Generalized Rough Membership Functions

As pointed out in [9], there are at least two views which can be used to interpret the rough set theory, operator-oriented view and set-oriented view. The operator-oriented view was proposed in [2] providing the generalization of lower and upper approximation operators in the presence of α-coverings of the universe. In this section, we provide the set-oriented view based on the notion of rough membership functions. In this case, rough membership functions of an element will be expressed into three values, minimum, maximum and average depending on similarity classes that cover the element. First of all, from weak fuzzy similarity relations and conditional probability relations, coverings of the universe can be defined and interpreted. The standard concept of rough membership functions can thus be generalized based on coverings of universe.

Definition 5. *Let U be a non-empty universe, and R be a conditional probability relation on U. For any element $x \in U$, $R_s^\alpha(x)$ and $R_p^\alpha(x)$ are defined as the set* *of elements that support x and the set of elements that are supported by x, respectively, to a degree of at least $\alpha \in [0,1]$, as follows:*

$$R_s^\alpha(x) = \{y \in U \mid R(x,y) \geq \alpha\},$$
$$R_p^\alpha(x) = \{y \in U \mid R(y,x) \geq \alpha\}.$$

The set $R_s^\alpha(x)$ can also be interpreted as consisting of elements that are similar to x, while $R_p^\alpha(x)$ consisting of elements to which x is similar. By the reflexivity, it follows that we can construct two coverings of the universe, $\{R_s^\alpha(x) \mid x \in U\}$ and $\{R_p^\alpha(x) \mid x \in U\}$. By extending standard rough membership function, we obtain three values of generalized rough membership function.

Definition 6. *Let $A \subseteq U$ be a crisp set, where U is a non-empty universe, and let R be a conditional probability relation on U. $R_s^\alpha(x) = \{y \in U|R(x,y) \geq \alpha\}$ denotes similarity class of x with α-cut, where $\alpha \in [0,1]$. $\mu_A^m(y)^\alpha, \mu_A^M(y)^\alpha$ and $\mu_A^*(y)^\alpha$ are defined as minimum, maximum and average rough membership functions of y with α-cut in the presence of set A, respectively as follows:*

$$\mu_A^m(y)^\alpha = \min\left\{\frac{|R_s^\alpha(x) \cap A|}{|R_s^\alpha(x)|}|x \in U, y \in R_s^\alpha(x)\right\},$$
$$\mu_A^M(y)^\alpha = \max\left\{\frac{|R_s^\alpha(x) \cap A|}{|R_s^\alpha(x)|}|x \in U, y \in R_s^\alpha(x)\right\},$$
$$\mu_A^*(y)^\alpha = \text{avg}\left\{\frac{|R_s^\alpha(x) \cap A|}{|R_s^\alpha(x)|}|x \in U, y \in R_s^\alpha(x)\right\}.$$

The above definition generalizes the concept of rough membership functions [7] and concretizes definition of generalized rough membership functions based on a covering proposed in [8]. In this case, the minimum, the maximum and the average equations may be assumed as the most pessimistic, the most optimistic and the balanced view in defining rough membership functions. The minimum rough membership function of y is determined by a set, $R_s^\alpha(x)$ to which y belongs, which has the smallest overlap with A compared to the cardinality of the set relatively. On the other hand the maximum rough membership function is determined by a set, $R_s^\alpha(x)$ to which y belongs, which has the largest overlap with A compared to the cardinality of the set relatively. The average rough membership function depends on the average of every set, $R_s^\alpha(x)$ to which y belongs. The relationships of the three rough membership functions can be expressed by:

$$\mu_A^m(y)^\alpha \leq \mu_A^*(y)^\alpha \leq \mu_A^M(y)^\alpha.$$

Moreover, the three rough membership functions may take varied values in calculation depending on the value of α. The minimum, maximum and average rough membership functions satisfy some properties as follows: for $A, B \subseteq U$ are crisp sets, where U is a non-empty universe,

(gr0) $\mu_U^m(x)^\alpha = \mu_U^*(x)^\alpha = \mu_U^M(x)^\alpha = 1,$

(gr1) $\mu_\emptyset^m(x)^\alpha = \mu_\emptyset^*(x)^\alpha = \mu_\emptyset^M(x)^\alpha = 0,$

(gr2) $[\forall R_s^\alpha(x), y \in R_s^\alpha(x) \Leftrightarrow z \in R_s^\alpha(x)] \Rightarrow [\mu_A^m(y)^\alpha = \mu_A^m(z)^\alpha,$
$\mu_A^*(y)^\alpha = \mu_A^*(z)^\alpha, \mu_A^M(y)^\alpha = \mu_A^M(z)^\alpha],$
(gr3) $y, z \in R_s^\alpha(x) \Rightarrow [\mu_A^m(y)^\alpha \neq 0 \Rightarrow \mu_A^m(z)^\alpha \neq 0, \mu_A^m(y)^\alpha = 1 \Rightarrow \mu_A^m(z)^\alpha = 1],$
(gr4) $y \in A \Rightarrow \mu_A^m(y)^\alpha > 0,$
(gr5) $\mu_A^M(y)^\alpha = 1 \Rightarrow y \in A,$
(gr6) $A \subseteq B \Rightarrow [\mu_A^m(y)^\alpha \leq \mu_B^m(y)^\alpha, \mu_A^*(y)^\alpha \leq \mu_B^*(y)^\alpha, \mu_A^M(y)^\alpha \leq \mu_B^M(y)^\alpha],$
(gr7) $A \neq \emptyset \Rightarrow \mu_A^m(x)^0 = \mu_A^*(x)^0 = \mu_A^M(x)^0 = \dfrac{|A|}{|U|} = \mathrm{P}(A).$

(gr0) and (gr1) show the boundaries condition of set, U and $\emptyset$. (gr2) and (gr3) indicate that two similar elements in coverings should have similar rough membership functions. (gr4) and (gr5) can be used to prove the real member of a given crisp set. (gr6) shows that the consistency of inclusive sets should have the same characteristics in comparison between their rough membership functions. If α is equal to 0, all rough membership functions of all elements are equal to the probability of a given non-empty crisp set in the universe as shown in (gr7).

Related to set-theoretic operators, $\neg$, $\cap$, and $\cup$, the rough membership functions satisfy some properties such as:

(g0) $\mu_{\neg A}^m(x)^\alpha = 1 - \mu_A^M(x)^\alpha,$
(g1) $\mu_{\neg A}^M(x)^\alpha = 1 - \mu_A^m(x)^\alpha,$
(g2) $\mu_{\neg A}^*(x)^\alpha = 1 - \mu_A^*(x)^\alpha,$
(g3) $\max(0, \mu_A^m(x)^\alpha + \mu_B^m(x)^\alpha - \mu_{A\cup B}^M(x)^\alpha)$
$\leq \mu_{A\cap B}^m(x)^\alpha \leq \min(\mu_A^m(x)^\alpha, \mu_B^m(x)^\alpha),$
(g4) $\max(\mu_A^M(x)^\alpha, \mu_B^M(x)^\alpha) \leq \mu_{A\cup B}^M(x)^\alpha$
$\leq \min(1, \mu_A^M(x)^\alpha + \mu_B^M(x)^\alpha - \mu_{A\cap B}^m(x)^\alpha),$
(g5) $\mu_{A\cup B}^*(x)^\alpha = \mu_A^*(x)^\alpha + \mu_B^*(x)^\alpha - \mu_{A\cap B}^*(x)^\alpha.$

By definition, generalized rough membership function provides four regions of $A \subseteq U$ as defined as follows:

1. Very positive region of A: $vpos(A) = \{x \in U | \mu_A^m(x) = 1\}$,
2. Positive region of A: $pos(A) = \{x \in U | \mu_A^m(x) > 0\}$,
3. Ambiguous region of A: $bnd(A) = \{x \in U | \mu_A^m(x) = 0, \mu_A^M(x) > 0\}$,
4. Negative region of A: $neg(A) = \{x \in U | \mu_A^M(x) = 0\}$.

It is necessary to denote some properties such that $vpos(A) \subseteq pos(A)$, $x \in A \Longrightarrow x \in pos(A)$, $x \in bnd(A) \Longrightarrow x \notin A$, $pos(A) \cap bnd(A) \cap neg(A) = \emptyset$ and $pos(A) \cup bnd(A) \cup neg(A) = U$. Also, one can defines $pos(A) - vpos(A)$ as a boundary of positive region or gives a special attention to the region in which $\mu_A^M(x) = 1$ as a part of positive region.

Covering of the universe in Definition 5 as a generalization of disjoint partition is considered as a crisp covering. Both crisp covering and disjoint partition are regarded as *crisp granularity*. However, when we consider a given set A be a fuzzy set on U instead of a crisp set and covering of the universe be a fuzzy covering instead of a crisp covering, we need to define a more generalized rough membership function of Definition 6. Here, fuzzy covering generalizes crisp covering. In this case, crisp covering can be constructed by applying α-level set of fuzzy covering. Fuzzy covering might be consider as a case of *fuzzy granularity* in which similarity classes as a basis of constructing the covering are regarded as fuzzy sets and defined as follows.

Definition 7. *Let U be a non-empty universe, and R be a (fuzzy) conditional probability relation on U. For any element $x \in U$, $R_s(x)$ and $R_p(x)$ are regarded as fuzzy sets and defined as the set that supports x and the set supported by x, respectively by:*

$$\mu_{R_s(x)}(y) = R(x,y), \quad y \in U,$$
$$\mu_{R_p(x)}(y) = R(y,x), \quad y \in U,$$

where $\mu_{R_s(x)}(y)$ and $\mu_{R_p(x)}(y)$ are grades of membership of y in $R_s(x)$ and $R_p(x)$, respectively.

Similarly, a more generalized rough membership function is also defined into three values, minimum, maximum and average.

Definition 8. *Let A be a fuzzy set on U, where U is a non-empty universe, and let R be a (fuzzy) conditional probability relation on U. $\mu_A^m(y), \mu_A^M(y)$ and $\mu_A^*(y)$ are defined as minimum, maximum and average rough membership functions of y in the presence of fuzzy set A, respectively as follows:*

$$\mu_A^m(y) = \min\left\{\frac{|R_s(x) \cap A|}{|R_s(x)|} | x \in U, \mu_{R_s(x)}(y) > 0\right\},$$
$$\mu_A^M(y) = \max\left\{\frac{|R_s(x) \cap A|}{|R_s(x)|} | x \in U, \mu_{R_s(x)}(y) > 0\right\},$$
$$\mu_A^*(y) = \text{avg}\left\{\frac{|R_s(x) \cap A|}{|R_s(x)|} | x \in U, \mu_{R_s(x)}(y) > 0\right\},$$

where $|R_s(x) \cap A| = \sum_{z \in U} \min\{\mu_{R_s(x)}(z), \mu_A(z)\}$ *and* $|R_s(x)| = \sum_{z \in U} \mu_{R_s(x)}(z)$.

Intersection is defined by minimum function in order to obtain property of reflexivity, although there are some operations of t-norm that might be used.

4 Illustrative Example

Let us illustrate the above concepts by using binary information table as shown in Table 1. Given $X = \{O_2, O_4, O_7, O_8\}$, where $X \subset U$, and $\alpha = 0.75$. Thus we just consider to similarity classes in which minimum degree of similarity concerning relationships between elements is equal to 0.75. By Definition 3, 4 and 5, we construct similarity classes of all elements in U by:

Table 1. Binary Information Table

Element	a_1	a_2	a_3	a_4	a_5	a_6	a_7	a_8
O_1	0	0	1	0	1	0	0	0
O_2	1	1	0	1	0	0	1	0
O_3	0	0	1	1	0	0	1	1
O_4	0	1	0	1	0	1	0	1
O_5	1	0	1	1	0	0	1	0

Element	a_1	a_2	a_3	a_4	a_5	a_6	a_7	a_8
O_6	0	0	1	0	1	0	1	0
O_7	0	1	1	0	0	0	1	0
O_8	1	1	0	0	0	0	1	1
O_9	0	1	0	1	1	0	1	0
O_{10}	0	1	0	0	0	1	1	0

Note: $O_1 = \{a_3, a_5\}$, $O_2 = \{a_1, a_2, a_4, a_7\}$, $O_3 = \{a_3, a_4, a_7, a_8\}$, atc.

$$
\begin{array}{ll}
R_s^{0.75}(O_1) = \{O_1\}, & R_s^{0.75}(O_6) = \{O_1, O_6\}, \\
R_s^{0.75}(O_2) = \{O_2, O_5, O_8, O_9\}, & R_s^{0.75}(O_7) = \{O_7\}, \\
R_s^{0.75}(O_3) = \{O_3, O_5\}, & R_s^{0.75}(O_8) = \{O_2, O_8\}, \\
R_s^{0.75}(O_4) = \{O_4\}, & R_s^{0.75}(O_9) = \{O_2, O_9\}, \\
R_s^{0.75}(O_5) = \{O_2, O_3, O_5\}, & R_s^{0.75}(O_{10}) = \{O_{10}\},
\end{array}
$$

By Definition 6, we calculate minimum, maximum, and average rough membership functions of element O_3, for instance. In this case, there are two similarity classes to which O_3 belongs. They are $R_s^{0.75}(O_3)$ and $R_s^{0.75}(O_5)$. The results are given by: $\mu_X^m(O_3)^{0.75} = 0$, $\mu_X^M(O_3)^{0.75} = 1/3$, $\mu_X^*(O_3)^{0.75} = 1/6$.
By rough membership function of all elements, four regions of X are given as:

$$
\begin{array}{ll}
vpos(X) = \{O_4, O_7\}, & bnd(X) = \{O_3, O_5\}, \\
pos(X) = \{O_2, O_4, O_7, O_8, O_9\}, & neg(X) = \{O_1, O_6, O_{10}\}.
\end{array}
$$

5 Conclusions

In this paper, we introduce the notion of weak fuzzy similarity relations. Two examples of such relations, conditional probability relations and fuzzy conditional probability relations, are suggested for the construction and interpreting coverings of the universe. Based on such covering, rough membership functions were generalized and defined into three values, minimum, maximum and average. Their properties are examined.

References

1. Intan, R., Mukaidono, M., 'Conditional Probability Relations in Fuzzy Relational Database ', *Proceedings of RSCTC'00*, (2000), pp.213-222.
2. Intan, R., Mukaidono, M.,Yao, Y.Y.,'Generalization of Rough Sets with α-coverings of the Universe Induced by Conditional Probability Relations', *Proceedings of International Workshop on Rough Sets and Granular Computing*, (2001), pp.173-176.
3. Intan, R., Mukaidono, M., 'Fuzzy Functional Dependency and Its Application to Approximate Querying', *Proceedings of IDEAS'00*, (2000), pp.47-54.
4. Komorowski, J., Pawlak, Z., Polkowski, L., Skowron, A., 'Rough Sets: A Tutorial', (1999).
5. Klir, G.J., Yuan, B., *Fuzzy Sets and Fuzzy Logic: Theory and Applications*,(Prentice Hall, New Jersey, 1995).

A Class of Quantitative-Qualitative Measures of Directed-Divergence

H.C. Taneja

Department Of Applied Sciences,
C.R.State College Of Engineering,
Murthal [Sonepat] – 131039, India

Abstract. A class of quantitative-qualitative measure of directed-divergence which do not require the condition of absolute continuity of the probability distribution envolved has been introduced and studied.

1 Introduction

Let $X = (x_1,x_2,\ldots,x_n)$ be a discrete random variable and $P = (p_1,p_2,\ldots,p_n)$, $0 \le p_i \le 1$, $\Sigma p_i = 1$ and $Q = (q_1,q_2,\ldots,q_n)$, $0 \le q_i \le 1$, $\Sigma q_i = 1$ be two probability distributions associated with X, and let $U = (u_1,u_2,\ldots,u_n)$, $u_i \ge 0$ be the utility distribution, where u_i is the utility of the outcome x_i of X for an observer with respect to some specified goal. The utility u_i of the outcome x_i is independent of the probabilities p_i or q_i and depends only on the qualitative characteristics of the physical system taken into account [6].

A quantitative-qualitative measure of relative information (or, directed-divergence) as suggested by Taneja and Tuteja [8] is given by

$$I(P/Q;U) = \sum_{i=1}^{n} u_i p_i \log(p_i / q_i), u_i \ge 0, 0 \le p_i, q_i \le 1 \tag{1.1}$$

The logarithm base 2 is used throughout this correspondence unless otherwise stated.

When the utilities are ignored, that is $u_i = 1$ for each i, clearly (1.1) reduces to the measure

$$I(P/Q) = \sum_{i=1}^{n} p_i \log(p_i / q_i), 0 \le p_i, q_i \le 1 \tag{1.2}$$

the Kullback's measure of directed-divergence [4].

The detailed properties and different characterizations and generalizations of the measure (1.1) have been studied by various authors [1,7,8].

It can be seen that the measure (1.1) is not symmetric in its probability distributions P and Q. To obtain a symmetric measure we define

$$J(P/Q;U) = I(P/Q;U) + I(Q/P;U)$$

N.R. Pal and M. Sugeno (Eds.): AFSS 2002, LNAI 2275, pp. 136–140, 2002.

$$= \sum_{i=1}^{n} u_i (p_i - q_i) \log(p_i / q_i) \tag{1.3}$$

as the quantitative-qualitative J-divergence measure.

When the utilities are ignored, (1.3) reduces to

$$J(P/Q) = \sum_{i=1}^{n} (p_i - q_i) \log(p_i / q_i) \tag{1.4}$$

the J-divergence defined by Jeffreys [3].

Clearly (1.1) and (1.3) have many properties in common.

It should be noted that the measure *I(P/Q ; U)* is undefined if, $q_i = 0$ and $p_i \geq 0$ for any $x_i \in X$. This means that the distribution P has to be absolutely continuous with respect to the distribution *Q* for *I(P/Q ; U)* to be defined. Similarly *J(P/Q ; U)* requires that *P* and *Q* be absolutely continuous with respect to each other. This is one of the problems with these divergence measures. To overcome this difficulty, in this communication we introduce a class of new quantitative-qualitative measures of directed divergence which do not require the condition of absolute continuity of the probability distributions involved.

2 Quantitative-Qualitative Measures of Directed-Divergence

Let $$P = (p_1, p_2, \ldots, p_n), 0 \leq p_i \leq 1, \sum p_i = 1;$$

$$Q = (q_1, q_2, \ldots, q_n), 0 \leq q_i \leq 1, \sum q_i = 1;$$

and

$$U = (u_1, u_2, \ldots, u_n), u_i \geq 0$$

be respectively the probability distributions and utility distribution associated with the random variable $X = (x_1, x_2, \ldots, x_n)$. We define

$$K(P/Q;U) = \sum_{i=1}^{n} u_i p_i \log(p_i /(p_i/2 + q_i/2)) \tag{2.1}$$

as a new quantitative-qualitative measure of directed-divergence.

When the utilities are ignored the measure (2.1) reduces to

$$K(P/Q) = \sum_{i=1}^{n} p_i \log(p_i /(p_i/2 + q_i/2)) \tag{2.2}$$

a measure of directed-divergence defined by J.Lin [5].

The measure *K(P/Q;U)* is well defined and is independent of the value of p_i, q_i and u_i for any i, in the sense that here q_i may be equal to zero for some specific i, independent of p_i. Also this measure is closely related to *I(P/Q;U)*, the measure (1.1) and it can be seen very easily that

$$K(P/Q;U) = I\{P/(P+Q)/2;U\} \tag{2.3}$$

where $P+Q = \{p_1+q_1, p_2+q_2, \ldots, p_n+q_n\}$.

Next we consider the following result:

Theorem 2.1
The measure *K(P/Q;U)* is bounded by the measure *I(P/Q;U)*, that is,

$$K(P/Q;U) \le (1/2)I(P/Q;U) \tag{2.4}$$

Proof : For p, q $\ge$ 0, we have

$$(p+q)/2 \ge \sqrt{pq}$$

Thus
$$K(P/Q;U) = \sum_{i=1}^{n} u_i p_i \log(p_i/(p_i/2+q_i/2))$$

$$\le \sum_{i=1}^{n} u_i p_i \log(p_i/\sqrt{p_i q_i})$$

$$= (1/2)\sum_{i=1}^{n} u_i p_i \log(p_i/q_i)$$

$$= (1/2)I(P/Q;U)$$

This completes the proof.

The measure *K(P/Q;U)* is not a symmetric measure. We define a symmetric measure based on *K(P/Q;U)* as

$$L(P/Q;U) = K(P/Q;U) + K(Q/P;U)$$

$$= \sum_{i=1}^{n} u_i p_i \log(p_i/(p_i/2+q_i/2)) + \sum_{i=1}^{n} u_i q_i \log(q_i/(p_i/2+q_i/2))$$

$$= \sum_{i=1}^{n} (u_i p_i + u_i q_i)\{p_i/(p_i+q_i)\log(2p_i/(p_i+q_i)) + q_i/(p_i+q_i)\log(2q_i/(p_i+q_i))\}$$

$$= \sum_{i=1}^{n} (u_i p_i + u_i q_i)\{1 - H(p_i/(p_i+q_i), q_i/(p_i+q_i))\} \tag{2.5}$$

We can see very easily that *L(P/Q;U)* satisfies the following

$$L(P/Q;U) \le (1/2)J(P/Q;U) \tag{2.6}$$

a relation equivalent to (2.4)

Also we know that for any $0 \le$ p, q $\le$ 1, p+q = 1,
H(p.q) $\ge$ 2 min (p,q), ref. [2,p,521] and also

min(p,q) = (1/2) {1- |p-q| } ; thus

$$1 - H(p,q) \le |p-q| \tag{2.7}$$

Using (2.7) in (2.5), we have

$$L(P/Q;U) \le \sum_{i=1}^{n} u_i |p_i - q_i|$$
$$= V(P/Q;U), \text{ say };$$

where we define *V(P/Q;U)* as the quantitative-qualitative variational distance between the two probability distributions $P = (p_1, p_2, \ldots, p_n)$ and $Q = (q_1, q_2, \ldots, q_n)$ associated with the random variable $X=(x_1, x_2, \ldots, x_n)$ when $U = (u_1, u_2, \ldots, u_n)$ is the associated utility distribution.

Thus we have the following results:

Theorem 2.2

The measure *L(P/Q;U)* is bounded by the measure *V(P/Q;U)*, that is,

$$L(P/Q;U) \le V(P/Q;U) \tag{2.8}$$

When the utilities are ignored, (2.8) reduces to

$$L(P/Q) \le V(P/Q) \tag{2.9}$$

Where

$$L(P/Q) = \sum_{i=1}^{n} (p_i + q_i)\{1 - H(p_i/(p_i+q_i), q_i/(p_i+q_i))\}$$

and $$V(P/Q) = \sum_{i=1}^{n} |p_i - q_i|$$

a result derived by J.Lin [5].

Further, since for $0 \le p, q \le 1$, we have $0 \le H(p,q) \le 1$; thus from (2.5), we have

$$L(P/Q;U) \le \sum_{i=1}^{n} u_i (p_i + q_i) \tag{2.10}$$

which, when the utilities are ignored, reduces to

$$L(P/Q) \le 2$$

a bound derived by J. Lin [5].

An interpretation of *L(P/Q;U)* in terms of 'useful' entropy may be given as follow: We have

$$L(P/Q;U) = \sum_{i=1}^{n} u_i p_i \log(2p_i/(p_i+q_i)) + \sum_{i=1}^{n} u_i q_i \log(2q_i/(p_i+q_i))$$

$$= -2\sum_{i=1}^{n} u_i((p_i+q_i)/2)\log((p_i+q_i)/2) + \sum_{i=1}^{n} u_i p_i \log p_i + \sum_{i=1}^{n} u_i q_i \log q_i$$

$$= 2H((P+Q)/2;U) - H(P;U) - H(Q;U)$$

where *H(P;U)* is the Longo's 'useful' entropy function [6]. This provides one possible physical interpretation of *L(P/Q;U)*, the quantitative-qualitative L-divergence measure. It may be interesting to further investigate the different properties, characterizations and generalizations of the measures introduced.

References

1. Bhatia P.K. (1991), 'Symmetry and quantitative-qualitative measures and their applications to coding', Ph.D. Thesis submitted to M D University, Rohtak.
2. Gallager R.G. (1968), 'Information theory and reliable communication', Wiley, New York.
3. Jeffreys H. (1946), 'An invariant form for the prior probability in estimation problems', In: Proc. Roy. Soc. Lon. Ser A, pp 453-461.
4. Kullback S. (1959), 'Information theory and statistics', Wiley, New York.
5. Lin J. (1991), 'Divergence Measures based on the Shannon entropy', IEEE Tran. Inf. Th. 37, 145-151.
6. Longo G. (1972), 'Quantitative-Qualitative Measures of Information', Springer, New York.
7. Taneja H.C. (1984), 'On the Quantitative-Qualitative measure of relative information', Information Sciences, 33: 223-227.
8. Taneja H.C. (1984), 'Characterization of quantitative-qualitative measure of relative information', Information Sciences, 33: 217-222.

(α, β) Reduction of Decision Table: A Rough Approach*

Shrabonti Ghosh and S.S. Alam

Department of Mathematics,
Indian Institute of Technology,
Kharagpur - 721302, India
alam@maths.iitkgp.ernet.in

Abstract. In this paper we generalise Pawlak's rough approach for simplifying a decision table in an information system. We consider an information system where attribute values are not always quantitative, rather subjective having vague or imprecise meanings. Some objects may have attribute values which are almost identical i.e., they can't be distinguished clearly by the attributes. Considering this observation we present a generalised method for reduction of decision table for different choice values of α and β, α being for condition attributes and β for decision attributes where $\alpha, \beta \in [\,0, 1\,]$. For $\alpha = 1$ and $\beta = 1$, the method reduces to Pawlak's method.

1 Introduction and Preliminaries

The issue of knowledge representation is of primary importance in current research in AI. Intuitively, knowledge can be perceived as a body of information about some parts of reality, which constitute our domain of interest. The knowledge representation system can be perceived as a data table, columns of which are labelled by attributes, rows are labelled by objects and each row represents a piece of information about the corresponding object. The data table can be obtained as a result of measurements, observations or it represents knowledge of an agent or a group of agents. In this paper we will consider a special and important class of KR-system, called decision tables, which play an important role in many applications. A decision table [1],[3],[4] and [6] is a kind of prescription which specifies what decisions should be undertaken when some conditions are satisfied. Most decision problems can be formulated employing decision table formalism; therefore, this tool is particularly useful in decision making. In the present paper we consider Pawlak's method [4],[6] for reducing a decision table in such a manner that the same decisions can be there with a smaller number of conditions. This kind of simplification eliminates the need for checking the unnecessary conditions, or in some applications, for performing expensive tests to arrive at a conclusion which eventually could be acheived by simpler means. We

* The authors are thankful to the referees for their valuable comments and suggestions for correction to the first version of this paper.

N.R. Pal and M. Sugeno (Eds.): AFSS 2002, LNAI 2275, pp. 141–147, 2002.

generalise Pawlak's method [2],[4] with the concept of "almost indiscernibility relation" [2] instead of "indiscernibility relation", using two choice values $\alpha, \beta \in [\,0, 1\,]$, α being for condition attributes and β for decision attributes. The following definitions and preliminaries are required in the sequel of our work and hence presented in brief.

Definition 1. *Let μ be a fuzzy set of X. The α-cut of μ is the crisp set μ_α defined by*

$$\mu_\alpha = \{x \in X: \mu(x) \geq \alpha\}.$$

Definition 2. *Let a be an attribute and D_a be its domain. A fuzzy relation S_a on D_a (i.e., a fuzzy set of $D_a \times D_a$) is said to be a fuzzy proximity relation if (i) $S_a(a_1, a_1) = 1$ and (ii) $S_a(a_1, a_2) = S_a(a_2, a_1)$, where $a_1, a_2 \in D_a$.*

Definition 3. *[2] Let S_a be a fuzzy proximity relation on D_a, the domain of the attribute a. Two elements $a_1, a_2 \in D_a$ are said to be α-identical if either (i) $(a_1, a_2) \in \alpha$-cut of S_a (denoted by $a_1 c_\alpha a_2$) or (ii) there exist a chain $z_1, z_2, z_3, \ldots, z_n$ in D_a such that $a_1 c_\alpha z_1 c_\alpha \ldots c_\alpha z_n c_\alpha a_2$. If a_1 and a_2 are α-identical, with respect to S_a, we write this relation using notation $a_1 I_\alpha a_2$.*

Definition 4. *[4] An Information System or a KR-system is a pair S = (U, A) where U - is a nonempty, finite set called the universe, A - is a finite set of attributes, $V = \bigcup_{a \in A} V_a$, where V_a is called the domain of attribute a, $f: U \times A \rightarrow V$ is an information function such that $f(x, a) \in V_a$ for every $a \in A$ and $x \in U$.*
Let $x \in U$. The function $f_x: A \rightarrow V$, such that $f_x(a) = f(x, a)$ for every $a \in A$ will be called information on x in S.

Definition 5. *[2] Let S = (U, A) be an information system. Let $P \subseteq A$. Consider the fuzzy proximity relations S_a, $\forall\, a \in P$. For a chosen $\alpha \in [0, 1]$ we denote by $P(\alpha)$ a binary relation over U defined by $x, y \in P(\alpha)$ iff $f_x(a)\, I_\alpha\, f_y(a)$, $\forall\, a \in P$. It can be proved that the binary relation $P(\alpha)$ is an equivalence relation over U. Also, we notice that $P(\alpha)$ is not exactly the indiscernibility relation IND(P) defined by Pawlak; rather it can be viewed as an "almost indiscernibility relation" over U. Here also $[x]_P$ denotes an equivalence class of the relation $P(\alpha)$ containing an object x, i.e., $[x]_P$ is an abbreviation of $[x]_{P(\alpha)}$. For $\alpha = 1$, $P(\alpha)$ reduces to the indiscernibility relation IND(P). In this sense, $P(\alpha)$ generalizes IND (P).*

Definition 6. *[2] Let S = (U, A) be an information system. Consider a level value $\alpha \in [0, 1]$. The subset of attributes P is α-independent, if for every $Q \subset P$, $P(\alpha) \neq Q(\alpha)$. Otherwise subset P is α-dependent. A subset $P \subseteq Q \subseteq A$ is a α-reduct of Q, if P is α-independent subset of Q and $P(\alpha)=Q(\alpha)$.*

Definition 7. *[2] Let $S = (U, A)$ be an information system. Let $P \subseteq A$ and $a \in P$. Consider a level value $\alpha \in [0, 1]$. An attribute a is α-superflous in P if $Q(\alpha) = P(\alpha)$ where $Q = P - \{a\}$. Otherwise the attribute a is α-indispensible in P.*

Definition 8. *[4] Decision tables can be defined in terms of KR-systems as follows.Let $K = (U, A)$ be a knowledge representation system and let $C, D \subseteq A$ be two subsets of attributes, called condition and decision attributes, respectively. With every $x \in U$ we associate a function d_x: $A \rightarrow V$, such that $d_x(a) = a(x)$, for every $a \in C \bigcup D$; the function d_x will be called a decision rule. We denote this decision table by $T = (U, C, D)$. Pawlak simplified the decision tables of an information system based on the indiscernibility relation. In the next section we analyse the same type of problems based on "almost indiscernibility relation" $P(\alpha)$ for different values of $\alpha \in [0, 1]$.*

2 (α, β) Reduction of a Decision Table

In this section we present our method of reduction of a decision table.

Definition 9. *Let C be an α-reduct of the set P of all condition attributes in a decision table $T = (U, P, Q)$ where $\alpha \in [\ 0,\ 1\ ]$. Then the decision table $T_\alpha = (U, C, Q)$ is called an α-table of T.*

2.1 Reduction of Decision Table

We now present an algorithm for simplification of a decision table: -
Algorithm: -

1. Choose properly a set of fuzzy proximity relations S_a, S_b, S_c,.......for each attributes a, b, c, ...
2. Choose a value $\alpha \in [\ 0, 1\ ]$.
3. Compute the α-reducts of condition attributes which is equivalent to elimination of some columns from the decision table, and write the α-table corresponding to one α-reduct.
4. Choose a value of $\beta \in [\ 0, 1\ ]$.
5. For every decision rule, compute the value reducts corresponding to β.
6. Eliminate duplicate rows.
7. Eliminate superfluous values of attributes, and obtain the final table known as the minimal (α, β) solution.

We will study the algorithm by an example below:

Let us consider the following decision table (which is attitude for choosing an aircraft):

Table 1.

U	a	b	c	d	e	f
1	Average	Stable	Long	Low	Very Slow	Negative (immediate response)
2	Not Good	Unstable	Medium	High	Slow	Neutral
3	Very Good	Medium	Short	Very High	Very Quick	Positive (immediate response)
4	Average	Stable	Medium	Average	Very Slow	Negative (after little thought)
5	Very Good	Unstable	Short	Low	Very Slow	Neutral
6	Good	Medium	Long	High	Quick	Positive (after little thought)
7	Very Good	Medium	Long	Low	Quick	Positive (after little thought)
8	Good	Unstable	Short	High	Very Slow	Negative (after little thought)

where a, b, c, d, e are condition attributes and f is decision attribute characterised by a - speed, b - stability limit, c - how long it can fly, d - height, e - how quick can it turn, f - attitude. For the sake of simplicity we denote the attitude "Negative (immediate response)" by N_1, the attitude "Negative (after little thought)" by N_2, "Neutral" by N, "Positive (immediate response)" by P_1 and "Positive (after little thought)" by P_2.
Now consider the following fuzzy proximity relations (Table 2 to Table 7) of the attributes a,b,c,d,e and f, respectively as:

Fuzzy Proximity Relations S_a

	Not Good	Average	Very Good	Good
Not Good	1.0	0.3	0.0	O.1
Average	0.3	1.0	0.4	0.8
Very Good	0.0	0.4	1.0	0.5
Good	0.1	0.8	0.5	1.0

Fuzzy Proximity Relations S_b

	Stable	Medium	Unstable
Stable	1.0	0.4	0.1
Medium	0.4	1.0	0.7
Unstable	0.1	0.7	1.0

Fuzzy Proximity Relations S_c

	Long	Medium	Short
Long	1.0	0.8	0.4
Medium	0.8	1.0	0.5
Short	0.4	0.5	1.0

Fuzzy Proximity Relations S_d

	High	Low	Average	Very High
High	1.0	0.3	0.5	0.7
Low	0.3	1.0	0.8	0.1
Average	0.5	0.8	1.0	0.2
Very High	0.7	0.1	0.2	1.0

Fuzzy Proximity Relations S_e

	Slow	Very Slow	Quick	Very Quick
Slow	1.0	0.4	0.3	0.1
Very Slow	0.4	1.0	0.2	0.0
Quick	0.3	0.2	1.0	0.7
Very Quick	0.1	0.0	0.7	1.0

Fuzzy Proximity Relations S_f

	N_1	N_2	P_1	P_2	N
N_1	1.0	0.7	0.0	0.0	0.4
N_2	0.7	1.0	0.0	0.0	0.5
P_1	0.0	0.0	1.0	0.8	0.4
P_2	0.0	0.0	0.8	1.0	0.5
N	0.4	0.5	0.4	0.5	1.0

For $\alpha = 0.8$ we compute the following:

$$\{a,b,c\}^* = \{\{1,4\},\{2\},\{3\},\{5\},\{6\},\{7\},\{8\}\}$$
$$\{a,b,d\}^* = \{\{1,4\},\{2\},\{3\},\{5\},\{6\},\{7\},\{8\}\}$$
$$\{a,b,e\}^* = \{\{1,4\},\{2\},\{3\},\{5\},\{6\},\{7\},\{8\}\}$$
$$\{a,c,d\}^* = \{\{1,4\},\{2\},\{3\},\{5\},\{6\},\{7\},\{8\}\}$$
$$\{a,c,e\}^* = \{\{1,4\},\{2\},\{3\},\{5\},\{6\},\{7\},\{8\}\}$$
$$\{a,d,e\}^* = \{\{1,4\},\{2\},\{3\},\{5\},\{6\},\{7\},\{8\}\}$$
$$\{b,c,d\}^* = \{\{1,4\},\{2\},\{3\},\{5\},\{6\},\{7\},\{8\}\}$$
$$\{c,d,e\}^* = \{\{1,4\},\{2,6\},\{3\},\{5\},\{7\},\{8\}\}$$
$$\{a,b,c,d,e\}^* = \{\{1,4\},\{2\},\{3\},\{5\},\{6\},\{7\},\{8\}\}$$

We observe that each of the sets of condition attributes {d, e}, {c, e}, {c, d}, {b, e}, {b, d}, {b, c}, {a, e}, {a, b} are α-superfluous or these attributes are α-dispensible attributes. Consequently we can remove the columns d, e or c, e or c, d or b, e or b, d or b, c or a, e or a, b from the table. Removing columns c, d from the table, we get the reduced table as

Table 2. α-table for $\alpha = 0.8$

U	a	b	e	f
1	Average	Stable	Very Slow	N_1
2	Not Good	Unstable	Slow	N
3	Very Good	Medium	Very Quick	P_1
4	Average	Stable	Very Slow	N_2
5	Very Good	Unstable	Very Slow	N
6	Good	Medium	Quick	P_2
7	Very Good	Medium	Quick	P_2
8	Good	Unstable	Very Slow	N_2

Now we choose a value of β, say $\beta = 0.7$. In the next step, we compute the value reducts for every decision rule. As an example, let us compute value reducts for the first decision rule of the decision table.
In order to compute α-reducts of the family F = $\{[1]_a, [1]_b, [1]_e\}$, i.e., {{ 1, 4, 6, 8}, {1, 4}, {1, 4, 5, 8 }}, (the abbreviation $[1]_a$ stands as per Definition 5), we have to find all subfamilies G $\subseteq$ F such that $\bigcap$ G $\subseteq [1]_f$, i.e., {1, 4, 8}, (where $[1]_f$ is the equivalence class of the almost indiscernibility relation f (β) corresponding to the rule 1). There are three sub-families of F given by

$$[1]_a \bigcap [1]_b = \{1,4,6,8\} \bigcap \{1,4\} = \{1,4\}$$
$$[1]_a \bigcap [1]_e = \{1,4,6,8\} \bigcap \{1,4,5,8\} = \{1,4,8\}$$
$$[1]_b \bigcap [1]_e = \{1,4\} \bigcap \{1,4,5,8\} = \{1,4\}$$

and all three of them are reducts of the family F. Hence we have three value reducts a(1) = Average, b(1) = Stable and a(1) = Average, e(1) = Very Slow, b(1) = Stable, e(1) = Very Slow. This means that the attribute values of attributes a, b and a, e and b, e are characteristic for decision class 1 and do not occur in any other decision classes in the decision table. In the next table, we list value reducts for all decision rules.

Table 3.

U	a	b	e	f
1	Average	Stable	×	N_1
$1'$	Average	×	Very Slow	N_1
$1''$	×	Stable	Very Slow	N_1
—	—	—	—	—
2	Not Good	Unstable	×	N
$2'$	Not Good	×	Slow	N
$2''$	×	Unstable	Slow	N
—	—	—	—	—
3	Very Good	×	Very Quick	P_1
$3'$	Very Good	Medium	×	P_1
$3''$	×	Medium	Very Quick	P_1
—	—	—	—	—
4	Average	×	Very Slow	N_2
$4'$	×	Stable	Very Slow	N_2
$4''$	Average	Stable	×	N_2

U	a	b	e	f
5	Very Good	Unstable	×	N
$5'$	Very Good	×	Very Slow	N
—	—	—	—	—
6	Good	×	Quick	P_2
$6'$	Good	Medium	×	P_2
$6''$	×	Medium	Quick	P_2
—	—	—	—	—
7	×	Medium	Quick	P_2
$7'$	Very Good	×	Quick	P_2
$7''$	Very Good	Medium	×	P_2
—	—	—	—	—
8	Good	×	Very Slow	N_2
$8'$	Good	Unstable	×	N_2

We observe from the above table that, for decision rules 1, 2, 3, 4, 6, 7 we have only three value reducts of condition attributes for each decision rule. Each of the decision rules 5 and 8 has two value reducts of condition attributes. Hence each of the decision rules 1, 2, 3, 4, 6, 7 has three reducts and each of the rules 5, 8 has two reducts. Thus there are $27 \times 27 \times 4 = 2916$ (not necessarily different) solutions to our problem.
One such solution is presented in the table below:

Table 4.

U	a	b	e	f
1	Average	Stable	×	N_1
2	Not Good	×	Slow	N
3	×	Medium	Very Quick	P_1
4	×	Stable	Very Slow	N_2
5	Very Good	×	Very Slow	N
6	Good	×	Quick	P_2
7	×	Medium	Quick	P_2
8	Good	Unstable	×	N_2

Table 5.

U	a	b	e	f
1	Average	Stable	×	N_1
2	Not Good	Unstable	×	N
3	Very Good	Medium	×	P_1
4	Average	Stable	×	N_2
5	Very Good	Unstable	×	N
6	×	Medium	Quick	P_2
7	×	Medium	Quick	P_2
8	Good	×	Very Slow	N_2

Another solution is shown in Table 5. Because decision rules 1, 4 and 6, 7 are identical in Table 5, we can represent our table in the form:

Table 6.

U	a	b	e	f
1, 4	Average	Stable	×	N_1
2	Not Good	Unstable	×	N
3	Very Good	Medium	×	P_1
5	Very Good	Unstable	×	N
6, 7	×	Medium	Quick	P_2
8	Good	×	Very Slow	N_2

Table 7.

U	a	b	e	f
1	Average	Stable	×	N_1
2	Not Good	Unstable	×	N
3	Very Good	Medium	×	P_1
4	Very Good	Unstable	×	N
5	×	Medium	Quick	P_2
6	Good	×	Very Slow	N_2

In fact, enumeration of decision rules is not essential, so we can enumerate them arbitrarily and we get as a final result in the Table 13. This solution is referred to as minimal (α, β) solution for $\alpha = 0.8$, $\beta = 0.7$.
Similarly considering different values of α and β we get different solutions.

3 Conclusion

Simplification of decision tables is of primary importance in many applications. After simplifying when we get a reduced decision table, in that table the same decision can be based on a smaller number of conditions. This kind of simplification eliminates the need for checking unnecessary conditions, and ultimately we get a decision table containing only those values of condition attributes which are necessary to make decisions. In this work, Pawlak's method of reduction of decision table is generalised for different choice values of α and β, α being for condition attributes and β for decision attributes when α, $\beta \in [0,1]$. For $\alpha = 1$ and $\beta = 1$, the method reduces to Pawlak's method.

References

1. Boryczka, M. (1989), Optimization of Decision Tables Using Rough Sets,*Bull. Polish Acad. Sci. Tech.*, **37** 321-332.
2. De, S.K., Biswas, R., Roy,A. R. (1999), Finding dependency of attributes in an information system, *The Journal of Fuzzy Math* **7** 335-344.
3. Kowalczyk, A. and Szymanski, J. (1989), Rough Simplification of Decision Tables, *Bull. Polish Acad. Sci. Tech.* **37** 359-374.
4. Pawlak, Z.(1991), ROUGH SETS, Theoretical Aspects of Reasoning about Data, *Kluwer Academic Publishers.*
5. Pawlak, Z. (1982), Rough Sets, *International Journal of Information and Computer Sciences* **11** 341- 356.
6. Pawlak, Z. (1985), Rough Sets and Decision Tables, *Lecture Notes in Computer Sciences 208*, Springer Verlag, 186-196.

Wavelet Transform Based Fuzzy Inference System for Power Quality Classification

Arvind K. Tiwari and K.K. Shukla

Dept. Computer Engineering, Institute of Technology,
Banaras Hindu University, Varanasi- 221 005, India
aktiwari@banars.ernet.in

Abstract. The paper presents a hybrid scheme using a Discrete Wavelet Transform and a Fuzzy Expert System for feature extraction and classification. The signal under test (electrical current or voltage for Power Quality study) is processed through a DWT decomposition block to generate the feature extraction curve. The DWT Level and Energy information from the feature extraction curve is then passed through a diagnostic module that computes the truth-value of the signal combination and determines the class to which the signal belongs. The proposed scheme is much simpler and powerful than currently available PQ classification schemes.

1 Introduction

The application of Fuzzy knowledge based approach is fast proliferating in a large number of engineering disciplines. The ability of this approach to closely emulate the behavior of a human expert allows automation of monitoring and control processes operating in vague and uncertain environments. Several Fuzzy Inference mechanisms are available for design of such systems. This paper presents a new approach to power quality monitoring that combines the time-scale localization capability of wavelet based feature extraction with human-like problem solving using Fuzzy Inference System. Fuzzy logic, which is the logic on which fuzzy control is based, is much closer in spirit to human thinking and natural language than the traditional logical system. Primarily, it provides an effective means of capturing the approximate, inexact nature of the real world. In essence, the fuzzy logic controller (FLC) provides an algorithm, which can convert the linguistic control strategy based on expert knowledge into an automatic control strategy.

A number of PQ classification methods have been proposed [1], [2], [3], [4], [5], [6], [7], [8]. The approach seems appropriate in detecting and identifying a particular type of disturbance, and are justified on purely intuitive ground. However, none of them present a methodology that can be used to classify or measure different Power Quality problems. In this paper, application of DWT, as a powerful preprocessing tool for detecting, and localizing PQ issues have been investigated. Preprocessing with orthogonal wavelets can be used in connection to fuzzy logic to express fuzzy rules.

N.R. Pal and M. Sugeno (Eds.): AFSS 2002, LNAI 2275, pp. 148–155, 2002.

Multiresolution analysis simplifies, for instance, considerably the design of a FLC containing rules in the form of

If "signal level" is low ..... then ....

Fuzzy Inferencing for classification and quantification of power signal disturbances has been investigated. The proposed technique will deal with the problem not only in time domain or frequency domain, but in a wavelet domain, which covers both time and frequency domains simultaneously. The paper is organized as follows. In section 2 we present a general introduction to Discrete Wavelet Transform. Application of DWT for monitoring PQ issues has been dealt in section 3 presents in brief the theory of Fuzzy Inference and its use for classification issues. Section 4 covers in detail the results related to PQ classification with DWT-Fuzzy Inference system. Finally a conclusion and references are presented in section 5 and 6.

2 Discrete Wavelet Transform

2.1 Multiresolution Signal Decomposition and Its Implementation

The Multiresolution Signal Decomposition (MSD) technique decomposes a given signal x(t) into its detailed and smoothed versions. A complete treatment of the theory can be found in [10].

Let $x_0(n)$ be a discrete-time signal recorded from a physical measuring device. From the MSD technique, the decomposed signal at scale j are $c_j(n)$ and $d_j(n)$, where $c_j(n)$ is the smooth version of the original signal, and $d_j(n)$ is the detailed representation of the original signal $x_0(n)$ in the form of wavelet transform coefficients. They are defined as

$$c_j(n) = \sum_k h(k-2n)x_{j-1}(k) \quad (1)$$

$$d_j(n) = \sum_k g(k-2n)x_{j-1}(k) \quad (2)$$

where, h(n) and g(n) are the associated filter coefficients. Fig. 1 best describes, implementation of a two level decomposition and reconstruction system the MSD technique.

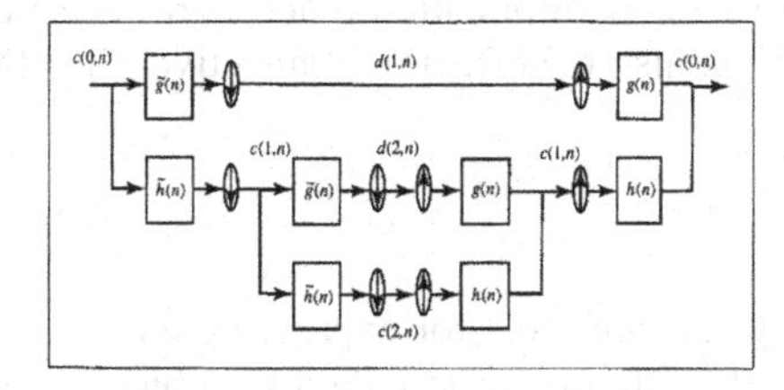

Fig. 1. A two level decomposition and reconstruction system

2.2 DWT in Feature Detection, Extraction and Localization

In this section, we summaries the application of the wavelet transform in detecting and extracting power quality disturbance features. In present investigation we utilize a dyadic-orthonormal wavelet transform with Daubechies' wavelet with a four - coefficient filter [9], [13]. We perform a 12-scale signal decomposition to ensure that all disturbance features in both high and low frequency are extracted. The first scale signal has frequency range of $f/2 - f/4$, where f is the sampling frequency of the time domain disturbance signal. The second, third, fourth, fifth and higher level signals have frequency ranges of $f/4 - f/8$, $f/8 - f/16$, $f/16 - f/32$, $f/32 - f/64$ **respectively.**

The aforementioned wavelet transforms response to signal behavior is related to the vanishing moment of the analyzing wavelet $\psi(t)$ given below:

$$\int_{-\infty}^{\infty} t^n \psi(t) = 0 \tag{3}$$

Where n = 0,1,2,3, ...***N-1***. The above expression is called a vanishing moment of order ***N***. Note that the order of the vanishing moment of an analyzing wavelet $\psi(t)$ is at least equal to one, i.e. ***N*** =1.

As far as detection and localization is concerned, the first finer decomposition levels of the distorted signal are normally sufficient. The other resolution levels are used to extract more information that can help in **PQ** classification. Classification of **PQ** problems in the present investigation is based upon **Parseval's Theorem**; which sates that if the used wavelets form an orthonormal basis and satisfies the admissibility condition [11], then the energy of the distorted signal is equal to the energy in each of the expansion coefficients i.e.:

$$\int |f(t)|^2 = \sum_{k=-\infty} |c(k)|^2 + \sum_{j=0}^{\infty} \sum_{k=-\infty}^{\infty} |d_j|^2 \tag{4}$$

3 Fuzzy Logic in PQ Classification

In general, a fuzzy control rule is a fuzzy relation, which is expressed as a fuzzy implication. The choice of a fuzzy implication function reflects not only the intuitive criteria for implication but also the effect of connective also [14], [15].

3.1 Fuzzy Logic

Fuzzy Logic and Fuzzy sets are tools for expressing and operating on knowledge that is imprecise, or where the interpretation is highly subjective and depends strongly on context or human opinion. For any crisp set C it is possible to define a characteristics function $\mu_c = U \rightarrow \{0,1\}$. In fuzzy set, the characteristics function is generalized to a membership function that assigns to every $u \in U$ a value from the unit interval [0,1] instead from the two elements set {0,1} [12].

3.2 Fuzzy Logic in the Frequency Domain

Preprocessing with orthogonal wavelets can be used in connection to fuzzy logic to express fuzzy rules in the frequency domain. Multiresolution analysis simplifies, for instance, considerably the design of a fuzzy controller containing rules in the frequency domain of the form

if „signal frequency " is low .... then ...

The method uses the fact that a Multiresolution analysis with orthogonal wavelets corresponds to the iterative filtering of the signal with quadrature mirror filters (QMF) [11]. Orthogonal wavelet decomposition fulfills therefore the power complementary condition. A QMF that satisfies the power complementary property is known as perfect reconstruction quadrature mirror filter (PRQMF). Also, this means that the frequency windows H_m corresponding to the first high pass filters and the nth low pass filter H_{low} satisfy the condition:

$$\sum_{m=1..n} |H_m(\omega)|^2 + |H_{low}(\omega)|^2 = 1 \tag{5}$$

The different domain frequency windows $|H_m(\omega)|^2$ can be used as membership functions. The degree of membership of a variable to the frequency window is given by

$$\mu(H_m) = \sum_{1} (d_m)^2 \ / (\sum_{m=1}^{n} (d_m)^2 + (c_m)^2) \tag{6}$$

The main advantage of this scheme is its simplicity. The wavelet coefficients may be used for other purposes such as signal denoising or for removing an offset. Many different functions can be realized at a reasonable computing power.

4 Detection, Localization and Classification

Using the localization property gained from the finer resolution levels, a time-frequency plot of the distorted signal is generated [13] in Matlab. The classification rules proposed in [13] are simple for the operator to detect, localize, and classify different power quality problems based on the proposed 'feature extraction curve'. The variation in three different zones of the feature extraction curve; peak, lower left part and lower right provides sufficient information for classifying different PQ problems. However, looking onto feature extraction curve many a time it will be difficult for operator to distinguish between different PQ problems impended in the signal simultaneously and also to predict about the magnitude of particular problems. To overcome the limitations we propose a Fuzzy Expert System based on Mamdani type fuzzy implication Fig. 2. Fig. 3 a. represents DWT based Decomposition of pure sinusoidal supply of magnitude 1.0 p.u. and Fig. 3b represents generalized PQ feature extraction curve of sinusoidal supply system based on procedure described in section 2.3 Fig. 4 represents change in DWT decomposition and feature extraction curve due to presence of sudden surge in supply system, which detection and localization is easily performed with DWT Decomposition plots whereas its classification could be done with the help of Feature extraction curve.

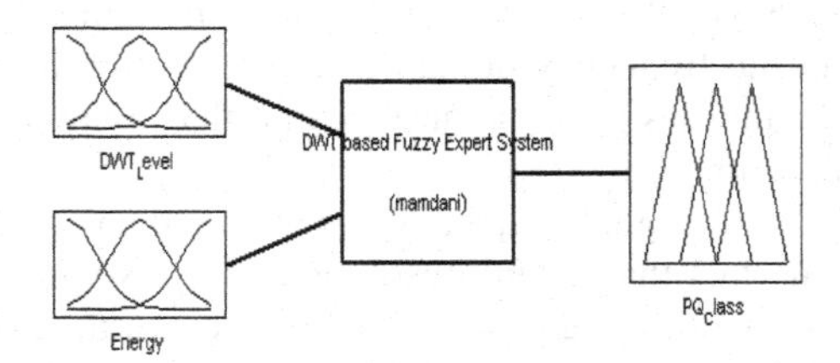

Fig. 2. Fuzzy Expert System based on Mamdani type fuzzy implication

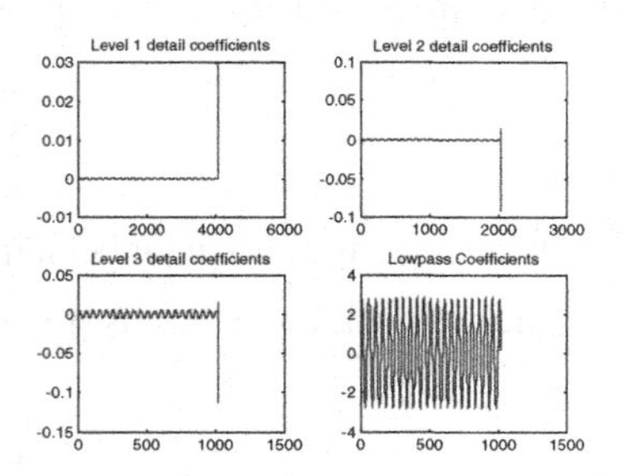

Normal Frequency Sag & Swell Zone

Low Frequency Disturbance Zone

High Frequency Disturbance Zone

Fig. 3a. DWT based Decomposition of pure sinusoidal supply of magnitude 1.0 p.u.

Fig. 3b. Generalized PQ feature extraction curve

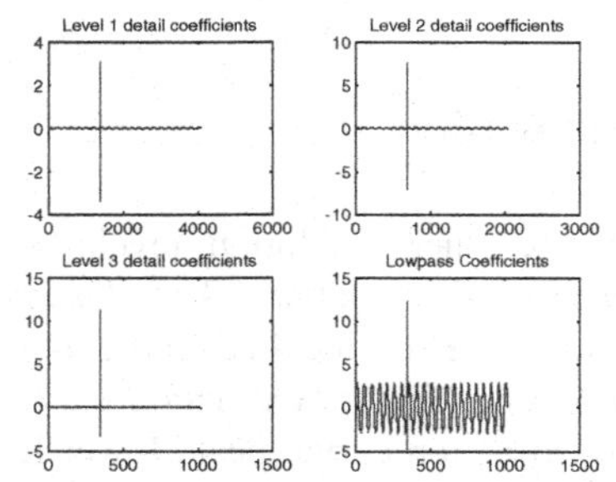

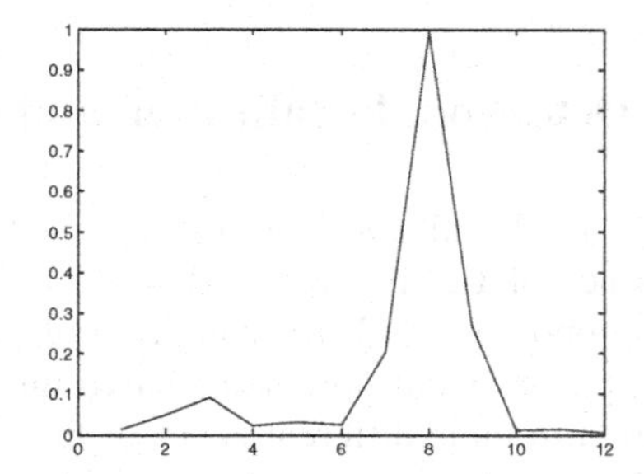

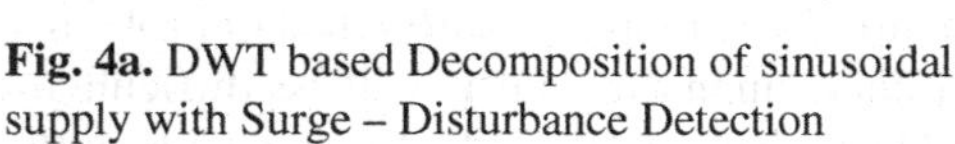

Fig. 4a. DWT based Decomposition of sinusoidal supply with Surge – Disturbance Detection

Fig. 4b. PQ feature extraction curve with Surge -Disturbance Classification

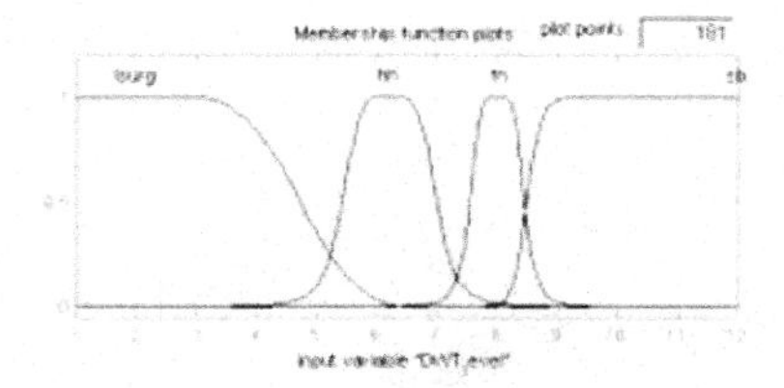

Fig. 5. Membership Function for DWT levels of Feature extraction Curve

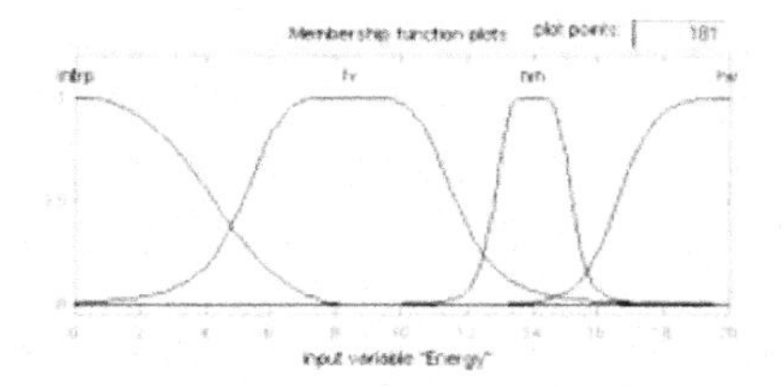

Fig. 6. Membership Function for Energy in Feature extraction Curve

4.1 Application of Fuzzy Expert System in PQ Classification

For classifying the PQ problems, four fuzzy sets are chosen from the DWT levels (lv) designated as surge (fast transient/surges), hh (higher order harmonic), fn (fundamental waveform), sb (sub harmonic component) Fig. 5. In a similar way four fuzzy sets are chosen for the amplitude of the feature extraction curve (en), designated as intrp (interruption/outage), lv (lower peak), nm (peak corresponding to fundamental) and hv (higher peak) Fig. 6.

The rule base for the fuzzy decision support system is listed below as:

Rule 1	If (DWT_Level is surge) and (Energy is hv)	then (PQ is surge)
Rule 2	If (DWT_Level is hh) and (Energy is lv)	then (PQ is higher harmonic)
Rule 3	If (DWT_Level is sb) and (Energy is lv)	then (PQ is sub harmonic)
Rule 4	: If (DWT_Level is fn) and (Energy is hv)	then (PQ is swell)
Rule 5	If (DWT_Level is fn) and (Energy is intrp)	then (PQ is outage)
Rule 6	If (DWT_Level is fn) and (Energy is lv)	then (PQ is sag)
Rule 7	If (DWT_Level is fn) and (Energy is nm)	then (PQ is good)
Rule 8	If (DWT_Level is hh) and (Energy is hv)	then (PQ is higher harmonic)
Rule9	If (DWT_Level is sb) and (Energy is hv)	then (PQ is sub harmonic)

The above rules have been obtained in consultation with an expert power system engineer and refined after several trials. The Fuzzy inferencing is done using the maximum product compositional rule of inference. If $\alpha_1, \alpha_2, \alpha_3, \ldots \alpha_7$ are the firing strength of the rule base for each category of the transient PQ disturbance (Normal, Sag, Swell, Higher Harmonic, Sub Harmonic, Outage and Surge), the output is obtained as Fig. 7;

$$\mu_0 = \alpha_1 OR \alpha 2 OR \alpha 3 OR \alpha 4 OR \alpha 5 OR \alpha 6 OR \alpha 7$$

$$= \max(\alpha_1, \alpha_2, \alpha_3, \alpha_4, \alpha_5, \alpha_6, \alpha_7) \quad (7)$$

From the Fig. 8 & Table 1 it is observed that each category of waveform is successfully classified as the output from the fuzzy expert system and shows truth-value of particular class that suddenly rises in most cases in comparison to normal waveform. Hence it is obvious that the proposed approach is computationally simple in comparison to ANN based and Fuzzy expert system with Fourier Linear Combiner based approaches [2-8].

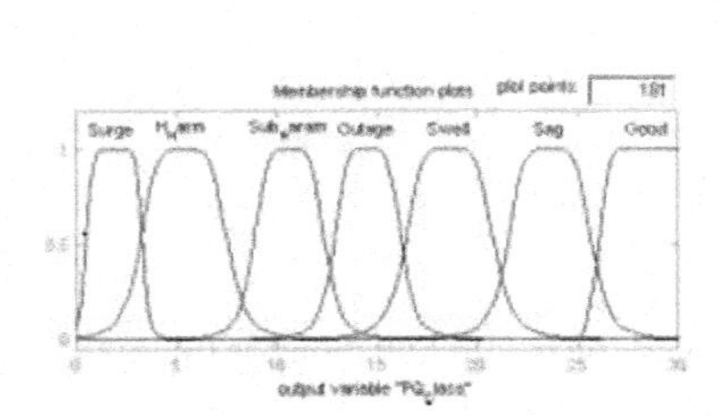

Fig. 7. Membership Function for PQ classes

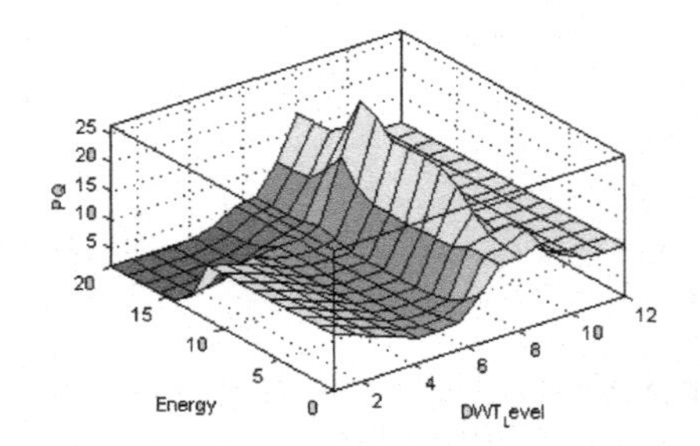

Fig. 8. Surface View of the Fuzzy Inference Systems

Table 1. PQ classification; Output of Model & Corresponding PQ Problem

Output of Fuzzy Model	Corresponding PQ
1.93	Surge
5.44	Higher Harmonic
10.6	Sub Harmonic
14.5	Outage
18.1	Swell
22.7	Sag
26.4	Good

5 Conclusion

The paper presents a new approach for the classification of Power Quality disturbances using a Discrete Wavelet Transform and Fuzzy Expert System. The DWT separates PQ problems that overlap in time and frequency. And presents a time frequency picture of the distorted signal under test. The obtained feature extraction curve presents a simple classification tool to detect, localize and classify different PQ problems. Further, the feature extraction curve is used as input to the Fuzzy Expert System for classification of different PQ problems. The Fuzzy Expert System yields a robust and accurate classification scheme for a variety of simulated waveforms. A power system expert has evaluated the system and the approach is found to emulate the reasoning of human expert satisfactorily and is computationally much simpler than currently available tools. Significant events in a real system give rise to characteristic waveforms containing features like sag, swell, surge, interruption, higher harmonics and sub harmonics etc. These features are the ones already used in our work to test the system. We are therefore confident that the performance figures reported by us are realistic and will be maintained when the system is applied in the field However, the techniques presented are equally applicable to a wide range of application areas like remote sensing, medical diagnosis, power system, speech analysis etc

References

1. R.C. Dugan, M.F. McGranaghan, H.W. Beaty, Electric Power Systems Quality, McGraw-Hill, 1996.
2. N. Kandil, et al., "Fault identification in an AC-DC transmission system using neural networks", IEEE Transactions on Power System, vol. 7, no. 2, pp. 812-819, May 1992.
3. A.K. Ghosh and D.L. Lubkeman, "The classification of Power System Disturbances Waveforms Using a Neural Approach", IEEE Transactions on Power Delivery, vol. 10, no. 1, pp. 109-115, January 1995.
4. P. Pillay and Bhattacharjee, "Power Quality Assessment via wavelet transform analysis", IEEE Transactions on Power System, vol. 11, no. 4, Nov. 1996.
5. G.T. Heydt and A.W. Gali, "Transient Power Quality Problems Analysed using Wavelets", IEEE Transactions on Power Delivery, vol. 12, no. 2, April 1997.
6. Surya Santoso, E.J. Powers, and P. Hofman, "Power Quality Assessment via wavelet transform analysis", IEEE Transactions on Power Delevery, vol. 11, no. 2, April 1996.
7. P.K. Dash, S. Mishra, M.M.A. Salma, and A.C. Liew, "Classification of Power System Disturbances Using a Fuzzy Expert System and a Fourier Linear Combiner.
8. O. Rioul and M. Vetterli, "Wavelets and signal processing", IEEE signal processing magazine, October 1991, pp.14-38.
9. S. Mallat, " A theory for Multiresolution signal decomposition: the wavelet representation", IEEE transaction on Pattern Recognition and Mach. Intel, vol. 11, pp.674-693, July 1989.
10. R.R. Rao and A.S. Bopardikar, "WAVELET TRANSFORMS: Introduction to Theory and Applications"
11. M.M. Marcos and W.R. Anis Ibrahim, "Electric Power Quality and Artificial Intelligence: Overview and Applicability", IEEE Spectrum, vol. 19, no. 6, June 1999.
12. D. Driankov, H. Hellendoorn and M. Reinfrank, "An Introduction to Fuzzy Control", Narosa Publishing House.
13. Arvind K. Tiwari, K.K. Shukla, "Discrete Wavelet Transform based Power Quality Classification", Accepted for publication in conference proceeding and presentation in international conference "Energy Automation and Information Technology – EAIT2001" to be held at IIT Kharagpur, Dec. 7-9, 2001.
14. C.C. Lee, "Fuzzy Logic in control Systems: Fuzzy Logic controller- Part I", IEEE Transaction on systems, man and cybernetics, vol. 20, no. 2, pp. 404-418, March/April 1990.
15. C.C. Lee, "Fuzzy Logic in control Systems: Fuzzy Logic controller- Part II", IEEE Transaction on systems, man and cybernetics, vol. 20, no. 2, pp. 419-435, March/April 1990.

A Fuzzy-Neural Technique for Flashover Diagnosis of Winding Insulation in Transformers

Abhinandan De and Nirmalendu Chatterjee

Electrical Engineering Department, Jadavpur University, Calcutta - 700032, India
adel@vsnl.net
nirmalendu@ieee.org

Abstract. A Fuzzy-Neural pattern recognition technique for insulation flashover diagnosis in oil filled power transformers has been described in the paper. Determination of exact nature and location of internal insulation flashover during impulse testing of power transformers is of practical importance to the manufacturer as well as designers. The presently available diagnostic techniques more or less depend on expertise of the test personnel and in many cases not beyond ambiguity and controversy. The new technique described in the paper relies on high discrimination power and excellent generalization ability of Fuzzy-Neural systems in complex pattern classification problem and overcomes the limitations of conventional diagnostic methods. The technique applied to winding model of typical high voltage transformers exhibited high diagnostic accuracy by successful detection and discrimination of flashovers of different nature and site of occurrence in the winding.

1 Introduction

As a major apparatus in power system, the power transformer has a vital role in system operation. All large power transformers have to pass successfully through stipulated impulse tests to qualify for actual field operation and determination of any internal insulation failure of winding during impulse testing of a power transformer is critically important. Unfortunately, when an internal flashover is suspected, there is no general method yet available for determination of nature of the flashover and detection of flashover location in the winding other than complete dismantling of the whole affected wining, followed by rigorous manual inspection, which involves wastage of time and money. Further, the conventional flashover detection techniques are highly dependent on the knowledge and skill of individual test personnel and in most cases raises controversy. Application of Artificial Intelligence can overcome many of the limitations of the conventional flashover diagnostic techniques. Intelligent systems have been successfully applied to other flashover diagnosis problems in power system. Two different approaches have been made to solve such problems, symbolic expert system approaches [1] and neural network or fuzzy-neural network approaches [2]. Expert systems have been criticised for their requirement of 'large-set knowledge acquisition' and huge effort to maintain the knowledge base. ANNs and Fuzzy-Neural Networks offer a simple and more robust solution to the flashover diagnosis problem due to their high adaptability and generalization ability.

N.R. Pal and M. Sugeno (Eds.): AFSS 2002, LNAI 2275, pp. 156–162, 2002.

1.1 Proposed Flashover Diagnosis Technique

The proposed method employs a Fuzzy-Neural system which attempts to recognize the frequency responses of the winding of typical high voltage transformers under normal and different faulty condition of winding insulation and learns to establish correlation between the nature and physical location of occurrence of internal insulation flashover and its associated frequency response. The frequency response of a transformer winding is obtained by transforming the digitally recorded time domain signals of winding neutral current and applied impulse voltage during test into frequency domain signals using a Fast Fourier Transformation (FFT) routine. The neutral current harmonics are subsequently divided by respective voltage harmonics to obtain frequency response of the winding admittance (transfer function). The integrity of the winding insulation is determined by comparing the frequency response patterns of the winding obtained at full and reduced test voltage. Internal insulation flashover of different natures and at different locations of occurrences generate distinct patterns in the transfer function amplitude-frequency spectrum. Analysis of such spectra patterns by researchers [3,4] has revealed that longitudinal flashover of insulation or series type insulation flashover located close to the high voltage line terminal of a winding is reflected by the presence of new poles at high frequency band in the amplitude-frequency plot of the transfer function, whereas, series flashover occurring near the ground terminal induces new poles in the low frequency region of the spectrum. For transverse flashover of insulation or shunt type flashovers increase in the pole height of the major resonances are observed rather than formation of new poles with a predominant effect for the line end shunt flashovers and less prominent effect for shunt flashovers near the ground terminal of the winding. Thus, for a particular type of winding, it is not difficult to establish a correlation between the distortion of the transfer function and the associated nature and location of the flashover. Also, it is not impossible to generalize this relationship for other similar type of winding designs.

Although detection of failure may be possible by simple superposition and comparison of several transfer function amplitude-frequency plots, in practical cases, frequency spectrum comparison based on simple human eye estimation may not be effective to detect very delicate and minute difference in the plots. More over, identification of flashover nature and location from the available plots require extensive study and expertise, and once again leads to controversy among the experts. This task of recognition and classification of large set of visually identical spectrum patterns, however, can be successfully handled by Fuzzy-Neural pattern classifiers. If the frequency response of a typical high voltage transformer winding with deliberately simulated flashovers at various locations are presented as set of input patterns to such a pattern classifier, the task is to identify the patterns with "similar distinctive feature" and group them together, the number of such groups and the feature of each group, however, are not explicitly stated. The pattern recognizer is intended to discover significant patterns or features from a set of amplitude-frequency spectra patterns of the winding and classify the patterns intelligently.

1.2 The Fuzzy-Neural Pattern Recognition Algorithm

The Fuzzy ARTMAP algorithm employed in the present work is same as that developed by Carpenter *et al* [5]. The system incorporates two fuzzy ART modules ART_a and ART_b, that are linked together via an inter-ART module F^{ab}, called a *map field* as in Fig.1. The map field is used to form predictive association between categories and to realize the *match tracking rule* whereby, the vigilance parameter of ART_a increases in response to a predictive mismatch at ART_b. Match tracking recognizes category structure so that predictive error is not repeated on subsequent presentation of the input.

ART_a and ART_b. Inputs to ART_a and ART_b are in the complement code form: for ART_a input $I = A = (a, a^c)$; and for ART_b input $I = B = (b, b^c)$. Variables in ART_a and ART_b are designated by superscripts *a* or *b*. For ART_a, $x^a \equiv (x^a_1,\ldots, x^a_{2Ma})$ denotes the F^a_1 output vector; $y^a \equiv (y^a_1,\ldots, y^a_{2Na})$ denotes the F^a_2 output vector; and $w^a_j \equiv (w^a_{j,1},\ldots, w^a_{j,2Ma})$ denotes the *j*th ART_a weight vector. For ART_b, $x^b \equiv (x^b_1,\ldots, x^b_{2Mb})$ denotes the F^b_1 output vector; $y^b \equiv (y^b_1,\ldots, y^b_{2Nb})$ denotes the F^b_2 output vector, and $w^b_k \equiv (w^b_{k,1},\ldots, w^b_{k,2Mb})$ denotes the *k*th ART_b weight vector. For the map field, $x^{ab} \equiv (x^{ab}_1,\ldots, x^{ab}_{Nb})$ denotes the F^{ab} output vector and $w^{ab}_j \equiv (w^{ab}_{j1},\ldots, w^{ab}_{jNb})$ denotes the weight vector from *j*th F^a_2 node to F^{ab}. Components of vectors x^a, y^a, x^b, y^b and x^{ab} are reset to zero between input presentations.

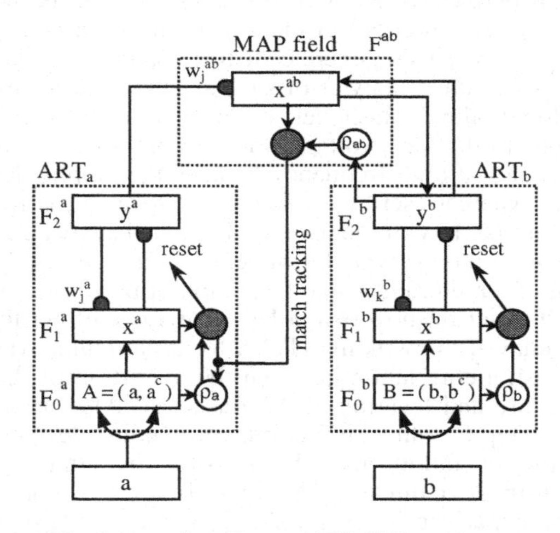

Fig. 1. A typical Fuzzy ARTMAP architecture

Map Field Activation. Map field F^{ab}, is activated when one of the ART_a or ART_b categories becomes active. When the *J*th F^a_2 node is chosen, then its weights w_J^{ab} activate F^{ab}. If the node *K* in F^b_2 is active, then node *K* in F^{ab} is activated by 1-to-1 pathways between F^b_2 and F^{ab}. If both ART_a and ART_b are active, then F^{ab} becomes active only if ART_a predicts the same category as ART_b via the weights w_J^{ab}.The F^{ab} output vector x^{ab} obeys the following:

$$\mathbf{x}^{ab} = \begin{cases} \mathbf{y}^b \wedge \mathbf{w}_J^{ab} & \text{if the } J\text{th } F^a_2 \text{ node is active and } F^b_2 \text{ is active.} \\ \mathbf{w}_J^{ab} & \text{if the } J\text{th } F^a_2 \text{ node is active and } F^b_2 \text{ is inactive} \\ \mathbf{y}^b & \text{if } F^a_2 \text{ is inactive and } F^b_2 \text{ is active} \\ 0 & \text{if } F^a_2 \text{ is inactive and } F^b_2 \text{ is inactive} \end{cases} \quad (1)$$

where, $\wedge$ is the fuzzy OR operator.
From (1), $\mathbf{x}^{ab} = 0$ if the prediction $\mathbf{w}_J^{ab}$ is disconfirmed by $\mathbf{y}^b$. Even such a mismatch triggers an ART_a search for a better category, as follows.

Match Tracking. At the start of each input presentation the ART_a vigilance parameter ρ_a equals a baseline vigilance, ρ'_a. The map field vigilance parameter is ρ_{ab}.

$$\text{If } |\mathbf{x}^{ab}| < \rho_{ab} |\mathbf{y}^b| \quad (2)$$

then, ρ_a is increased until it is slightly larger than $|A \wedge w_J^{ab}|\,|A|^{-1}$, where A is the input to F^a_1, in complement coding form. Then,

$$|x^a| = |A \wedge w_J^a| < \rho_a |A| \tag{3}$$

where, J is the index of the active F^a_2 node. When this occurs, ART_a search leads either to activation of another F^a_2 node J with

$$|x^a| = |A \wedge w_J^a| \geq \rho_a |A| \tag{4}$$

and

$$|x^{ab}| = |y^b \wedge w_J^{ab}| \geq \rho_{ab} |y^b| \tag{5}$$

or, if no such node exists, to the shutdown of F^a_2 for the remainder of the input presentation.

Map Field Learning. Learning rule determines how the map field weights w^{ab}_{jk} change through time, as follows. Weights w^{ab}_{jk} in $F^a_2 \rightarrow F^{ab}$ paths initially satisfy

$$w^{ab}_{jk}(0) = 1. \tag{6}$$

During resonance with the ART_a category J active, w_J^{ab} approaches the map field vector x^{ab}. With fast learning, once J learns to predict the ART_b category K, that association is permanent; i.e., $w_{JK}^{ab} = 1$ for all time.

2 Application of the Diagnostic Technique to Model Winding

As a case study, response of a high voltage 3-phase power transformer model has been considered with the high voltage winding comprising of 80 number disk coils subjected to standard 1.2/50 μs lightning impulse voltage as per standard test specification. Artificial winding insulation flashovers simulating two most frequently occurring flashover types, namely i) the longitudinal insulation flashover between adjacent disk coils of winding and ii) the transverse insulation flashover between a coil and nearby objects at ground potential have been deliberately created at different points of the winding. The typical Fuzzy ARTMAP network illustrated in Fig.1 has been used as the basic pattern recognizer to recognize the responses of the winding effected by flashovers of different natures and different physical locations of occurrence in the winding. The efforts have been concentrated on identification of different flashover natures and zone of occurrence of such flashovers in the winding. The idea is then to classify these insulation flashover patterns into different pre-designated flashover categories.

2.1 Classification of Training Exemplars

At the first stage of learning a known set of flashover patterns were presented to the network as training exemplars. These training exemplars were chosen to represent some typical flashover patterns in respect of nature of flashover and location of occurrence. The network was trained to classify these training exemplars into some predetermined fixed number of fault categories. The baseline vigilance parameter in the training algorithm was continuously adjusted during training phase to optimize the rigidity of the classification process. An optimum choice of the baseline vigilance parameter classified the training exemplar set into nine classes representing one 'no-flashover' class and eight flashover categories respectively, where each of the training exemplar was made to represent a distinct flashover category with predetermined class boundary.

2.2 Classification of Various Flashover Patterns

After the initial phase of training with the exemplar set is completed and the category designations are obtained, new flashover patterns are presented to the trained network for further enhancement of learning. Flashover patterns for both longitudinal and transverse type flashovers are presented for faults at various locations in the winding. Table 1 shows a flash over pattern classification example, where fault patterns corresponding to longitudinal as well as transverse type flashovers in two arbitrarily chosen disk coils from each of the four winding zones have been illustrated. The classification and grouping of different fault patterns is made by observing the similarity in array indices of output neurons, activated by various input fault patterns.

Table 1. Illustration of flashover pattern classification

Input Pattern No.	Fault Characteristics			Array Index of the Active Neuron	Designated Class Label for the Fault Pattern
	Nature of Insulation Flashover	Location of Insulation Flashover	Faulty Disk No.		
1	No Flashover	None	None	2	1
2	Longitudinal Flashover	Within 0 - 25% of winding length i.e. in between Disk no. 1 - 20	7	11	2
3			13	11	
4		Within 25 - 50% of winding length i.e. in between Disk no. 21 - 40	26	27	3
5			32	27	
6		Within 50 - 75% of winding length i.e. in between Disk no. 41 - 60	46	34	4
7			55	34	
8		Within 75 - 100% of winding length i.e. in between Disk no. 61 - 80	67	43	5
9			74	43	
10	Transverse Flashover	Within 0 - 25% of winding length i.e. in between Disk no. 1 - 20	6	97	6
11			11	97	
12		Within 25 - 50% of winding length i.e. in between Disk no. 21 - 40	25	81	7
13			33	81	
14		Within 50 - 75% of winding length i.e. in between Disk no. 41 - 60	45	70	8
15			54	70	
16		Within 75 - 100% of winding length i.e. in between Disk no. 61 - 80	66	57	9
17			73	57	

2.3 Accuracy of the Flashover Diagnosis Technique

On successful completion of training and enhancement of learning, an evaluative study of flashover diagnostic accuracy of the proposed technique was made by presenting new flashover patterns to the system with simulated flashovers at new locations which were not in the earlier training input set. The system could correctly recognize these new patterns and classify them according to topological similarity with the pre-existing patterns. In general, it was observed that, identification and discrimination of flashover nature is comparatively easier for line end flashovers, that is, flashovers within 0-25% length of the winding from line terminal as longitudinal and transverse flashovers within these zones produced vastly different spectrum patterns. Whereas, discrimination of 'series' and 'shunt' type flashovers was most difficult and sometimes inaccurate in 75-100% winding length near the winding neutral, as they generated closely similar spectrum patterns. Indication of occurrence of a flashover however, was accurate all the time due to large topological dissimilarity between the responses of the winding under normal and flashover conditions. For rest of the winding zones, the system exhibited good diagnostic accuracy by identification and correct indication of flashover nature and location.

Percentage accuracy of the diagnostic system was evaluated by repeating the experiment on 20 different sample sets of flashover patterns, similar to the example set presented in Table 1. The percentage accuracy has been computed based on ratio of the actual number of accurately classified fault patterns to the total number of presented fault patterns in 20 trials. Summary of these classification results has been presented in Table 2.

Table 2. Summary of the flashover diagnosis results

Zone of flashover in the winding	Percentage diagnostic accuracy	
	Flashover-nature identification accuracy	Flashover-zone identification accuracy
0-25%	100%	95%
25-50%	95%	85%
50-75%	90%	80%
75-100%	80%	70%

As presented in Table 2, the tested model yielded high diagnostic accuracy in flashover-nature identification for flashovers within 0-50% length of winding. Flashover-zone identification accuracy was very high for flashovers within 0-25% length of winding and fairly good in the 25-50% winding zone. In 75-100% winding zone the flashover-zone identification accuracy was poor but acceptable and flashover-nature identification accuracy was fairly good. As the spectrum pattern corresponding to no-flashover condition of winding is largely different from any of the flashover patterns, the classifier experienced no difficulty in recognizing no-flashover conditions, yielding 100% diagnostic accuracy in detecting no-flashover condition of winding.

3 Conclusion

A Fuzzy-Neural technique for insulation flashover diagnosis in oil filled power transformers has been described in the paper. This new approach overcomes many of the limitations of conventional impulse test methods. The method employs a Fuzzy-Neural pattern recognition algorithm to recognise the frequency responses of the winding admittance of typical high voltage transformers under normal and different flashover condition of winding insulation and learns to establish correlation between the nature and physical location of occurrence of internal insulation flashover in a transformer winding and its associated frequency response. The proposed diagnostic method could efficiently recognise and discriminate internal insulation failures of different nature and different location of occurrence in the winding and yielded high diagnostic accuracy when tested on a model winding of high voltage oil filled power transformer.

References

1. Fukui, C., Kawakami, J.: An Expert System for Fault Section Estimation using Information from Protective Relays and Circuit Breakers. IEEE Trans. on Power Delivery, Vol. 1, No. 4 (1986) 83-90
2. Aggarwal, Raj K., Xuan, Q.Y., Johns, Allan T.: A Novel Approach to Fault Diagnosis in Multicircuit Transmission Lines Using Fuzzy ARTmap Neural Networks. IEEE Trans. on Neural Networks, Vol.10, No.5 (1999) 1214 - 1221
3. Malewski, R., Poulin, B.: Impulse Testing of Power Transformers using Transfer Function Method. IEEE Trans. on Power Delivery, Vol.3, No.2 (1988) 476-489
4. De, A., Chatterjee, N.: Frequency Spectrum Analysis: a Reliable method for Impulse Fault Diagnosis in Transformers. Proc. of All India conference on Application of Innovative Technology in Modern Substation Equipment, The Institution of Engineers (I), Calcutta (2000)
5. Carpenter, G.A., Grossberg, S., Markuzon, N., Reynolds, J.H., Rosen, D.B.: Fuzzy ARTMAP: A neural network architecture for incremental supervised learning of analog multidimensional maps. IEEE Trans. on Neural Networks, Vol.3 (1992) 698-712

Feature Identification for Fuzzy Logic Based Adaptive Kalman Filtering

Abhik Mukherjee, Partha Pratim Adhikari, and P.K. Nandi

Dept of Computer Science and Technology
Bengal Engineering College (DU)
P.O. Botanic Garden, Howrah 711 103, India
{abhik,partha,pkn}@becs.ac.in

Abstract. Standard approaches to fuzzy logic based adaptive Kalman filter use features based on adjustment of noise statistics according to performance of plant and sensor noise sources. Availability of this information is limited to specific domains. Here the Kalman gain computed by conventional Kalman filter is modified using online estimate of measurement residuals which is always available. Arguments are given for qualitative relationship between the residuals and Kalman gain tuning. This fuzzy logic based scheme is computationally simple and hence fast.

1 Introduction

Kalman filter (KF) was presented by R.E.Kalman [1] in the year 1960 as a tool for estimation. It is a set of mathematical equations that provides an efficient computational (recursive) solution of the least-squares method [2]. The filter supports estimations of past, present, and even future states [3].

The conventional KF estimate is based on apriori knowledge [4] about system dynamics. It assumes that both the measurement noise and plant noise are Gaussian white with zero mean which is reasonable in most cases. However, there is possibility of mismatch in system modelling as well as noise statistics in actual application. Hence, there is always scope for making the estimator adaptive with respect to the exact environment. A lot of analytical research on adaptive KF has been done to address this goal. Works like [5,6] are aimed at overcoming the limitations of the original work on several aspects. Studies of these works reveal the scope of applying fuzzy logic based techniques to achieve adaptation. The fuzzy logic based scheme should be able to achieve indirectly the same kind of adaptiveness as the analytical schemes do through direct computations.

Fuzzy logic has been used for designing adaptive KF [7,8]. Work devoted to design of adaptive estimator for global positioning system (GPS) using fuzzy logic can be found in [9]. In these works the authors have considered some domain specific performance measure as input features. In the present work, it has been argued that a fuzzy logic based scheme can be designed for adaptive KF, in the lines of generalized analytical approaches, which can work across all domains.

An overview of KF is presented in Section 2.1, followed by a preliminary discussion of fuzzy logic based control systems in Section 2.2. These sections provide

N.R. Pal and M. Sugeno (Eds.): AFSS 2002, LNAI 2275, pp. 163–170, 2002.

the necessary background. The arguments for feature identification is detailed in Section 3 along with a brief outline of the overall design. Some simulation results are presented in Section 4 to establish the relevance of the work.

2 Preliminaries

2.1 Kalman Filter Algorithm

The system model considered here is the linear, discrete, stochastic sequence described by the following equations:

$$X_{k+1} = \Phi_k X_k + W_k$$
$$Z_k = H_k X_k + V_k$$

where, X_k - State vector, Z_k - Measurement vector, Φ_k - State transition matrix, H_k - Measurement sensitivity matrix, W_k - Process noise vector, V_k - Measurement noise vector.
The stochastic disturbance vectors W_k and V_k are treated as zero mean Gaussian noise sequences with the following properties:

$$E\left[W_i W_j^T\right] = Q_i \delta_{ij}, \qquad E\left[V_i V_j^T\right] = R_i \delta_{ij}$$

where Q_i and R_i are plant and measurement noise covariances respectively. The standard KF equations can be written as follows:

Time Update	
Project the state ahead	Project the error covariance ahead
$\hat{X}_{k+1}(-) = \Phi_k \hat{X}_k(+)$	$P_{k+1}(-) = \Phi_k P_k(+) \Phi_k^T + Q_k$

Measurement Update		
Compute the Kalman gain	Update estimate with measurement	Update error covariance
$K_k = P_k(\blacksquare) H_k^T (H_k P_k(-) H_k^T + R_k)^{-1}$	$\hat{X}_k(+) = \hat{X}_k(-) + K_k[Z_k - H_k \hat{X}_k(-)]$	$P_k(+) = (I - K_k H_k) P_k(-)$

In this context, $r_k = Z_k - H_k \hat{X}_k(-)$ is called the measurement residual.

2.2 Fuzzy Control Systems

The design of controller involves the realization of physical transfer functions to effect a change of control variable depending on the output error. In many real life systems, the design of controller is not feasible mainly due to lack of exact analytical knowledge of the physical model. Fuzzy logic helps in such situations for design of a robust controller by using simple rules depicting the nature of relationship between the output error and the control variable. In absence of exact analytical model, such descriptions depict more or less the physical realities of the controller requirements.

A complete treatment of the theory of fuzzy control can be found in [10]. Computationally feasible features depicting the output error are chosen. These are partitioned into several overlapping fuzzy sets, with predefined membership functions. A computed crisp value of feature (sometimes normalized) can belong to each of these fuzzy sets with a fuzzy membership value computed using the membership functions. A rulebase is formulated with these fuzzy sets as antecedents, the consequent being some fuzzy set of the control variable. The rules are processed using fuzzy set operations to obtain fuzzy values for the consequent set. From these fuzzy values, a unique crisp value is computed for the control variable through defuzzification. Often this value is denormalized to obtain the actual change of control variable necessary.

3 Proposed Scheme

The general approach for developing fuzzy based adaptive KF is to tune the noise covariances on the basis of some performance metrics of estimation. When plant and sensor noise sources can be identified, their degradation or improvement can be monitored to tune the noise covariances online. In [9], the distance travelled by the vehicle between GPS updates and the geometrical dilution of precision (GDOP value) of the receiver are contributors to sensor noise and rules are formulated to tune the measurement noise covariance R_k depending on these input features. The process noise covariance P_k, Q_k are influenced by the estimation performance and hence are adjusted online through a rulebase involving such performance metrics as features. Since these covariances are used in the KF algorithm, the scheme becomes adaptive.

However such models are limited to specific domains subject to the availability of such information. In most real life situations, measurement residuals are the only available information, whereby identifying Q_k, R_k and modelling mismatch separately is a difficult proposition. In the present work proper insight has been developed into the physical aspects of residuals. Appropriate logic is sought through which the change in residuals would translate into changes of Kalman gain. These arguments are presented in details in this section. It also includes an outline of the overall design methodology for the fuzzy based scheme. As a whole this work provides some general heuristic formulations for an adaptive KF.

3.1 Analytical Viewpoint

Common analytical approaches to adaptive KF involve online updating of measurement and process noise covariances [11]. When the estimator is able to compute these noise statistics and use their latest online values for the estimation, the KF can adapt to the real time fluctuations of noise statistics. According to Mayers and Tapley [11],

$$\begin{aligned}\hat{R}_k &= \frac{1}{(N-1)}\sum_{k=1}^{N}\left[(r_k-\hat{r}_k)(r_k-\hat{r}_k)^T - \frac{(N-1)}{N}H_kP_k(-)H_k^T\right]\\ &= \sigma_r^2 - \frac{1}{N}\sum_{k=1}^{N}H_kP_k(-)H_k^T \end{aligned} \tag{1}$$

This clearly indicates that residuals are involved in computation of R_k. Similar expressions for Q_k, the plant noise covariance, can also be derived. These noise covariances are involved in the computation of Kalman gain (refer to Section 2.1). From this fact it can be argued that online updating of measurement and plant noise statistics based on measurement residuals have an indirect impact on the Kalman gain. Thus the residuals may be considered for online adhoc tuning of the Kalman gain. Hence such tuning should make the filter adaptive.

In the Kalman gain computation equation presented in Section 2.1, the numerator involves the error covariance corresponding to the non-measurable state variables $(P_k(-)H_k^T)$. The denominator contains error covariance terms of the state variables for which measurements are available $(H_k P_k(-)H_k^T)$ along with measurement noise covariance (R_k). Thus the qualitative relationship between the residuals and the Kalman gain can be explained by decoupling the Kalman gain into two parts K_M and K_S where $K_{k_{m \times n}} = \begin{bmatrix} K_{S_{(m-n)\times n}} \\ K_{M_{n \times n}} \end{bmatrix}$. Here n is the number of measurable state variables out of m total states.

When sensor noise is high, error covariance R_k should increase. This in turn implies that reliability on present measured data has to be reduced. This can be achieved through reduction in K_M values. The Kalman gains corresponding to the remaining state variables, for which measurement is not available, behave in a different way. Greater plant error covariance pertaining to non-measurable state variables implies that recent estimation quality is poor. So KF should respond by giving more importance to presently available data. This can be achieved by increasing the corresponding K_S gain components.

3.2 Feature Selection

Individual residual values are the contributors to the continuously changing noise covariance. The present motivation is to capture the changes in noise covariance through properties of residuals. Proportional, Integral and Derivative of error do represent any error function well in time domain for physical systems and forms the basis of designing PID controllers. It can be argued that proper definitions of P, I and D of measurement residuals can adequately represent in time domain the characteristics of residuals in respect of tuning the Kalman gain in an adhoc manner. Definitions of P, I and D considered here are:

- absolute value of current residual, $P = |r_k|$.
- accumulated residual over last w iterations, $I = \Sigma_{k-w+1}^{k} |r_k|$.
- rate of change of residual over last w iterations, $D = (|r_k| - |r_{k-w+1}|)$.

P represents the absolute value of current residual. Irrespective of its sign, for large values of residual, K_M values should be reduced and K_S values should be increased with respect to the Kalman gains computed by conventional KF.

I represents the accumulated residual on a moving window. I therefore reflects the trend of convergence of the estimator. The Kalman gain tuning would be similar to the rules formulated for P. However I influences the adaptation from a long term point of view. I can therefore account for an overall bias in estimation due to erratic model of plant and measurement noise.

D represents the rate of change of absolute value of residual over a few (say, 5) iterations. A positive value of D implies that the estimation is tending to go more and more off the mark, since the present residual value has become higher than the previous one. Adaptation to this condition calls for less emphasis on the present measurement, which can be achieved by decreasing the Kalman gain K_M and leaving K_S intact. When D is negative, the estimation is tending to become better. Under such situations it would be best to leave all the gains unchanged with respect to D, at least temporarily.

Having identified the proper features, their fuzzy sets can be defined. The following linguistic variables are considered for the input features and output change of Kalman gain:

Variable	Linguistic grades	Remarks
P	ZO PS PM PL	-ve values do not arise
I	ZO PS PM PL	-ve values do not arise
D	ZO PS PM PL	-ve values included in ZO
ΔK_M	NL NM NS ZO	+ve values not considered
ΔK_S	ZO PS PM PL	-ve values not considered

ZO – Zero or no change, PS/NS – +ve/-ve small, PM/NM – +ve/-ve medium, PL/NL – +ve/-ve large

The input features are normalized with respect to global maximum and minimum values, which set the design criterion and margin of performance of the system. The output of the fuzzy system is simply a small percentage value indicating the adjustment necessary for the Kalman gain. Absolute value of Kalman gain is meant in all cases - thus the increase in case of negative K_k values are to be interpreted accordingly.

3.3 Design Overview

Fuzzification. Different types of membership functions can be considered for each linguistic variable. The choice depends on the nature of distribution of the feature values. In this case the noise involved with residuals are stochastic in nature. Symmetric triangular membership functions have been used for the fuzzy sets in order to accommodate the Gaussian distribution with zero mean for the noise associated with residuals. Bell shaped functions can also be used but triangular ones are preferred for simplicity. The process of fuzzification converts crisp values of P, I and D into fuzzy membership values belonging to each of the linguistic variables.

Rulebase Design. The rulebase consists of fuzzy theoretic expressions and are obtained based on the discussions in Section 3.2. Rulebases used are shown (see next page) separately for the two types of gain components.

Defuzzification. The height method of defuzzification has been used. Given a set of values of P,I,D, let α_i be the firing strength (degree of fulfilment) of the i-th rule. The firing strength can be computed using some T-norm, although it is not relevant here as there is only one clause in the antecedent. The defuzzified value ΔK is then computed as $\Delta K = \frac{\sum_{i=1}^{N} \alpha_i u_i}{\sum_{i=1}^{N} \alpha_i}$, where u_i is the peak (i.e. the point with membership value of unity) of the consequent of the rule and N is the number of rules. Note that, for ΔK_S, $N = 8$ and for ΔK_M, $N = 12$.

In case of Kalman gain matrix, each gain must be considered separately. Since a Kalman gain element depends on all the measurement residuals, the rulebase has to incorporate features of each residual. Defuzzification takes care of the combined effect of all the residuals. Finally the gain fraction is denormalized. Defuzzified value of Kalman gain is a fractional value and is to be interpreted as a percentage of gain with respect to actual gain computed by conventional KF.

Rules for K_S	**Rules for K_M**
1. IF P is PS THEN ΔK_S is PS	1. IF P is PS THEN ΔK_M is NS
2. IF P is PM THEN ΔK_S is PM	2. IF P is PM THEN ΔK_M is NM
3. IF P is PL THEN ΔK_S is PL	3. IF P is PL THEN ΔK_M is NL
4. IF P is ZO THEN ΔK_S is ZO	4. IF P is ZO THEN ΔK_M is ZO
5. IF I is PS THEN ΔK_S is PS	5. IF I is PS THEN ΔK_M is NS
6. IF I is PM THEN ΔK_S is PM	6. IF I is PM THEN ΔK_M is NM
7. IF I is PL THEN ΔK_S is PL	7. IF I is PL THEN ΔK_M is NL
8. IF I is ZO THEN ΔK_S is ZO	8. IF I is ZO THEN ΔK_M is ZO
	9. IF D is PS THEN ΔK_M is NS
	10. IF D is PM THEN ΔK_M is NM
	11. IF D is PL THEN ΔK_M is NL
	12. IF D is ZO THEN ΔK_M is ZO

4 Preliminary Simulation Results

The domain for studying the simulation of the adaptive KF is the transfer alignment [12] for the inertial navigation system (INS) of a moving vehicle. In this system the estimation of the misalignment angles from velocity and position error measurement, corrupted by the sensor and measurement noise, is attempted by implementing the adaptive estimator. The estimated states can be utilized for alignment of navigation axes of Inertial Measurement Unit (IMU).

The present simulation studies have been directed towards identification of proper variables for designing fuzzy logic based KF. The features of the measurement residual have been studied for different types of noise. Common noise types include zero mean Gaussian, random bias and random walk. Random bias and random walk are correlated and coloured noise. The Kalman gain variation has been studied for an analytical model of R_k adaptive KF. Figure 1 shows the nature of P, I and D of the measurement residual under random walk noise condition in case of conventional KF. It must be noted here that P and I have

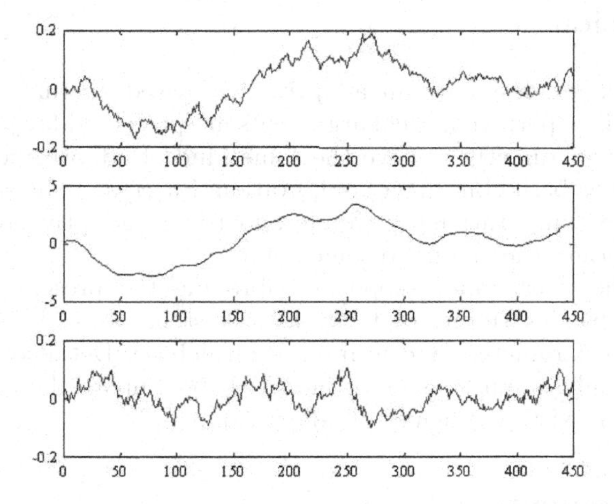

Fig. 1. Plot of P, I and D of measurement residual with iterations

been plotted with actual sign for better understanding. Here different definitions of P, I and D were considered and the best revelation of the characteristics of residuals was obtained in the manner in which it is defined in Section 3.

Figure 2 shows how the values change with the iterations for a typical gain component. Gain of conventional KF is compared with the gain values obtained in case of R_k adaptive analytically designed KF discussed in Section 3.1. The gains of both the estimators are of the same order of magnitude. This reveals the prospects of tuning the gains directly as has been argued in the present work.

This analysis forms part of the preliminary studies required for designing the fuzzy logic based adaptive KF. These results support the concept of the proposed fuzzy adaptive filter. However the full design depends on proper tuning of parameters, which would involve rigorous simulation studies.

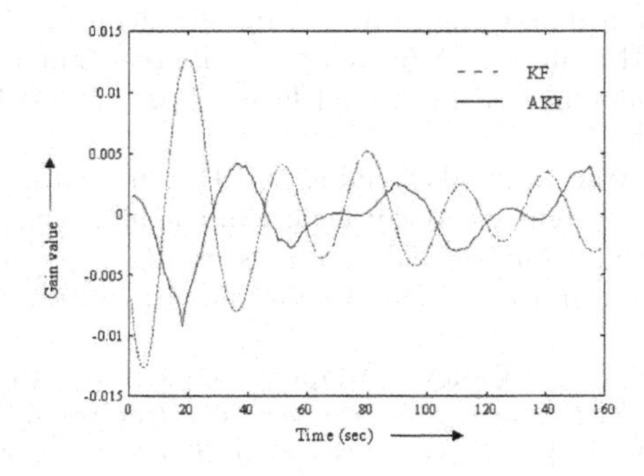

Fig. 2. Comparative study of change of Kalman gain for a typical state variable

5 Conclusion

The framework for designing an adaptive KF based on fuzzy logic has been discussed in this paper. Analytical arguments are presented for justifying the use of fuzzy logic to indirectly achieve the same kind of adaptiveness as in case of adaptive schemes based on direct computation. Success of the scheme therefore would result in computationally simple and robust adaptive KF. Preliminary results substantiate the claims to some extent.

There is, however, much scope of improving the present work. Different schemes available for tuning of fuzzy logic systems must be applied for setting the design parameters and refining the rulebase. Detailed domain specific performance analysis must also be conducted to compare the fuzzy logic based scheme to other existing schemes of adaptation.

Acknowledgment

The motivation of the work was derived from a project on Kalman filtering sponsored by DRDO, India. The authors also thank the anonymous reviewers for their comments.

References

1. R. E. Kalman, "A new approach to linear filtering and prediction theory," *Journal of Basic Eng , vol 82*, pp. 34–45, Mar 1960.
2. R. G. Brown and Y. C. Hwang, *Introduction to Random Signals and Applied Kalman Filtering.* New York: John Wiley and Sons, 1997.
3. G. Welch and G. Bishop, "An introduction to kalman filter," (http://www.cs.unc.edu/ welch/kalman/kalmanIntro.html).
4. A. Gelb, *Applied Optimal Estimation.* MIT press, 1974.
5. A. H. Jazwinski, "Adaptive filtering," *Automatica, vol. 5*, pp. 475–485, 1969.
6. R. K. Mehra, "Approaches to adaptive filtering," *IEEE Trans. on Automatic Control, vol. AC-17*, pp. 693–698, Oct 1972.
7. J. Lalk, "Intelligent adaption of kalman filters using fuzzy logic," (Proceedings of third IEEE conference on Computational Intelligence), pp. 744–749, 1994.
8. Y. H. Lho and J. H. Painter, "A fuzzy tuned adaptive kalman filter," (Proceedings of third IEEE conference on Industrial Fuzzy Control and Intelligence systems), pp. 144–148, 1993.
9. K. Kobayashi, K. Cheok, K. Watanabe, and F. Munekata, "Accurate differential global positioning system via fuzzy logic kalman filter sensor fusion technique," *IEEE Transactions on Industrial Electronics, vol 45 3*, pp. 510–518, June 1998.
10. D. Driankov, H. Hellendron, and M. Reinfrank, *An Introduction to Fuzzy Control.* Narosa, 1997.
11. K. A. Mayers and B. D. Tapley, "Adaptive sequential estimation with unknown noise statistics," *IEEE Trans. on Automatic Control*, pp. 520–523, 1976.
12. M. Kayton and W. R. Fried, *Avionics Navigation System, 2nd ed.* John Wiley & Sons, 1997.

Soft-Biometrics: Soft-Computing Technologies for Biometric-Applications

Katrin Franke[1] and Javier Ruiz-del-Solar[2]

[1] Dept. of Pattern Recognition, Fraunhofer IPK, Berlin, Germany
katrin.franke@ipk.fhg.de
[2] Dept. of Electrical Engeneering, Universidad de Chile, Santigo, Chile
jruizd@cec.uchile.cl

Abstract. Biometrics, the computer-based validation of persons' identity, is becoming more and more essential due to the increasing demand for high-security systems. A biometric system testifies the authenticity of a specific physiological or behavioral characteristic possessed by a user. New requirements over actual biometric systems as robustness, higher recognition rates, tolerance for imprecision and uncertainty, and flexibility call for the use of new computing technologies. In this context soft-computing is increasingly being used in the development of biometric applications. Soft-Biometrics correspond to a new emerging paradigm that consists in the use of soft-computing technologies for the development of biometric applications. The aim of this paper is to motivate discussions on application of soft-computing approaches in specific biometric measurements. The feasibility of soft-computing as a tool-set for biometric applications should be investigated.

1 Introduction

Biometrics offers new perspectives in high-security applications while supporting natural, user-friendly and fast authentication. Biometric identification considers individual physiological characteristics and/or typical behavioral patterns of a person to validate their authenticity. Compared to established methods of person identification, employing PIN-codes, passwords, magnet- or smart cards, biometric characteristics offer the following advantages:

- They are significant for each individual,
- They are always available,
- They cannot be transferred to another person,
- They cannot be forgotten or stolen,
- They always vary[1].

Although, there was a strong growth in biometric technologies during the past years [1], the introduction of biometrics into mass market applications, like telecommunication or computer-security, was comparable weak [2].

[1] Rem.: The presentation of two 100% identical feature sets indicates fraud.

N.R. Pal and M. Sugeno (Eds.): AFSS 2002, LNAI 2275, pp. 171–177, 2002.

From our point of view there are three main aspects responsible for the current situation. First, soft- as well as hardware (sensor) technologies are still under development and testing, although, black sheep promising 100 percent recognition rates. Second, there is a lack of standardization and interchange of biometric systems; basically, such systems are proprietary. And third, there are only few large-scale reference projects that gave evidence of the usability and acceptance of biometrics into real worlds applications.

The work presented in this paper will contribute to the first aspect by employing soft-computing approaches to improve algorithms of biometric analysis. The aim is to provide tool-sets that are able to handle natural variations being sticking in biometrics. Also, the tool-sets should be tractable, robust and of low costs. So, the authors studied soft-computing approaches and their feasibility into biometric measurements.

Section 2 gives a short overview on biometrics where section 3 introduces soft-computing. Section 4 deals with the introduction of soft-computing into biometric application and claims the paradigm of soft-biometrics. A realized application examples, in particular for signature verification, will be described in section 5. Finally, section 6 concludes the presented work and provides some practical hints.

2 Biometrics

Biometric systems comprise the following components: data acquisition and preprocessing; feature extraction and coding; computation of reference data and validation. The systems compare an actual recorded characteristic of a person with a pre-registered characteristic of the same or another person. Thereby, it has to be decided between *identification* (1 to many comparison) and *verification* (1 to 1 comparison). Then, the matching rate of the both characteristics is used to validate, whether the person is what they claim to be. The procedures seem to be equivalently to the traditional methods using PIN or ID-number. However, the main difference is founded by the fact that in biometrics an absolute statement *identical / not identical* cannot be given. For instance a credit card has exact that number "1234 5678 9101" or not, contrary, a biometric feature varies naturally at any acquisition.

Biometric technologies will be divided into approaches utilizing *physiological* characteristics, also referred as *passive* features, and approaches using *behavioral* characteristics that are *active* features. Behavioral characteristics, used e.g. in speaker recognition, signature verification or key-stroke analysis are always variable. On the other hand physiological characteristics employed e.g. in hand, fingerprint, face, retina or iris recognition, are more or less stabile. Variations may be caused by injuries, illness as well as variations during acquisition.

Each biometric system has to be able to handle diverse variation by using "tolerance-mechanisms". Also, it should be possible to adjust a statement about a person's identity gradually with a certain probability, and, it should allow for tuning a system not to reject a person falsely or to accept another person without permission.

Due to the variability of the biometric characteristics, a resulting error rate cannot be easily assigned. However, adapted algorithms that are able to handle inaccuracies and uncertainty might slow down resulting error rates.

Biometric approaches have to solve the two-class problem *person accepted* or *person rejected*. So, the performance of biometric systems is measured with two basic rates: False acceptation rate (FAR) is the number of falsely accepted individuals; False rejection rate (FRR) is the number of falsely rejected individuals [3].

3 Soft-Computing

Since the early days of Artificial Intelligence scientists and engineers have been searching for new computational paradigms capable of solving real-world problems efficiently. Soft-Computing (SC) is one of such paradigms that has emerged in the recent past as a collection of several models of computation, which work synergistically and provide the capability of flexible information processing. The principal constituents of SC are fuzzy logic, neural networks, evolutionary computing, probabilistic reasoning, chaotic theory and parts of machine learning theory. SC is more than a melange of these disciplines, it is a partnership, in which each of the partners contributes a distinct methodology for addressing problems in its domain. In this perspective, these disciplines are complementary more than competitive.

4 Soft-Biometrics: Soft-Computing and Biometrics

SC being able to consider variations and uncertainty is suitable for biometric measurements due to the following reasons:

- Biometric features do not have an absolute "ground truth" and they will hardly reach this. Biometric features always vary!
- Derivations from the "ideal" biometric characteristic are difficult or even unable to describe analytically.
- High accuracy within the measurement may cause inflexibility and the loss of generalization ability.

SC is increasingly being used in biometric systems whereas biometrics employing SC approaches are referred as *soft-biometrics*.

The general biometric system whose block-diagram is shown in figure 1 is made of a pre-processing module (PP); a feature extraction and coding module (FE/C); a reference determination and/or classifier generator module (RD/CG); an analysis and validation module (AV) and a result fusion module (RF). The PP-module comprises diverse methods that treat recorded data in such a way that significant features can be extracted easily. The FE/C-module includes

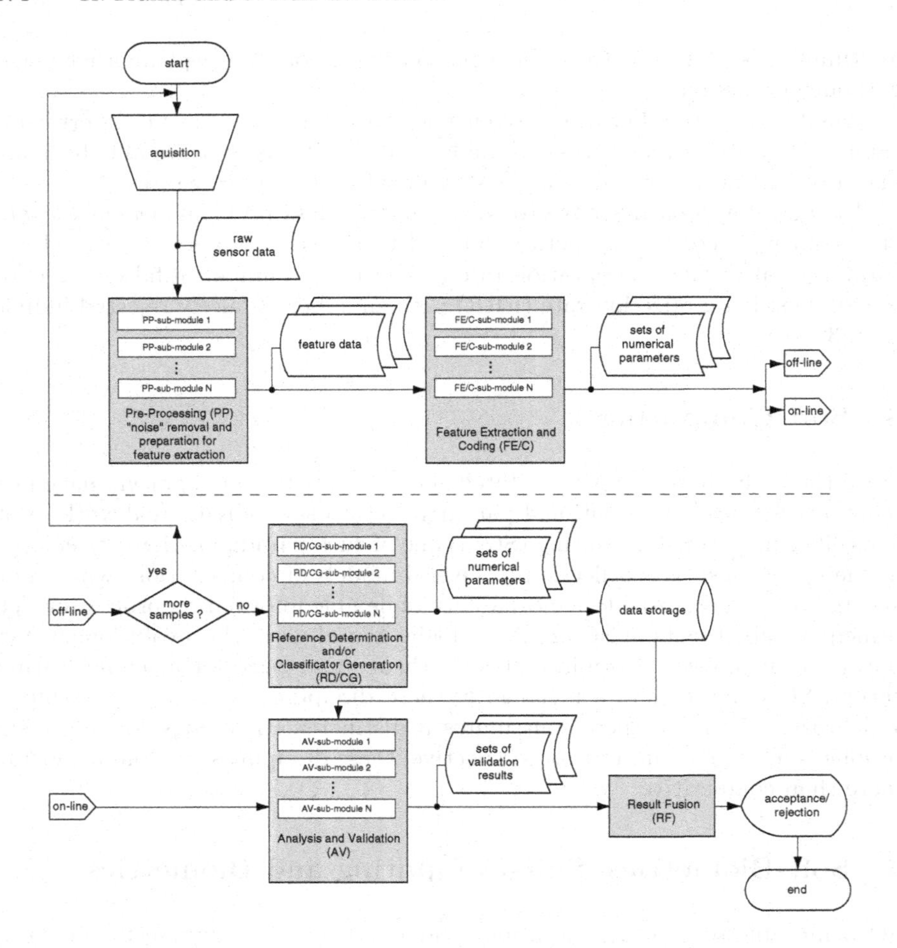

Fig. 1. Block-diagram of a biometric system in general

methods that convert treated input data into numerical parameters, which represent specific aspects of a biometric characteristic. Within the RD/CG-module the numerical parameters are used to determine reference/template-data or to generate classifiers. It is only employed in the off-line phase during enrollment/training. The AV-module is activated during the on-line phase to analyze and to validate the numerical parameters of a questioned (also called sample) characteristic. Last but not least the RF-module combines different outputs, in case there is more than one AV-module, to decide whether a person is what they claim to be.

SC can be introduced into any component/module of a biometric system. The application of SC as classifiers or decision-ruler is widely spread [4], whereas, SC in pre-processing and feature extraction is sparsely used. From our point of view the application of SC in biometrics has to be decided individually. Since SC is always data-driven, the available data has to be analyzed to decide in detail whether it is useful to employ SC or not.

To give an example. In fingerprint identification for the matching of e.g. 14 minutia the application of SC is overpowered, contrary in static signature verification, here, e.g. the number and shape of signature strokes varies. (Rem.: In case there are two almost identical sets of strokes it indicates fraud.) Consequently, SC approaches might be employed if the biometric features are of high complexity, less accurate and an analytical description is time consuming or almost impossible (see figure 2).

On the other hand it does not make sense to feed SC approaches with any available detail of a biometric characteristic. As in the traditional approach features that are considered during validation have to be significant. Here, SC can support the selection of typical details of a biometric characteristic.

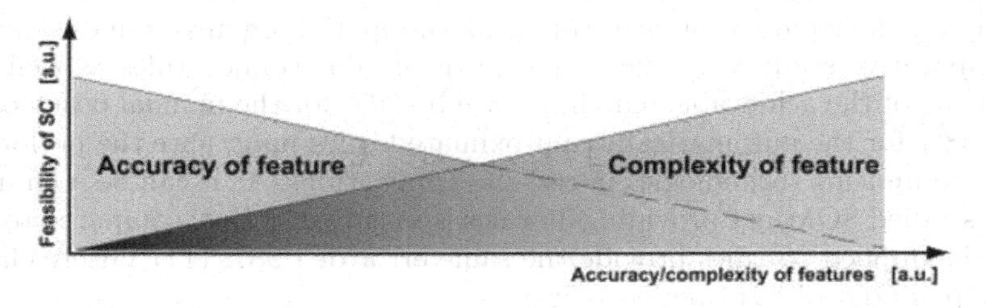

Fig. 2. Feasibility of soft-computing depending on complexity and accuracy of the biometric features

5 Application Example: Signature Verification

In signature verification artificial neuronal networks (ANN) were mainly employed as trainable classifiers [5],[4]. Until now there are only few approaches that use other SC approaches (e.g. fuzzy logic (FL) and/or evolutionary computation (EC)), too [6].

During the past years we have developed the *SIC Natura* system [7], [8] for signature verification that includes SC methods in the document-PP-module, in the RD/CG-module and within the AV-module.

For pre-processing, in particular for the elimination of textured backgrounds on paper documents, EC was considered [9], [8]. During a preparation phase a probe and a goal image has to be presented to the *LUCIFER* system, which generates morphological images-processing-filters for further usage (see figure 3).

To derive fuzzy matching rules a neuro fuzzy approach proposed by Kosko [10] is employed within the RD/CG module [11]. Thereby, questioned and reference signatures are threaded morphologically to derive such called regions. Then, human labeled region-training-sets are used to select and to adapt fuzzy rules for upcoming region matching within the AV-module.

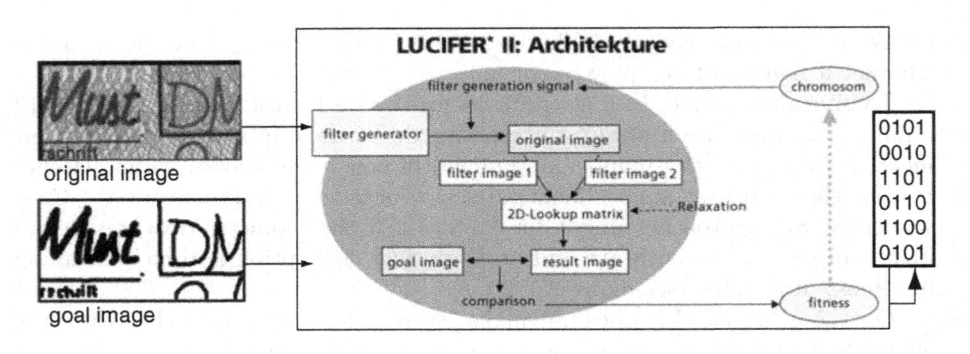

Fig. 3. LUCIFER-module for background filter generation

Here, soft-computing technologies are employed to derive a sophisticated schema for the region approach validating the spatial organization of signatures. A comparative study with the first empirically determined rules as well as the increasing of the achievable matching rate by 20% for the manual designed rules up to 98% for the automatically approximated rules punctuate the performance of soft-computing technologies in real-life applications. As it can be seen in table 1, the studied SOM-net providing 35 rules is mostly effective, compared to the 45 rules the applied 35 rules provide the same error of 1.98% (117 falsely classified region pairs out of 5914 region pairs).

Table 1. Result for SOM-architectures providing different numbers of adapted fuzzy rules. The total number of samples was 2929 for training and 5914 for testing

Net nodes	Number of rules	MSE 1. stage	MSE 2. stage	Err. train. data	Err. all data
3×8	19	0.0327	0.0286	129	255
3×12	29	0.0281	0.0217	98	202
3×16	35	0.0216	0.0171	55	117
3×24	54	0.0180	0.0131	45	117

Since there is a bundle of feature extraction and validation methods within the *SIC Natura* system further work will be devoted to the fusion of sub-module-results by using fuzzy fusion (e.g. Fuzzy Integral [12]).

6 Conclusions

SC approaches can be employed within all components of the biometric system, like for data pre-processing itself or for designing of adapted pre-processing filters, also, for the extraction of significant features, for reference determination, as classifiers or as decision-ruler as well as for result fusion. SC has to be used with care in biometric applications. Take a look at the kind of biometric data that are available to make your final decision.

Acknowledgement

The authors would like to thanks Mario Köppen, Aureli Soria-Frisch, Jan Schneider and Anita Stellmacher for inspiring discussions about soft-computing and biometrics.

References

1. Elsevier Advanced Technology: Biometrics technology today (01/1997-02/2000)
2. Newham, E., Bunney, C., C.Mearns: Biometrics report (1999)
3. Tele Trust - AG 6: Biometrische Identifikation - Bewertungskriterien zur Vergleichbarkeit biometrischer Verfahren (1998)
4. Schneider, J., Franke, K., Nickolay, B.: Konzeptstudie - Biometrische Authentifikation. Technical report, Fraunhofer IPK Berlin (2000)
5. Leclerc, F., Plamondon, R.: Automatic signature verification: the state of the art-1989-1993. International Journal of Pattern Recognition and Artificial Intelligence **8** (1994) 643–660
6. Yang, X., Furuhashi, T., Obata, K., Uchikawa, Y.: Constructing a high performance signature verification system using a GA method. In: Proceedings 2nd New Zealand International Two-Stream Conference on Artificial Neural Networks and Expert Systems, Dunedin, New Zealand (1995) 170–173
7. Franke, K., Köppen, M., Nickolay, B., Unger, S.: Machbarkeits- und Konzeptstudie Automaisches System zur Unterschriftenverifikation. Technical report, Fraunhofer IPK Berlin (1996)
8. Franke, K., Köppen, M.: A computer-based system to support forensic studies on handwritten documents. International Journal on Document Analysis and Recognition **3** (2001) 218–231
9. Franke, K., Köppen, M.: Towards an universal approach to background removal in images of bankchecks. In: Proceedings 6th International Workshop on Frontiers in Handwriting Recognition (IWFHR), Tajon, Korea (1998)
10. Kosko, B.: Fuzzy Engineering. Prentice Hall Internationall, Inc. (1997)
11. Franke, K., Zhang, Y.N., Köppen, M.: (Static signature verification employing a kosko-neuro-fuzzy approach) submitted to the International Conference on Fuzzy Systems (AFSS) 2002.
12. Grabisch, M., Nicolas, J.M.: Classification by fuzzy integral: Performance and tests. Fuzzy Sets and Systems **65** (1994) 255–271

Comparative Study between Different Eigenspace-Based Approaches for Face Recognition

Pablo Navarrete and Javier Ruiz-del-Solar

Department of Electrical Engineering, Universidad de Chile, Chile
{pnavarre,jruizd}@cec.uchile.cl

Abstract. Different eigenspace-based approaches have been proposed for the recognition of faces. They differ mostly in the kind of projection method been used and in the similarity matching criterion employed. The aim of this paper is to present a comparative study between some of these different approaches. This study considers theoretical aspects as well as simulations performed using a face database with a few number of classes.

1 Introduction

Among the most successful approaches used in face recognition we can mention *eigenspace-based* methods, which are mostly derived from the *Eigenface*-algorithm. These methods project the input faces onto a dimensional reduced space where the recognition is carried out, performing a holistic analysis of the faces. Different eigenspace-based approaches have been proposed. They differ mostly in the kind of projection/decomposition method been used and in the similarity matching criterion employed. The aim of this paper is to present a comparative study between some of these different approaches. The comparison considers the use of three different projection methods (Principal Component Analysis, Fisher Linear Discriminant and Evolutionary Pursuit) and five different similarity matching criteria (Euclidean-, Cosines- and Mahalanobis-distance, Self-Organizing Map and Fuzzy Feature Contrast). The pre-processing aspects of these approaches (normalization, illumination invariance, geometrical invariance, etc.) are not going to be addressed in this study. It should be noted that a previous comparative study that does not include the Fuzzy Feature Contrast method was presented in [4]. The mentioned methods are described in section 2, and the comparative study is presented in section 3.

2 Eigenspace-Based Approaches

Eigenspace-based approaches approximate the face vectors (face images) with lower dimensional feature vectors. The main supposition behind this procedure is that the face space (given by the feature vectors) has a lower dimension than the image space (given by the number of pixels in the image), and that the recognition of the faces can be performed in this reduced space. These approaches consider an off-line phase or

N.R. Pal and M. Sugeno (Eds.): AFSS 2002, LNAI 2275, pp. 178–184, 2002.

training, where the face database is created and the *projection matrix*, the one that achieve the dimensional reduction, is obtained from all the database face images. In the off-line phase are also calculated the *mean face* and the reduced representation of each database image. These representations are the ones to be used in the recognition process.

2.1 General Approach

Figure 1 shows the block diagram of a generic eigenspace-based face recognition system. A preprocessing module transforms the face image into a unitary vector and then performs a subtraction of the mean face ($\overline{\mathbf{x}}$). After that, the resulting vector, **x,** is projected using the projection matrix $\mathbf{W} \in R^{N \times m}$ that depends on the eigenspace method been used (see section 2.2). This projection corresponds to a dimensional reduction of the input, starting with vectors $\mathbf{x}$ in R^N (with N the image vector dimension) and obtaining projected vectors $\mathbf{q}$ in R^m with $m<N$ (usually $m<<N$). The *Similarity Matching* module compares the similarity of the reduced representation of the query face vector $\mathbf{q}$ with the reduced vectors $\mathbf{p}^k \in R^m$ that represent the faces in the database. By using a given criterion of similarity (see section 2.3), this module determines the most similar vector $\mathbf{p}^k$ in the database. The class of this vector is the result of the recognition process, i.e. the identity of the face. In addition, a *Rejection System* for unknown faces is used if the similarity matching measure is not good enough (see description in [1]).

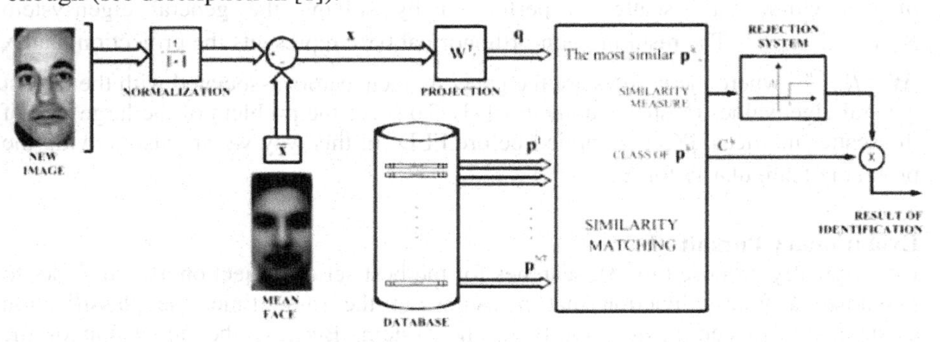

Fig. 1. Block diagram of a given eigen-space face recognition system.

2.2 Projection/Decomposition Methods

Principal Components Analysis - PCA

PCA is a general method to identify the principal differences between signals and after that to make a dimensional reduction of them. Let $\mathbf{X} = \left[(\mathbf{x}^1 - \overline{\mathbf{x}})(\mathbf{x}^2 - \overline{\mathbf{x}})\cdots(\mathbf{x}^{NT} - \overline{\mathbf{x}})\right]$ be the matrix of the normalized training vectors. $\mathbf{x}^j$ represents the normalized j image vector, $\mathbf{x}$ is the mean face image and NT is the number of training images. Then, $\mathbf{R} = \mathbf{X}\mathbf{X}^{\mathrm{T}}$ will be the correlation matrix estimator. The eigenvectors of $\mathbf{R}$ represent a special basis in the image space, and the eigenvalues are the projection variance on each of this axes (the Eigenfaces). Therefore PCA will chose only the eigenvectors of $\mathbf{R}$ associated with the higher variance and in this way will reduce the dimension of the training images. Also PCA give us the projection matrix $\mathbf{W} \in R^{N \times m}$ for reducing every image that follows the same statistical pattern. Computational aspects of the implementation of this method are explained in [4].

Fisher Linear Discriminant - FLD

FLD searches for the projection axes on which the face images of different classes are far from each other, and at the same time where the face images of the same class are close from each other. In a similar way of PCA using the $\mathbf{R}$ matrix, FLD uses two scatter matrices, $\mathbf{S}_b$ and $\mathbf{S}_w$, for representing the separation between the individual class means respect to the global mean face, and the separation between vectors of each class respect to their own class mean, respectively:

$$\mathbf{S}_b = \sum_{i=1}^{NC} \mathrm{P}(C_i)\left(\mathbf{m}^{(i)} - \mathbf{m}\right)\left(\mathbf{m}^{(i)} - \mathbf{m}\right)^{\mathrm{T}}; \quad \mathbf{S}_w = \sum_{i=1}^{NC} \mathrm{P}(C_i)\,\mathrm{E}\left[\left(\mathbf{x}^{(i)} - \mathbf{m}^{(i)}\right)\left(\mathbf{x}^{(i)} - \mathbf{m}^{(i)}\right)^{\mathrm{T}}\right] \quad (1)$$

where $\mathbf{m}$ is the global mean vector, $\mathrm{P}(C_i)$ are the occurrence probabilities associated to each class C_i, $\mathbf{m}^{(i)}$ are the average vectors of C_i, and $\mathbf{x}^{(i)}$ are the vectors associated to C_i. The maximization of the between class scatter and the minimization of the within class scatter is performed by solving the general eigensystem $\mathbf{S}_b\,\mathbf{w}^k = \lambda_k\,\mathbf{S}_w\,\mathbf{w}^k$. The resulting non-orthonormal base represents the projection matrix $\mathbf{W} \in R^{N \times m}$, where the m rows are the general eigenvectors associated with the largest general eigenvalues (Fisher Parameters [4]). To solve the problem of the large size of the scatter matrices, PCA is applied before FLD. In this way we are also solving the problem of singularity for $\mathbf{S}_w$.

Evolutionary Pursuit - EP

EP, originally proposed in [3], searches for the best set of projection axes in order to maximize a fitness function that measures, at the same time, the classification accuracy and generalization ability of the system. Because the dimension of the solution-space of this problem is very large, it is solved using Genetic Algorithms. In order to obtain the EP-faces an initial dimensional reduction is first performed using PCA, and then a Whitening Transformation is applied (equivalent to a Mahalanobis metric system, see 2.3). In the Whitened-PCA space are performed several rotations between pair of axes and then a subset of them is selected. Each rotation is coded using a chromosome representation. In this representation each chromosome represents a certain projection system. To evaluate this system the following fitness function is used:

$$\zeta(\alpha_k, a_i) = \zeta_a(\alpha_k, a_i) + \lambda\,\zeta_s(\alpha_k, a_i), \quad (2)$$

where $\zeta_a(\alpha_k, a_i)$ measures the accuracy, $\zeta_s(\alpha_k, a_i)$ measures the generalization ability, and λ is a positive constant (see definitions in [3]).

2.3 Similarity Matching Methods

Euclidean Distance

$$d(\mathbf{x},\mathbf{y}) = \sqrt{(\mathbf{x}-\mathbf{y})^{\mathrm{T}}(\mathbf{x}-\mathbf{y})}\ . \tag{3}$$

Cosine Distance

$$\cos(\mathbf{x},\mathbf{y}) = \frac{\mathbf{x}^{\mathrm{T}}\mathbf{y}}{\|\mathbf{x}\|\|\mathbf{y}\|}\ . \tag{4}$$

Mahalanobis Distance

$$d(\mathbf{x},\mathbf{y}) = (\mathbf{x}-\mathbf{y})^{\mathrm{T}}\,\mathbf{R}^{-1}\,(\mathbf{x}-\mathbf{y})\ ; \qquad \mathbf{R}\text{: correlation matrix.} \tag{5}$$

From a geometrical point of view this distance has a scaling effect in the image space. Taking into consideration the face image subset, directions in which a greater variance exist are compressed and directions in which a smaller variance exist are expanded. It can be proved that in the PCA space the Mahalanobis distance is equivalent to the Euclidean distance, weighting each component by the inverse correspondent eigenvalue (see demonstration in [4]), and it is often called Whitening (PCA) Transformation.

SOM Clustering
Self-Organizing Maps (SOMs) are used as associative networks to match the projected query face with the corresponding projected database faces. The use of a SOM to implement this module improves the generalization ability of the system. The SOM approach uses reference vectors $\mathbf{m}_i$ to approximate the probability distribution of the faces in a 2D map [2]. In the training phase of the SOM a clustering of the reduced face vectors is carried out. Thereafter the SOM is transformed in an associative network by labeling all its nodes. Both procedures are explained in [4].

Fuzzy Feature Contrast – FFC

$$S(\mathbf{x},\mathbf{y}) = \sum_{i=1}^{m} \min\{\mu_i(\mathbf{x}),\mu_i(\mathbf{y})\} - \alpha\sum_{i=1}^{m} \max\{\mu_i(\mathbf{x})-\mu_i(\mathbf{y}),0\} - \beta\sum_{i=1}^{m} \max\{\mu_i(\mathbf{y})-\mu_i(\mathbf{x}),0\} \tag{6}$$

where $\mu_i(\mathbf{x})$ is a membership function associated with the i-component of vector $\mathbf{x} \in R^m$. This similarity measure, originaly proposed in [5], is a fuzzy implementation of the Feature Contrast model from Tversky. The first sum measure the common features (intersection) and the others represent the distinctive features (difference in the two possible ways). The positive parameters α and β adjust the contrast of the three kind of features. By chosing $\alpha \neq \beta$ it is possible to introduce asymetries between the subject-referent comparison. This model considers that all the features are independent, and that can be assumed in PCA and WPCA, but not in FLD and EP. In our implementation we normalize each feature of PCA (in WPCA it is not necessary) and we chosed $\mu_{i\,(\mathbf{x})}$ linear between –1 and 1, with x_i normalized.

3 Comparison among the Approaches

In order to test the described methods we have made several simulations based in the Yale University - Face Image Database. We use 150 images of 15 different classes. Then we preprocessed the images by masking them in windows of 100 x 200 pixels placing the several face features in the same relative places. In table 1 we show the results of several simulations using different kind of representations and similarity matching methods. For each simulation we used a fixed number of training images, using the same type of images per class, according with the Yale database specification. In order to obtain representative results we take the average of 20 different set of images for each fixed number of training images. All the images not used for training are used for testing.

We can see that the best models always are obtained with the Fisher representation, and the difference against the other representations decrease when the number of training images per class decrease, showing that the FLD discrimination ability strongly depends on the number of training images per class. The best results are almost always obtained with FLD- cosine. The systems that seem to be as efficient as FLD-cosine are SOM and Withening-cosine.
The best results using FFC were obtained employing an asymmetric subject-referent comparison: α=0.5 β=5. This means that in the question "how is the subject face similar to the referent face?" the answer focus more on the features of the referent (the unknown face). The generalization ability of the systems is not well measured in our simulations because the number of selected axes is about the same of the number of classes (15). That affects the FLD representation method as well as the FFC and SOM similarity matching methods. For this reason in future works we want to perform our comparative study on a larger database, like FERET. We think that this will improve the relative recognition ability of the methods being affected for the small number of classes.

Another important issue is the computational cost of the training processes. In PCA this computational cost is mainly due to the process of determining **R**, $O(NT^2 \cdot N)$, and solving the eigensystem, $O(NT^3)$. If we suppose that the number of training images NT is much smaller than the number of pixels per image N, then the computational cost of PCA is just the cost of determining **R**, $O(NT^2 \cdot N)$. In our implementation of FLD we requires previously the computation of PCA to reduce the vectors dimension to m_1 ($m_1 < NT$), and the additional cost is due to the process of determining the scatter matrices, $O(m_1^2 \cdot NC)$, and solving the general eigensystem,

$O(m_1^3)$. Nevertheless the additional cost in FLD is usually much smaller than the PCA initial cost. Finally EP requires much more computations because this process must iterate until a given criterion is accomplished. The computational cost of on-line operation is mainly given by the comparisons with database vectors, $O(NT \cdot m)$, except when the SOM-based similarity measure is used, $O(\text{(number of nodes)} \cdot m)$. The numerical stability for the different methods depends mostly of the numerical algorithms used for solving eigensystems. Either in PCA or FLD this is not a critical problem because always involves the management of symmetric matrices.

Table 1. Mean recognition rates using different numbers of training images per class, and taking the average of 20 different training sets. The small numbers are the standard deviation of each recognition rate.

	im./class	axes	Euclidean	cos(·)	SOM	FFC	Whitening Euclidean	Whitening cos(·)	Whitening SOM	Whitening FFC
PCA	**6**	56	87.9 6.2	86.0 6.8	84.6 7.0	77.1 10.1	64.7 9.4	79.3 11.6	64.7 10.5	77.1 10.1
FISHER		17	91.5 6.6	91.6 6.5	90.3 6.7	83.9 9.3	91.9 5.8	92.6 5.6	92.1 6.2	85.6 8.3
E.P.		16	81.2 9.0	85.3 8.7	83.7 9.8	77.2 8.0	-	-	-	-
PCA	**5**	34	88.7 3.8	87.1 4.1	86.0 5.1	78.5 8.1	69.5 8.9	83.2 9.0	66.1 10.5	78.5 8.1
FISHER		15	92.2 5.7	91.7 6.2	90.3 6.4	85.1 9.1	92.3 4.7	92.4 5.7	92.1 5.3	85.4 8.5
E.P.		13	84.1 5.7	87.7 6.6	86.7 7.6	78.7 6.8	-	-	-	-
PCA	**4**	46	87.3 3.9	86.7 3.9	84.8 3.6	77.6 5.2	72.9 5.5	84.4 5.6	66.7 6.5	77.6 5.2
FISHER		18	90.3 4.5	91.1 5.0	90.3 4.4	84.4 5.9	90.4 4.2	91.0 4.4	90.1 4.7	82.9 5.7
E.P.		18	83.6 4.6	86.9 4.7	85.0 5.0	74.7 6.0	-	-	-	-
PCA	**3**	35	86.6 4.0	85.4 3.9	82.0 5.6	77.9 4.6	75.0 5.6	84.8 5.4	67.4 6.9	77.9 4.6
FISHER		15	89.0 3.6	90.4 4.0	87.4 4.0	80.7 6.3	88.9 3.1	89.9 3.9	88.7 3.9	81.5 3.4
E.P.		14	81.1 4.3	86.9 3.7	82.5 3.7	75.9 4.4	-	-	-	-
PCA	**2**	26	82.7 5.9	80.8 5.9	76.2 7.9	71.1 5.9	75.6 4.9	82.1 4.6	60.8 7.3	71.1 5.9
FISHER		15	81.5 5.6	82.2 5.8	79.4 5.8	69.3 8.6	80.7 4.7	82.8 4.9	78.8 5.8	73.6 6.2
E.P.		14	77.8 5.6	81.2 5.3	76.0 7.3	70.0 7.4	-	-	-	

Acknowledgements

This research was supported by the DID (U. de Chile) under Project ENL-2001/11 and by the join "Program of Scientific Cooperation" of CONICYT (Chile) and BMBF (Germany).

References

1. Golfarelli M., Maio D., and Maltoni D., "On the Error-Reject Trade-Off in Biometric Verification Systems", *IEEE Trans. Pattern Analysis and Machine Intelligence*, vol. 19, no. 7, pp. 786-796, July 1997.
2. Kohonen T., "Self-Organized Maps", 1997.
3. Liu C., and Wechsler H., "Evolutionary Pursuit and Its Application to Face Recognition", *IEEE Trans. Pattern Analysis and Machine Intelligence*, vol. 22, no. 6, pp. 570-582, June 2000.
4. Navarrete P., and Ruiz del Solar J., "Eigenspace-based Recognition of Faces: Comparisons and a new Approach", *Proc. of the Int. Conf. on Image Analysis and Processing ICIAP 2001*, pp. 42-47, Sept. 26-28, Palermo, Italy.
5. Santini S., and Jain R., "Similarity Measures", *IEEE Trans. Pattern Analysis and Machine Intelligence*, vol. 21, no. 9, pp. 871-883, September 1999.

Static Signature Verification Employing a Kosko-Neuro-fuzzy Approach

Katrin Franke, Yu-Nong Zhang, and Mario Köppen

Dept. of Pattern Recognition, Fraunhofer IPK, Berlin, Germany,
katrin.franke@ipk.fhg.de

Abstract. To overcome difficulties in transferring "classical" handwriting examination methods into computer algorithms a hybrid neuronal system, proposed by Kosko [1], was employed to derive rules for signature region matching. The segmentation of signatures, written on paper documents, into regions will be presented and the two stage fuzzy rule learning, for finding and tuning the fuzzy rules will be discussed. By using the neuro-fuzzy approach [1] a region matching performance of 98% was achieved.

1 Introduction

The *SIC Natura* system [2], [3] for static and pseudo-dynamic signature verification provides the frame for the studies described below. It is being applied for validating whether a questioned signature, written on a piece of paper, is from the same person as a reference already stored in the database. Also, new references might be prepared for database storage[1].

From our point of view the challenge in developing signature validation systems is based on the knowledge transfer, going from the human handwriting examiner to computer algorithms. An analytical, mathematical formulation of this acquired and consolidated knowledge is very time-consuming and in many cases almost impossible. Here, soft-computing technologies provide a capable tool-sets to deal with [4].

The presented region approach is a special analysis method of the *SIC Natura* system. Contrary to former works [2], [3] the presented approach focuses on the validation of the spatial organization of a signature, which is only one aspect of the whole signature verification procedure. The proposed regions support a better representation of the spatial organization of the signature strokes than such called identity grids [5] being equidistant and fix for each writer. Moreover, soft-computing technologies are employed to evaluate the proposed region approach, in particular to derive a more sophisticated validation schema.

A known neuro-fuzzy approach will be employed as a tool-set during system design. Thereby, questioned signature regions were presented to a human examiner who has to label the regions with their validation result. The labeled

[1] Due to the planned application there is the restriction of using just one reference signature per writer.

N.R. Pal and M. Sugeno (Eds.): AFSS 2002, LNAI 2275, pp. 185–190, 2002.

regions are used to determine adapted matching rules employing soft-computing technologies, in particular a Kosko-Neuro-Fuzzy approach [1].

A comparative study with the first empirically determined rules as well as the increasing of the achievable matching rate by 20% for the manual designed rules up to 98% for the automatically approximated rules punctuate the performance of soft-computing technologies in real-life applications.

In section 2 the region approach will be presented. And, section 3 discusses the application of the Kosko-Neuro-Fuzzy approach [1] for deriving region matching rules.

2 The Region Approach

The segmentation of signatures into regions will be proposed. Also, starting from the basic region matching-algorithm the application of the Kosko-Neuro-Fuzzy approach will be motivated.

Reasons for using regions: The region matching is a macroscopic investigation of the spatial organization of a signature. As it can be seen in figure 1, regions are bloated contour-strokes that link local stroke characteristics to global stroke distribution[2].

Studying the significance of single signature strokes [2], we found out that there are many similar strokes for different writers. In particular, short strokes with a low curvature are very common, e.g. in letters "m", "n" and "u". Looking into detail we also recognized that the segmentation of the writing space varies. The main advantage of bloated stroke (the regions) is that they became unique by considering their neighborhoods.

Production of signature region algorithm: For obtaining spatial organization of a signature, the scanned signature image is brought into the normalized form of a binary image (the way to do this depends on the document context [3], and will not be presented here). The region production is determined from a single parameter δ, which gives an offset direction.

1. In the binary image, black-white transitions in the direction δ are set to black, all other pixels to white (this gives the so-called *shadow image* [6] in direction δ). It has to be noted that thus the binary signature image is represented by a one-pixel thick contour, decomposed into a number of separated strokes. The single strokes are now the regions R_i of level $n = 0$.
2. Design a signature mask M by eroding the signature image (i.e. expanding black) with a full-sized 7×7 structuring element five times (see [7] for a description of morphological operations).
3. Set region R_i of level n as region R_i of level $(n+1)$. Also, assign each neighbor of region R_i of level n, which does not already belong to another region R_j of level $(n + 1)$ and is within the signature mask M, to region R_i of level $(n + 1)$.
4. Repeat step 3 until there are no pixels left within the signature mask M.

[2] In the average a signature can be represented by 30 to 70 regions.

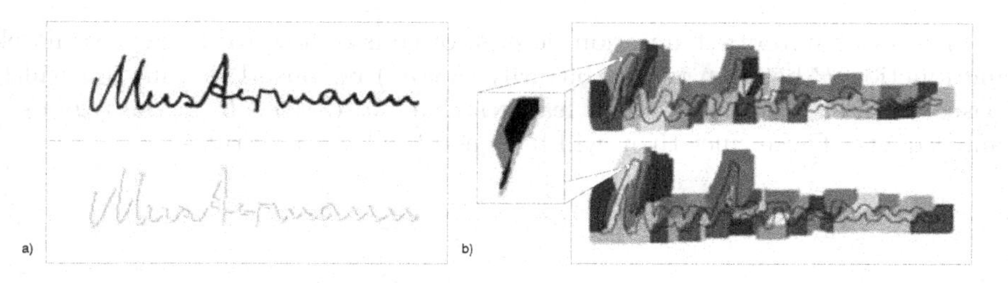

Fig. 1. Region production and matching: a) binary signature image and corresponding shadow image; b) regarding their centers of gravity, regions of reference and pattern image were laid one upon the other

For convenience, each region is marked within the image with a separate color. Also, for better performance, three shadow images for directions $\delta_1 = (1, -1)$, $\delta_2 = (1, 0)$ and $\delta_3 = (1, 1)$ were fused before starting step 3 (compare figure 1).

Region matching algorithm: Once the regions of two images I_1 and I_2 are obtained, the goal of the region matching algorithm is to compute a measure of overall region similarity. The algorithm presumes a matching function $P(R_i, R_j)$ between two single regions R_i from image I_1 and R_j from image I_2 with result values from $[0, 1]$ to be given. In the following, image I_1 is considered to be the given reference image, and I_2 the questioned pattern image.

1. Scale the pattern image onto the same size as the reference image. This will give a linear transformation L of each pixel of the pattern image into the reference image. And, compute all centers-of-gravity (COGs) of all regions in both images. Set the matching degree value to $E = 0$. Also, create an array of flags, the *reference usage*, with one flag set to `false` for each region of the reference image.
2. For one region R_i of the pattern image, transform its COG to a point of the reference image by applying L. Select the five nearest COGs of regions in the reference image. If the distance to the nearest COG is larger than a value d_{max}, evaluate the matching with 0 and go to step 5.
3. Among the five nearest COGs, select the one with the corresponding region area (number of pixels belonging to that region) closest to the area of R_i. This be region R_j of the reference image.
4. Evaluate the matching for R_i by computing $P(R_i, R_j)$ and add this value to E. Also, set the flag of R_j in the reference usage array to `true`.
5. Repeat steps 2 to 4 for each region of the pattern image.
6. Divide the matching value E by the number of regions in the pattern image (thus getting a value r_1 between 0 and 1), and also compute the sum of all regions of the reference image with corresponding flag `true` in the reference usage array. Divide this sum by the total area of all regions in the reference image, thus giving a second value r_2 between 0 and 1. The matching between both images is computed as weighted sum of r_1 and r_2. For this application, r_1 was weighted with 0.8 and r_2 with 0.2. These weights were determined empirically since r_1 measures the overall matching and r_2 the reference usage.

The most important question here is, of course, how to design the matching function P between two (arbitrarily shaped) regions. This function will be based on rules with intervals of features that are derived from the regions as antecedents. Those rules then, will be generalized to fuzzy rules.

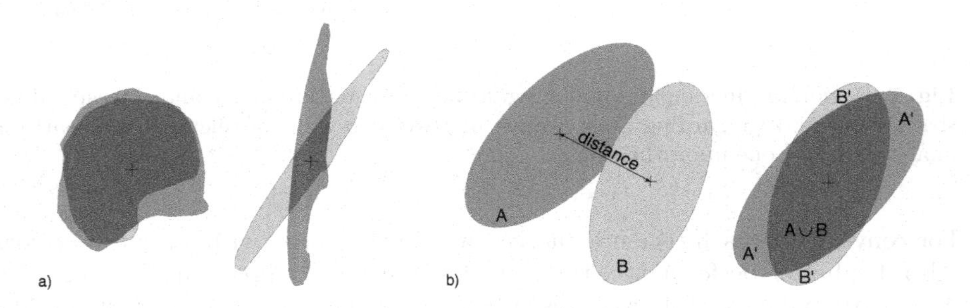

Fig. 2. Region and their features

The analysis of signatures in general as well as the matching of regions requires tolerance mechanisms, since signing is a behavioral characteristic of a human and underlies permanent variations. To design a matching function P the basic characteristics of regions as well as their variations have to be studied. There are various types of regions. Mainly, they can be categorized in compact and stretched regions (see figure 2). Looking at the intra-individual region variations we recognized that slant variations are most relevant for the stretched regions, even if their shape is almost stable. Finally, we decided to design adapted matching rules that take primary region characteristics into account.

A subset of 254 region pairs was analyzed to empirically determine numerical parameters and six rules for the region matching function $P(R_i, R_j)$. Thereby, the region pair (R_i, R_j) is described by the numerical parameters: area for of the reference region, $p_1 = A(R_j)$; relatively to the area of the pattern region, $p_2 = A(R_j)/A(R_i)$; the intersection of both regions, $p_3 = A(R_j) \cap A(R_i)$; the difference set for the reference, $p_4 = A(R_j) - p_3$ and for the pattern, $p_5 = A(R_i) - p_3$; as well as the distance d between the COGs of both regions, $p_6 = d$.

The application of the first designed matching function P yield to the fact that the rules were to rough and marginal cases led to false classifications. One solution could be the refinement of rules, the other one a stepwise extension of the rule-basis. Very quickly we recognized that the complexity of the input data is not suitable for further manual adapting of the matching rules by investigating primary region characteristics. It has to be considered that the region sizes vary between few up to more than thousand image points (pixels). Also, the numerical parameters span a six-dimensional vector space. Finally, to avoid rule explosion and the loss of generalization ability, we decided to consider soft-computing approaches for finding and tuning the adapted region matching function $P(R_i, R_j)$.

3 Fuzzy Rule Learning

The manually designed rules for region matching can be generalized to fuzzy rules. The rules approximate the mapping function P of (region) input data $X = [p_1, ..., p_6]$ to (validation) output data Y. The rules might not only be derived by observing humans action. Under the pre-condition that there is a representative dataset $[X, Y]$ the rules might be also determined automatically by fuzzy function approximation. Due to the reported performance, the hybrid neuronal system proposed by Kosko [1] is employed. The basic principle follows a geometric approach, which assumes fuzzy rules as point in an n-dimensional vector space. The hybrid neuronal system combines unsupervised and supervised neuronal learning. Thereby, the first neuronal system uses unsupervised learning to find the covariance matrices of synaptic vectors. It picks the first set of rules, which approximate the mapping function P, based on the statistics of the training data $[X, Y]$. The second neuronal system uses supervised learning to tune the rules with stochastic gradient descent. In our approach we employed a SOM-net [8] [9] for the first learning stage. In the second we used gradient decent with permanent adaptation of the learning constant.

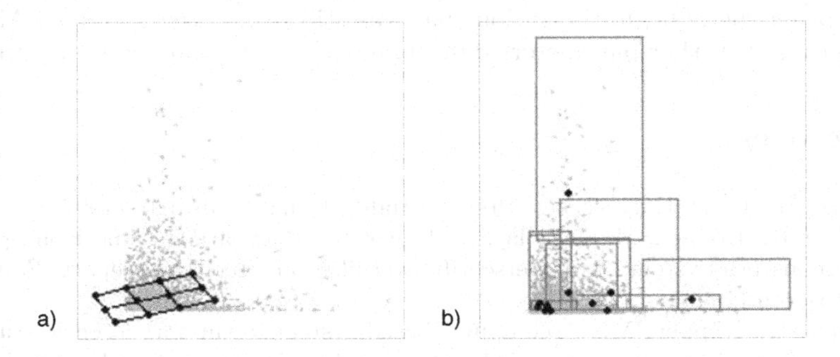

Fig. 3. Rule finding by applying 3 × 4 SOM-net: feature projection with a) linear initialized SOM-nodes, b) trained SOM-nodes and corresponding patches that represent fuzzy rules

For training and testing we provided 5914 region pairs that were manually labeled by a human examiner. Although, the region pairs were take from genuine signatures, there were pairs that could not be labeled non-ambiguously or that has to be labeled as non-identically. In this cases the matching value were $P(R_i, R_j) = 0.5$ resp. $P(R_i, R_j) = 0.0$, otherwise the matching value were set to $P(R_i, R_j) = 1.0$. Then, the mapping function P can be approximated by $(p_{1_{ij}}, ..., p_{6_{ij}}, P(R_i, R_j))$. Out of the overall labeled region pairs we took 2929 pairs, without considering any specific criteria, for finding and tuning the fuzzy rules.

As mentioned before, it is quite difficult to predict the required number of rules. Since there are three matching values $[0.0, 0.5, 1.0]$ possible, we decided to

Table 1. Result for SOM-architectures providing different numbers of adapted fuzzy rules. The total number of samples was 2929 for training and 5914 for testing.

Net nodes	Number of rules	MSE 1. stage	MSE 2. stage	Err. train. data	Err. all data
3×8	19	0.0327	0.0286	129	255
3×12	29	0.0281	0.0217	98	202
3×16	35	0.0216	0.0171	55	117
3×24	54	0.0180	0.0131	45	117

investigate SOM-architectures with $3 \times K$ nodes. In figure 3 can be seen how the input data are clustered by the founded rules of a 3×4 SOM-net.

The results for different SOM-architectures as well as the corresponding mean square errors (MSE) are given in table 1. As it can be seen, the SOM-net providing 35 rules is mostly effective. Compared to the 45 rules the applied 35 rules provide the same error of 1.98% (117 falsely classified region pairs out of 5914 region pairs). Contrary, due to a lower amount of rules (10 rules less) less computational effort can be expected.

The achieved results of 98.02% matching rate by applying the 35 rules tune very optimistic. However, the implemented approach might be further improved by adapting the SOM-initialization that was still linear (see figure 3). Also, the application of further parameters describing the region pairs could be studied.

References

1. Kosko, B.: Fuzzy Engineering. Prentice Hall Internationall, Inc. (1997)
2. Franke, K., Köppen, M., Nickolay, B., Unger, S.: Machbarkeits- und Konzeptstudie Automaisches System zur Unterschriftenverifikation. Technical report, Fraunhofer IPK Berlin (1996)
3. Franke, K., Köppen, M.: A computer-based system to support forensic studies on handwritten documents. International Journal on Document Analysis and Recognition **3** (2001) 218–231
4. Franke, K., del Solar, J.R.: (Soft-biometrics: Soft-computing technologies for biometric-applications) submitted to the International Conference on Fuzzy Systems (AFSS) 2002.
5. Murshed, N., Sabourin, R., Bortolozzi, F.: A cognitive approach to off-line signature verification. International Journal of Pattern Recognition and Artificial Intelligence **11** (1997) 801–825
6. Watanabe, S., Furuhashi, T., Obata, K., Uchikawa, Y.: A study on feature extraction using fuzzy net for off-line signautre recognition. In: Proceedings of 1993 International Joint Conference on Neuronal Networks. (1993) 2857–2860
7. Serra, J.: Image Analysis and Mathematical Morphology. Academic Press, London (1982)
8. Kohonen, T.: Self-Organizing maps. Springer (1995)
9. Kohonen, T.: SOM-PAK, Self-Organizing Map Program Package, Helsinki University of Technology. (1995)

A Control Analysis of Neuronal Information Processing: A Study of Electrophysiological Experimentation and Non-equilibrium Information Theory

Prasun K. Roy[1,3,*], John P. Miller[2], and D. Dutta Majumder[3,4]

[1] Lishman Brain Unit, Maudsley Hospital, University of London, London SE5 8AZ, UK
[2] Centre for Computational Biology, Montana State University, Bozeman, MT-59717, USA
[3] Indian Statistical Institute, ECSU, 203 Barrackpore Trunk Road, Calcutta-35, India*
[4] Institute of Cybernetics Systems & Information Technology, 155 Asokgarh, Calcutta-35, India
pk_roy@doctors.org.uk, jpm@cns.montana.edu, icsitddm@vsnl.net

Abstract. A model of information transmission across a neuron is delineated in terms of source (stimulus)-encoder-channel-decoder-behaviour (response). From cybernetic analysis of experimental data, we perform frequency/time domain and stability analyses and obtain the Bode, Nichols and Nyquist plots, Root locus plane, transfer function and response equation, all confirmed by data. We consider a new paradigm of information theory based on non-equilibrium dynamics of fluc-tuation, organization and information (Nicolis-Prigogine), that is the counterpart of Shannon-Boltzmann approach to information-entropy based on equilibrial dyna-mics. The Prigogine theorem of minimum entropy production and Rosen's prin-ciple of optimum design were observed to characterize neural transmission in a particular test neuron operating near optimal sensitivity regime. Using Nyquist theorem and generalized temperature concept, we compute a non-equilibrial ent-ropy production and neurodynamic temperature equivalent during neural informa-tion processing. A trans-information/temperature plot implies an order-disorder Bose transition and zero neurodynamic entropy (near 0^0N) as informational analog of third law of thermodynamics (near 0^0K). Neural applications are explored.

1 Introduction

Within the last decade the emerging field of computational neuroscience has provided much insight into neuronal function, specially with regard to information transmission and communication. Many researchers (including one of the authors [1,2]) have generated extensive experimental data on neuronal information processing, using various types of stimulus/response paradigms and analyzing the data with the tool of information theory. From a very different starting point, Wiener and others have developed the field of cybernetics as an integrated synthesis of control, communication and computation [C^3] techniques, for study of information and regulation in biological systems. Numerous workers (as two of the authors) have shown how cybernetics provide a unitary theory of activity and dynamics of a wide variety of physical, biological or neurological systems [3,4]. In this paper we explore a unification of control theoretic or "cybernetic" approaches with information theoretic approaches.

Neuroscientists have generally used the classical model of information theory, namely the Shannon formalism consonant with the Boltzmann statistical mechanics

N.R. Pal and M. Sugeno (Eds.): AFSS 2002, LNAI 2275, pp. 191–203, 2002.

of equilibrium thermodynamics of the 1870s, in their analyses of neural systems. The basic information parameter within the context of this formalism is *entropy*. However, the newer discipline of non-equilibrium thermodynamics, developed in the 1970s by Prigogine and Eigen, shows how organization and information can be actuated in living systems -- whether neuronal, cellular or genetic -- in the form of novel functional structures, called dissipative structures [5-7]. Lately Hofkirchner, Nicolis, Avramescu, Baddeley and others [9-12] have used this non-equilibrium approach to information theory, to obtain insights into information processing in biological and cognitive systems, the basic parameter here being the *entropy production* and *information progression*. In the present study, we consider the application of non-equilibrium information theory and entropy production to neural functioning and elucidate an approach towards studying optimal neural functioning, neurothermodynamics and neurodynamical temperature. For exploring the concept of neurodynamic temperature and neural trans-information, we quantify the variation of gain and phase with frequency (i.e. Bode, Nyquist and Nichols plots.) For investigating neuronal optimality we use Root locus plane representations and Input-Output analysis. We shall first present a study of neural information processing using tools of control theory and input-output analysis. We shall then use these findings to proceed towards an informational neurothermodynamic approach.

2 Control Systems Analysis

Schema 1 shows our neural information processing pathway, the *source* is an external stimulus. We proceed to explore the system response in frequency and time domains.

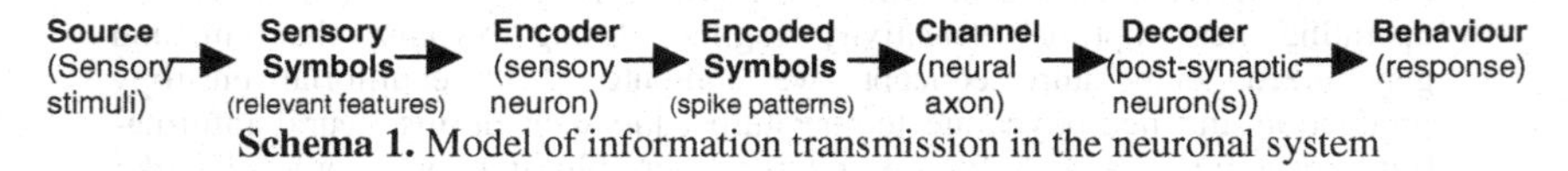

Schema 1. Model of information transmission in the neuronal system

Bode Plots of the Neurone. The specimen neurone considered here is a primary mechanosensory interneurone identified as "10-3" within the cercal sensory system of an insect, the cricket, *Acheta domestica*. This interneuron is sensitive to air currents in the animal's immediate environment. All experimental recordings to be discussed here is taken from two reports published previously [1,2]. For these studies, the air current stimuli were generated by movement of loudspeakers. The responses of the neurone were recorded with intracellular micro-electrodes. Experimental neuro-physiological data is available to characterize the relationship between Stimulus (air current) and Response (neural spike train) within the stimulus frequency band of 5-400 Hz. For the analysis presented in those reports, the metric used for characterizing the stimulus-response sensitivity was the stimulus-response coherence (i.e., the cross-correlation of the stimulus waveform and the spike train response, normalized by the autocorrelation of the stimulus and the autocorrelation of the response).

Coherence is a measure that can be related to the mutual information between stimulus and response, assuming a linear encoding scheme. This particular neurone encoded significant information only within the 5-80 Hz band (bandwidth mean ≈ 40 Hz) [1]. Data is presented on stimulus-response coherence (denoted here as the gain "g") versus frequency f for the cell (fig 5 of ref.[1]), and on the frequency f versus change of time shift (ΔT) of the cell's peak response relative to stimulus frequency component that elicited that response (fig 3 'inset', *ibid*). The stimulus-response coherence data were obtained for 10 different frequencies: 10 Hz, 20 Hz,....90 Hz, 100 Hz. From the time shift ΔT and f, we can calculate the phase shift $\Delta\omega$. With this data we can plot the ω–f and g-f relations in a log scale of frequency, thus computing the Bode diagrams of the neurone. These plots are shown in fig. 1a and b, where the dashed lines denote curve of closest fit. Coherence (denoted as gain g) tends to zero after 90 Hz onwards. The Bode plots characterize the frequency encoding sensitivity of the neurone under varying stimulation, assuming a linear encoding scheme.

Nyquist Trajectory and Stability Principle. For each of the 10 frequency values above, we can further construct a polar plot of gain and phase shift as Nyquist figure for the neurone (Fig.2a). Note that as frequency increases, the Nyquist curve moves in an ovoid arcuate trajectory that approaches the origin. The Nyquist critical point at –1 gain and 180^0 phase is outside the trajectory, thus indicating that the neural system is stable. A system can be made unstable by maneuvering the Nyquist trajectory so that it encloses the critical point. Such a transition may be achieved by introducing de-compensation by means of a superimposed transfer function or synaptic delays.

Nichols Plot and Root Locus Plane. Instead of using the polar graph of the Nyquist figure, we can also illustrate the gain-phase relation using orthogonal coordinates (Nichols plot, fig 2b). A Nichols plot is useful in examining the system behaviour of a population of neurones, since this plot (in contrast to Nyquist plots) permits a simple graphical addition of the curves of individual neurones. To get a further idea of the behaviour of the neuronal system, we need to plot the imaginary frequency vs. real frequency, in the root locus plane. This plot is shown in fig. 3a. For each of the sampling frequencies f_n=10 Hz, 20 Hz,...100 Hz with their respective phase shifts ($\Delta\omega_n$) of the phasors, we readily calculate the corresponding components of real and imaginary frequency. The root locus plane can furnish the stability dynamics and complex RLC parameters of the neural circuit, as immittance, impedance and reactance. In fig. 2b, the root locus plane shows an anticlockwise ovoid trajectory as frequency rises. Ovoid root locus representation has been noted earlier by Stark for the myoneural system [13], though not for a neuroinformational system as ours.

3 Input-Output Analysis

We now proceed to describe the system function through input-output characteristics. Specifically, we study the transient response of the neurone to a step change in the amplitude of an input stimulus [2]. For that study, the air current stimuli were band-passed white noise waveforms. Neural responses were recorded for step changes in the RMS velocity of the white noise air current stimulus waveforms. Air currents having RMS velocities of 0.25, 0.5, 1, 2, 4 and 8 units [1 unit = air current of 250 $(cm/s)^2$] were presented. It was observed that there was threshold effect: only stimulus intensities above about 3 units produce an observable step response change in spike rate that is different from the pre-stimulus values [fig 3b, curve G (4 units stimulus) and curve H (8 units)]. Weaker stimuli had no significant effect on spike discharge. The two higher stimuli caused immediate responses: the spike rate increased and then exponentially decreased with time (neglecting perturbations). Similar exponential decreases of neural responses to step function inputs are found in invertebrates (e.g., stimulation of femoral spine antenna of cockroach [15]) and in vertebrates (e.g., feather response to step input of ambient temperature in birds [16]).

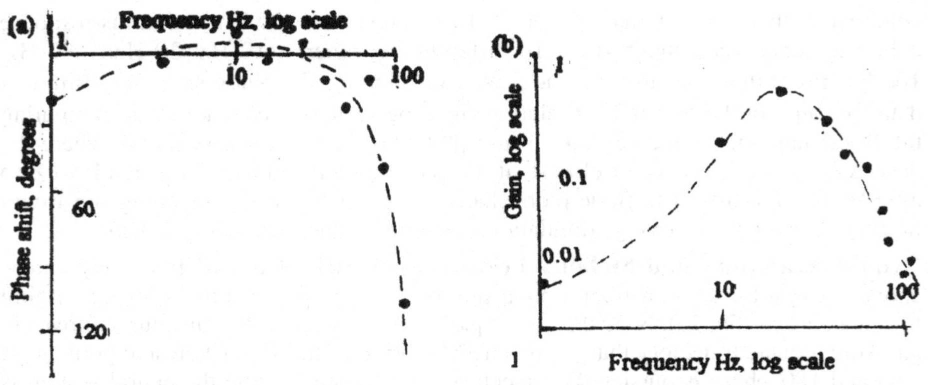

Fig. 1a,b. Bode diagrams of the neuronal system, showing relation of phase (linear scale), gain (log scale) and frequency (log scale).

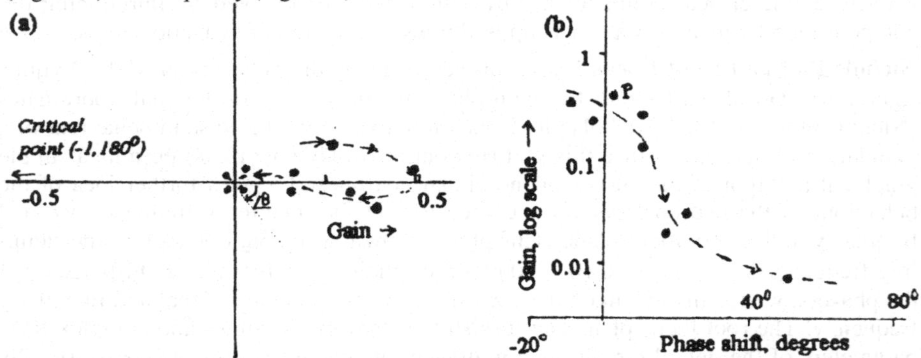

Fig. 2a,b. Nyquist and Nichols plots of phase versus gain, in polar and orthogonal coordinates respectively. Stability implied by Nyquist critical point N not being enclosed by the trajectory.

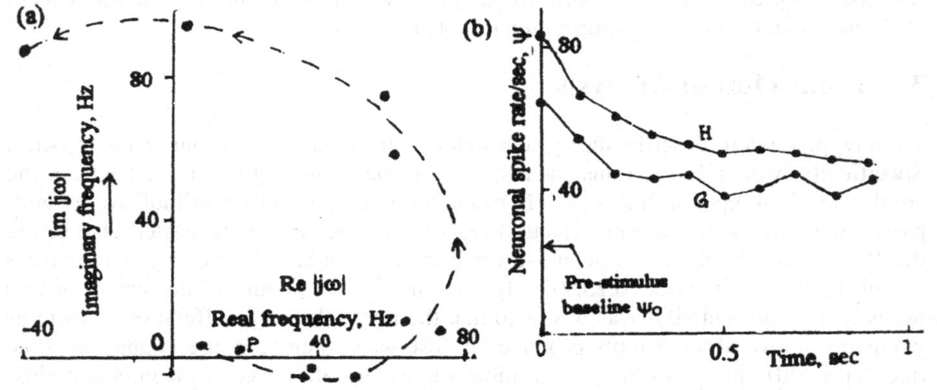

Fig. 3a. Root locus plane plots the neural Real frequency Re $|j\infty|$ and Imaginary frequency Im $|j\infty|$.

Fig. 3b. Steps response of neurone with pre-stimulus baseline activity $S_0 \approx 24$ spikes/sec. Similar exponentially decreasing step response found in invertebrates and vertebrates (see text).

Let S be neural spike rate at time *t* (due to stimulus), and S_0 be the pre-stimulus spike rate. The effective spike rate induced by the stimulus is $\Delta S = S - S_0$ (fig 3b). The general equation relating *t* with the effective spike rate S would be $\Delta S = C\exp(-at)$, i.e., $S = S_0 + C\exp(-a\,t) = S_0[1 + k\exp(-a\,t)]$. Here $k = C/S_0$, and S_0 is approximately constant for various curves: $S_0 \approx 24$. Thus $k = C/24$. Plotting the experimental points of curve G in semilog axes, i.e., $ln\ \Delta S$ versus *t* yields the curve in fig 4a. The resultant points are nearly collinear. A straight line of best fit is drawn as an estimate, from which we obtain C=33.78, k=1.41 and the neural step response equation of curve G becomes:

$$S_{est} = 24\ [1 + 1.41\exp\ (-0.92\,t)]\ . \quad (1)$$

Response Equation: Transfer Function. Let us assess the predictive power of this control system model. The time constant (or system constant) describing this neuronal system is 1/a, i.e., 1.087 sec. System constant remains constant during various stimuli. Define the stimulus intensity of curve G as M units (M=4, see above). Now present a stimulus with intensity of N units, i.e., 10 log N decibels [w.r.t. air current of 250 $(cm/s)^2$], the new C value of response is $[(log\ N)/(log\ M)]$ times the earlier C value of 33.78 [i.e. $C_{new} = 33.78(\log N/\log 4) = 56.11\log N$, so $k = C_{new}/24 = 2.34\log N$]. The log factors come from Weber-Fechner law of sensory neurophysics, stating that response is proportional to logarithm of stimulus. Thus the general response equation S_{est} and transfer function $[H(s)=Y(s)/X(s)]$ of our neuronal system to stimulus intensity N, is:

$$S_{est} = 24\ [1 + 2.34\log N\exp\ (-0.92\,t)]\ . \quad (2)$$

Transfer Function: $$H(s) = [24/s + 56.11\ log\ N\{1/(s + 0.92)\}]/s. \quad (3)$$

The transfer function can also be obtained from the Bode plots, using the values of low frequency gain and high frequency slope of gain curve. Let us now consider a stimulus of 8 units. Setting N=8 in eq. 2, we predict a response:

$$S_{est} = 24\ [1 + 2.11\exp(-0.92)]; \qquad \text{i.e.,} \qquad ln\ [S_{est} - 24] = 3.9 - 0.92\,t\ . \quad (4)$$

In fig. 4a we plot eq. 4 and the ten experimental points of fig. 3b. Note close correspondence between our estimated equation and experimental data. The utility of this approach is to assure that a control theory model is reliable for experimental analysis.

4 Non-equilibrium Information Theory: *Entropy Production*

Having explored a neural control approach, we may now proceed to non-equilibrium aspects. The basic concept of non-equilibrium dynamics is that stochastic fluctuation, inherent in any system, can lead to energy output, entropy production and information dissipation, as the level of organization, activation and information processing rises [9-11]. Nicolis has applied the non-classical non-equilibrium information theory model to analyze informational aspects of neuronal systems [10]. The origin of non-equilibrium dynamics can be traced to Einstein's well known stochastic fluctuation equation linking entropy change and probability of stochastic fluctuation [17]:

$$p = C\ exp\ (\Delta S/k); \qquad \text{i.e.,} \qquad p = c\ \ exp\ (\Delta\, I/\,k).$$

in an information theory framework. This can be taken to be a starting equation for non-classical information theory, as the Shannon-Boltzman equation is the basis of classical information theory. Important concepts are the Nyquist-Hartley generalized information entropy concept and the Nyquist theorem, which relates the energy dissipation power *(P)* of a system with the frequency *(f)* and the intensity of fluctuation (T_E) as characterized by the concept of generalized temperature [18]:

$$P = k\, T_E\, df\ . \tag{5}$$

Neurodynamic Temperature. The postulate of the temperature concept in physics is embodied in Zeroth Law of Thermodynamics, a version of which states that *there is a parameter (i.e., the temperature scale) that estimates the degree of hotness or thermal excitation of a physical object* [19]. To describe an informational counterpart of temperature, we introduce the concept of generalized temperature, e.g., the concept of equivalent temperature used in information transmission systems. It is defined by the amount of equivalent fluctuations or thermal noise that will be actuated in an element by increased temperature, which will produce a similar effect to the electrical fluctuation in question. The 'equivalent temperature' T_E (in a different scale of 'equivalent degrees absolute') can be found out from the fluctuation's noise parameter F [20]:

$$T_E = (F\text{-}1)\, T_0\ . \tag{6}$$

Here T_0 is the temperature of any reference context of the system. We normalize T_0=1, for reasons to be explained below. The noise parameter F can be characterized by the noise:signal ratio, i.e., reciprocal of signal:noise ratio, viz. 1/SNR [20]. For our neuronal system [1], the coherence (here called the gain "g") = [SNR/(1+ SNR)], whence F or [1/SNR] = [(1-g)/g]. Thus, from the gain values at different frequencies (Bode plot: fig.1b) we can calculate the different T_E values at those frequencies. Thence, from eq. 5, we calculate the energy dissipation power at any frequency f_1 (w.r.t. a reference frequency f_2) by calculating r, the ratio of the powers P_1 and P_2 at the two frequencies f_1 and f_2. In the ratio the k and df will cancel out.

Reference frequency is chosen to be 15 Hz, since the particular interneurone under consideration becomes maximally sensitive at 15 Hz [1]. The power ratio r can be expressed as R decibels = 10 $\log_{10} r$. Since neurons encode significant information up to 80 Hz (fig 1b), we consider the points of f = 10 Hz, 20 Hz,...80 Hz. Thence we plot R against f to obtain the graph shown in fig. 4b. It shows that energy dissipation curve is minimum at approximately 20 Hz. Note that gain is maximum at 20 Hz (fig. 1b), implying that the neuronal system is transmitting information most efficiently at that frequency. This minimum energy dissipation behaviour appears to be an illustration of the well known Prigogine-Nicolis Principle of non-equilibrium dynamics, viz. Principle of Minimum energy dissipation or entropy production [5-7]:

Prigogine-Nicolis Principle of Minimum Entropy Production: The rate of energy dissipation or entropy production of a system at the stationary or optimal state, is a minimum.

Optimal Design. The principle of minimum entropy production is related to Rosen's principle of optimal design and Pontryagin's principle of optimal control, and is a basic principle of biological functioning, whether in animal cell, plant cell or microorganism. It is logical that this principle should be applicable to neural cell functioning. Indeed, the principle has wide application, e.g. a typical insect as fruit-fly furnishes a minima graph like fig. 4b when energy dissipation is plotted against myoneural frequency, as gauged by wing movement frequency or velocity of flying. Energy dissipation attains a minimum value at an optimal flying velocity [7], the velocity being optimal as the insect can cover maximum distance at that velocity. The Prigogine principle clarifies that the cricket sensory information processing and the fruit-fly motor information processing, can operate most economically with the minimum entropic cost, at the corresponding optimal frequencies.

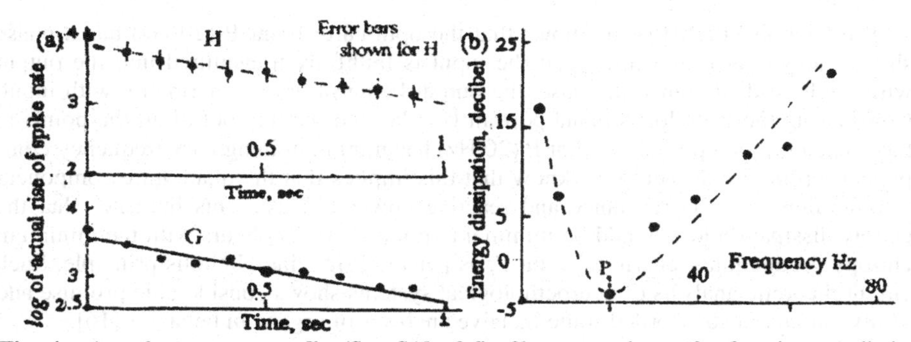

Fig. 4a. Actual step response [ln $(S - S_0)$] of fig 3b: neuronal transfer function, predictive equation.

Fig 4b. Optimality in energy dissipation: Prigogine's Principle of Minimum entropy production.

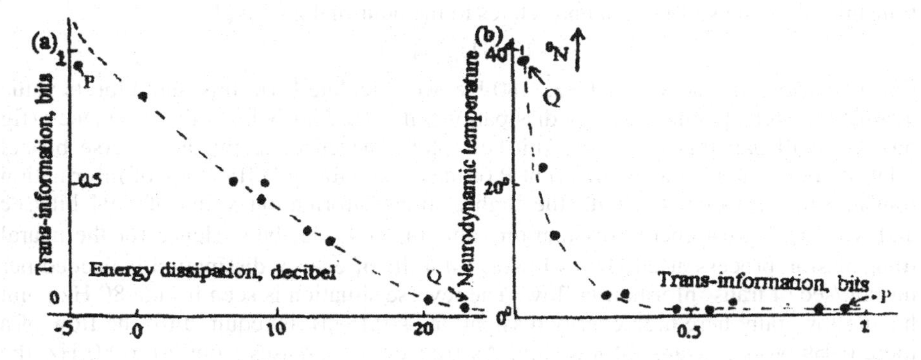

Fig 5a. Bit-Bel plot of neurone: Trans-information (bits) and energy dissipation (bels).

Fig 5b. Neurodynamic temperature equivalent ^{0}N vs. trans-information: Third thermodynamic law. An informational version of Third law of Thermodynamics

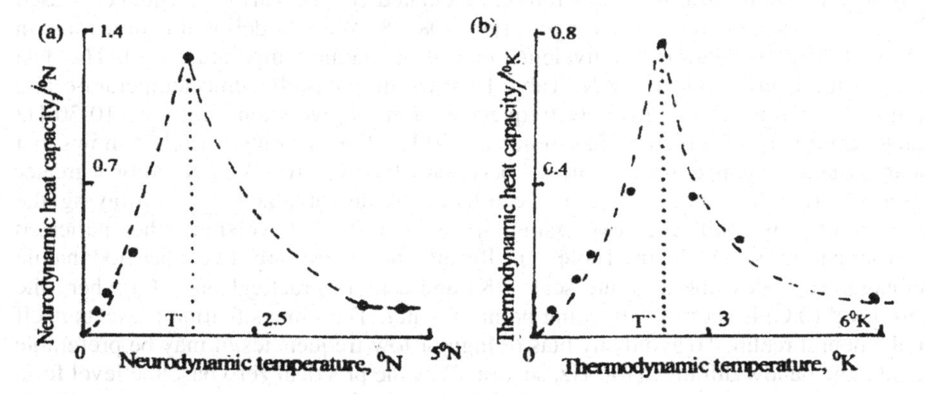

Fig. 6a. Neurodynamic thermal capacity, second order transition and absolute zero behaviour.

Fig 6b. Thermodynamic thermal capacity, second order transition and absolute zero behaviour. The transition temperature T' and T" for the two Lambda (Λ) like points are shown.

The Nichols chart (fig.2b) shows that the optimality point P (20 Hz) has a phase shift of about zero, indicating that the input is faithfully transmitted into the output with the least disturbance in phase, the output being in unison or fidelity with input. Considering the Root locus plane portrait (fig.3a), we see that out of all the points in the trajectory, the optimality point P (20 Hz) has minimum imaginary frequency component. From circuit theory we know that this implies that the quadrature component is minimum, i.e., the reactance and reactive power is least, This indicates that the energy dissipation loss would be minimal at point P. This concurs with the minimum entropy production scenario of point P, as per the Prigogine-Nicholis principle. Such circuit theoretic analysis of neuroethological systems shows considerable promise and utility, and has also afforded some incisive analyses of organism behavior [16].

Bit-Bel Plot. Overall neuronal encoding accuracy can be gauged by the Trans-information parameter I_T, which is the mutual information between the stimulus waveform and the spike train responses. The I_T (in bits/sec) is the information flow transmitted by the spike-train, and relates to the neuronal gain g [1]:

$$I_T = -\log_2 [1 - g] .$$

For the various frequencies 10 Hz,...80 Hz, we calculate I_T (in bits) and plot its value against the corresponding energy dissipation value (R deci-bels) at that frequency (fig 5a). We will call this curve the "bit-bel" plot. The curve shows an inverse bit-bel relation. For instance the 20 Hz point P of maximal gain and efficiency of information transmission is associated with the highest trans-information value of 0.94 bits/sec and with the lowest energy dissipation value of –4.44 decibels. Hence for the neural transmission process at 20 Hz, as low as –4.4 dB of energy dissipation is needed per unit bit/sec of trans-information flow. The reverse situation is seen for the 80 Hz point having low gain: here high energy dissipation (+20.15 dB) is required for the flow of a mere 0.059 bits/s of trans-information. As frequency increases from 20 to 80 Hz, the locus of the point moves down the curve, showing a decrease in the efficacy.

Temperature–Trans-information Plot: Third Law of Thermodynamics. The neural equivalent absolute temperature, calculated for the various frequencies used for these experimental studies, ranged from 0.08^0-38^0. We will define this new scale in terms of "degrees neuronal equivalent", or neurodynamic temperature (^{0}N). The data thus varies from 0.08^0N to 38^0N. Table-I shows the neurodynamic temperature and trans-information for the different frequencies. Let us give attention to the 10-30 Hz range, centred around the optimal frequency 20 Hz. For this range, Table I shows that neurodynamic temperature varies between 0.08^0N to 1.57^0N. For greater discrimination it is better to use a centigrade scale, obtained by multiplying the temperature by 100 and expressing it in a re-scaled version (the italicized neurodynamic scale ^0N in Table 1). Recall that there are two thermodynamic temperature scales: the absolute scale (^{0}K) and a more practical one (^{0}C) where the zero level (0^0C) is set to the freezing point of water. The same situation presents itself in the neural realm. To avoid any bias of high or low frequencies, it may be preferable to take the bandwidth mean (40 Hz, section 2) as the practical zero baseline level for a new *practical* neural temperature scale ^0P. That is, the neural temperature at 40 Hz state (112 ^0N, Table 1) would be defined as 0 ^0P while that of the neural absolute zero point (0^0N) would become -112 ^0P. Thus the parallelism:

Thermodynamic scale: $x.\ ^0\mathrm{K} = (x - 273)\ ^0\mathrm{C}$; *Neurodynamic scale*: $y\ ^0N = (y - 112)\ ^0P$

Fig 5b plots the temperature ^{0}N versus trans-information I_T for frequencies 10 Hz, 20 Hz,...80 Hz. From fig 5a and fig. 5b we can construct a temperature–frequency plot (not shown) which shows that 1^{0}N is the temperature characterizing 34 Hz of neural spike frequency. This 1^{0}N temperature is the unit and normalization factor T_0 in eq 6. Note that as neurodynamic temperature tends to zero (fig 5b), the curve appears to approach the horizontal axis asymptotically, displaying the following behaviour:

The Absolute Zero Principle: The neurone's trans-information parameter asymptotically tends to very high values, as the neurodynamic temperature approaches 0^{0}N.

As trans-information is maximized (point P, fig 5b), the efficacy of neuronal informa-tion transmission increases. As maximization of mutual information is associated with maximization of negentropy and minimization of uncertainty [22], we may say that entropy will minimize toward zero. Hence it can be construed *that entropy tends to zero as the neurodynamic temperature of the neural system approaches 0^{0}N.* Clearly this statement is the neurodynamic counterpart of Third Law of Thermodynamics, which states that as temperature of a system tends to 0^{0}K, the entropy tends to zero.

From an information theoretic approach, Mandelbrot elaborates why thermody-namic laws are applicable to general dynamic systems, whether biological, communi-cational or cognitive [23]. We have endeavoured to extend this view to neural system.

Table 1. Neurodynamic temperature, Trans-information & Heat capacity at different frequency

Freq., Hz	5 (P_1)	10 (P_2)	15 (P_3)	20 (P_4)	30 (P_5)	40 (P_6)	50 (P_7)	60 (P_8)	70(P_9)
T-info., Bit	0.10	0.47	0.86	0.94	0.61	0.56	0.26	0.23	0.04
Temp., ^{0}N	12.81	1. 57	0.22	0.08	0. 90	1.12	4.06	4.67	38. 0
Temp., ^0P	1169	45	-90	-104	-81	0	294	355	3692
Ht.cap, c	1. 37	1. 30	0.24	0.15	0.44	0.72	0.21	0.16	4. 5

5 Exploring the Neurodynamic Absolute Zero

An analysis of system behaviour, temperature or absolute zero is provided by Debye-Einstein model of conduction of energy and signal across a thermodynamic system:

The Debye-Einstein specific heat postulate: This indicates that as temperature approaches absolute zero, the thermal capacity tends to zero.

Heat capacity 'c' is given by $\Delta E/\Delta T$ where E and T denote energy output and temperature. We explore whether a Debye-type analog occurs in the neurodynamic system. The strength of energy output is characterized by energy remitted per unit frequency [18,14]. For each of *n* frequency columns in Table 1, we calculate the energy remitted intensity, i.e. energy dissipation per hertz (say E_n), using data of fig 4b. From Table 1, the temperature increases in the sequence of columns: P_4, P_3, P_5, P_6, P_2, P_7, P_1, P_9. For each of these successive stages of temperature increase, we calculate the consecutive temperature differences ΔT_n.

For these stages we also compute the ΔE_n for each stage. The quotient gives the neuroinformational thermal capacity corresponding to the different frequencies and temperature. In Table 1 and fig. 6a we show the numerical value of thermal capacity against the corresponding neurodynamic temperature (in ^{0}N); comparatively low temperatures have been shown for emphasis ($T<5^0N$). We can call this informational counterpart of specific heat capacity by a new term "specific information", taking a cue from Nicolis [10]. Fig 6a shows that neuroinformational heat capacity and temperature jointly approach zero, illustrating the neural counterpart of the Debye principle. Thus his approach may be applied to neural analysis for obtaining fresh insights. The equivalent situation for a thermodynamic system is shown in fig 6b; this is the well known second-order "order-disorder" transition, with the transition temperature T`` and heat capacity curve displaying the Λ (lambda) behaviour [14]. Thus, from fig. 6:

Thermodynamic transition temperature = 2.1^0K; *Neurodynamic transition temperature* = 1.6^0N

The thermodynamic system becomes much ordered when cooled. Experimental examples of Λ transition are Bose-Einstein condensates. A Λ-type behaviour is not unexpected in neurodynamics, where neural function would attain an ordered state as neurodynamic temperature and activation decreases. The rise of ordered functioning is experimentally demonstrated by the minimization of amplitude and phase errors in 15-40 Hz range [1], corresponding to low temperatures ($T<1.5^0N$). Note that concepts of critical temperature transition and Bose entities have been respectively applied to neural and symbolic transmissions [8, 12, 21], but in different contexts. Our reliance on the Debye approach is further consolidated by its success to analyse other types of biological information systems, as nucleic macromolecules. Flory and Lifschitz have used the Debye-Einstein approach of statistical thermodynamics of quasi-particles and condensation, to develop a precise understanding of molecular bioinformational interactions, thus showing that second-order transition, entropy minimization, transition temperature and theta point occur in macromolecular informational systems [24].

6 Neurothermodynamics: Toward a Rigorous Basis

Our thermodynamic energy dissipation model of neural information may be given a rigorous basis by newer experimental findings on neurometabolism [12, 25], whereby calorimetric energy expenditure of neural processing is currently available in terms of the ubiquitous biological currency of energy: 7,000,000 ATP molecules (4.2 micro-erg) per trans-information bit. These values may be nearly universal from insect neurones to the human brain [12], and so our findings on isolated neurones could also be applicable across species in general. Neural energy dissipation or entropy production originates primarily from oxygen consumption and cellular metabolism for neural Na/K/proton exchange pumps and neurotransmitter uptake. The neurothermodynamic approach, based on ATP energy dissipation, has the potential of offering unifying insights across different levels of functional organization, from molecular neurochemistry, through neural transmission/neural circuits, to brain and behaviour.

Findings from Bioenergetics of Information Transmission. Based on nonlinear analysis of bioenergetics processes, the current approach to frequency optimality, coherence and information transmission in biological systems is the Frohlich-Davydov (F-D) approach, applicable to cellular systems universally, ranging from unicellulars and yeast to neurones and mammalian cells [28, 29]. This is based on non-equilibrium dynamic principles and on Debye model of energy/information transfer in biothermo-dynamic systems via discrete quasi-particle type wave states as phonon or soliton. Our neuro-thermodynamic approach is complimentary and integrative with the prevalent F-D approach. The F-D analysis uses the concept of cellular excitation and transmission through quasi-particle type wave-states as phonon and soliton displaying Bose-Einstein transition. The F-D Principle indicates that [28]:

> *Frohlich-Davidov Principle of Coherent Transmission:* A system with energy input, may, when considered as a non-equilibrium dynamic system, exhibit sensitivity between input frequencies ω_1 and ω_2; and an excitation-transmission mode at an intermediate frequency ω' can become coherently activated, if the energy input rate exceeds a threshold value Φ.

Our neural experimental study (Sec. 2) illustrates the F-D Principle. Experiments show that there is a critical energy input threshold (Φ, RMS power 9.5×10^{-6} cm/s^2) for transmission to occur [1]. The minimum and maximum frequencies of neuronal sensitivity are 5 Hz (ω_1) and 70 Hz (ω_2) [1], and coherent optimal excitation occurs at 20 Hz (ω'). The Λ behaviour is a manifestation of Bose condensates [14], and the neuro-dynamic Λ-type behaviour (fig 6a) experimentally implies that neural function may be usefully analysed in terms of Bose-Einstein type quasi-transition and its consequential Frohlich-Davydov approach to bioenergetics which uses the concepts of Bose entities as phonons or solitons. The F-D bio-transmission model delineates that the coherent excitations, mediated through Bose-like quasi-entities as phonons and solitons, will undergo Bose-Einstein type condensed transition at lower temperature, and the system will attain an ordered single state with very low energy level [28, 29]. In our study we see a similar process: the neuronal optimality of 20 Hz occurs at low neurodynamic temperature (0.08^0N), with a low energy level (0.36 dB); the neurone can be taken to approximate an ordered single state as entropy attains zero level.

Technological Implications. A basic property of Frohlich model is mode softening or broadening where the coherent modal frequency becomes a broadened optimum zone, instead of a sharp frequency peak. This zone could be due to the biomolecular constituents which would induce a Raman shifting and broadening of the frequency, e.g. C=O bonds in α-helices enable Raman shifts in solitons [28]. Such a spectral broadening around the peak of 20 Hz is illustrated in our neural example in fig 4b. Neural information processing has been insightfully analysed in terms of soliton-phonon transmission, Ca^{++}-coupled axonal sol-gel transition process and Frohlich coherence in hydrophobic pockets [26, 27]. The model shows that the Frohlich process can actuate a neuronal automata, thereby producing a computational system with complex information transmission modes at the exact velocity of action potentials. Evidence indicates that Frohlich pumping of phonons and solitons induces Bose cooperativity within neuronal structure and frequency can undergo Raman shift.

Our neurothermodynamic coherence approach appear to be corroborated independently by recent experimental electronics findings. Like our neural example sketched above, a similar optimality process is found to occur in engineering information transmission systems, like fibre cables using solitonic information transmission. Here Frohlich coherent entities as solitons can undergo Raman-like shifting, called "pulse broadening of Raman Bosons" (PBRB) [30]. This process is of critical importance in efficient optimality of fibre transmission [31]. Thus we see that comparative study of Raman soliton process in neuronal transmission may probe important insights and suggestions for telecommunication and computer engineering.

7 Conclusion: Future Prospects

We have demonstrated that Neurothermodynamics and Non-equilibrium Information Theory---the newly emerging paradigm of nonclassical information processing---have potentiality to furnish valuable insights into neural function: from synapses to the brain. An example is the proposed neurodynamic equivalent temperature scale characterizing neural information processing. Indeed all the thermodynamic laws can be said to have informational counterparts in the neurodynamic domain: from zeroth law to fourth law (Nicolis-Prigogine principle). The first law (conservation law) in the neurocomputational realm is conservation of emitted and received informational quanta (encoding and decoding spikes). The zeroth law, second law (Shannon principle), third law and fourth law are already outlined above.

The lowest neurodynamic temperature, obtained in our experiment is 8^0N (at 20 Hz). Future experiments are needed to show how near one can approach the neural absolute zero and isolate any neurodynamic analogues to cooperative thermodynamic phenomena, such as zero point energy and biological superconduction or zero-noise information transmission [8, 28, 29]. The latter could be gauged by experimentally finding the neuronal impedance (sec. 4) abruptly falling to zero near 0^0N . Investigation of system behaviour around thermodynamic absolute zero enabled a crucial advance in physical sciences, hence experimental study of neural parameters around neurodynamic absolute zero may not only hold much potentiality for neuroscience, but also for new bionics or information technology area designed on neural concepts.

Acknowledgements. For initiating collaboration, thanks are due to US National Inst. of Mental Health, National Brain Research Centre-India, and especially to Drs V Ravindranath, NBRC, Richard Nakamura, NIMH, AR Thakur, Jadavpur University. PKR thanks Drs Walter Freeman, UCBerkeley and RE Shaw, Einstein Inst/UConnecticut for discussions on neurocontrol theory.

References

1. Theunissen, F., et al: Information theoretic analysis, J. Neurophysiol. 75 (1996) 1345-1364
2. Clague, H., et al: Effects of adaptation on neural coding, J. Neurophysiol. 77 (1997) 207-220
3. Dutta Majumder, D.: Cybernetics and systems: A unitary science, Kybernetes 8 (1979) 7-15
4. Roy, P., Majumder, D.: A biothermodynamic study, J. Intelligent Systems 10 (2000) 57-104
5. Prigogine, I.: Nobel Lecture: Chemistry - 1977, Nobel Foundation, Stockholm (1978)
6. Nicolis, G., Prigogine, I.: Self-organization in non-equilibrium systems, Wiley, N.Y. (1977)
7. Zotin, A.: Thermodynamic bases of biological processes, W. de Gruyter, N. Y. (1990)
8. Little, W. Persistent states in the brain, Mathematical Biosciences, 19 (1974) 101-120
9. Hofkirchner, H: The quest for unified theory of information, Gordon-Breach, London (1997)

10. Nicolis, J.: Information processing, In: Basar, E.: Synergetics of brain, Springer, NY (1983)
11. Avramescu, A.: Coherent informational energy, J. Documentation 36 (1980) 293-312
12. Baddeley, R., Hancock, P.: Information theory and the brain, C.U.P., Cambridge (2000)
13. Stark, L: Neurological control systems analysis: Bio-engineering, Plenum, N.Y. (1978)
14. Pippard, Sir B.: Elements of Classical Thermodynamics, C.U.P., Cambridge (1991)
15. Pringle, J, Wilson, V: Response to harmonic stimulus, J. Exp. Biol. 29 (1952) 220-234
16. Mcfarland, D.: Feedback mechanisms in animal behaviour, Academic Press, London (1981)
17. Einstein, A.: Die molukular-kinetischen theorie, Annalen der Physik 17 (1905) 549-560
18. Dorf, R.: The electrical engineering handbook, CRC Press, Boca Raton (1993)
19. Halliday, D., Resnick, R.: Physics – Part I, Toppan-Wiley, Tokyo (1983)
20. Buckingham, M.: Noise in electronic devices and systems, Halstead Press, N. Y. (1995)
21. Herdan G.: The advanced theory of choice in language, Mouton, The Hague (1989)
22. Brillouin, L.: Science and information theory, Academic Press, New York (1966)
23. Mandelbrot, M: Thermostatistics, In: Cherry C: Information theory, Butterworths, NY (1976)
24. Flory, P.: Nobel Lecture: Chemistry - 1974, Nobel Foundation, Stockholm (1975)
25. Sarpeshkar, R.: Analog vs. digital neurobiology, Neural Computation 10 (1998) 1601-38
26. Sataric, M.: Energy transfer mechanism involving soliton, Nanobiology 1 (1992) 445-56
27. 27. Rasmussen, S.: Computational connectionism in neurones, Physica-D 42 (1990) 428-49
28. Frohlich, H., Kremer, F: Coherent excitations in biological systems, Springer, Berlin (1983)
29. Davidov, A.: Quantum physics and biology, Springer, Berlin (1986)
30. Agrawal, G: High capacity Raman soliton systems, Optical Experiments, 9(2) (2001) 66-73
31. Corney, J: Noise limit: Raman effect in soliton propagation, Opt.Comm.,140 (1997) 215-17

Modeling of Nonlinear Systems by Employing Self-Organization and Evaluation – SOR Network –

Takeshi Yamakawa and Keiichi Horio

Department of Brain Science and Engineering
Graduate School of Life Science and Systems Engineering
Kyushu Institute of Technology
680-4 Kawazu, Iizuka, Fukuoka 820-8502, Japan
yamakawa@brain.kyutech.ac.jp, horio@tsuge98.ces.kyutech.ac.jp

Abstract. In this paper, an adaptive self-organizing relationship (ASOR) network, which is the extension of the self-organizing relationship (SOR) network proposed by the authors, is proposed. The SOR network can obtain the desired input/output relationship of a target system by using the input/output vector pairs and their evaluations. In order to add the ability of adaptation to the SOR network, the new algorithm that the learning rate and the area of the neighborhood are adjusted according to need is employed. The ASOR network can adapt to the change of the desired input/output relationship of the target system. The effectiveness of the proposed ASOR network is verified by applying it to design of the control system of the DC motor whose load changes with time.

1 Introduction

The self-organizing map (SOM) neural network was developed by Teuvo Kohonen during the period 1979-1982[1][2]. It consists of the input layer and the competitive layer. The competitive layer contains processing units that have the weight vectors. During the learning, applying the input vectors to the input layer, the weight vectors of the units, that are defined as the winner unit and the neighboring units, are updated to be attracted to the input vectors. So the distribution of the weight vectors after the learning indicates that of the input vectors.

By using the feature of the SOM, we proposed the self-organizing relationship (SOR) network which consists of the input layer, the output layer and the competitive layer [3][4]. The SOR network is established to approximate the desired relationship between input and output of a system. The input and output vector pairs of the target system are employed as the learning vectors, and are applied to the input and output layer, respectively. Furthermore, the modified Kohonen's self-organizing learning law, in which the weight vectors are updated to be attracted to or to be repulsed from the learning vector, is employed. In the modified learning law, the evaluations of the learning vectors are used in

N.R. Pal and M. Sugeno (Eds.): AFSS 2002, LNAI 2275, pp. 204–213, 2002.

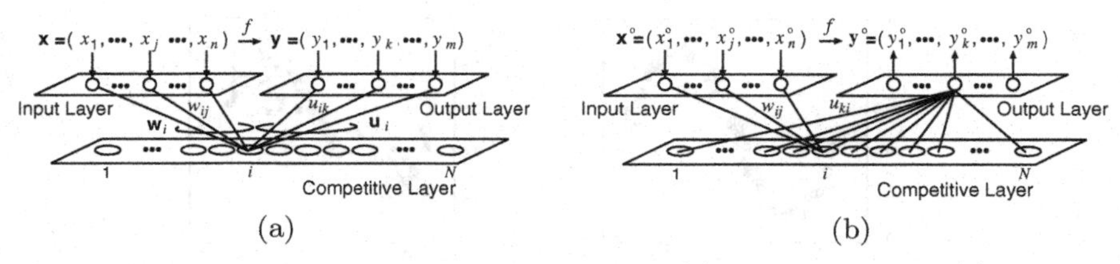

Fig. 1. The self-organizing relationship network. (a) Learning mode. (b) Execution mode.

order to decide the direction of the update (attraction or repulsion). There are many systems that the correct input/output relationship are not available but the input/output relationship can be evaluated by evaluation function or intuition of the user and so on. The SOR network can approximate desired I/O relationships of the target system by employing the modified learning law. The validity of the SOR network is verified by applying it to design of the control system[3][4]. However, the learning rate and the area of the neighborhood have to decrease as learning progresses, and it is difficult to apply it to the system whose desired input/output relationship changes dynamically.

In this paper, we propose the adaptive SOR (ASOR) network in which the learning rate and the area of the neighborhood are adjusted according to need. The learning rate and the area of the neighborhood are determined by the necessity measure which represents the quality of the current state of the network. The convergence and the restart of the learning can be controlled by using the necessity measure, and the network adapt to the change of the desired input/output relationship of the target system.

The effectiveness and the validity of the proposed ASOR network is verified by applying it to design of the adaptive control system.

2 Self-Organizing Relationship Network

The SOR network consists of the input layer, the output layer and the competitive layer, in which n, m, and N units are included, respectively, as shown in Fig. 1. The i-th unit in the competitive layer is connected to the units in the input layer and the output layer with weight vector $\mathbf{w}_i = (w_{i1}, \cdots, w_{in})$ and $\mathbf{u}_i = (u_{i1}, \cdots, u_{im})$, respectively. The network can be established by learning in order to approximate the desired function $\mathbf{y} = f(\mathbf{x})$.

2.1 Learning Mode of SOR Network

The random input/output vector pair $(\mathbf{x}, \mathbf{y})$ is applied, as the learning vector, to the input and the output layer together with the evaluation E for the input/output vector pair. The evaluation E may be assigned by the network designer, given by the intuition of the user or obtained by examining the system under test. The value of E is positive or negative in accordance with judgment of the designer, preference of the user or score of examination. The positive E

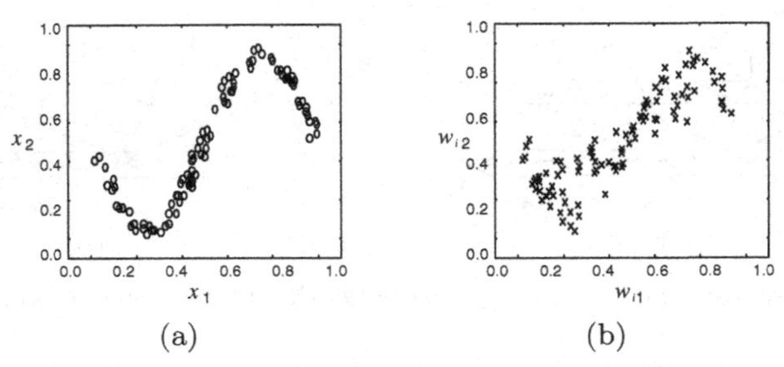

Fig. 2. The input vectors for the SOM and the weight vectors of the SOM after the learning. (a)There are 100 input vectors $\mathbf{x} = (x_1, x_2)$ for the SOM. (b)The weight vectors of the SOM after the learning.

causes the self-organization of attraction to the learning vector and the negative one does that of repulsion from the learning vector. The self-organizing network is named SOR network by employing this key idea.

Fig. 2(a) shows 100 of the vectors which are the input vectors for the SOM. In the learning of the SOM, the input vector $\mathbf{x}$ is applied to the input layer, and the c-th unit which has the closest weight vector $\mathbf{w}_c$ to the input vector is defined as the winner unit. The units that are located within the neighborhood of the winner unit are defined as the neighboring units. $\Delta\mathbf{w}_i$ calculated by Eq. (1) is added to the old weight vectors $\mathbf{w}_i$ of the winner unit and neighboring units in order to obtain the new weight:

$$\Delta\mathbf{w}_i = \alpha(t) \cdot (\mathbf{x} - \mathbf{w}_i), \tag{1}$$

where $\alpha(t)$ is learning rate which decreases with time. The distribution of the weight vectors after the learning is shown in Fig. 2(b). Here, number of the learning iteration is 1000, and the initial value of the learning rate $\alpha(0)$ is 0.2. It is indicated that the distribution of the weight vectors does not represent that of the input vectors, so the input vectors are not suitable for the Kohonen's learning law.

On the other hand, the SOR network employ the learning vectors which cause self-organization of repulsion from the input vector, so the distribution of the weight vectors after the learning indicates the distribution of the input vectors. Fig. 3(a) shows 80 of the learning vectors which are given positive evaluation E (=1.0), and 20 of ones which are given negative evaluation E (=-1.0). In the learning of the SOR network, the learning vector $\mathbf{I} = (\mathbf{x}, \mathbf{y}) = (x_1, y_1)$ is applied to the input and output layers, the c-th unit in the competitive layer, which has the closest weight vector $\mathbf{v}_c = (\mathbf{w}_c, \mathbf{u}_c) = (w_{c1}, u_{c1})$ to the learning vector, is defined as the winner unit. The units that are located within the neighborhood of the winner unit are defined as the neighboring units. $\Delta\mathbf{v}_i$ calculated by Eq. (2) is added to the old weight vectors $\mathbf{v}_i$ of the winner unit and neighboring units in order to obtain the new weight:

$$\Delta\mathbf{v}_i = \begin{cases} \alpha(t) \cdot E \cdot (\mathbf{I} - \mathbf{v}_i) & E \geq 0 \\ \beta(t) \cdot E \cdot \exp(- \parallel \mathbf{I} - \mathbf{v}_i \parallel) \cdot \mathbf{sgn}(\mathbf{I} - \mathbf{v}_i) & E < 0, \end{cases} \tag{2}$$

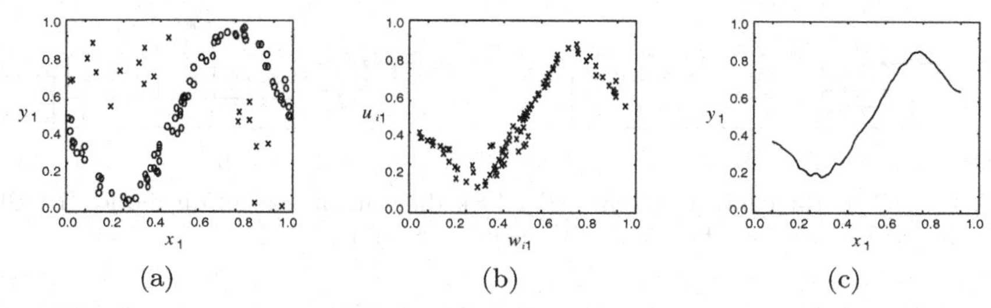

Fig. 3. The learning vectors for the SOR network, the weight vectors of the SOR network after the learning and the I/O characteristics generated by the SOR network. (a)The learning vectors for the SOR network ($\cdots E = 1.0$, $\times \cdots E = -1.0$). (b)The weight vectors of the SOR after the learning. (c)Input/output characteristics generated by the SOR network after the learning.

where $\alpha(t)$ and $\beta(t)$ are learning rates which decrease with time, and $\mathbf{sgn}(\cdot)$ is sign function. In other words, when the evaluation E is positive or negative, the weight vectors of the winner unit and the neighboring units are attracted to or repulsed from the learning vector $\mathbf{I}$, respectively. Fig. 3(b) shows the weight vectors of the SOR network after the learning. Here, number of the learning iteration is 1000, and the initial value of the learning rate $\alpha(0)$ and $\beta(0)$ are 0.2 and 0.1, respectively. It is known that the distribution of the weight vectors of the SOR network after the learning show that of the learning vectors which have the positive evaluation E. It is verified that the modified learning algorithm, which employ the self-organization of attraction and repulsion, is very useful to obtain the reasonable distribution of the learning vectors.

2.2 Execution Mode of SOR Network

After the learning, the SOR network is ready to use as the input/output relationship generator. This operation is referred to as the execution mode and it is illustrated in Fig. 1(b). The actual input vector $\mathbf{x}^o$ is applied to the input layer, and the output z_i of the i-th unit in the competitive layer is calculated by:

$$z_i = \exp\left(-\frac{\| \mathbf{x}^o - \mathbf{w}_i \|}{\gamma}\right), \tag{3}$$

where γ is a constant representing fuzziness of similarity. z_i represents the similarity measure between the weight vector $\mathbf{w}_i$ and the actual input vector $\mathbf{x}^o$. The output y_k^o of the k-th unit in the output layer is calculated by:

$$y_k^o = \sum_{i=1}^{N} z_i u_{ki} \; / \; \sum_{i=1}^{N} z_i, \tag{4}$$

where u_{ki} is a weight from the i-th unit in the competitive layer to k-th unit in the output layer and it is equal to u_{ik} obtained in the learning mode. The output of the network $\mathbf{y}^o = (y_1^o, \cdots, y_k^o, \cdots, y_m^o)$ represents the weighted average

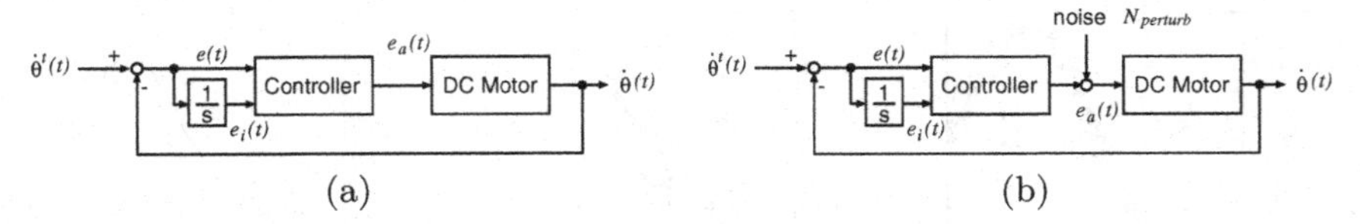

Fig. 4. DC motor control system. (a) Block diagram in execution mode. (b) Block diagram with noisy signal employed to achieve learning.

of $\mathbf{u}_i$ by similarity measure z_i. Fig. 3(c) shows the input/output characteristics generated in the execution mode of the SOR network. Here, γ is 0.025. It is clarified that the input/output characteristics of the SOR network is reasonable.

3 Adaptive Self-Organizing Relationship Network

The SOR network can extract the desired input/output relationship of the target system. However the learning rate and the area of the neighborhood decrease as learning progresses, and it is difficult to apply the SOR network to the system whose desired input/output relationship changes dynamically.

In this paper, the adaptive SOR (ASOR) network, in which the learning rate and the area of the neighborhood are adjusted according to need, is proposed. In the operation of the ASOR network, the execution mode and the learning mode are carried out repeatedly. The input vector is applied, and the output vector is calculated in the execution mode. The input/output relationship is evaluated, and the learning of the ASOR network is carried out by using the input/output vector pair and its evaluation. In order to adjust the learning rate and the area of the neighborhood the necessity measure of the learning *nec* is employed. The necessity measure at the step t is given by:

$$nec(t) = nec(t-1) + \eta E(t) \tag{5}$$

where $E(t)$ is the evaluation value of the input/output relationship generated in the execution mode at the step t. η is a parameter and given by:

$$\eta = \begin{cases} \eta_+ & 0 \leq E(t) \\ \eta_- & 0 > E(t), \end{cases} \tag{6}$$

where η_+ and η_- are negative parameters which decide the change rate of the $nec(t)$. When the evaluation of the input/output relationship is bad $(0 > E(t))$, in other words the state of the network is not desired, the necessity measure of the learning increases. On the other hand, when the evaluation of the input/output relationship is good $(0 \leq E(t))$, the necessity measure of the learning decreases. The learning rates $\alpha(t)$ and $\beta(t)$ and the area of the neighborhood $N_c(t)$ at the step t are calculated based on the necessity measure by:

$$\alpha(t) = \alpha_0 \cdot nec(t), \tag{7}$$

$$\beta(t) = nec(t-1) - nec(t), \tag{8}$$

$$N_c(t) = N_{c0} \cdot nec(t), \tag{9}$$

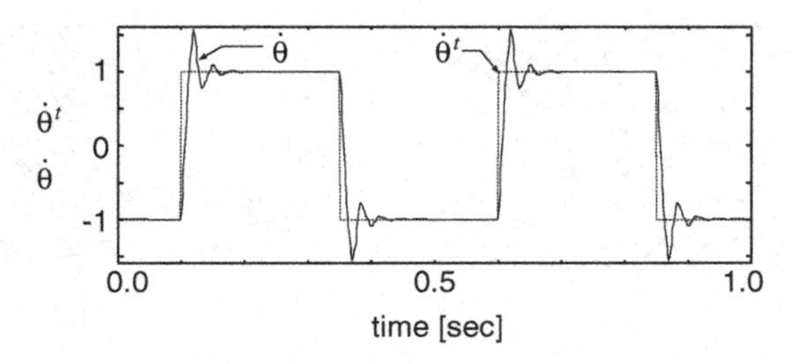

Fig. 5. The speed of the motor shaft $\dot{\theta}(t)$ and its target value $\dot{\theta}^t(t)$ when the DC motor is controlled by the PI controller.

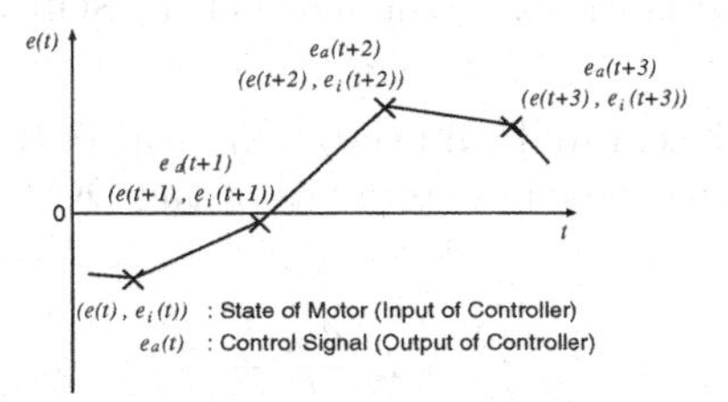

Fig. 6. How to evaluate the input/output relationship of the controller.

where α_0 and N_{c0} are parameters which decide the learning rate and the area of the neighborhood. The learning rate $\alpha(t)$ and area of the neighborhood $N_c(t)$ become large when the necessity measure of the learning is large. In order to compensate the stability of the learning, the learning rate $\beta(t)$ becomes large when the necessity measure of the learning is decreasing. When the necessity measure is large, in other words the state of the network is not desired, the desired input/output relationship can not be generated in the execution mode. The perturbation noise is added to the output of the network to search for the good input/output relationship. The range of the perturbation noise is given by:

$$N_{perturb}(t) = N_{perturb0} nec(t), \tag{10}$$

where $N_{perturb0}$ is a parameter which decide the range of the perturbation noise. When the necessity measure of the learning is large the range of the noise has to be large.

Using this algorithm, the ASOR network can adapt to the change of the desired input/output relationship of the target system.

4 Experimental Results

In order to verify the effectiveness of the proposed ASOR network, we apply it to design of the controller of the DC motor whose load changes with time. Fig. 4(a) shows the feedback control system of the DC motor. In Fig. 4, $\dot{\theta}(t)$ and $\dot{\theta}^t(t)$ are speed of the motor shaft and its target value, respectively. The controller is the system with two inputs and one output. The inputs are error $e(t)$ between

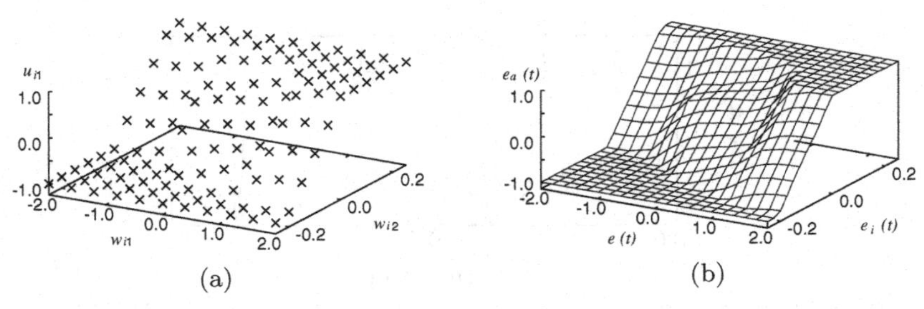

Fig. 7. The weight vectors and the input/output characteristics after the learning. (a) The weight vectors of the SOR network after the learning. (b) The input/output characteristics generated in the execution mode of the SOR network.

$\dot{\theta}(t)$ and $\dot{\theta}^t(t)$ and its integrated value $e_i(t)$, and the output is the supply voltage to the DC motor $e_a(t)$. The characteristics of the DC motor used here is given by:

$$e_a = L_a \frac{di_a}{dt} + R_a i_a + e_b, \tag{11}$$

$$e_b = K_b \dot{\theta}, \tag{12}$$

$$T = K_T i_a, \tag{13}$$

$$J_m \ddot{\theta} + b_m \dot{\theta} = T, \tag{14}$$

where e_a, L_a, i_a, R_a and e_b represent supply voltage, armature inductance, armature current, armature resistance and counter electromotive force, respectively. θ, T, J_m and b_m represent position of the motor shaft, motor torque, moment of inertia and coefficient of friction, respectively. K_b and K_T represent coefficients. In the experiments, L_a, R_a, J_m, b_m, K_b and K_T are 8.6×10^{-3}, 3.2, 3.0×10^{-5}, 3.0×10^{-6}, 6.0×10^{-2} and 1.7×10^{-2}, respectively.

Fig. 5 shows the response of the $\dot{\theta}(t)$ and $\dot{\theta}^t(t)$ when the DC motor is controlled by the PI controller. The output of the PI controller $e_{a,PI}(t)$ is calculated by:

$$e_{a,PI}(t) = K_P e(t) + K_I e_i(t), \tag{15}$$

where K_P and K_I are proportional and integral parameters, and they are decided to be 1.2 and 9.4, respectively, by trial and error. It is known that the response of the speed of the motor shaft has large overshoot and it vibrates before steady state.

4.1 Experiments of SOR Controller

The control system is designed by using the SOR network. At first the learning vectors and their evaluations have to be obtained. For the purpose, the noisy signal is added to the output of the PI controller $e_{a,PI}(t)$ as shown in Fig. 4(b), and the response of the motor is observed. The typical response of the error $e(t)$ is shown in Fig. 6. At time step t, under the input signals of the controller $e(t)$ and $e_i(t)$ the perturbation noise is added to the output of the controller to be

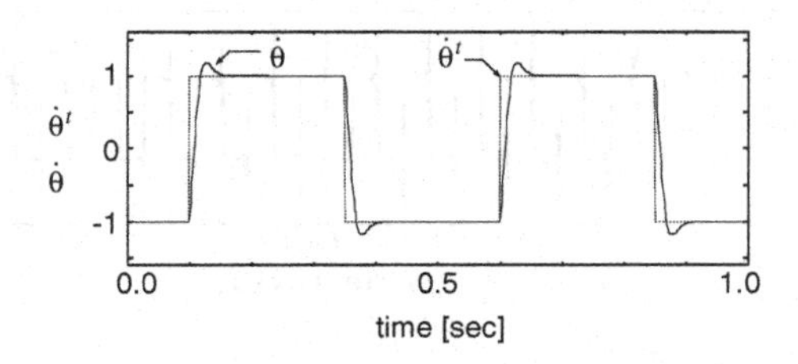

Fig. 8. The speed of the motor shaft $\dot{\theta}(t)$ and its target value $\dot{\theta}^t(t)$ when the DC motor is controlled by the SOR network.

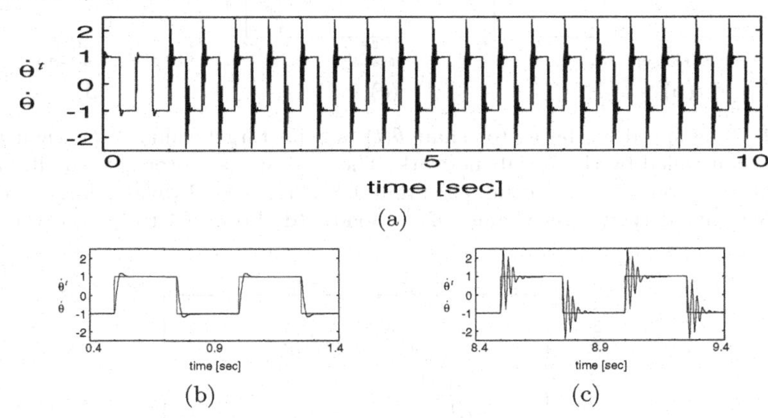

Fig. 9. The speed of the motor shaft $\dot{\theta}(t)$ and its target value $\dot{\theta}^t(t)$ when the DC motor is controlled by the conventional SOR network. The load of the motor gets smaller at time 2 second. (a) From 0 to 10 second. (b) From 0.4 to 1.4 second (load is large). (c) From 8.4 to 9.4 second (load is small).

$e_a(t)$ which is fed to the DC motor. This voltage supply $e_a(t)$ causes $e(t+1)$ and $e_i(t+1)$ at the next step $t+1$. The evaluation of the input/output relationship $E(t)$ at time step t can be calculated by:

$$E(t) = \sqrt{e(t)^2} - \sqrt{e(t+1)^2}. \tag{16}$$

This equation means that the evaluation $E(t)$ becomes positive when the error decreases at next step. On the other hand, $E(t)$ becomes negative when the error increases at next step.

The learning of the SOR network is achieved by using the learning vectors and their evaluations obtained above. In the learning, the number of units in the competitive layer is 121 (11×11), and the number of learning iteration is 1000. The initial value of the learning rate $\alpha(0)$ and $\beta(0)$ are 0.2 and 0.1, respectively. The weight vectors $\mathbf{w}_i$ are fixed, and only weight vectors $\mathbf{u}_i$ are updated. Fig. 7(a) shows the weight vectors after the learning, and Fig. 7(b) shows the input/output characteristics generated in the execution mode. The fuzziness parameter γ in Eq.(3) is 3.0. The SOR network whose weight vectors are fixed is used as the controller of the system as shown in Fig. 4(a), and Fig. 8 shows response of the

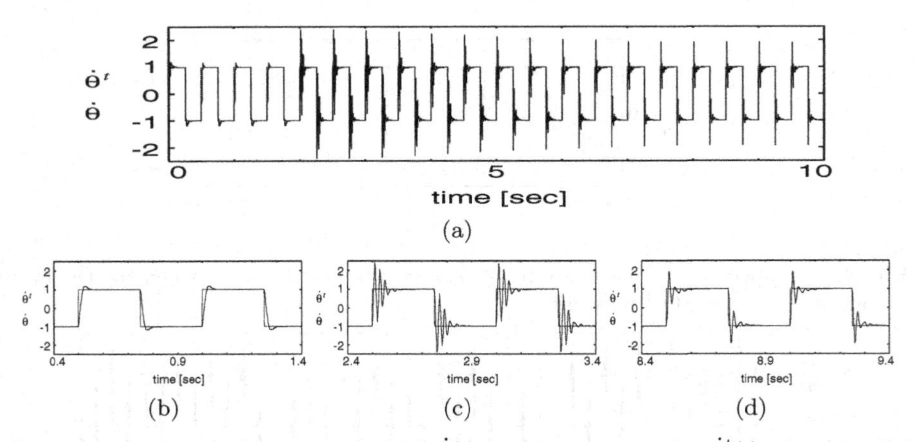

Fig. 10. The speed of the motor shaft $\dot{\theta}(t)$ and its target value $\dot{\theta}^t(t)$ when the DC motor is controlled by the ASOR network. The load of the motor gets smaller at time 2 second. (a) From 0 to 10 second. (b) From 0.4 to 1.4 second (load is large). (c) From 2.4 to 3.4 second (just after change of the load). (d) From 8.4 to 9.4 second (load is small).

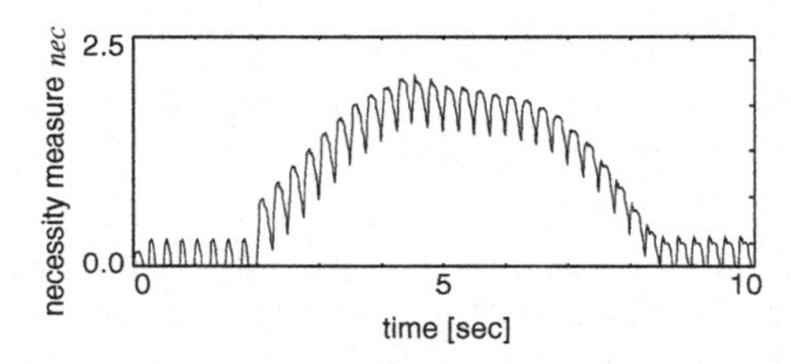

Fig. 11. The necessity measure of the learning $nec(t)$.

$\dot{\theta}(t)$ and $\dot{\theta}^t(t)$. The response of $\dot{\theta}(t)$ controlled by the SOR network has smaller overshoot compared to that controlled by the PI controller (Fig. 5), and it does not vibrate before steady state.

4.2 Experiments of ASOR Controller

In this section, the experiments, in which the load of the motor changes with time, is described. When the DC motor is controlled by the SOR network whose weight vectors are obtained above experiment and fixed, the speed of the motor shaft $\dot{\theta}(t)$ and its target value $\dot{\theta}^t$ are shown in Fig. 9(a). Here, the load gets smaller ($Jm = 1.0 \times 10^{-5}$) at time 2 second. Fig. 9(b) and (c) show the $\dot{\theta}(t)$ and $\dot{\theta}^t(t)$ from 0.4 to 1.4 second, and from 8.4 to 9.4 second, respectively. You can see that the response of $\dot{\theta}(t)$ after the change of the load has a large overshoot and takes a long time to converge.

The experiment in which the ASOR network is used as the controller is described. In the experiment, η_+ and η_- in Eq.(6) are 0.01 and 0.05, respectively. α_0, N_{c0} and $N_{perturb0}$ in Eq.(7), (9) and (10) are 1.0, 10.0 and 1.0, respectively.

Fig. 10(a) shows the response of the $\dot{\theta}(t)$ and $\dot{\theta}^t(t)$ controlled by the ASOR network whose initial weight vectors are that obtained above (Fig. 7(a)). Here, the load gets smaller (Jm : from 3.0×10^{-5} to 1.0×10^{-5}) at time 2 second. Fig. 10(b), (c) and (d) show the response of the $\dot{\theta}(t)$ and $\dot{\theta}^t(t)$ from 0.4 to 1.4 second, from 2.4 to 3.4 second, and from 8.4 to 9.4 second, respectively. Fig. 11 shows the necessity measure of the learning.

During the large load, the response of the $\dot{\theta}(t)$ is reasonable and the necessity measure of the learning $nec(t)$ is not so large. After the change of the load, the response of the $\dot{\theta}(t)$ has a large overshoot and vibrates, and $nec(t)$ increases. The learning rate $\alpha(t)$ and area of the neighborhood $N_c(t)$ also increases, and the learning of the ASOR network is achieved. The response of the $\dot{\theta}(t)$ just after the change of the load (Fig. 10(c)) is very similar to that controlled by the conventional SOR network (Fig. 9(c)), because the adaptation is not advanced. The overshoot of the $\dot{\theta}(t)$ becomes smaller with progress of the adaptation. After the elapse of enough time the response of the $\dot{\theta}(t)$ (Fig. 10(d)) has not so large overshoot and does not take a long time to converge. It indicates that the proposed adaptation algorithm, in which the learning rates and the area of the neighborhood are adjusted according to need, is effective to apply the SOR network to the system whose desired input/output relationship changes with time.

5 Conclusions

In this paper, the adaptive self-organizing relationship (ASOR) network, in which the weight vectors are updated based on the necessity of the learning, is proposed. The ASOR network is applied to design of the controller of the DC motor. The necessity of the learning is calculated in reference to the evaluation of the response of the motor. The learning of the ASOR network is advanced when the necessity measure of the learning is large. Thus the convergence and restart of the learning can be controlled by the necessity measure. The effectiveness and validity of the proposed ASOR network is verified by applying it to design of the control system of the DC motor whose load changes with time.

References

1. T. Kohonen, "Self-organized formation of topologically correct feature maps," *Biol. Cybern.*, Vol.43, pp.59-69, 1982.
2. T.Kohonen, *Self-organization and associative memory*, Second edition, Berlin: Springer-Verlag, 1988.
3. T. Yamakawa and K. Horio, "New Design Method of Fuzzy Logic Controller Using Self-Organizing Relationship," *Proceedings of IIZUKA'98*, pp.155-158, Iizuka, Japan, Oct. 16-20, 1998.
4. T. Yamakawa and K. Horio, "Self-Organizing Relationship (SOR) Network," *IEICE Trans. on Fundamentals of Electronics, Communications and Computer Science*, Vol.E82-A, No.8, pp.1674-1678, 1999.

Characterization of Non-linear Cellular Automata Model for Pattern Recognition

Niloy Ganguly[1], Pradipta Maji[2], Arijit Das[2],
Biplab K. Sikdar[2], and P. Pal Chaudhuri[2]

[1] Computer centre, IISWBM, Calcutta, India 700073
n_ganguly@hotmail.com
[2] Department of Computer Science & Technology, Bengal Engineering College (D U), Howrah, India 711103
{pradipta@,arij@,biplab@,ppc@ppc.}becs.ac.in

Abstract. This paper establishes the non-linear Cellular Automata (*CA*) as a powerful pattern recognizer. The special class of *CA*, referred to as *GMACA* (*Generalized Multiple Attractor Cellular Automata*), is employed for the design. The desired *CA* model, evolved through an efficient implementation of genetic algorithm, are found to be at the *edge of chaos*.

1 Introduction

Pattern recognition is the study as to how the machines can learn to distinguish patterns of interest from their background. In conventional approach of pattern recognition, the machine compares the given input pattern with each of the stored patterns and identifies the closest match. The search requires the time proportionate to the number of patterns learnt/stored.

The *Associative Memory* provides solution to the problem where the time to recognize a pattern is independent of the number of patterns stored. The design divides the entire state space into some pivotal points. The states close to a pivotal point get associated with that specific point. Identification of a pattern, distorted due to noise, amounts to traversing the transient path from the given pattern to the closest pivotal point. As a result, the process of recognition becomes independent of the number of patterns learnt/stored.

Since early 80's the model of associative memory has attracted considerable interest among the researchers [1]. Both sparsely connected machine (Cellular Automata) [2] and densely connected network (Neural Net) [1] have been explored to solve this problem of pattern recognition. The seminal work of Hopfield [3] made a breakthrough by modeling a *recurrent*, *asynchronous*, *neural net* as an *associative memory* system. However, the complex structure of neural net model has partially restricted its application. Search for alternative model around the simple structure of Cellular Automata (*CA*) continued [2,4].

The design of *CA* based associative memory is mostly concentrated around uniform *CA* [2,4]. This structure restricts the *CA* to evolve as a general purpose pattern recognizer [2]. Some works on non-uniform *CA* has also been reported

N.R. Pal and M. Sugeno (Eds.): AFSS 2002, LNAI 2275, pp. 214–220, 2002.

in [5,6]. However, the evolution of CA as a general purpose pattern recognizer is still to be explored.

In this paper, we propose the evolution of CA based associative memory for pattern recognition. The more general class of CA referred to as *Generalized Multiple Attractor CA* is employed for the design. It displays very encouraging result in pattern recognition. The memorizing capacity of $GMACA$, as established in this paper, found to be better than that of conventional Hopfield Net. We employ genetic algorithm (GA) to arrive at the desired $GMACA$ configurations.

The evolved CA rule space is then subjected to extensive study. The study categorizes the CA rule by conducting an in-depth analysis of its local and global parameters. The diverse parameters, λ [7] and $\mathcal{Z}$ parameter [8], entropy, mutual information etc., developed over the years to characterize the CA, display consistency for the CA rule space evolved for the design. The $GMACA$ that exhibit maximum memorizing potential are found to be in between order and the chaos - that is, lie at the *edge of chaos* [9]. It confirms the general belief that the complex computation occurs at the *edge of chaos* [7,9].

In order to present the underlying principle of CA based associative memory, in *Section II* we present an overview of CA. The $GMACA$ based associative memory and its application for pattern recognition is next outlined in *Section III.* In *Section III*, we also report the experimental observations that prove the potential of CA based model of associative memory, ideally suited for pattern recognition. Characterization of the proposed model is provided in *Section IV.*

2 Cellular Automata Preliminaries

A *Cellular Automaton* (CA) is a discrete system which evolves in discrete space and time. It consists of a large number of cells, organized in the form of a lattice. Each cell acts as a single processor and communicates only with the neighboring cells. The updation of state (next state) for a cell depends on its own state and the states of its neighboring cells and is usually specified in the form of a *rule table* [10]. If the same rule is applied to all the cells, then the CA is referred to as $uniform$ CA, else it is a *hybrid* CA. In this paper we will discuss only 3-neighborhood (left neighbor, self and right neighbor) one dimensional CA and each CA cell having only two states - 0 or 1.

A $GMACA$ is a *hybrid* CA that employs non-linear rules with AND/OR logic and can perform pattern recognition task. *Fig.1* illustrates the state space of a 4-cell hybrid $GMACA$ with rule vector $< 98, 236, 226, 107 >$.

In the present work, we have concentrated on the non-linear $GMACA$ and its attractive feature to memorize the large number of patterns. An extensive study to characterize the $GMACA$ is reported.

3 $GMACA$ for Pattern Recognition

The pattern recognizer is trained to get familiarized with some specific pattern set $\mathcal{P} = \{\mathcal{P}_1, \cdots, \mathcal{P}_i, \cdots \mathcal{P}_k\}$ to memorize $\mathcal{P}$ and can recognize an incoming pat-

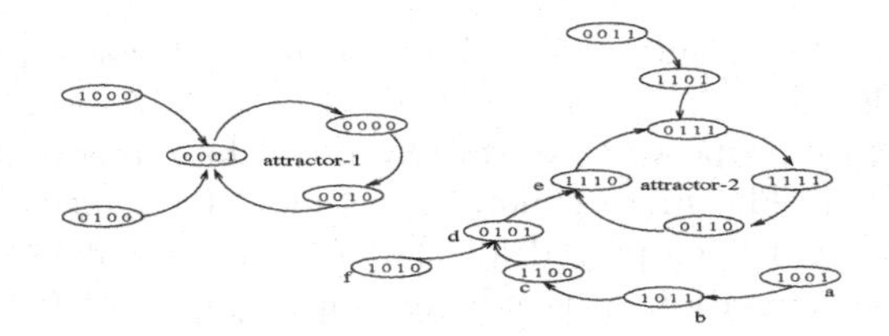

Fig. 1. State space of a GMACA based 4-bit pattern recognizer

tern $\mathcal{P}_i$ even if the pattern is corrupted with limited noise. When the new pattern $\acute{\mathcal{P}}_i$ is input to the system, the patern recognizer identifies it as $\mathcal{P}_i$, where $\acute{\mathcal{P}}_i$ is the closest match to $\mathcal{P}_i$. If the entire pattern is viewed as a binary string, then the hamming distance between $\acute{\mathcal{P}}_i$ and $\mathcal{P}_i$ is the least among all $\mathcal{P}$'s. The hamming distance between $\acute{\mathcal{P}}_i$ and $\mathcal{P}_i$ is the measure of noise.

A *GMACA*, having its state space distributed into disjoint basins (*Fig.1*) with transient and attractor states, models an associative memory. The entire state space around some pivotal points can be viewed as the state space generated by a *GMACA* with its attractors and transient states. Memorizing the set of pivotal points $\{\mathcal{P}_1, \cdots, \mathcal{P}_i, \cdots \mathcal{P}_k\}$ is equivalent to the design of a *CA* with the pivotal points as the attractor states. Any other transient point $\acute{\mathcal{P}}_i$, in close vicinity of a pivotal point $\mathcal{P}_i$, can be considered as a pattern with some noise. The correct output $\mathcal{P}_i$ can be produced in time proportionate to the state traversal of the *GMACA*.

3.1 Design Goals

The *GMACA* which can perform pattern recognition for a design should maintain the following two relations:

R1: Each attractor basin of the *GMACA* should contain one and only one pattern ($\mathcal{P}_i$) to be learnt in its attractor cycle; and

R2: The hamming distance of each state $\mathcal{S}_i \in \mathcal{P}_i$-basin with $\mathcal{P}_i$ is lesser than the hamming distance of $\mathcal{S}_i$ with any other $\mathcal{P}$s.

Fig.1 illustrates a 4-cell *GMACA* that maintains both R1 and R2. It learns two patterns, $\mathcal{P}_1 = 0000$ and $\mathcal{P}_2 = 1111$. The state $\acute{\mathcal{P}} = 1110$ has the hamming distances 3 and 1 with $\mathcal{P}_1$ & P_2 respectively. If $\acute{\mathcal{P}}$ is given as the input to be recognized, then the recognizer designed with the *GMACA* of *Fig.1* is loaded with $\acute{\mathcal{P}}= 1110$. The *GMACA* returns the desired pattern $\mathcal{P}_2$ after two time steps.

The next subsection describes the *GA* based searching scheme implemented to evolve the *GMACA*, tuned for a set of patterns to be recognized.

3.2 Evolving *GMACA* Based Associative Memory

The aim of this design is to evolve the *GMACA* (rule vector) that can perform pattern recognition task. This subsection describes the *GA* based solution to evolve the *GMACA* rules with the desired functionality.

In our model, the GA starts with an initial population (IP) of 50 chromosomes (CA rules). Each CA rule evolves over several generations under selection (elitist model), crossover, and mutation. The evolution process is controlled by the fitness function $\mathcal{F}(C_r)$.

Fitness function: The fitness $\mathcal{F}(C_r)$ of a particular chromosome C_r (CA rule) in a population is determined by the hamming distance between the attractor $\acute{\mathcal{P}}_i$, evolved for a state from the run (generation), and the desired attractor $\mathcal{P}_i$. The chromosome is run with 300 randomly chosen initial configurations (ICs) and fitness of CA is determined by averaging the fitness for each individual IC.

Let us assume that, a chromosome C_r is run for the maximum iterations ($\mathcal{L}_{max}$) for an IC and it reaches to a state $\acute{\mathcal{P}}_i$. If $\acute{\mathcal{P}}_i \notin$ any attractor cycle, that is, it is still a *transient state*, then the fitness value of C_r is considered as zero. On the other hand, if $\acute{\mathcal{P}}_i \in$ an attractor cycle containing $\mathcal{P}_i$, then the fitness of C_r is $\frac{n-|\mathcal{P}_i-\acute{\mathcal{P}}_i|}{n}$, where n is the length of the CA rule (chromosome). But, if any attractor other than $\mathcal{P}_i$ exits in that attractor cycle, then fitness of C_r is also considered as zero. Therefore, fitness function

$$\mathcal{F}(C_r) = \frac{1}{k}\sum_{i=1}^{k}\frac{n- \mid \mathcal{P}_i - \acute{\mathcal{P}}_i \mid}{n} \tag{1}$$

where k is the number of random ICs.

Selection, Crossover and Mutation: From the exhaustive experimentation, we have set the ratio of selection, mutation and crossover to be performed on the present population (PP) to form the next generation population (NP). To construct NP, 10% of the best CA rules (elite rules) of PP are selected without any modifications. The 80% chromosomes of NP are constructed from crossover operations among the chromosomes of PP. The rest 10 % population of NP are generated out of mutations of the elite rules. We employ single point crossover between randomly chosen pairs of rules and the single point mutation.

3.3 Performance of $GMACA$ Based Model

The experiments to evolve desired n-cell $GMACA$ for different values of n are carried out. For each n, 15 different sets of patterns to be trained are selected randomly. The number of patterns to be learnt is then progressively increased.

Table 1 demonstrates the pattern recognition capability of a $GMACA$ based design. *Column II* of *Table 1* depicts the maximum number of patterns that an n-cell $GMACA$ can memorize -that is, the number of patterns for which the GA has obtained 100% fit rules. The average number of generations required to converge the GA are noted in *Column III.* The results of Hopfield Net are provided in *Column IV.* The experimentation clearly indicates that the $GMACA$ have much higher storage capacity in comparison to Hopfield Net.

Table 1. Performance of *GMACA* based pattern recognizer

Pattern size (n)	*CA* based Patt. Recog		No of patterns memorized in Hopfield net
	No of patterns memorized	No of gene^n required	
10	4	128	2
15	4	66	2
20	5	172	3
25	6	210	4
30	7	280	5
35	8	340	5
40	9	410	6
45	10	485	7

4 Characterization of *GMACA* Rule Space

In this section, we make an extensive study on the derived *GMACA*. The study is directed towards the categorization of *GMACA* rules that performs the complex computations in pattern recognition. The study has been performed (i) locally by exploring *CA* rule table, and (ii) globally by exploring the dynamical behavior of the space-temporal patterns generated by the *CA* set.

4.1 Exploring *CA* Rule Table

The *CA* rule space can be characterized by evaluating the parameters - λ [7] and $\mathcal{Z}$ [8]. Different values of these parameters correspond to different *CA* dynamics.

Characterization of the uniform *CA* state transition, based on the value of λ, is reported in [7]. As the λ value is shifted from 0 to 0·5 the average behavior of the *CA* passes through the transitions : homogeneous $\rightarrow$ periodic $\rightarrow$ complex $\rightarrow$ chaotic. The *CA* behaves in the reverse order when the λ value is incremented from 0·5 to 1. The range of λ for which the rule exhibit complex function is referred to as λ_c. All complex computations are likely to occur near λ_c [7]. The region is referred to as the *edge of chaos*. For the hybrid *CA* domain we introduce the parameter λ_{av}, the average of λ values for different *CA* cells. In *Table 2*, we provide the λ_{av} values for different fit *GMACA* rules, evolved for different n. *Column II* indicates the resulted mean of λ_{av} above $0 \cdot 5$ while the *Column IV* indicates mean value below $0 \cdot 5$. Corresponding standard deviations are noted in *Column III* and *V* respectively.

The value of $\mathcal{Z}$ varies from 0 to 1. The $\mathcal{Z}$ value close to 1 indicates chaotic behavior of the *CA*, while $\mathcal{Z} = 0$ indicates order [8]. Any intermediate value of $\mathcal{Z}$ identifies the complex *CA* rules. The $\mathcal{Z}$ parameters of evolved *GMACA* rules for different values of n are also reported in *Table 2*. *Column VI* depicts the mean value of $\mathcal{Z}$ and *Column VII* contains the standard deviation in $\mathcal{Z}$ parameter to arrive at the desired *GMACA* rules.

Table 2 illustrates that the λ_{av} of the evolved *GMACA* are clustered around in the areas that are roughly equidistance from 0·5 (from 0·433 to 0·494 and from

Table 2. Parameters characterizing rule space

CA	$\lambda_{av} > 0.5$		$\lambda_{av} < 0.5$		$\mathcal{Z}$ parameter	
size (n)	Mean	Std. Devi^n	Mean	Std. Devi^n	Mean	Std. Devi^n
10	0.537	0.020	0.472	0.018	0.618	0.033
15	0.524	0.021	0.458	0.025	0.642	0.037
20	0.512	0.005	0.487	0.007	0.621	0.012
25	0.518	0.007	0.486	0.006	0.622	0.009
30	0.509	0.005	0.465	0.025	0.610	0.029
35	0.511	0.021	0.481	0.018	0.627	0.022
40	0.523	0.011	0.476	0.019	0.633	0.018
45	0.521	0.013	0.489	0.009	0.639	0.033

0·503 to 0·557) which corresponds to the region at *edge of chaos* [9]. Similarly, the $\mathcal{Z}$ parameter values for the evolved *GMACA* are found intermediate between 0 and 1. It indicates the *GMACA* rules as the complex rules [8].

4.2 Space Temporal Study

Dynamical behavior of space-time patterns generated by the *GMACA* is an another way to characterize the *CA* rule space [9,7,10]. The macroscopic measurements of *CA* dynamics like *entropy* [9], *mutual information* are used to classify the *CA* rules.

Entropy is the measure of randomness of a system. The maximum entropy (close to 1) of a system signifies *chaotic behavior.* The values shown in the *Columns II* and *III* of *Table 3* are the mean and standard deviation of entropy, computed for the evolved *GMACA* rules. The high entropy and low standard deviation indicate that the evolved *GMACA* rules are at the *edge of chaos* [9].

Table 3. Space temporal study to categorize *CA* rule space

CA	Entropy		Mutual
size (n)	Mean	Std. Devi^n	Information
10	0.813	0.024	0.729
15	0.921	0.033	0.740
20	0.859	0.037	0.638
25	0.903	0.021	0.829
30	0.867	0.030	0.688
35	0.878	0.031	0.724
40	0.857	0.052	0.713
45	0.839	0.021	0.683

Mutual information measures the correlation between patterns generated at a fixed time interval. If a pattern $\mathcal{P}_1$ is the copy of $\mathcal{P}_2$, then mutual information between $\mathcal{P}_1$ and $\mathcal{P}_2$ is 1. Whereas, the mutual information between two statistically independent patterns is 0. The ordered *CA* rules do not create spatial structures, in effect generate pattern set with low mutual information. On the other hand, the complex *CA* rules create highly correlated structures. To measure the mutual information, we select patterns generated from the evolved *GMACA* rule, separated by the window of size 6. The mutual information value corresponding to an evolved *GMACA* rule is noted in *Column V* of *Table 3.* All the values in *Column V* are found to be very high.

All the study performed in this section, points that the *GMACA* evolved for pattern recognition, belong to the same class and are most likely to be found at the *edge of chaos.*

5 Conclusion

This paper has established the Cellular Automata as a powerful machine in designing the pattern recognition tool. The pattern recognizer is designed with the hybrid *GMACA* based associative memory. It is proved experimentally that the storage capacity of *GMACA* based associative memory is far better than that of Hopfield Net and can train much more number of patterns. A new parameter λ_{av} extends the Langton's observation in the domain of hybrid *CA*. The consistent values of different parameters of the *GMACA* rule space ensure that the *GMACA* rules which can perform complex computation of pattern recognition are most likely to be occurred at the *edge of chaos.*

References

1. J. Hertz, A. Krogh and R. G. Palmer, *"Introduction to the theory of Neural computation"*, Santa Fe institute studies in the sciences of complexity, Addison Wesley, 1991.
2. M. Chady and R. Poli, *"Evolution of Cellular-automaton-based Associative Memories"*, Technical Report no. CSRP-97-15, May 1997.
3. J. J. Hopfield, *"Pattern Recognition computation using action potential timings for stimulus representations"*, Nature, 376: 33-36; 1995.
4. E. Jen, *"Invariant strings and Pattern Recognizing properties of 1D CA"*, Journal of statistical physics, 43, 1986.
5. P. Pal Chaudhuri, D Roy Chowdhury , S. Nandi and S. Chatterjee, *"Additive Cellular Automata, Theory and Applications, VOL. 1"*, IEEE Computer Society Press, Los Alamitos, California.
6. M. Sipper, *"Co-evolving Non-Uniform Cellular Automata to Perform Computations"*, Complex Systems, 7: 89-130, 1993.
7. W. Li, N. H. Packard and C. G. Langton, *"Transition Phenomena in Cellular Automata Rule Space"*, PhysicaD, 45; 1990.
8. Wuensche, A., and M. J. Lesser, *"The Global Dynamics of Cellular Automata"*, Santa Fe Institute Studies in the Science of Complexity, Addison-Wesley, 1992.
9. M. Mitchell, P. T. Hraber and J. P. Crutchfield, *"Revisiting the Edge of Chaos: Evolving Cellular Automata to Perform Computations"*, Complex Systems, 7:89-130, 1993.
10. S. Wolfram, *"Theory and application of Cellular Automata"*, World Scientific, 1986.

On Convergence of a Neural Network Model Computing MSC

Swapan K. Parui and Amitava Datta

Indian Statistical Institute,
203 B. T. Road, Calcutta - 700 035
amitava@isical.ac.in,
WWW Home Page: http://www.isical.ac.in/~amitava

Abstract. A self-organizing neural network model that computes the centre and radius of the minimum circle spanning a given finite planar set is proposed by Datta [8]. Here we mathematically prove that the model converges to the desired centre of the minimum spanning circle.

1 Introduction

The problem of finding the minimum spanning circle (MSC) or the smallest enclosing circle for n given points in the Euclidean plane is very well-known and various techniques to solve this problem are developed over a long span of time [1]–[7]. The MSC has applications in optimization, pattern recognition, image analysis, statistical estimation and many other areas. A neural network formulation of this problem, using a self-organizing model, has been proposed by Datta [8]. In the present paper, we work out rigorous analysis of this model and show that the learning process converges. The model yields the centre and the radius of the desired minimum circle with the learning coefficient α satisfying a certain condition similar to that imposed on stochastic approximation processes.

2 The MSC Model

Let $S = \{P_1, P_2, ..., P_n\}$ be a set of n given points in the Euclidean plane. The problem is to find the centre and radius of the smallest circle such that no point of S falls outside the circle. Such a circle is called the minimum spanning circle. The problem can be stated as: Find the point $W = (w_1, w_2)$ so that

$$\max_j \|P_j - W\| \tag{1}$$

is minimized over all choices of W in the plane. In other words, we compute

$$r = \min_{(w_1,w_2)} \max_j \{((w_1 - a_j)^2 + (w_2 - b_j)^2)^{\frac{1}{2}}\} \tag{2}$$

where (a_j, b_j) is the co-ordinates of $P_j, j = 1, 2, \ldots, n$. The radius of the MSC is r and the optimum (w_1, w_2) is the centre.

N.R. Pal and M. Sugeno (Eds.): AFSS 2002, LNAI 2275, pp. 221–227, 2002.

The MSC-model [8] uses the principle of simple competitive learning [9], and learns the position of the centre of the MSC from the input without any supervision. An array of processors interconnected among themselves, forming a network, is used. The $2D$ co-ordinates in S are treated as the input signals. Assign one processor to each point of S. Another processor π stores a weight vector $W = (w_1, w_2)$. This weight vector is updated iteratively in the self-organization process. The weight vector here represents a position in the plane where the processor π can be thought to be located.

Assign the initial value of the weight vector W at random[1]. At iteration t we find the farthest point (w.r.t. Euclidean distance) from $W(t)$. Let $P_k = (a_k, b_k)$ be the farthest point, $k \in \{1, 2, ..., n\}$. That is,

$$\|P_k - W(t)\| = \max_j \|P_j - W(t)\| \tag{3}$$

Then the weight vector W is updated as follows:

$$W(t+1) = W(t) + \alpha(t)(P_k - W(t)) \tag{4}$$

The weight update process is continued repeatedly with $\alpha(t) \to 0$ as $t \to \infty$ enabling the process to stabilize when the difference between the weight vectors at two successive iterations is negligible. For details of the model, see [8].

3 Convergence of MSC-Model

We shall now prove that the weight vector $W(t)$ converges to the centre of the minimum spanning circle so that the centre can be obtained with any given level of accuracy.

Definition 1. For the set S, the *farthest point Voronoi diagram* (FPVD) is a partition of the plane defined by the convex regions V_r, $1 \leq r \leq n$ where

$$V_r = \{P : \|P - P_r\| > \|P - P_t\| \;\; \forall\, t \neq r\} \tag{5}$$

V_r is the farthest point Voronoi polygon corresponding to P_r, $1 \leq r \leq n$. Farthest point Voronoi diagrams are shown in Fig.1. The proof is divided into four cases:

Case-1: If $S = \{P_1\}$ then $W^* = P_1$ where, W^* is the limiting value of $W(t)$.

Case-2: If $S = \{P_1, P_2\}$ then $W^* = \frac{1}{2}(P_1 + P_2)$.

Case-3: If $S = \{P_1, P_2, P_3\}$ then W^* is either (a) the midpoint of one of the sides of the triangle $P_1P_2P_3$ or (b) the common FPV vertex of the three FPV polygons V_1, V_2, V_3.

Case-4: If $S = \{P_1, P_2, \ldots, P_n\}$ then

(a) If the MSC is determined by two points P_i, P_j then $W^* = \frac{1}{2}(P_i + P_j)$.

(b) If the MSC is determined by three points P_i, P_j, P_k then W^* = the common FPV vertex of the three FPV polygons V_i, V_j, V_k.

---

[1] In the original model [8], the centroid of points is taken as the initial value.

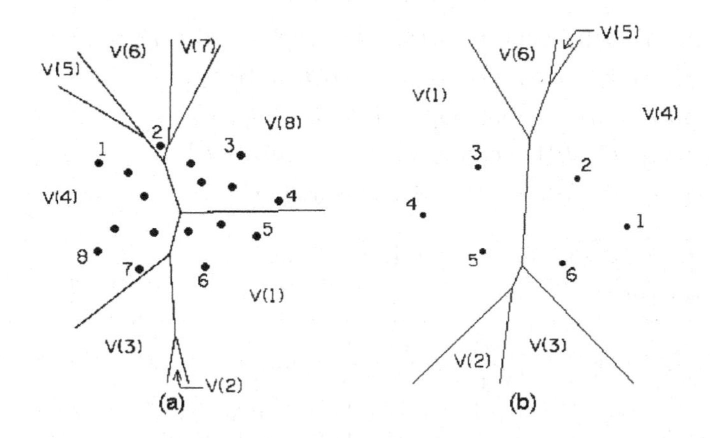

Fig. 1. FPVD for two planar sets. MSC is determined by (a) three (b) two points.

Before going into the proof, we state below a few known results ([2,10,11]).

Result 1. If there is a circle C passing through three points, P_i, P_j and P_k such that they do not lie on any open semicircumference of C and if C contains all the points, then C is the MSC.

Result 2. If the MSC passes through exactly two points P_i and P_j, then the line segment P_iP_j forms a diameter of the MSC.

Result 3. If no two points of S form a diameter of the MSC, then the MSC passes through at least three points which do not lie on any open semicircumference of the MSC.

Result 4. The MSC is unique.

Let $X(t) \in S$, $t = 0, 1, 2, \ldots$ be such that $\|X(t) - W(t)\| = \max_i \|P_i - W(t)\|$, $0 < \alpha(t) < 1$ for all t and $\alpha(t)$ decreases to 0 as t tends to ∞ such that $\sum_{t=0}^{\infty} \alpha(t) = \infty$.
Note that the points of S come from a bounded region. Hence, $\|W(0) - P_i\| < K$ for all i, for some $K > 0$. It is easy to see that $\|W(t) - P_i\| < K \; \forall \; i, \forall \; t$.

Lemma 1. $\prod_{t=0}^{\infty}[1 - \alpha(t)] = 0$.

Proof. Since $0 < \alpha(t) < 1$, for all t, $\sum_{t=0}^{\infty} \alpha(t) = \infty$ iff $\prod_{t=0}^{\infty}[1 - \alpha(t)] = 0$ ([12]). ∎

We shall now prove a result on $\alpha(t)$. For any given t_0, let $S(0) = \alpha(t_0)$ and $S(t+1) = [1 - \alpha(t_0 + t + 1)]S(t) + \alpha(t_0 + t + 1)$ for $t = 0, 1, 2, \ldots$.
Note that

$$S(1) = \alpha(t_0)[1 - \alpha(t_0 + 1)] + \alpha(t_0 + 1)$$
$$S(2) = \alpha(t_0)[1 - \alpha(t_0 + 1)][1 - \alpha(t_0 + 2)] + \alpha(t_0 + 1)[1 - \alpha(t_0 + 2)] + \alpha(t_0 + 2)$$

$$
\begin{aligned}
S(t) = {} & \alpha(t_0)[1-\alpha(t_0+1)][1-\alpha(t_0+2)]\ldots[1-\alpha(t_0+t)] \\
& +\alpha(t_0+1)[1-\alpha(t_0+2)]\ldots[1-\alpha(t_0+t)] \\
& +\alpha(t_0+2)[1-\alpha(t_0+3)]\ldots[1-\alpha(t_0+t)]+\ldots \\
& +\alpha(t_0+t-2)[1-\alpha(t_0+t-1)][1-\alpha(t_0+t)] \\
& +\alpha(t_0+t-1)[1-\alpha(t_0+t)]+\alpha(t_0+t)
\end{aligned}
$$

Lemma 2. $S(t)$ goes to 1 as t goes to ∞.
Proof. It is easy to see that

$$
\begin{aligned}
1-S(1) &= [1-\alpha(t_0)][1-\alpha(t_0+1)] \\
1-S(2) &= [1-S(1)][1-\alpha(t_0+2)] \\
&= [1-\alpha(t_0)][1-\alpha(t_0+1)][1-\alpha(t_0+2)] \\
1-S(t) &= [1-\alpha(t_0)][1-\alpha(t_0+1)][1-\alpha(t_0+2)]\ldots[1-\alpha(t_0+t)]
\end{aligned}
$$

From Lemma 1, $\prod_{t=t_0}^{\infty}[1-\alpha(t)]=0$. Hence Lemma 2. ■

Applying Eqn.(4) repeatedly we get

$$
\begin{aligned}
W(t_0+t+1) = {} & [1-\alpha(t_0)][1-\alpha(t_0+1)][1-\alpha(t_0+2)]\ldots[1-\alpha(t_0+t)]W(t_0) \\
& +\alpha(t_0)[1-\alpha(t_0+1)][1-\alpha(t_0+2)]\ldots[1-\alpha(t_0+t)]X(t_0) \\
& +\alpha(t_0+1)[1-\alpha(t_0+2)]\ldots[1-\alpha(t_0+t)]X(t_0+1) \\
& +\alpha(t_0+2)[1-\alpha(t_0+3)]\ldots[1-\alpha(t_0+t)]X(t_0+2)+\ldots \\
& +\alpha(t_0+t-2)[1-\alpha(t_0+t-1)][1-\alpha(t_0+t)]X(t_0+t-2) \\
& +\alpha(t_0+t-1)[1-\alpha(t_0+t)]X(t_0+t-1) \\
& +\alpha(t_0+t)X(t_0+t)
\end{aligned}
$$

Case 1: $n = 1$.
Here, $X(t) = P_1$ for all t. Hence, from above, we get

$$W(t_0+t+1) = [1-S(t)]\,W(t_0) + S(t)\,P_1$$

Lemma 3. For Case 1, $W(t_0+t)$ goes to P_1 as t goes to ∞.
Case 2: $n = 2$. The centre of the MSC is $\frac{1}{2}(P_1+P_2)$.
Lemma 4. For Case 2, $W(t_0+t)$ goes to $(P_1+P_2)/2$ as t goes to ∞.
Proof. It is based on Proposition 1 and Proposition 2 below.
Proposition 1. $W(t_0+t+1)$ can be made arbitrarily close to the straight line L passing through P_1 and P_2.
Note that

$$
\begin{aligned}
& \|P_1-W(t_0+t+1)\|+\|P_2-W(t_0+t+1)\| \\
\leq {} & [1-\alpha(t_0)][1-\alpha(t_0+1)][1-\alpha(t_0+2)]\ldots[1-\alpha(t_0+t)][\|P_1-W(t_0)\|
\end{aligned}
$$

$$
\begin{aligned}
&+\|P_2 - W(t_0)\|] \\
&+\alpha(t_0)[1-\alpha(t_0+1)][1-\alpha(t_0+2)]\dots[1-\alpha(t_0+t)][\|P_1 - X(t_0)\| \\
&+\|P_2 - X(t_0)\|] \\
&+\alpha(t_0+1)[1-\alpha(t_0+2)]\dots[1-\alpha(t_0+t)][\|P_1 - X(t_0+1)\| \\
&+\|P_2 - X(t_0+1)\|] \\
&+\alpha(t_0+2)[1-\alpha(t_0+3)]\dots[1-\alpha(t_0+t)][\|P_1 - X(t_0+2)\| \\
&+\|P_2 - X(t_0+2)\|] + \dots \\
&+\alpha(t_0+t-2)[1-\alpha(t_0+t-1)][1-\alpha(t_0+t)][\|P_1 - X(t_0+t-2)\| \\
&+\|P_2 - X(t_0+t-2)\|] \\
&+\alpha(t_0+t-1)[1-\alpha(t_0+t)][\|P_1 - X(t_0+t-1)\| + \|P_2 - X(t_0+t-1)\|] \\
&+\alpha(t_0+t)[\|P_1 - X(t_0+t)\| + \|P_2 - X(t_0+t)\|] \\
= {} &[1-\alpha(t_0)][1-\alpha(t_0+1)][1-\alpha(t_0+2)]\dots[1-\alpha(t_0+t)]\ [\|P_1 - W(t_0)\| \\
&+\|P_2 - W(t_0)\|] \\
&+\ \alpha(t_0)[1-\alpha(t_0+1)][1-\alpha(t_0+2)]\dots[1-\alpha(t_0+t)]\ \|P_1 - P_2\|] \\
&+\ \alpha(t_0+1)[1-\alpha(t_0+2)]\dots[1-\alpha(t_0+t)]\ \|P_1 - P_2\| \\
&+\ \alpha(t_0+2)[1-\alpha(t_0+3)]\dots[1-\alpha(t_0+t)]\ [\|P_1 - P_2\| + \dots \\
&+\ \alpha(t_0+t-2)[1-\alpha(t_0+t-1)][1-\alpha(t_0+t)]\ [\|P_1 - P_2\| \\
&+\ \alpha(t_0+t-1)[1-\alpha(t_0+t)]\ \|P_1 - P_2\| \\
&+\ \alpha(t_0+t)\ \|P_1 - P_2\| \\
= {} &[1-S(t)]\ [\|P_1 - W(t_0)\| + \|P_2 - W(t_0)\|]\ +\ S(t)\ \|P_1 - P_2\|
\end{aligned}
$$

Now, $S(t)$ goes to 1 as t tends to ∞. So, $W(t_0+t+1)$ gets arbitrarily close to the line segment joining P_1 and P_2. Hence Proposition 1. ■

Proposition 2. $W(t_0+t+1)$ can be made arbitrarily close to the perpendicular bisector L_1 of the line segment P_1P_2.

Let H_1 and H_2 be the two half planes defined by L_1 such that P_i belongs to H_i $(i = 1, 2)$. Without loss of generality, let $W(t_0)$ lie in H_1. From Lemma 3, we know $W(t_0 + t)$ will belong to H_2 for some t. Suppose $W(t_0 + 1), W(t_0 + 2), \dots, W(t_0 + t_1 - 1)$ belong to H_1 and $W(t_0 + t_1)$ belongs to H_2. Now, the perpendicular distance between $W(t_0+t_1)$ and L_1 is less than or equal to $\|W(t_0+t_1-1) - W(t_0+t_1)\|$. (Fig.2)

Note that $\|W(t_0+1) - W(t_0)\| = \alpha(t_0)\|X(t_0) - W(t_0)\| \leq \alpha(t_0)K$.

Since $\alpha(t) \to 0$ as $t \to \infty$, for any $\delta > 0$ we can take t_0 to be sufficiently large so that $\|W(t_0+1) - W(t_0)\| < \delta$ and in general, $\|W(t_0+t) - W(t_0+t-1)\| < \delta$ for all $t > 0$. Now, $W(t_0 + t_1 + t)$ will belong to H_1 for some t. Suppose, $W(t_0+t_1), \dots, W(t_0+t_1+t_2-1)$ belong to H_2 and $W(t_0+t_1+t_2)$ belongs to H_1. From Fig.2(a), it is clear that the distance between $W(t_0 + t_1 + t)$ and L_1 is less than δ for all $t > 0$. Hence Proposition 2. ■

Case 3. Suppose $n = 3$. Consider the farthest point Voronoi diagram (FPVD) of the points. The only vertex of the FPVD of P_1, P_2, P_3 may lie either (a) outside

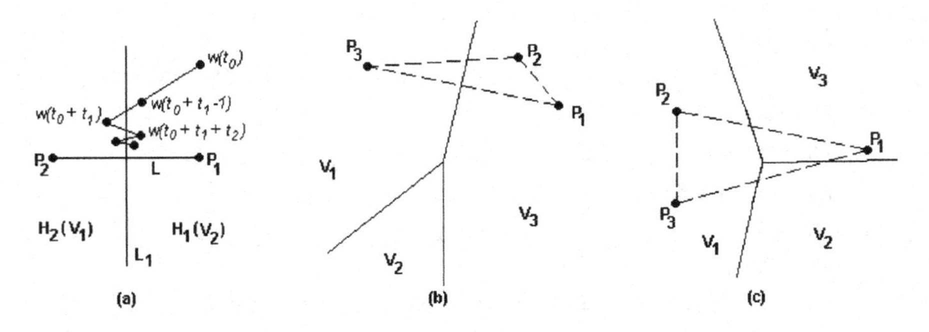

Fig. 2. The trajectories of $W(t)$. (a) Case 2, (b) Case 3(a), (c) Case 3(b). The FPVD is denoted by solid lines and the triangle formed by the input points is denoted by dashed lines.

(Fig.2(b)) or (b) inside (Fig.2(c)) the triangle $\Delta P_1P_2P_3$ formed by P_1, P_2, P_3. The FPV polygon corresponding to the point P_i is V_i $(i = 1, 2, 3)$.

Case 3(a). In the former case, for some large t_0, $W(t)$ will lie outside V_2 for all $t > t_0$. From Case-2, it is then clear that $W(t)$ converges to $(P_1 + P_3)/2$ which is the centre of the MSC.

Case 3(b). In the latter case, consider the three perpendicular bisectors L_{ij} of the line segments P_iP_j $(1 \leq i \neq j \leq 3)$. From the arguments given in the proof of Proposition 2 in Case-2, $W(t)$ will be arbitrarily close to each of the three L_{ij}'s. Thus $W(t)$ converges to the intersection point of the three bisectors which is the FPV vertex where the FPV polygons V_1, V_2, V_3 meet. Here, the FPV vertex is the centre of the MSC. Note that for any large t_0 and for any i, there will be a $t_i > t_0$ such that $W(t_i)$ lies in V_i. In other words, each V_i will be visited by $W(t)$ infinitely many times.

Case 4. Suppose $n > 3$.
We claim that W^*, the limiting value of $W(t)$, will lie either (a) on an open edge (an edge excluding the vertex points) or (b) on a vertex of the FPV diagram of the point set S. Suppose not. Suppose, W^* lies in the interior of some V_p. Then, after some time, $W(t)$'s will always lie in the interior of V_p in which case, the limiting value of $W(t)$ will be P_p (from Case 1). It is a contradiction since V_p cannot contain P_p. Hence we have two subcases:

Case 4(a). The limiting value W^* lies on an open FPV edge. Suppose this edge is the common edge of the two FPV polygons V_i and V_j. That means, for sufficiently large t_0, $W(t)$ will lie either in V_i or in V_j for all $t > t_0$. Moreover, for any t' there will exist t_i, $t_j > t'$ such that $W(t_i)$ and $W(t_j)$ are in V_i and V_j respectively. In other words, both V_i and V_j (and only they) will be visited by $W(t)$ infinitely many times. This case is similar to Case 2 and hence $W^* = \frac{1}{2}(P_i + P_j)$ (see Fig.1). Now construct a circle C, centred at W^* passing through P_i, P_j. Obviously, P_iP_j forms the diameter of the circle C and since P_i,

P_j are the farthest points from W^*, C contains all the points of S. Hence, by Results 1–4, C is the MSC.

Case 4(b). The limiting value of W^* lies on an FPV vertex. Suppose W^* lies on the common FPV vertex v_{ijk} of three FPV polygons say, V_i, V_j and V_k. That means, for sufficiently large t_0, $W(t)$ will lie either in V_i or V_j or V_k for all $t > t_0$ and each of V_i, V_j, V_k will be visited by $W(t)$ infinitely many times. Then it will be similar to Case 3(b). Note that W^* will be equidistant from P_i, P_j and P_k. Construct a circle C, centred at W^* and passing through P_i, P_j and P_k. We claim that P_i, P_j and P_k do not lie on an open semicircumference of C. Suppose not. Then v_{ijk} falls outside the triangle $\Delta P_i P_j P_k$. Hence, from Case 3(a), W^* cannot lie on v_{ijk} (see Fig.2(b)). This is a contradiction since W^* lies on v_{ijk}. Moreover, since P_i, P_j and P_k are the farthest points from v_{ijk}, C contains all the points of S. Hence C is the MSC from Results 1–4.

4 Conclusions

Datta [8] proposed a neural network model for the Minimum Spanning Circle problem. The model was found to give quite satisfactory results on the basis of computer simulations. It was also conjectured that the model could be straightway implemented for any higher dimension without additional time complexity. In the present article, we mathematically analyze the MSC-model for $2D$ case and prove that it produces the desired MSC after convergence.

References

1. D. W. Hearn and J. Vijay. Efficient algorithms for the (weighted) minimum circle problem. Operation Research **30** (1982) 777-795.
2. R. C. Melville. An implementation study of two algorithms for the minimum spanning circle problem. *Computational Geomerty* (G. T. Toussaint, Ed.), North-Holland, (1985) 267-294.
3. N. Megiddo. Linear time algorithms for linear programming in R3 and related problems. SIAM J. of Computing **12** (1983) 759-776.
4. R. Skyum. A simple algorithm for smallest enclosing circle. Infor. Proc. Letters **37** (1991) 121-125.
5. R. D. Smallwood. Minimax detection station placement. Operation Research **13** (1965) 636-646.
6. J. J. Sylvester. A question in the geometry of situation. Quart. J. Math. **1** (1857) 79-?.
7. G. Toussaint and B. Bhattacharya. On geometric algorithms that use the farthest-point Voronoi diagram. *Tech. Rept. SOCS-81.3*, McGill Univ. School of Computer Sci., 1981.
8. A. Datta. Computing minimum spanning circle by self-organization. Neurocomputing **13** (1996) 75-83.
9. K. Mehrotra, C. K. Mohan and S. Ranka. *Elements of Artificial Neural Networks*, MIT Press, 1997.
10. F. P. Preparata and M. I. Shamos. *Computational Geometry: An Introduction*, New York: Springer-Verlag, 1985.
11. M. Shamos. *Problems in Computational Geometry*, Carnegie-Mellon Univ., 1977.
12. T. M. Apostol. *Mathematical Analysis*, second edition, Addison-Wesley, 1979.

Recognition of Handprinted Bangla Numerals Using Neural Network Models

U. Bhattacharya[1], T.K. Das[1], A. Datta[2], S.K. Parui[1], and B.B. Chaudhuri[1]

[1] Computer Vision and Pattern Recognition Unit,
Indian Statistical Institute, 203, B. T. Road, Calcutta - 35, India
{ujjwal,swapan,bbc}@isical.ac.in
[2] Computer and Statistical Service Center,
Indian Statistical Institute, 203, B. T. Road, Calcutta - 35, India
amitava@isical.ac.in

Abstract. This paper proposes an automatic recognition scheme for handprinted Bangla (an Indian script) numerals using neural network models. A Topology Adaptive Self Organizing Neural Network is first used to extract from a numeral pattern a skeletal shape that is represented as a graph. Certain features like loops, junctions etc. present in the graph are considered to classify a numeral into a smaller group. If the group is a singleton, the recognition is done. Otherwise, multilayer perceptron networks are used to classify different numerals uniquely. The system is trained using a sample data set of 1880 numerals and we obtained 90.56% correct recognition rate on a test set of another 3440 samples. The proposed scheme is sufficiently robust with respect to considerable object noise.

1 Introduction

Automatic recognition of optical characters, in particular handwritten characters, is a challenging problem. Though this research was initially for English characters only, during last few years a need for such research was felt in different regional languages. Machine recognition of handwritten English (Arabic) ZIP codes has been implemented with reasonable success in USA. Considerable research for recognition of handwritten numerals [8], [11] has been conducted in other scripts also. This paper is concerned with the recognition of handwritten numerals in Bangla, the second-most popular language and script in the Indian subcontinent and the fifth-most popular language in the world.

Handwritten character recognition, because of large variability of the input, is much more difficult than printed character recognition. The variability is caused by variation of shapes resulting from writing habit, style, education, mood etc., as well as factors such as writing instrument, writing surface and scanning quality. To the best of our knowledge, there are only few pieces of published work on recognition of handwritten Bangla numerals [4], [9].

It has been noted that the success of a character recognition technique is predominantly decided by the choice of distinguishing features [7], [10]. The set of features should be insensitive to distortion, human style variation and scaling

N.R. Pal and M. Sugeno (Eds.): AFSS 2002, LNAI 2275, pp. 228–235, 2002.

of the characters. To simplify the recognition scheme, many existing methods assume some constraints on handwriting with respect to tilt, size, relative positions, stroke connections, distortions etc. In this paper we consider numerals written inside fixed rectangular boxes.

During last few years, neural network (NN) based classification models have gained immense popularity. They are useful in designing high accuracy systems [5] because they perform well in the presence of noise and are adaptive in nature. Another advantage is the parallel nature of NN algorithms. In this paper we propose an approach to Bangla numeral recognition using NN models. We first use a topology adaptive self-organizing neural network (TASONN) model to obtain the graph representation of the input character. Then we consider structural features of this graph describing the topology of the character along with a hierarchical tree classifier to classify handprinted Bangla characters into a few subclasses. To further classify the characters belonging to such a subclass we use multilayer perceptron classifiers (MLP) along with different sets of features obtained from the graph representation.

2 Overview of NN Models

In the first phase of the proposed method we use the unsupervised TASONN while in the final phase the supervised MLP model is used.

2.1 Topology Adaptive Self-Organizing Neural Network Model

In a TASONN model [3], initially there is no processor (nodes), that is, the set of processors is null. New processors are added to the network depending on the distribution of the input points. The input vectors here are the positional coordinates of the object (numeral) pixels. The network is evolved through learning and provides a vector skeleton of the numeral pattern, which essentially provides a planar straight line graph.

In the present implementation of this TASONN model, initially there is a small set of processors (nodes) and the weight vectors associated to different processors are adaptively updated during learning until it converges with respect to the current set of processors. A new processor is added in the region of maximum density of input pixels and further adaptation of the weight vectors is done. This process converges when the number of processors reaches a certain preassigned value.

2.2 Multi-layer Perceptron Neural Network Model

Each MLP network here consists of an input layer, one hidden layer and an output layer. The weights along the connections are determined through training of the network. This training may be achieved by the well-known backpropagation algorithm [6]. This algorithm performs a gradient descent on some mean-square error surface and it usually converges very slow in many real life applications [2].

Zero	One	Two	Three	Four	Five	Six	Seven	Eight	Nine
০	১	২	৩	৪	৫	৬	৭	৮	৯

Fig. 1. Ideal samples of Bangla numerals

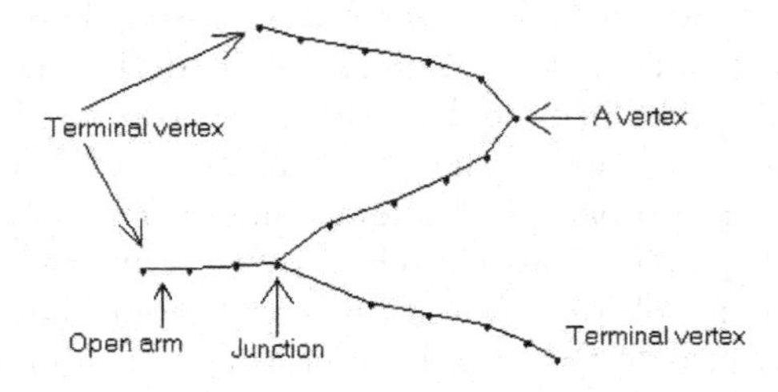

Fig. 2. Relevant parts of the graph of character two

However, the speed of convergence may be largely improved by suitably selecting its learning rate. For the present problem, we consider backpropagation algorithm with self-adaptive learning rates [1]. Here different learning rates are considered for individual weights and these are updated during learning.

3 Selection of Features and Recognition Scheme

The set of Bangla numerals is shown in Fig. 1. Let us now describe several definitions and notations related to the graph obtained by TASONN from such numerals.

Definition 1. A *junction* is a vertex of a graph having more than two neighbours.

Definition 2. A *terminal vertex* is a vertex which has only one neighbour.

Definition 3. The *lowest vertex* is the vertex situated below all other vertices.

Definition 4. The *lowest terminal vertex* is the terminal vertex situated vertically below all other terminal vertices.

Definition 5. An *open arm* ia a link between a terminal vertex and its neighbour.

Definition 6. The *right open arm* is the link between the rightmost terminal vertex and its neighbour.

Definition 7. The *cycle volume* is the proportion of the number of vertices forming the cycle to the total number of vertices forming the graph.

Definition 8. The *character height* is the height of the smallest rectangle enclosing the graph.

Definition 9. The *cycle centroid* is the centroid of vertices forming the cycle.

Similarly, we can define other relevant parts of a graph like *highest vertex*, *second highest vertex*, *highest terminal vertex*, *second highest terminal vertex*, *rightmost vertex*, *rightmost terminal vertex*, *left open arm* and *character width*. Some relevants parts of the graph corresponding to the image of a numeric character are shown in Fig.2.

Now let us introduce a few notations as listed in the table below.

Notation	Description
$HD1$	The horizontal distance between the junction and lowest terminal vertex normalized with respect to character width; (If there exists more than one junction, we consider the lowest such junction).
$HD2$	Horizontal distance between the highest terminal vertex and the lowest terminal vertex normalized with respect to character width; (If there exists only one terminal vertex, this distance is assumed to be zero).
$HD3$	Horizontal distance between the highest terminal vertex and the second highest terminal vertex normalized w.r.t. character width; (If there exists only one terminal vertex, this distance is assumed to be zero).
Vol	Cycle volume
$SgnM$	This variable takes the value -1 if the cycle centroid is situated to the left of the junction; else, it takes the value +1. If any one of the cycle or junction is not unique, then we consider the lowest such objects.
$DistR$	Dist. between the rightmost vertex and the rightmost terminal vertex.
$GradR$	Slope of right open arm.
$GradL$	Slope of left open arm.

We further introduce notations like $VD1$, $VD2$ and $VD3$ representing vertical distances similar to $HD1$, $HD2$ and $HD3$ respectively.

In the proposed method, several features of the graph obtained by TASONN have been considered for the recognition purpose. The most important ones are the existence of junction(s), cycle(s), the change in direction (clockwise or anticlockwise) along the graph and the number of terminal node(s). These four features are used for classifying the numerals in a class hierarchy (Fig.3). This class hierarchy is constructed manually on the basis of the training set. Each node of the tree represents a particular group of Bangla numerals. The choice of features and decision rules at each non-terminal node of this tree structure are as follows:

Root Node (Level 0):
Features: Existence of loop and junction.
Decision Rule: If no loop and no junction then Class A; If loop exists but no junction then class B; If no loop but junction exists then Class C; If both loop and junction exist then Class D.

Features and decision rules for the non-terminal nodes in this level are as follows:

Nodes at level 1:
Node representing Class A:
Features: Any change in the direction (clockwise/anticlockwise) while traversing the graph starting from a terminal vertex, then visiting its neighbour, then visiting the other neighbour of that neighbour and so on? (This feature is considered in cases when there is no junction).

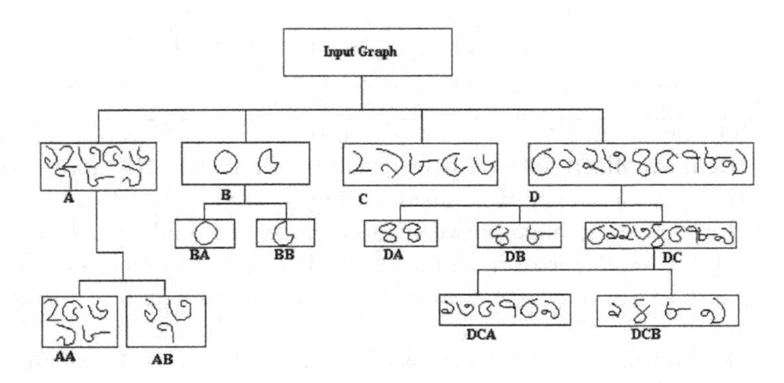

Fig. 3. Hierarchical tree classifier

Decision Rule: If the direction changes, then subclass AA. Else, subclass AB.

Node representing Class B:
Features & Decision Rule: These are similar to the node representing class A and the subclasses are respectively BA consisting only the numeral five and BB consisting only the numeral zero.

Node representing Class D:
Features: Number of loops and number of junctions.
Decision Rule: If the number of loops is two, then subclass DA; If number of loop is one and number of junctions is two, then subclass DB; If both the number of loop and number of junction are one, then subclass DC.

Node representing subclass DC (Level 2):
Features: Number of terminal vertices.
Decision Rule: If this number is one then subclass DCA. Else, subclass DCB.

The list of leaf nodes of our tree classifier are the following: C in level 1, AA, AB, BA, BB, DA and DB in level 2, DCA and DCB in the lowest level. Among these BA, BB and DA represent a single numeral each, *viz.* five, zero and four respectively. So, for these three nodes no further recognition strategy is required. For the rest six nodes further recognition is obtained as follows. To classify a pattern in such a node as a unique numeral we use an MLP network. The architectures of these MLPs are different for different leaf nodes. For each of them we use a different set of features and the number of possible output classes is also different. The features, network architectures and number of output classes are listed in Table 1.

4 Experimental Results

In this section we present results taken on a set of 5320 handwritten Bangla numerals. These characters are read by 300 dpi HP scanner. Since the proposed method is invariant of the size of characters, we do not apply any normalization technique. We randomly choose 1880 samples (188 samples/per class), which is

Table 1. List of features & architectures for different MLP

Leaf node	Features	Architecture	Possible classes
C	$HD3, VD3, DistR, GradR, GradL$	$5 \times 6 \times 5$	5
AA	$HD2, VD2, VD3, GradR$	$4 \times 6 \times 5$	5
AB	$VD2, VD3, GradL$	$3 \times 3 \times 3$	3
DB	$HD2, VD2, GradR$	$3 \times 3 \times 2$	2
DCA	$Vol, VD1, VD3$	$3 \times 4 \times 6$	6
DCB	$HD1, HD2, SgnM$	$3 \times 4 \times 4$	4

Table 2. Confusion matrix obtained on the training set of 1880 samples

Recognized as →

Digit	১	২	৩	৪	৫	৬	৭	৮	৯	০
১	90.5	1.01	1.06	0	0	0	1.2	1.7	2.51	0
২	3.2	92.3	0	0	0	0	0	2.5	0	0
৩	0.51	0	91.4	0	0.5	2.8	0	0	0	2.1
৪	1.2	0.5	0	92.6	1.5	0	0	2.5	0	1.5
৫	0.5	0.35	0	0	90.8	1.5	0	0	0	2.7
৬	0	0	3.6	0	1.72	92.1	0	0	0	0
৭	3.02	0.5	1.2	0	0	0	93.1	0.7	0	0
৮	0	2.1	0	2.1	0	0	0	91.3	0	0
৯	2.91	0.4	0	0	0	0	0	0	93.03	0
০	0	0	1.5	0	3.2	0.4	0	0	0	93.51

Table 3. Confusion matrix obtained on the test set of 3440 samples

Recognized as →

Digit	১	২	৩	৪	৫	৬	৭	৮	৯	০
১	89.8	1.08	1.1	0	0	0	1.5	1.9	3.15	0
২	3.5	90.4	0.3	0	0	0	0	3.1	0	0
৩	1.1	0.1	90.6	0	0.85	3.2	0	0	0	2.6
৪	1.5	0.6	0.1	90.1	1.8	0	0	3.31	0	1.7
৫	0.6	0.4	0	0	89.03	1.7	0	0	0	3.1
৬	0	0	3.9	0	2.1	91.1	0	0	0	0
৭	3.7	0.8	1.5	0	0	0	91.4	0.9	0	0
৮	0	2.3	0.8	2.3	0	0	0	90.07	0.7	0
৯	4.2	1.2	0	0	0	0	0	0	91.01	0
০	0	0	2.6	0	4.1	1.02	0	0	0	92.1

about 35% of the whole data set, as the training set and the test set consists of remaining 3440 characters (344 samples/per class).

In the first phase of recognition, to obtain the graph representation of the character, TASONN is used with an initial number of processors which is $\min(10, 0.1N)$, where N is total number of pixels in the character. Further processors are added until total number of processors become $\max(15, .1N)$.

Based on the graph, so obtained, an appropriate class of the input character is decided and the class is determined using appropriate MLP. The training of the MLPs are performed by a modified backpropagation algorithm with learning rate = 0.01 and learning rate modification factor = 0.1. The maximum number of learning iterations = 5000. A character corresponding to which the difference between the largest and the second largest values in the output layer of the MLP is less than a certain threshold value, has been rejected by the proposed method.

Fig. 4. (*a*) Noisy input, (*b*) Output graph

Table 2 and Table 3 show classification results on training and test sets respectively in terms of confusion matrices. The figures in these tables are average values of different instances with 25 different independent random initializations. From the table it is noted that the correct classification rate is 91.8% on the training set and 90.56% on the test set. Table 2 and Table 3 give, respectively, the recognition result on the training set and on the test set. Numbers of misclassified samples are 107 and 242 corresponding the two different sets while those numbers of rejections are 47 and 83 respectively.

5 Conclusions

The proposed two-stage recognition system of Bangla numerals provides acceptable classification results. This method is based on two different neural network architectures which can minimize the use of heuristics. Whatever other methods (very few in numbers) available, are either largely heuristics based or not tested on a sufficiently large database.

The proposed method performs very well even when the scanned image of the character gets affected by noise. In Fig.4(a) a character (eight) affected by 50% random noise is shown and its graph representation (obtained by TASONN) is shown in Fig. 4(b). This representation can eliminate the effect of noise and so the recognition is also not affected. This fact also ensures that the method works even if there are small breaks on the contour of the characters.

References

1. Bhattacharya U. and Parui S. K.: Self-Adaptive Learning Rates in Backpropagation Algorithm Improve Its Function Approximation Performance. Proc. of the IEEE International Conference on Neural Networks, Australia (1995) 2784-2788.
2. Bhattacharya U., Chaudhuri B. B. and Parui S. K.: An MLP-based Texture Segmentation Method Without Selecting a Feature Set. Image and Vision Computing. 15 (1997), 937-948.
3. Datta A., Parui S. K. and Chaudhuri B. B.: Skeletonization by a Topology Adaptive Self-Organizing Neural Network, Pattern Recognition. 34 (2001) 617-629.
4. Dutta A., Chaudhuri S.: Bengali Alpha-Numeric Character Recognition Using Curvature Features. Pattern Recognition. 26 (1993) 1757-1770.
5. Garris M. D., Wilson C. L. and Blue J. L.: Neural Network-Based Systems for Handprint OCR Applications, IEEE Transactions on Image Processing, 7 (1998) 1097-1112.
6. Hecht-Nielson R.: Neurocomputing. New York: Addison-Wesley, 1990, Chapter 5.

7. Lam L., Suen C. Y.: Structural Classification and Relaxation Matching of Totally Unconstrained Handwritten ZIP-Code Numbers. Pattern Recognition. 21 (1988) 19-31.
8. Mori S. Suen C. Y., Yamamoto K.: Historical Review of OCR Research and Development. Proceedings of the IEEE. 80 (1992) 1029-1058.
9. Pal U. and Chaudhuri B. B.: Automatic Recognition of Unconstrained Off-line Bangla Hand-written Numerals, Advances in Multimodal Interfaces, Springer Verlag Lecture Notes on Computer Science (LNCS-1948), Eds. T. Tan, Y. Shi and W. Gao. (2000) 371-378.
10. Shimura M.: Multicategory Learning Classifiers for Character Reading, IEEE Trans. Syst. Man Cybern. 3 (1973) 74-85.
11. Suen C. Y.: Computer Recognition of Unconstrained Handwritten Numerals. Proceedings of the IEEE. 80 (1992) 1162-1180.

Optimal Synthesis Method for Binary Neural Network Using NETLA

Sang-Kyu Sung[1], Jong-Won Jung[1], Joon-Tark Lee[1], Woo-Jin Choi[2], and Seok-Jun Ji[3]

[1]School of Electrical, Electrical & Computer Engineering, Dong-A University, Korea
{sksungkor,jongwonj}@hanmail.net, jtlee@mail.donga.ac.kr
[2]Saracom. Co. Ltd
wjchoi@saracom.net
[3]Hanla Level. Co. Ltd
seokjunji@yahoo.co.kr

Abstract. This paper describes an optimal synthesis method of binary neural network for pattern recognition. Our object is to minimize the number of connections and the number of neurons in hidden layer by using a Newly Expanded and Truncated Learning Algorithm (NETLA) for the multilayered neural networks. The synthesis method in NETLA uses the Expanded Sum of Product (ESP) of the boolean expressions and is based on multilayer perceptron. It has an ability to optimize a given binary neural network in the binary space without any iterative learning as the conventional Error Back Propagation (EBP) algorithm. Furthermore, NETLA can reduce the number of the required neurons in hidden layer and the number of connections. Therefore, this learning algorithm can speed up training for pattern recognition problems. The superiority of NETLA to other learning algorithms is demonstrated by an application to the approximation problem of a circular region.

1 Introduction

The binary neural network based n-tuple method have been used in image processing for pig evisceration, scene analysis, and pattern recognition because of having many potential advantages in knowledge manipulation. However, the problems of real-time processing in binary neural network for huge quantities of data was not solved[1]. To solve the problem, various kinds of the learning algorithms have been proposed[3].

Recently, EBP algorithm has been applied to many binary to binary mapping problems. However, EBP algorithm applied to binary to binary mapping problems results in long training time and inefficient performance. Since the number of neurons in the input and output layer are determined by the dimensions of the input and output vectors, respectively, the abilities of three layer binary neural network depend on the number of neurons in the hidden layer and the number of connections. Therefore, one of the most important problems in application of three layer binary neural network is to determine the necessary number of neurons in the hidden layer. Recently, various learning algorithms has been proposed to determine the required number of neurons in the hidden layer for any binary to binary mapping. Firstly, learning algorithm

N.R. Pal and M. Sugeno (Eds.): AFSS 2002, LNAI 2275, pp. 236–244, 2002.

called expanded and truncated learning (ETL) was proposed to train a three layer binary neural network for the generation of binary to binary mapping. Its basic principles are to find a set of required separating hyperplanes through a geometrical analysis of given training inputs and to determine the number of neurons in the hidden layer. However, there are disadvantages that it requires the tedious sequential learning over all the patterns [5]. Secondly, in the MSP Term Grouping Algorithm(MTGA) the training patterns are represented only in Minimal Sum of Product(MSP) form of the boolean expressions satisfying the Unate property. However, there are disadvantages that the not-Unate terms are to be given as additional neurons in the hidden layer [6].

In this paper, a Newly Expanded and Truncated Algorithm (NETLA) is proposed which can be applied even for the case of not-Unate terms. Here the boolean expressions are represented as the Expanded Sum of Product (ESP) for optimal synthesis. Therefore, NETLA does not require any special necessity condition as MTGA's Unate property and can automatically determine the required number of neurons in the hidden layer. Furthermore, by finding redundancy part of input pattern, it can reduce the number of connections between input and hidden layer. Here, the proposed geometrical learning algorithm called NETLA is applied to train a three layer binary neural network and its superiority over other networks [4],[7] is demonstrated through a practical pattern recognition problem(problem of circular region approximation).

2 Preliminaries

2.1 The Basic Unit Structure of Binary Neural Network

Assume that the binary training input patterns can be separated by n-dimensional hyperplanes which is expressed as net function of Eq.(1) for n- inputs;

$$net(x,T) = w_1x_1 + w_2x_2 + \cdots + w_nx_n - T = 0 \tag{1}$$

where, x_i is the i-th input T is the threshold value and w_i is the connection weight between the i-th input and the neuron. In this case, the set for training inputs must be linearly separable (LS), and the n-dimensional hyperplanes can be established by n-inputs with hard-limiter activation function as Eq.(2).

$$y = \begin{cases} 1 : \sum_{i=1}^{n} w_i x_i - T \geq 0 \\ 0 : otherwise \end{cases} \tag{2}$$

Where, y is the output of neuron.

2.2 Decision on Weight and Threshold Values in the Hidden Layer

To decide the weight and threshold values, we used the Reference HyperSphere (RHS) method. RHS means separating hyperspheres which enclose all the training patterns and SHS (Separating HyperSphere) includes only the true patterns. Hyperplanes are obtained by intersecting two hyperspheres. By using RHS method, we can find the necessary hyperplanes into which the training inputs can be partitioned in the n-dimensional unit hypercube with center, $(1/2, 1/2, \cdots, 1/2)$ and radius, $\sqrt{n}/2$.

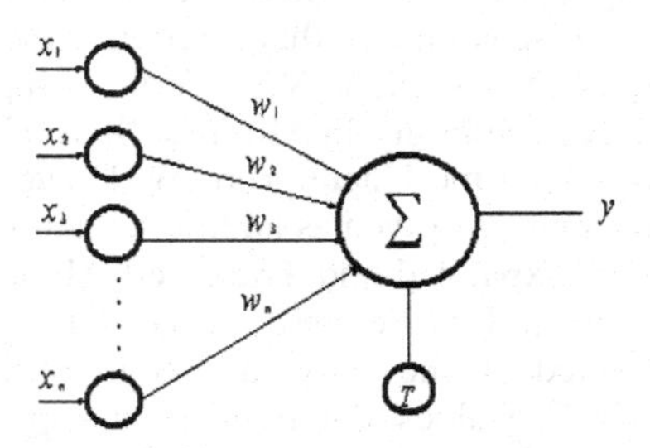

Fig. 1. The basic unit structure of binary neural network

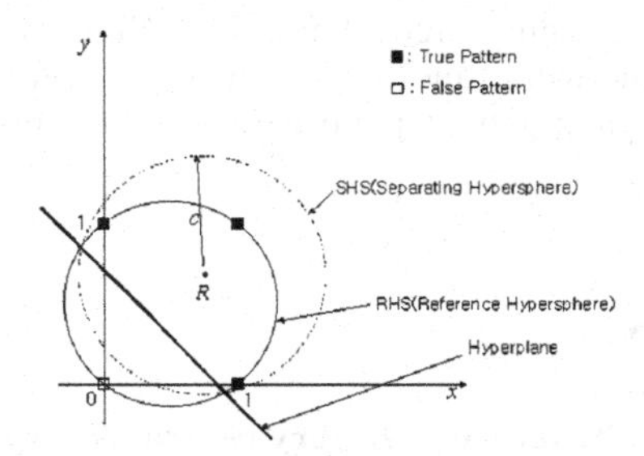

Fig. 2. Example of RHS method for two dimension

The equation for the RHS is written as Eq.(3).

$$(x_1 - \frac{1}{2})^2 + (x_2 - \frac{1}{2})^2 + \cdots + (x_n - \frac{1}{2})^2 = \frac{n}{4} \tag{3}$$

We assume the SHS with its center, $(c_1/c_0, c_2/c_0, \cdots, c_i/c_0)$ and radius, R, inside the SHS is written as Eq.(4).

$$(x_1 - \frac{c_1}{c_0})^2 + (x_2 - \frac{c_2}{c_0})^2 + \cdots + (x_n - \frac{c_i}{c_0})^2 = R^2 \tag{4}$$

Therefore, when these two n-dimensional hyperspheres intersect, the n-dimensional hyperplanes can be found. The hyperplane equation obtained by subtracting Eq.(4) from Eq.(3) and multiplying by c_0 as follows

$$(2c_1 - c_0)x_1 + (2c_2 - c_0)x_2 + \cdots + (2c_n - c_0)x_n - T = 0 \tag{5}$$

where, $(c_1 \cdots c_n)$ are each binary variables of quantized values. c_0 is total number in set of included true vertices. These two classes of training input can be separated by an n-dimensional hyperplane as Eq.(6)

$$\sum_{i=1}^{n} w_i x_i - T = 0 \tag{6}$$

If the Eq.(5) is equal to Eq.(6), then each of w_i and T are $2c_i - c_0$ and $R^2 c_0 - \sum_{i=1}^{n} c_i^2 / c_0$. Therefore, the weight values can be calculated as

$$w_i = 2c_i - c_0 \tag{7}$$

And threshold values calculated as

$$T = R^2 c_0 - \sum_{i=1}^{n} c_i^2 / c_0 \tag{8}$$

Where, number of hyperplanes is equal to number of neurons in the hidden layer.

2.3 Example of Decision on Weight and Threshold Values

Suppose, a set of n-bit training input vectors is given and the desired binary output is assigned to each training input vector. Also, let there be two classes of training input vectors (i.e., vertices) which can be separated by an n- dimensional hyperplane as in Eq.(1).

where, $T = [\frac{t_{\min} + f_{\max}}{2}]$ and $[x]$ is the smallest integer greater than or equal to x.

If $t_{\min} > f_{\max}$, then there exists a separating hyperplane.

If $t_{\min} \le f_{\max}$,then there does not exist a separating hyperplane.

Let us consider a function of three input variables, $f = (x_1, x_2, x_3)$. If inputs are {000, 010, 011, 111}, then $f = (x_1, x_2, x_3)$ produces '1'. If inputs are {001, 100}, then $f = (x_1, x_2, x_3)$ produces '0'. The inputs {101, 110} are don't care. With the training inputs {000, 010, 011, 111, 001, 100}, $f = (x_1, x_2, x_3)$ produces '1' or '0'. And, let $t_{\min}$ be the minimum value of $\sum_{i=1}^{n}(2c_i - c_0)v_t^i$ among all the vertices in (set of included true vertices) SITV, and $f_{\max}$ be the maximum of $\sum_{i=1}^{n}(2c_i - c_0)v_r^i$ among the rest vertices. Where, v_t^i is i-th bit of the true vertex in the set and v_r^i is i-th bit of the training vertex in the set. For this example, $t_{\min} = \min[-3x_1 + x_2 - x_3]$ for set of included true vertices {000, 010, 011}, and $t_{\min} = 0$. $f_{\max} = \max[-3x_1 + x_2 - x_3]$ for vertices {001, 100 111}, thus $f_{\max} = -1$. Since $t_{\min} > f_{\max}$ and $T = 0$, the hyperplane $-3x_1 + x_2 - x_3 = 0$ separates the vertices in the set of included true vertices {000, 010, 011} from the rest vertices. For the given example, it turns out that the set of included true vertices are {000, 010, 011}. If all true vertices are included in the set of included true vertices, then the given function is a linearly separable function and only one neuron is required for the function. Since, set of included true vertices of the first neuron includes only {000, 010, 011}, the remaining vertices are converted to expand into the first hypersphere. That is, the false vertices {001, 100} are converted into true vertices, and the remaining true vertex {111} is converted into a false vertex. Choose one true vertex, say {001} and test if the new vertex can be added to set of included true vertices. It turns out that set of included true vertices includes all currently

declared true vertices {000, 010, 011, 001, 100}. Therefore, two neurons are required in the hidden layer. The second required hyperplane is Eq.(5)
Where, $c_0 = 5$, $c_1 = 1$, $c_2 = 2$, and $c_3 = 2$. That is, $-3x_1 - x_2 - x_3 - T = 0$. Hence, $t_{min} = -3$ and $f_{max} = -5$. Since $t_{min} > f_{max}$ and $T = -4$, the required hyperplane is $-3x_1 - x_2 - x_3 + 4 = 0$.
Therefore, the algorithm finds two separating hyperplanes, that is, two neurons in the hidden layer.

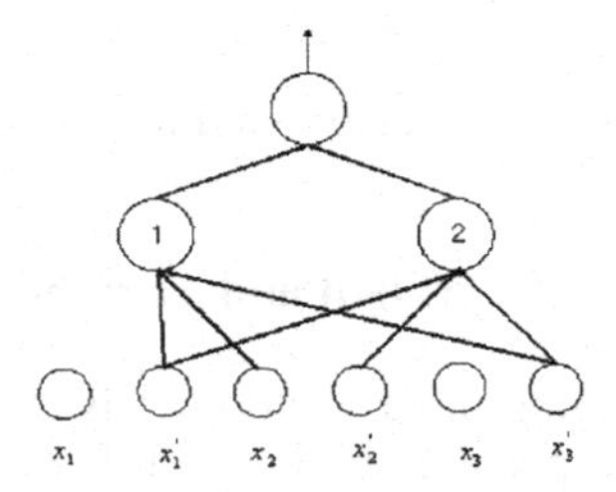

Fig. 3. The structure of three layer binary neural network for given example

Table 2. Weight and threshold values in hidden layer for given example

Layer	Unit	Weight			Threshold
Hidden	1	-6	2	-2	-1
	2	-3	-1	-1	-4

2.4 The Advantage and Synthesis Process of NETLA

The synthesis process for NETLA is shown in figure 4. NETLA can synthesize regardless of input pattern order for binary neural network. The advantages of NETLA are as follows:

i) Since the weight and threshold values can be directly calculated from false and true patterns, so iterative learning is not needed.

ii) No special necessity condition (like Unate property).

3 Simulations

ETL, MTGA and NETLA are applied to the approximation problem of one circular region and the results are stated in the following subsections. Through these applications, we show the reduction in number of hidden neurons and connections.

Consider the separation problem of a circular region in 2-dimensional space within a square with sides of length 8 with origin in the lower left corner. A circle of diameter 4 is placed within the square, at (4,4), and then the space is sampled with 64 grid points located at the center of 64 identical squares covering the large square. Of these points, 52 fall outside of the circle (desired output '0') , and 12 fall within the circle(desired output '1'). After two axes (cross and vertical) are quantized as 3 bits, resultant function before synthesis for the problemb is as follows.

$$f = (x_1, x_2, x_3, x_4, x_5, x_6) = x_1' x_2 x_4' x_5 x_6 + x_1' x_2 x_3 x_4' x_5 + x_1' x_2 x_3 x_4 x_5' + x_1' x_2 x_4 x_5' x_6$$
$$+ x_1 x_2' x_4 x_5 x_6' + x_1 x_2' x_3' x_4' x_5 + x_1 x_2' x_3' x_4 x_5' + x_1 x_2' x_4 x_5' x_6'$$

where, $(x_1 \cdots x_6)$: true pattern $(x_1' \cdots x_6')$: false pattern

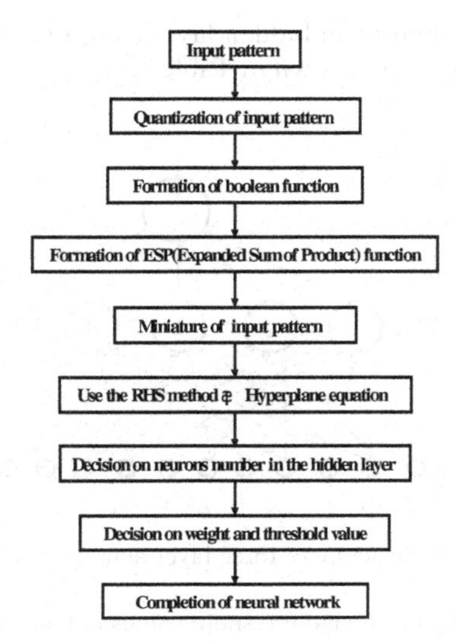

Fig. 4. The block diagram of NETLA

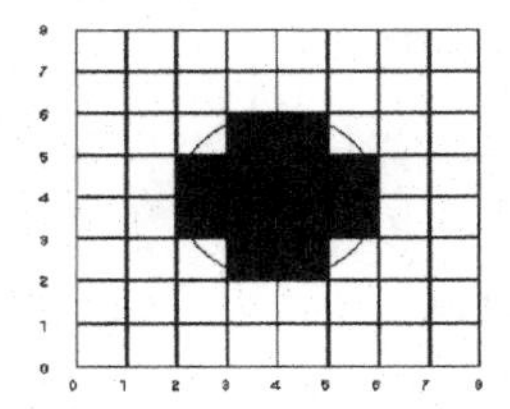

Fig. 5. Circular region obtained by 6-bit quantization

Consider the separation problem of a circular region in 2-dimensional space within a square with sides of length 8 with origin in the lower left corner. A circle of diameter 4 is placed within the square, at (4,4), and then the space is sampled with 64 grid points located at the center of 64 identical squares covering the large square. Of these points, 52 fall outside of the circle (desired output '0') , and 12 fall within the circle(desired output '1'). After two axes (cross and vertical) are quantized as 3 bits, resultant function before synthesis for the problemb is as follows.

$$f = (x_1, x_2, x_3, x_4, x_5, x_6) = x_1' x_2 x_4' x_5 x_6 + x_1' x_2 x_3 x_4' x_5 + x_1' x_2 x_3 x_4 x_5' + x_1' x_2 x_4 x_5' x_6$$
$$+ x_1 x_2' x_4 x_5 x_6' + x_1 x_2' x_3' x_4' x_5 + x_1 x_2' x_3' x_4 x_5' + x_1 x_2' x_4 x_5' x_6'$$

where, $(x_1 \cdots x_6)$: true pattern $(x_1' \cdots x_6')$: false pattern

3.1 Synthesis Result for ETL

The synthesis result for ETL is as follows

$$f = (x_1, x_2, x_3, x_4, x_5, x_6) = x_1' x_2 x_3 x_4' x_5 x_6 + x_1' x_2' x_3' x_4' x_6 + x_1' x_2' x_3' x_4' x_5 x_6 + x_1' x_2'$$

By ETL requires five neurons in hidden layer as in Fig. 6. The weights and threshold values in the hidden layer are shown in Table 3.

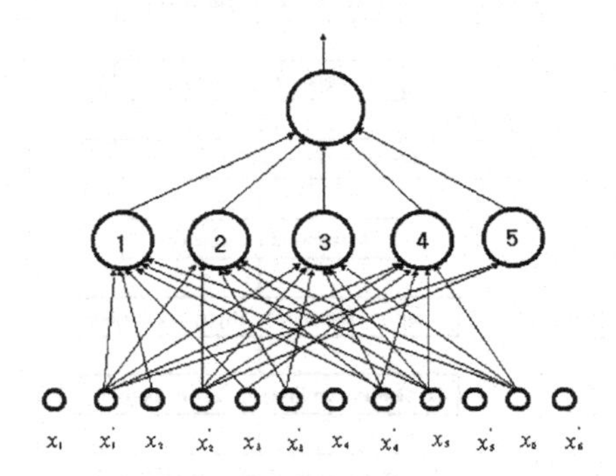

Fig. 6. Structure of three layer neural network for ETL

Table 3. Weight and threshold values in hidden layer for ETL

Layer	Unit	Weight						Threshold
Hidden	1	-3	3	1	-3	3	1	7
	2	-29	-3	-1	-3	3	1	-5
	3	-29	-3	-1	-3	3	1	-25
	4	-19	-13	-1	-3	3	1	-20
	5	-16	-16	0	0	0	0	-24

3.2 Synthesis Result for MTGA

MTGA synthesis method represents the function in the MSP form and then checks for the Unate's property. The Unate's property means that a complemented and uncomplemented variable x_i should not appear simultaneously in the MSP form of the function. The Unate property is explained using the following example. Assume two function $f_1(x_1,x_2,x_3)$ and $f_2(x_1,x_2,x_3)$

$$f_1(x_1,x_2,x_3)=x_1'x_2+x_2'x_3+x_1x_3 \text{ : not- Unate}$$

$$f_2(x_1,x_2,x_3)=x_1'x_2+x_2x_3+x_1'x_3 \text{ : Unate}$$

The synthesis result for MTGA is as follows

$$f=(x_1,x_2,x_3,x_4,x_5,x_6)=(x_1'x_2x_4'x_5x_6+x_1x_2'x_3x_4'x_5)+(x_1x_2'x_3x_4x_5'+x_1'x_2x_4'x_5x_6)$$
$$+(x_1x_2'x_4x_5x_6'+x_1x_2'x_3'x_4'x_5)+(x_1x_2'x_3'x_4x_5'+x_1x_2'x_4x_5'x_6')$$

MTGA requires four neurons in hidden layer as in Fig. 7. The weight and threshold values in hidden layer for MTGA are shown in Table 4.

Table 4. Weight and threshold values in hidden layer for MTGA

Layer	Unit	Weight						Threshold
Hidden	1	-3	3	1	-3	3	-1	6.5
	2	-3	-3	1	3	-3	1	5.5
	3	3	-3	-1	-3	3	1	5.5
	4	3	-3	-1	3	-3	1	4.5

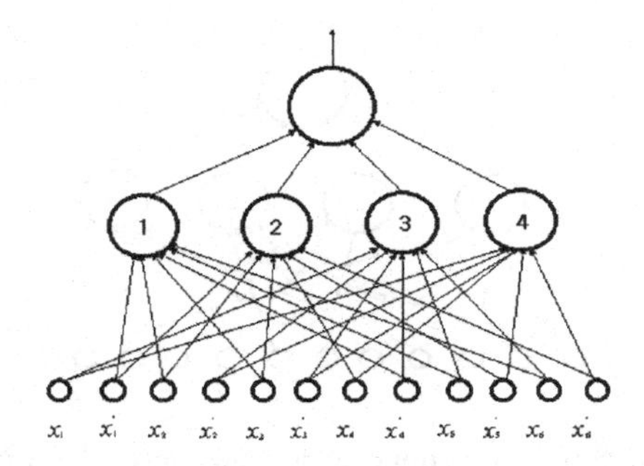

Fig. 7. Structure of three layer neural network for MTGA

3.3 Optimal Synthesis Result for the Proposed NETLA

Using NETLA the resultant synthesis will be as

$$f(x_1, x_2, x_3, x_4, x_5, x_6)$$
$$=(x_1'x_2x_3x_4 + x_1'x_2x_4x_6 + x_1x_2'x_3'x_4 + x_1x_2'x_4x_6')\cdot x_5' + (x_1'x_2x_4'x_6 + x_1'x_2x_3x_4' + x_1x_2'x_4'x_6' + x_1x_2'x_3'x_4')\cdot x_5$$

And each additional hyperplane for the 1st and the 2nd term is extended as follows

$(x_1'x_2x_3x_4 + x_1'x_2x_4x_6 + x_1x_2'x_3'x_4 + x_1x_2'x_4x_6')\cdot 1$ and $(x_1x_2'x_4x_6' + x_1x_2'x_3'x_4' + x_1'x_2x_4'x_6 + x_1'x_2x_3x_4')\cdot 1$

Therefore, we can obtain function as follows

$$f(x_1, x_2, x_3, x_4, x_5, x_6)$$
$$=(x_1'x_2x_3x_4 + x_1'x_2x_4x_6 + x_1x_2'x_3'x_4 + x_1x_2'x_4x_6')\cdot x_5' + (x_1'x_2x_3x_4 + x_1'x_2x_4x_6 + x_1x_2'x_3'x_4 + x_1x_2'x_4x_6')\cdot 1$$
$$+(x_1x_2'x_4'x_6' + x_1x_2'x_3'x_4' + x_1'x_2x_4'x_6 + x_1'x_2x_3x_4')\cdot x_5 + (x_1x_2'x_4x_6' + x_1x_2'x_3x_4' + x_1'x_2x_4'x_6 + x_1'x_2x_3x_4')\cdot 1$$

and then, $f(x_1, x_2, x_3, x_4, x_5, x_6) = (x_1'x_2x_3x_4 + x_1'x_2x_4x_6 + x_1x_2'x_3'x_4 + x_1x_2'x_4x_6')\cdot(x_5'+1)$

$$+(x_1x_2'x_4'x_6' + x_1x_2'x_3'x_4' + x_1'x_2x_4'x_6 + x_1'x_2x_3x_4')\cdot(x_5+1)$$

where, $(x_5'+1)$ and (x_5+1) become '1'.

Consequently, synthesis result function $f(x_1,x_2,x_3,x_4,x_5,x_6)$ is as follows

$$=(x_1'x_2x_3x_4 + x_1'x_2x_4x_6 + x_1x_2'x_3'x_4 + x_1x_2'x_4x_6') + (x_1'x_2x_4'x_6 + x_1'x_2x_3x_4' + x_1x_2'x_4'x_6' + x_1x_2'x_3'x_4')$$
$$=(x_1'x_2x_3x_4 + x_1'x_2x_4x_6) + (x_1x_2'x_3'x_4 + x_1x_2'x_4x_6') + (x_1'x_2x_4'x_6 + x_1'x_2x_3x_4') + (x_1x_2'x_4'x_6' + x_1x_2'x_3'x_4')$$

NETLA needs only four neurons in hidden layer as in Fig. 8. The weights and threshold values in hidden layer are shown in Table 5. Thus, we see that NETLA requires the minimum number of connections compared to ETL and MTGA. Also, in case of NETLA the number of neurons in hidden layer is less compared to ETL.

Table 5. Weight and threshold values in hidden layer for NETLA

Layer	Unit	Weight					Threshold
Hidden	1	-3	3	1	-3	-1	6.5
	2	-3	-3	-1	3	1	5.5
	3	-3	3	-1	3	1	5.5
	4	3	-3	-1	3	1	4.5

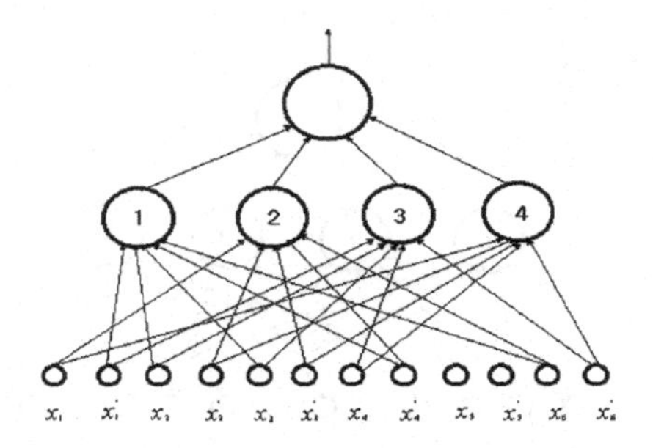

Fig. 8. Structure of three layer neural network for NETLA

4 Conclusions

Since the NETLA proposed in this paper can automatically calculate the weights and threshold values from all the true and false patterns, so it dose not require any iterative learning. The NETLA can reduce the number of the required neurons in hidden layer and connections. Also, training speed of NETLA is much faster than that of other learning algorithms for the generation of binary to binary mapping. The superiority of this NETLA to other algorithms was proved by the approximation problem of one circular region. In the future, we are going to apply this algorithm to fingerprint identification, digital image processing, etc.

References

1. A.W. Andersen, S.S. Christensen, T.M. Jorgensen.: An Active Vision System for Robot Guidance using a Low Cost Neural Network Board. In European Robotics and Intelligent Systems Conference. (1994) 480-488
2. P.L. Bartlett, T. Downs.: Using Random Weight to rain Multilayer Network of Hard-limiting Units. IEEE Trans. Neural Networks. (1992) 202-210
3. M.L. Brady, R. Rayhavan, J. Slawny.: Back Propagation fails to separate where Perceptrons Succeed. IEEE Trans. Circuits Systems. (1989) 665-674
4. S. Park, J.H. Kim, H. Chang.: A Learning Algorithm for Discrete Multilayer Perceptron. in Proc. Int. Sysp. Circuit Systems. Singapore(1991)
5. J.H. Kim, S.K. Park, Han, H. Oh, M.S. Han.: The Geometrical Learning of Binary Neural Network. IEEE Trans. Neural Networks. vol. 6.no. 1. (1995) 237-247
6. Z. Kohavi.: Switching and Finite Automata Theory. (1986) 2nd ed. McGraw-Hill
7. Donald L. Gray, Anthony N. Michel.: A Training Algorithm for Binary Feedforward Neural Networks. IEEE Trans. Neural Networks. (1992) 176-194

A Neural Network Based Seafloor Classification Using Acoustic Backscatter

Bishwajit Chakraborty

National Institute of Oceanography, Dona Paula, Goa: 403 004
bishwajt@darya.nio.org

Abstract. This paper presents a study results of the Artificial Neural Network (ANN) architectures [Self Organizing Map (SOM) and Multi-Layer Perceptron (MLP)] using single beam echosounding data. The single beam echosounder, operable at 12 kHz, has been used for backscatter data acquisitions from three distinctly different seafloor's from the the Arabian Sea. With some preprocessing of the snapshots, the performance of the SOM network is observed to be quite good. For unsupervised SOM network, only single snapshot is used for the training, and number of snapshots for subsequent testing of the network. Feature selection from ASCII data is an important component for an supervised MLP based network. Four selected features are used for training the the network. The test results of the MLP based network are also discussed in the text.

1 Introduction

Seafloor classification using remote high frequency acoustic system has been recognized as a useful tool. The seafloor's characteristics are extremely complicated due to variations of the many parameters at different scales. The parameters include sediment grain size, relief height at water- sediment interface and variations within the sediment matrix. Sound signal scattering is generally controlled by these factors, and seafloor classification using multi-parameter scattering model is generally employed [1, 2]. Use of backscatter model studies to identify finer scale bottom roughness (water - sediment interface and sediment volume roughness), is a technique presently developing[3]. Results of the model based studies considering parametric forms of seafloor roughness distribution [Rice PDF (Probability Density Function)] for single beam echosounding backscatter signal from the Arabian Sea shelf (areas between Mangalore and Kochi off the west coast of India) were developed from three varying seafloor sediments (sand, silty sand or clayey silt)[4]. For model based studies using echosounding backscatter data, it is difficult to acquire close grid seafloor ground truth information as a support for validation. In order to determine spatially close sampled sediment ground truth data[5], frequent time-consuming sediment sampling is necessary. Again, the analytical method pertaining to the determination of the sediment grain size distribution using sampled sediment, imposes an additional hindrance to seafloor

N.R. Pal and M. Sugeno (Eds.): AFSS 2002, LNAI 2275, pp. 245–250, 2002.

classification using modeling work. Hence, a methodology needs to be developed for real time online estimation of the seafloor sediment type distribution.

An Artificial Neural Network (ANN) based computational technique known as Kohonen's Self Organizing Map (SOM), has been applied recently[6]. SOM is a neural network architecture that exhibits the ability to discover the underlying patterns in the input data. The SOM based representation of the seafloor sediment classifier provides a nonparametric distribution of the seafloor types in real time, which in turn is also suitable for further parameter based scattering studies.

For a Multi-Layer Perceptron (MLP) based seafloor classifier, features need to be extracted from raw data. For each class of data, desired output must be established on the basis of ground truth information. Four acoustic feature parameters include bottom echo statistics and scattering models are generated from the ASCII echo data for training the network. These feature parameters are then fed to the neural network that performs the required classification. The testing is being performed, once the network is trained.

In this paper, a comprehensive presentation of the ANN based seafloor classifiers are made from three distinctly different seafloor regions. An application of the two ANN architectures like: SOM and MLP, are applied for seafloor classifications. In general, required training time is more for MLP than the SOM. However, for MLP architecture the classification time is less in comparison to SOM. This study reviews general performance comparisons between the employed scattering model and two basic neural network architectures for seafloor classification applications.

2 Survey Area and Echo Data

Survey regions between Mangalore and Kochi in the western continental shelf of India are chosen for backscatter data collection. The sediment samples were collected during the RV Gavesheni cruises 30 and 207. The detailed sediment distribution is known in the three areas where the echo signal data is collected. Sandy sediment bottom is located around 11°30' N and 75°06' E and the silty sand and clayey silt are located around 10°40' N, 75°25' E and 12°40' N, 74°40' E respectively. The water depths are varying from 45 m to 60 m in the three areas. The selected transmission pulse length is 1 ms. The echo waveform data was collected during ORV Sagar Kanya cruise 74.

The echo signal is acquired from the 12 kHz. deepsea echosounder (Honeywell-Elac) by using an interface. The echo signal is fed to the programmable delay circuit. The sounding trigger from the echosounder initializes the delay period. The delay circuit tracks the received echo and a gate is set for the digitization of the complete pulse length. The pulse is sampled at a frequency of 50 kHz. with the help of PCL-718 high performance data acquisition card installed in the Personal Computer (PC). The A/D (analog to digital) conversion is performed in the external pulse trigger mode whereby the conversion begins on the arrival of the external pulse which is generated by the delay circuit electronics. The output of PCL-718 is a 12 bit data which is stored in the PC hard disk in ASCII format.

Two hundred digital sample values are obtained from receiving signal of normal incidence beam output for each transmission ping. In the remaining section of the text this is reffered to as a snapshot. Echo waveform data from three different sedimentary regions are used for the application of the SOM and MLP architectures.

3 Self Organizing Map (SOM)

A SOM is characterized by the formation of a topographic map of the input patterns, in which the spatial locations of the neurons in the lattice correspond to inherent features of the input patterns. Certain pre-processing steps like feature extraction can be avoided in Kohonen's SOM. Once the classification network is designed for distinctly known sediment areas, identification of the seafloor sediment areas can be carried out in real time.

Here, the echo waveform data from three known seafloor sediments form three clusters in the SOM. As mentioned, we have assigned 200 samples from each snapshot as an input vector. A 200 x 1 input grid and initial 100 x 1 output grid were chosen for this study [6]. When an input vector belonging to a particular class is presented during the training, a reinforcement of only a few selected weights takes place. The detailed sequence of activities during training phase include: Network initialization, presentation of each input snapshot to the network, and distance calculation between the input and the weights associated with each output node. Also, minimum distance selection i.e., to designate the output node with the computed minimum distance to be the winning node is an important component. The essential weight updating for the designated output node and its neighbors are defined by the neighborhood size. New updated weights around the node for a neighborhood size for the next presentation of the same training snapshots are continued. An adaptation term (varies between 0 and 1, and decreases with the number of iterations) is defined as learning rate. It also allows to decrease the rate of weight updating. Here, the learning rate decreases as per the function, [$\sim 0.4/(t^{0.2})$], where t is the iteration number. The training iterations continue till the computed value of distance is less than the prespecified error value ($\sim 10^{-6}$). The training process attempts to cluster the nodes on the topological map to reflect the range of class types found in the training data (coarse mapping). Each time a new input class is applied to the network, the winning node must first be located. This identifies the region of the map that will have its weights updated.

Once training is completed, the network is tested using the entire data set. Individual data snapshots belonging to the three different classes are fed as input vectors and the winning neuron for each snapshot is ascertained. In our study the number of iterations to reach the prespecified error value (10^{-6}) ranged from 150 to 200. The neighborhood size N decreased in size with iterations in order to localize the area of maximum activity. The neuron neighborhoods vary from 58-68, 87-97, and 23-33 for sandy, silty sand, and clayey silt seafloor sediments respectively. The percentage of classification is reasonably good (82.5% and 71.25% respectively).

However poor classification is seen for silty sand (45%). Using raw data as input the classification efficiency of the SOM is not adequate, and some preprocessing is essential. A moving average technique is employed for snapshots belonging to three different sediment areas for an output neuron grid of size 100 x 1. A moving average of 50 snapshots show that the moving average technique provides improved classification. The classification percentages for sand, silty sand, and clayey silt sediment seafloor are found to have improvement (82.5%, 100%, and 100% respectively). The classification of three given seafloor sediment for a moving average of 35 and 20 snapshots are carried out. Improved classification i.e., 97.5%, 97.5%, and 100%, is obtained for the seafloor sediments of sand, silty sand, and clayey silt respectively for 35 snapshots. Degradation in the classification percentage for silty sand (87.5%) and clayey silt (63.75%) sediments for 20 moving average snapshots is observed. Results of the study emphasize that the moving average of 35 snapshots is ideal when an output grid of 100 neurons is used. Again, in order to bring about a reduction in the network's computational complexity, an output grid of smaller size is preferable. In this respect, output grids having a smaller number of neurons were selected while using the moving average of 35 snapshots. A comparison between the output neurons grid sizes of 100 x 1, 50 x 1, 25 x 1, and 15 x 1 for a 35 moving average snapshots indicate very interesting results. We observe that the classification percentage remains constant for the sandy, silty sand, and clayey silt sediment seafloor, i.e., results are found to be converging well. In this case, the percentage classifications are found to be obtained as: 97.5%, 97.5%, and 100% for the sand, silty sand, and clayey silt respectively. Hence, the output neuron grid size of 15 x 1 can be considered as most suitable when using a moving average of 35 snapshots. Excellent classification (~100 %) is found, while testing the network for three given sediment seafloor. The possibility of inaccurate classification is increased when the number of output neurons (15 x 1) is further reduced along with the number of snapshots for the moving average (less than 35 snapshots).

4 Multi-layer Perceptron (MLP)

Important configuration parameters of a MLP network are the number of hidden layers and the number of nodes in each hidden layer. Currently, there seems to be no simple method for determining the best possible network configuration for a given application. The number of nodes in the output layer remained the same as the number of classes to be recognized, i.e. three in this study. To determine the number of hidden neurons (h) one rough guideline is being chosen [7] [~ $(mn)^{0.5}$]. Therefore, for a 3-layer network with n input neurons and m output neurons, the decision to select hidden layer (h) neurons, can be made based on above rule. The number of neurons in the input and output layers are eight and three respectively for this study, which provide five neurons in the hidden layer.

For MLP network, feature selection is an important component. But there is no general rule for determining the best features for a given problem. The echo signal intensity variation can statistically provide either the variation of seafloor sediment materials or spatially varying roughness patterns due to geological processes. Therefore, the selected features should reflect the differences between the regions with homogeneous variation. We have examined several combinations of input feature vectors in order to achieve proper classification. Four features have been used to train MLP's and classify unknown vectors: received echo energy, pulse length, γ (computed using echo peak of the waveform in the Rice PDF study) [4], and kurtosis parameter (computed using the entire waveform). Firstly, the echo energy for each snapshot is considered to be one of the four features used in the training of the MLP based sediment classifier. The echo energy is being computed by summing the squares of individual sample values (number of sample values of digitized received signal, i.e., 200, used for echo energy calculation in this study). Secondly, the pulse length of the received signal is being considered as an important seafloor classification feature. Since, the received signal pulse length is a functional of the seafloor type, therefore, this selected feature will contribute significantly for classification. Thirdly, we chose γ, a parameter measures relative roughness or smoothness of the sea bottom (given in previous section of the paper). The last feature 'kurtosis' (k), also know as the 'coefficient of excess' [~ $(\mu_4 / \sigma^4) - 3$]. And, μ_4 and σ^2 are the fourth central moment and the variance of the snapshot respectively. Based on the point scattering model, the signal reverberation is resulting from random process and happens to be linear superposition from large number of individual point scatterers. The kurtosis (k) can be related to the number of scatterers (N) contributing to the reverberant return at any particular time. For a rectangular transmitted pulse, the relationship is reduced to: $N = 3 / k$. Therefore, assuming different seafloor types have a different number of dominant scatterers per unit area, the kurtosis of the received reverberation can form the basis for a classification procedure. However, real reverberation data often exhibit deviations from this normal model[8], and incorporation of non- Gaussian nature becomes eminent.

Four features are extracted from the training echo data. These are fed to the neural network as one set of input vector per training set. For network training, we have used five sets of input vectors (each set of four feature vectors is generated by ten snapshots for averaging) are used exclusively for training while the remaining were used for testing the network classification efficiency. Thus, with every run of the program, the testing or validation presentations appeared to the network as unknown data. Training over many epochs was sufficient to realize the assigned error goal for this study, and training algorithm employed is known as the Levenberg- Marquadt training algorithm. We have used *newff* command of the MATLAB[9]. After the *newff* command creates the network object *net* and initializes the weights and biases of the network, the network is ready for training. During training, the weights and biases of the network are iteratively adjusted to minimize the network performance function. The default performance function for feedforward networks is mean squared error between the network outputs and a target outputs. Five training sets of input vectors are necessary for batch mode training by invoking the function *train*. In batch mode, the weights and biases of the network are updated in the direction of the negative gradient of the performance function only after the entire training set has been applied to the network. The tan-sigmoidal transfer function was used for the input and hidden layers, while a purely linear transfer function was used for the output layer. The tested results reveal interesting implication of the MLP network for considered four feature parameters for seafloor classifications. Maximum success rate of 89% is observed for a combination of four features for three seafloor sediment areas. Again, performance wise for individual seafloor, success rate is found to be 100% for relatively less coarse (mixed) and fine grainsize sediments of silty sand and clayey silt respectively. Reduced success rate is found for the coarse grained sandy sediments (67%).

Acknowledgement

I acknowledge encouragement from Dr. Ehrlich Desa, Director, NIO while carrying out this work. Help from Dr. Vijay Kodagali, Scientist, NIO, is also thankfully acknowledged. NIO Contribution number 3707 is allotted to this paper.

References

1. Stanton, T.K.: Sonar Estimates of Seafloor Microroughness. J. Acoust. Soc. Am. **75** (1984) 809-818
2. Chakraborty, B.: Effects of Scattering due to Seafloor Microrelief on a Multifrequency-Sonar Seabed Profiler. J. Acoust. Soc. Am. **85** (1989) 1478-1481
3. Chakraborty, B., Schenke, H.W., Kodagali, V., Hagen,R.: Seabottom Characterization using Multibeam Echosounder Angular Backscatter: An Application of the Composite Roughness Theory. IEEE TGARS. **38** (2000) 2419-2422.
4. Chakraborty B., Pathak, D.: Sea Bottom Backscatter Studies in the Western Continental Shelf of India. Jour. Sound Vib. **219** (1999) 51-62
5. Nair, R.R., Hashimi,N.H., Rao,V.P.: Distribution and Dispersal of Clay Minerals on the Western Continental Shelf of India. Mar. Geol. **50** (1982) 1-9
6. Chakraborty, B., Kaustubha, R., Hegde, A., Pereira, A.: Acoustic Seafloor Sediment Classification using SOMs (IEEE TGARS - In-Press).
7. Masters, T.: Practical Neural Network Recipes in C++, Academic Press, San Diego (1993)
8. Alexandrou, D., de Moustier, C., Haralabaus, G.,: Evaluation and Verification of Bottom Acoustic Reverberation Statistics Predicted by the Point Scattering Model. J. Acous. Soc. Am. **91**(1992) 1403-1413
9. Matlab Neural Network Toolbox User's Guide (1998) Versions 5.3, Mathworks Inc.

Designing Rule-Based Classifiers with On-Line Feature Selection: A Neuro-fuzzy Approach

Debrup Chakraborty and Nikhil R. Pal

Indian Statistical Institute
Calcutta 700035, India
{debrup_r,nikhil}@isical.ac.in

Abstract. Most methods of classification either ignore feature analysis or do it in a separate phase, offline prior to the main classification task. This paper proposes a novel neuro-fuzzy scheme for classification with online feature selection. It is a four-layered feed-forward network for fuzzy rule based classification. The network learns the classification rules from the training data as well selects the important features. The rules learned by the network can be easily read from the network. The system is tested on both synthetic and real data and found to perform quite well.

1 Introduction

For pattern recognition task all features present in a data set may not have the same importance, i.e., some features may be redundant and also some may have derogatory influence on the classification task. Thus selection of a proper subset of features from the available set of features is important for designing of efficient classifiers. But most of the existing feature selection methods [2,9,10] perform feature analysis in a separate phase, offline with the main classification process. In [3] we discussed a methodology for simultaneous feature analysis and system identification in a 5-layered neuro-fuzzy framework. Here we modify the methodology in [3] for the classification task. This is a neural fuzzy system [4,6,7] for classification. The various layers of the network performs different functions of a fuzzy system, also it selects the important features and learns the rules required for the classification task.

2 The Network Structure

Let there be s input features $\mathbf{x} = (x_1, x_2, ..., x_s)^T$ and t classes $(c_1, c_2, ..., c_t)$. The proposed system will deal with fuzzy rules of the form, R_i : If x_1 is A_{1i} and x_2 is A_{2i} ..... and x_s is A_{si} then $\mathbf{x}$ belongs to class c_l with degree of certainty d_l, $(1 \leq l \leq t)$. Here A_{ji} is the i-th fuzzy set defined on the domain of x_j.

The neural-fuzzy system is realized using a four layered network as shown in Figure 1. The first layer is the input layer, the 2nd layer is the membership function and feature selection layer, the third layer is called the antecedent layer and the 4th layer is the output layer. The node functions with its inputs and

N.R. Pal and M. Sugeno (Eds.): AFSS 2002, LNAI 2275, pp. 251–259, 2002.

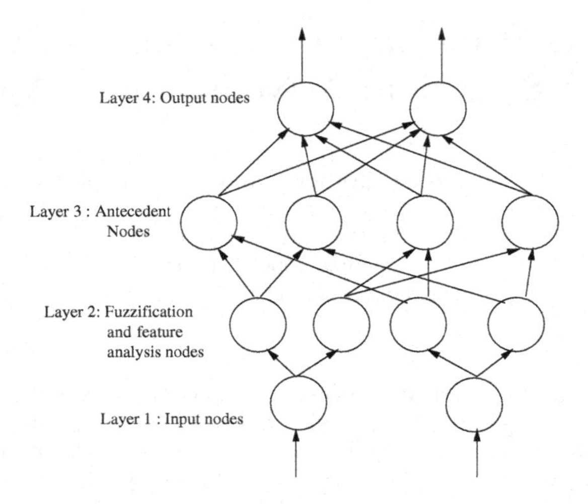

Fig. 1. The network structure.

outputs, are discussed next layer wise. We use suffixes p, n, m, l to denote respectively the suffixes of the nodes in layers 1 through 4 in order. The output of each node is denoted by z.

Layer 1: Layer 1 has N^1 (= the number of input features) nodes each representing an input linguistic variable. If x_p denotes the input to any node in layer 1 then the output of the node will be $z_p = x_p$.

Layer 2: This is the fuzzification and feature analysis layer. Each node in layer 2 represents a membership function of a linguistic value associated with an input linguistic variable. All connected weights between the nodes in layer 1 and layer 2 are unity. If there be N_i fuzzy sets associated with the ith feature and if there are s input features then the number of nodes in layer 2 would be $N^2 = \sum_{i=1}^{s} N_i$. The output of a node in layer 2 is computed by

$$\bar{z}_n = exp\{-\frac{(z_p - \mu_n)^2}{\sigma_n^2}\}. \tag{1}$$

Here μ_n and σ_n are the mean and spread respectively of a bell shaped function representing a term of the linguistic variable x_p associated to node n (n indicates the n-th term (fuzzy set) of the linguistic variable x_p) .

For feature selection, the output of this layer needs to be modified so that every indifferent feature x_p gets eliminated. If a linguistic variable x_p is not important for the classification task, then the values of x_p should not have any effect on the firing strength of the rules involving that input variable. Since for any $T - norm$, $T(1, \alpha) = \alpha$, $0 \leq \alpha \leq 1$ [7], this can be realized if an indifferent feature always generates a membership of unity in all its term sets. Thus we associate a modulator function f_n with each node n in layer 2 to modify the activation function as $z_n = f_n \cdot \bar{z}_n$. For an indifferent (or bad) feature we want all linguistic values defined on that feature to result in a membership value of 1.

So we model f_n as

$$f_n = (\bar{z}_n)^{-\lambda_p} \tag{2}$$

Here $\lambda_p \in [0,1]$ is a parameter associated with a particular *linguistic variable* x_p of which node n is a term. From (2) we see that when λ_p is nearly 1 then f_n is nearly $\frac{1}{\bar{z}_n}$, and when λ_p is nearly 0 then f_n is nearly 1. So for bad features λ_p should get large values (close to 1) and small values (close to 0) for good features. Thus, for a bad feature, the modulated membership value would be $f_n.\bar{z}_n \approx \frac{1}{\bar{z}_n}.\bar{z}_n \approx 1$ irrespective of the value of x_p; and for a good feature, it would be $f_n.\bar{z}_n \approx 1.\bar{z}_n \approx \bar{z}_n \approx$ the actual membership value. Since λ_p must take values between zero and one, we model λ_p by $e^{-\beta_p^2}$. Thus, the activation function of any node n in layer 2 would be as :

$$z_n = \bar{z}_n.f_n = \bar{z}_n^{(1-e^{-\beta_p^2})}. \tag{3}$$

The tunable parameter β_p can be learnt by back-propagation. When β_p^2 takes a large value then z_n tends to $\bar{z}_n$ and for small values of β_p^2, z_n tends to 1, thereby making the feature indifferent. Therefore, we want to make β_p^2 take large values for good features and small values for bad ones through the process of learning.

Layer 3: Each node in this antecedent layer represents an IF part of a fuzzy rule. The number of nodes in this layer is $N^3 = \prod_{i=1}^{s} N_i$. There are many operators ($T - norms$) for fuzzy intersection [7] of which min and product are quite popular. We use a soft version of min, which we call as $softmin$, defined as:

$$softmin(x_1, x_2, ..., x_n, q) = [(x_1^q + x_2^q + ... + x_n^q)/n]^{\frac{1}{q}}.$$

As $q \to -\infty$, $softmin$ tends to minimum of all x_i's where $i = 1, 2, .., n$. For our purpose we use $q = -12$ in all the examples reported. Note that, though $softmin$ is not a *T-norm* it satisfies the property of *identity* , as $q \to -\infty$, so, $softmin$ is compatible with our feature selection strategy. Thus, the output of the m-th node in layer 3 is

$$z_m = \left(\frac{\sum_{n \in P_m} z_n^q}{|P_m|}\right)^{\frac{1}{q}} \tag{4}$$

where P_m is the set of indices of the nodes in layer 2 connected to node m of layer 3 and $|P_m|$ denotes cardinality of P_m.

Layer 4: Each node in this output layer represents a class. If there are t classes then there will be $N^4 = t$ many nodes in layer 4. The nodes in this layer performs an OR operation, which combines the antecedents of layer 3 with the consequents. The nodes in layers 3 and 4 are fully connected. Let w_{lm} be the connection weight between node m of layer 3 and node l of layer 4. The weight w_{lm} represents the certainty factor of a fuzzy rule, which comprises the antecedent node m in layer 3 as the IF part and the output node l in layer 4 representing the THEN part. These weights are adjustable while learning the fuzzy rules. The OR operation is performed by some *s-norm* [7]. We use here the *max* operator. Thus, the output of the node l in layer 4 is computed by

$$z_l = max_{m \in P_l}(z_m w_{lm}), \tag{5}$$

where P_l represents the set of indices of the nodes in layer 3 connected to the node l of layer 4. Since the learnable certainty factors w_{lm}'s should be non-negative as well it should lie in $[0,1]$. So, we model w_{lm} by $e^{-g_{lm}^2}$. The g_{lm} is unrestricted in sign but the effective weight $w_{lm} = e^{-g_{lm}^2}$ will always lie in [0,1]. Therefore, the output of the l-th node in layer 4 will be

$$z_l = max_{m \in P_l}(z_m e^{-g_{lm}^2}). \tag{6}$$

Since it is enough to pick the node with the maximum of product of firing strength and certainty factor, for applications it is not necessary to force w_{lk} to lie in $[0,1]$. It is enough to make it non-negative. So we use $w_{lk} = g_{lm}^2$. Thus, (6) can be modified to

$$z_l = max_{m \in P_l}(z_m g_{lm}^2). \tag{7}$$

3 Learning of Feature Modulators and Rules

In the training phase, the concept of error backpropagation (EBP) is used to minimize the error function

$$e = \frac{1}{2}\sum_{i=1}^{N} E_i = \frac{1}{2}\sum_{i=1}^{N}\sum_{l=1}^{t}(y_{il} - z_{il})^2, \tag{8}$$

where t is the number of nodes in layer 4 and y_{il} and z_{il} are the target and actual outputs of node l in layer 4 for input data $\mathbf{x}_i$; $i = 1, 2, ..., N$. We use EBP for adjusting the learnable weights in layer 4 and the parameters β_p in layer 2. We use online update scheme and hence derive the learning rules using the instantaneous error function E_i.

The learnable parameters i.e. the feature modulators (β_p) and the certainty factor of the rules (g_{lm}) are updated according to the equations below:

$$g_{lm}(t+1) = g_{lm}(t) - \eta\,(\partial E/\partial g_{lm}) \tag{9}$$

$$\beta_p(t+1) = \beta_p(t) - \mu\,(\partial E/\partial \beta_p)\,. \tag{10}$$

In the above equations η and μ are learning coefficients.

4 Pruning the Network

4.1 Pruning Redundant Nodes

Let us consider a classification problem with n input features. So layer 1 of the network will have n nodes. Let the indices of these nodes be denoted by i $(i = 1\ to\ n)$. Let $\mathcal{N}_i$ be the set of indices of the nodes in layer 2, which represents the fuzzy sets on the feature represented by node i of layer 1. We also assume that p $(p < n)$ of the n features are indifferent/bad as dictated by the training. Let R be the set of indices of the nodes, which represents the p indifferent/bad features. Hence, any node with index i in layer 1 such that $i \in R$ is redundant. Also any node j in layer 2, where $j \in \mathcal{N}_i$ and $i \in R$ is also redundant. In our network construction, a node in layer 3, can be uniquely identified by its connections with the nodes in layer 2. We can indicate a node m in layer 3 as $S_m = [x_{m1}, x_{m2}, ..., x_{mn}]$ where $x_{mi} \in \mathcal{N}_i$. Consequently the number of nodes in layer 3 is $\prod_{i=1}^{n}(N_i)$ where $N_i = |\mathcal{N}_i|$. Now for any $i \in R$ we can group the nodes in layer 3 into N_i many groups, we call them G_{ik}, where $k = 1,\ 2, ...,\ N_i$. Every node in the kth group is connected to the kth fuzzy set on the ith feature. Let S_m be a node in layer 3, which belongs to the kth group, i.e., $S_m = [x_{m1}, x_{m2}, ..., x_{mn}] \in G_{ik}$. Then for every group G_{il}, $l \neq k$, $l = 1, 2, ..., N_i$, there exists exactly one node $S_q = [x_{q1}, x_{q2}, ..., x_{qn}]$, such that $x_{qj} = x_{mj}$, $\forall j \neq i$, $j = 1, 2, ..., n$, where $i \in R$ is a bad feature.

Thus, every group of nodes has identical connection structure with the nodes of layer 2 except for its connection to a node corresponding to the redundant feature i, and as per our construction that particular node produces an output membership value of 1, for all feature values. Hence, it is enough to keep only one of the N_i groups and the other $N_i - 1$ groups of nodes are redundant.

The crucial part of this pruning method lies in the determination of the set of redundant nodes in layer 1. For this we use the value of $1 - e^{-\beta_i^2}$ as an indicator. We have seen earlier, that for good features $1 - e^{-\beta_i^2}$ takes values close to 1 and for bad features it is close to 0. So we fix a small positive threshold th such that $i \in R$, if $\left(1 - e^{-\beta_i^2}\right) < th$. After pruning the network will be able to retain its performance more or less at the same level when th is small. But a few epochs of additional tuning of the weights may further improve the performance of the network.

Selection of th: We have used Gaussian membership functions for the input fuzzy sets, hence as per our formulation the output of the nodes in layer 2 can be represented by $z_n = (\bar{z}_n)^{\gamma_p}$, where $\bar{z}_n = exp\{-\frac{(z_p - \mu_n)^2}{\sigma_n^2}\}$, and $\gamma_p = 1 - e^{-\beta_p^2}$. If we consider $\sigma_n = \sqrt{2}\sigma_n{}'$, then we have,

$$\bar{z}_n = exp\{-(z_p - \mu_n)^2/(2\sigma'^2_n)\} \tag{11}$$

and

$$z_n = \left[exp\{-(z_p - \mu_n)^2/(2\sigma'^2_n)\}\right]^{\gamma_p}. \tag{12}$$

We know that 99% of the area under the membership function in (11) lies over the interval $[\mu_n - 3\sigma'_n, \mu_n + 3\sigma'_n]$. Consequently, the value of $\bar{z}_n$, beyond this interval would be negligibly small. For a bad/indifferent feature we want the modulated membership value z_n to be almost unity over the entire interval $[\mu_n - 3\sigma'_n, \mu_n + 3\sigma'_n]$. Therefore, we can safely choose that value of γ_p as the threshold th, which makes $z_n = c$ $(c \approx 1)$ at $z_p = \mu_n - 3\sigma'_n$ and at $\mu_n + 3\sigma'_n$. Thus, from (12) we obtain the threshold $th = -\frac{ln(c)}{4.5}$. Note that, for such a choice if $z_p \in (\mu_n - 3\sigma'_n, \mu_n + 3\sigma'_n)$, then $z_n \geq c$. In the present investigation we use $th = 0.05$ assuming $c = 0.8$.

4.2 Pruning Incompatible Rules

In our network, the nodes in layer 3 and layer 4 are fully connected and each link corresponds to a rule. The weight associated with each link is treated as the certainty factor of the corresponding rule. If there are t classes then layer 4 will have t nodes and there will be t rules with the same antecedent but different consequents, which are inherently inconsistent.

Suppose layer 3 has N^3 nodes (i.e., N^3 antecedents) and layer 4 has N^4 nodes (N^4 consequents or classes). Each node m in layer 3 is thus connected to N^4 nodes in layer 4. From the set of N^4 links we retain *only one* link with a node in layer 4 that has the highest certainty factor associated with it.

4.3 Pruning Zero Rules and Less Used Rules

After removal of the incompatible rules each node in layer 3 is connected with only one node in layer 4. Suppose node m in layer 3 which is connected to a node l in layer 4 has a very low weight $w_{lm} < \omega_{low}$ (we take $\omega_{low} = 0.001$). Then, the rule associated with the node pair m and l has a very low certainty factor which does not contribute significantly in the classification process. We call such rules as *zero rules* and are removed from the network.

Further, our network starts with all possible antecedents which cover the total hyperbox bounding the data. Hence there may (usually will) be rules which are never fired or fired by only a few data points. Such rules are not well supported by the training data and may result in bad generalization. We call such rules as *less used rules* and they should also be removed. For every antecedent node m in layer 3, we count the number of training data points t for which the firing strength of the antecedent clause is greater than a threshold $\alpha \in (0, 1)$. If t is less than $\gamma\%$ of the size of the training data then we can consider node m as well as the rule represented by m as inadequately supported by the training data. All such nodes m in layer 3 along with their links can be removed. In our simulations α is selected as 0.1 and γ as 2.

5 Simulation Results

The methodology is tested on three data sets of which one is a synthetic data set named ELONGATED [8] and other two are real data sets named IRIS [1] and PHONEME [5,11]. We divide each data set X into training (X_{tr}) and test (X_{te}) sets randomly, such that $X_{tr} \cup X_{te} = X$ and $X_{tr} \cap X_{te} = \phi$. The summary of the data sets used is given in Table 1.

For each feature we decided fuzzy sets, such that they span the total domain of the feature space with considerable amount of overlap in between them. The initial values of the certainty factor of the rules are all set to 1.0. The initial

Table 1. Summary of the data sets

Name	Total Size	Trng. Size	Test Size	No. of classes	No. of features
ELONGATED	1000	500	500	2	3
IRIS	150	100	50	3	4
PHONEME	5404	500	4904	2	5

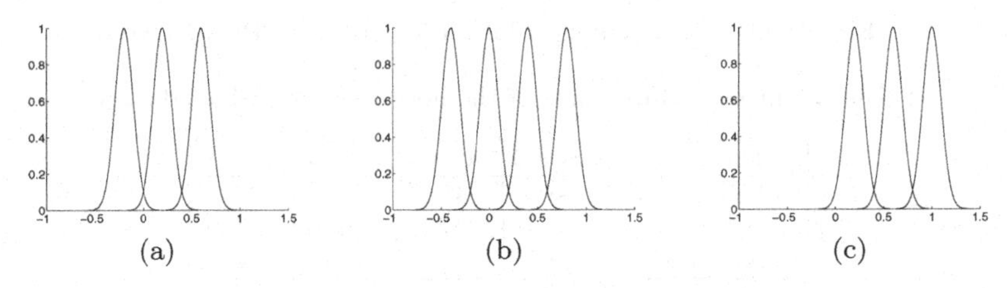

Fig. 2. Fuzzy sets used for ELONGATED: (a) feature 1 (b) feature 2 (c) feature 3

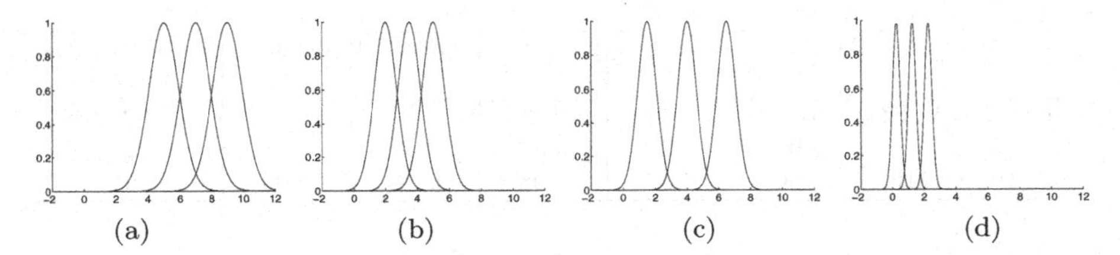

Fig. 3. Fuzzy sets used for IRIS: (a) feature 1 (b) feature 2 (c) feature 3 (d) feature 4

values of β_p's are set to 0.001, which means that initially the network considers all features to be equally unimportant. The learning parameters η and μ are differently chosen for different data sets. As stated earlier the threshold th is taken as 0.05 and α is taken as 0.1.

The training takes place in three phases. The phase 1 training is called the feature selection phase. This phase is terminated when the feature modulators stabilize. After phase 1 training, the redundant nodes (if any) as dictated by the β_p values are pruned and the network is trained for a few epochs, this is called the phase 2 training. After phase 2 training, the *incompatible rules*, the *zero rules* and the *less used rules* are removed and the network is again tuned for a few epochs, which we call the phase 3 training. We emphasize that in phase 2 and phase 3 only the certainty factors of the rules are updated.

For ELONGATED we used 3 fuzzy sets for features 1 and 3 and 4 fuzzy sets for feature 2. The fuzzy sets are shown in figure 2. For IRIS we used 3 fuzzy sets for each feature as shown in figure 3. For PHONEME we used 3 fuzzy sets for each feature but in this case the membership functions used for all features were the same (Figure 4). The initial architectures of the networks used for the different data sets are shown in Table 2. Table 3 shows the values of $1 - e^{-\beta_p^2}$ after phase 1 training. Thus the network selects 2 features from

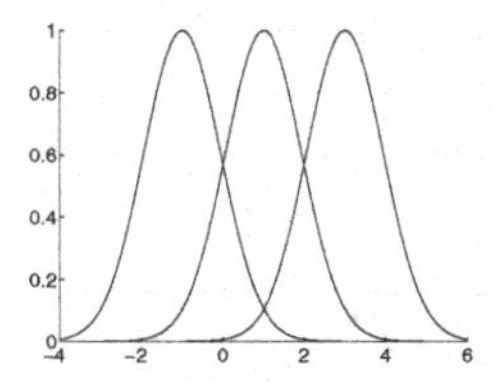

Fig. 4. The fuzzy sets used for all 5 features of PHONEME

Table 2. Initial architectures of the networks for various data sets

Data set	Layer 1	Layer 2	Layer 3	Layer 4
ELONGATED	3	10	36	2
IRIS	3	12	81	3
PHONEME	5	15	243	2

Table 3. $1 - e^{-\beta_p{}^2}$ values after phase 1 training

Data set	Feature 1	Feature 2	Feature 3	Feature 4	Feature 5	No of features selected
ELONGATED	0.51	0.44	0.01			2
IRIS	0.00	0.00	0.52	0.51		2
PHONEME	0.66	0.68	0.98	0.93	0.00	4

ELONGATED and IRIS and 4 features from PHONEME. In ELONGATED, the network starts with $36 \times 2 = 72$ rules. After pruning the redundant nodes the number of rules becomes $12 \times 2 = 24$. Next the incompatible rules were pruned to obtain 12 rules. Finally, 1 rule was found to be less used. Hence the number of rules represented by the final network was 11. In case of IRIS, pruning of redundant nodes reduced the number of rules from $81 \times 3 = 243$ to $9 \times 3 = 27$. The pruning of incompatible rules resulted in 9 rules out of which 4 were less used rules. So the final network for IRIS had 5 rules. In PHONEME the initial number of rules were $243 \times 2 = 486$. The removal of the redundant nodes and incompatible rules results in 81 rules out of which 32 rules were Zero rules and 3 were less used. Thus the final architecture represented 46 rules. Table 4 shows the rule reduction and the misclassifications for the data sets used.

6 Conclusion

A new scheme for fuzzy rule based classification in a neuro-fuzzy framework is presented. The novelty of the system lies in it's capacity to select good features online. The network described here is fully readable. Simulation results show that the system does a good job for all data sets tried. For further development of the methodology, some specialized tools of exploratory data analysis may be used to decide upon the number of input fuzzy sets. The parameters of the

Table 4. Performance of and rule reduction by the proposed system

Data set	Initial no. of rules	Final no. of rules	Misclassification on X_{tr}	Misclassification on X_{te}
ELONGATED	72	11	0 (0%)	0 (0%)
IRIS	243	5	3 (3%)	1(2%)
PHONEME	486	46	85(17%)	966 (19%)

input fuzzy sets can also be tuned using backpropagation. We believe, these will further improve the performance of the network.

References

1. Anderson, E.: The Irises of the Gaspe peninsula. Bulletin of the American IRIS Society **59** (1935) 2–5
2. De, R., Pal, N.R., Pal, S.K.: Feature analysis: neural network and fuzzy set theoretic approaches. Pattern Recognition **30** (1997) 1579–1590
3. Chakraborty, D., Pal, N.R.:Integrated feature analysis and fuzzy rule-based system identification in a neuro-fuzzy paradigm. IEEE Trans. on Syst. Man Cybern **31** (2001) 391–400
4. Keller, J., Yager, R., Tahani, H.: Neural network implementation of fuzzy logic. Fuzzy Sets and Systems **45** (1992) 1–12
5. Kuncheva, L.: Fuzzy Classifiers. Physica-Verlag (2000)
6. Lin, C.T., Lee, C.S.G.: Neural network based fuzzy logic control and decision system. IEEE Transactions on Computers **40** (1993) 1320–1335
7. Lin, C.T., Lee, C.S.G.: Neural Fuzzy Systems. Prentice Hall P T R, Upper Saddle River, NJ (1996)
8. Mao, J., Jain, A.K.: Artificial neural networks for feature extraction and multivariate data projection. IEEE Trans. on Neural Networks **6** (1995) 296–317
9. Pal, N.R.: Soft computing for feature analysis. Fuzzy Sets and Systems **103** (1999) 201–221
10. Ruck, D.W, Rogers, S.K., Kabrisky, M.: Feature selection using a multilayered perceptron. Journal of Neural Network Computing (1990) 40–48
11. *ftp://ftp.dice.ucl.ac.be/pub/neural-nets/ELENA*

Decomposed Neuro-fuzzy ARX Model

Marjan Golob and Boris Tovornik

Institute of Automation, Faculty of Electrical Engineering and Computer Science
University of Maribor, Smetanova 17, 2000 - Maribor, Slovenia
mgolob@uni-mb.si

Abstract. This paper explores a new approach for the modelling and identification of non-linear dynamic systems. A model, named *the Decomposed Neuro-Fuzzy Auto-Regressive* with *eXogenous* input model (DNFARX), based on decomposed structure of the fuzzy inference system, is proposed. An evolution of a neural network learning algorithm for the decomposed structure of the fuzzy inference system is suggested. A comparative study of the dynamic system modelling with conventional fuzzy inference system based models and the proposed model is presented for Box-Jenkins data set.

1 Introduction

The most common input-output model is non-linear *Auto Regressive with eXogenus* input (ARX) model [8]. In the discrete-time setting ARX model may be written as

$$y(k) = f(y(k\text{-}1),\ldots,y(k\text{-}n_y),u(k\text{-}\tau\text{-}1),\ldots,u(k\text{-}\tau\text{-}n_u)) \ , \tag{1}$$

where $f(\cdot)$ is a non-linear static transition function, $y(k\text{-}1),\ldots,y(k\text{-}n_y),u(k\text{-}\tau\text{-}1),\ldots,u(k\text{-}\tau\text{-}n_u)$ are the past model outputs and inputs, τ is an input time delay and n_y, n_u presents model order. The main task in system modelling is to determine the best function approximation for unknown non-linear function $f(\cdot)$. Fuzzy systems, neural network based systems and neuro-fuzzy systems can be used to approximate the non-linear state or input-output transition function.

An *Fuzzy Rule-Base System* (FRBS) presents two main components: the *Rule Base* (RB), representing the knowledge about the problem being described in the form of fuzzy if-then rules, and the *Fuzzy Inference System* (FIS) needed to obtain an output from FRBS. With symbolical notation [7] of the fuzzification end defuzzification operators the output of the general fuzzy model (GFM) is defined as follows:

$$\begin{aligned} y(k) = defuzz\{&fuzz(y(k\text{-}1)) \circ \ldots \circ fuzz(y(k\text{-}n_y)) \circ \\ &fuzz(u(k\text{-}\tau\text{-}1)) \circ \ldots \circ fuzz(u(k\text{-}\tau\text{-}n_u)) \circ R\} \ , \end{aligned} \tag{2}$$

where $\circ$ stands for a family of composition operators (i.e. *max-min, max-prod, sum-prod*) and R is relation, represents FIS behaviour that depends upon the rules in RB.

For example, a RB of a FRBS based fuzzy model of a dynamic system may consist of rules as example the l-th complex rule r_l

N.R. Pal and M. Sugeno (Eds.): AFSS 2002, LNAI 2275, pp. 260–266, 2002.

IF $y(k\text{-}1)$ is A_{l1} and … $y(k\text{-}n_y)$ is A_{lny} and $u(k\text{-}\tau\text{-}1)$ is B_{l1} and … $u(k\text{-}\tau\text{-}n_u)$ is B_{lnu} (3)
THEN $y(k)$ is A_{l0} .

l=1,2,…,p, A_{l0} … A_{lny} are fuzzy sets of the output universes of discourse $Y \in R$, B_{l1} … B_{lnu} are fuzzy sets of the output universes of discourse $U \in R$. Typically, rule consequence takes the following three form: A_{l0} is general fuzzy sets, A_{l0} is singleton fuzzy sets, or A_{l0} is linear function. The FIS of RBFS are usually of the relational (Mamdani) or functional (Sugeno) type [7]. A well-known neuro-fuzzy system for function approximation is the ANFIS model [10]. ANFIS is used to implement a Sugeno-type fuzzy system that uses a differentiable t-norm and differentiable membership functions. In proposed decomposed fuzzy ARX models we apply a fuzzy inference system which can be interpreted linguistically and uses a differentiable t-norm and membership functions.

The structure of this paper is as follows. After reporting about the dimensionality problem in dealing with high-dimensional input-output model problems and mentioning several methods to solve this problem, we propose, in section 2, the neuro-fuzzy version of the decomposed fuzzy ARX model where every fuzzy subsystem is realised with FIS based on bases fuzzy sets with derivative membership functions. In section 3, we simulate the neuro-fuzzy identifier for the Box and Jenkins's gas furnace data set [1] and we show the merit of the decomposed fuzzy ARX model. Performance comparisons of the proposed model with other related approaches are summarised by performance evaluation index computed from simulation results. Concluding remarks are finally made in section 4.

2 Decomposed Neuro-fuzzy ARX Model

Fuzzy or neuro-fuzzy systems, used in general fuzzy model (2) suffer from the curse of dimensionality, which leads to a rule space, and hence to computational and memory demands, exploding combinatorially with an increasing number of inputs. A large fuzzy rule base, generated form expert operators or by some learning or identification schemes, may contain redundant, weakly contributing, or outright inconsistent components. Moreover, in pursuit of good approximation, one may be tempted to overly assign the number of antecedent sets, thereby resulting in large fuzzy rule bases and problems in computation time and storage space. In [15] a method to reducing a fuzzy rule set by capturing a large extent of its input/output characteristics is introduced. The method is based on conducting singular value decomposition of the rule consequences and generating proper linear combinations of the original membership functions to form new ones for the reduced sets. The proposed approach can be applied regardless of the fuzzy paradigm adopted for the fuzzy rule base. The considerable reduction of the rules in the rule base is straightforward, but the rule structure is still stay complex (more then one fuzzy variable on the IF-part of fuzzy rules) and the dimensionality problem of the single-stage fuzzy reasoning process is still significant. Since the dimensionality problem is critical to the success of fuzzy systems in dealing with high-dimensional problems, there have been increasing attempts to address it through different ways. For the ANFIS model [10], Jang [11] has subsequently proposed an input selection method where only two most important inputs are considered and all the others are ignored. In paper [6], the structure of the rule premise (i.e., which inputs should appear in the premise part of a fuzzy rule) is optimised using a genetic algorithm based on a local cost function, which is consistent with the motivation to establish local optimal fuzzy rules. It is shown that optimisation of the rule structure cannot only reduce the rule complexity and improve the system performance, but also reveal the dependencies between the system inputs and the system output. As is well known, genetic algorithms are a very time-consuming searching processes and are sometimes computational prohibitive when a very high-dimensional problem is being considered.

Most of the existing fuzzy system or fuzzy neural network based models have been proposed to implement different types of single-stage fuzzy reasoning mechanisms. To address the dimensionality problem, fuzzy neural network modelling based on multistage fuzzy reasoning is pursued in paper [2] and two hierarchical network models, namely incremental type and aggregated type, are introduced.

In paper [5] two fuzzy models for SISO dynamic systems based on the decomposed compositional rule of inference have been presented. The simplification of the decomposed model structure have been used in both models. Two relational type of decomposed FIS with *max-min* composition and *max-prod* composition have been implemented and corresponding fuzzy identification algorithms have been proposed. Advantages of proposed models are; the reduction of number of rules in rule base, the simple composition of a fuzzy sets and a two-dimensional fuzzy relations. Code optimisation and hardware inference realisation is possible, and close relationship between structure of the proposed model and the discrete linear ARX model is evident.

With symbolical notation of the fuzzification operator end defuzzification operator and with respect to the simplified decomposed compositional rule of inference [5] the output of the decomposed fuzzy ARX model for SISO dynamic system with n_u+n_y two-dimensional fuzzy relations R_i is

$$y(k)=\frac{1}{n_u+n_y}\cdot\{defuzz\left[fuzz(y(k-1))\circ R_{a_1}\right]+\cdots+defuzz\left[fuzz\left(y(k-n_y)\right)\circ R_{a_{n_y}}\right]+ \quad (4)$$

$$defuzz\left[fuzz(u(k-\tau-1))\circ R_{b_1}\right]+\cdots+defuzz\left[fuzz(u(k-\tau-n_u))\circ R_{b_{n_u}}\right]\},$$

where R_{a_1} ... $R_{a_{n_y}}$ and R_{b_1} ... $R_{b_{n_u}}$ are relational matrices represents decomposed fuzzy subsystems behaviours. Note, that the dynamic structure is similar to the structure of a linear auto-regressive model, i.e. discrete linear ARX model [8]. Rule bases of a fuzzy subsystems are consist of simple rules, namely rules with only one antecedent variable and one fact in premise part of the rule.

In [5] RBFS subsystems of the decomposed fuzzy ARX model were realised with fuzzy relational equations. With fuzzy relational matrix identification based approach, the domains of the antecedent and the consequent variables are a priori partitioned into a specified number of membership functions of base fuzzy sets. The RB is then established to cover the antecedent space by using logical combinations of the antecedent terms. If no knowledge is available as to which variables cause the non-linearity of the system, all the antecedent variables are usually partitioned uniformly. However, the complexity of the system's behaviour is typically not uniform. Some operating regions can be well approximated by a single model, while other regions require rather fine partitioning. In order to obtain an efficient representation with as few rules as possible, the membership functions must be placed such that they capture the non-uniform behaviour of the system. This often requires that the process measurement are also used to form and optimize the membership functions.

In proposed neuro-fuzzy version (DNFARX) of the decomposed fuzzy ARX model (4) every fuzzy subsystem is realised with FIS based on bases fuzzy functions with derivative Gaussian membership functions as are proposed by Wang [14]. All of fuzzy subsystems are SISO fuzzy systems with one input and one output variable and simplification of original Wang fuzzy system with singleton fuzzifire, modified center of average defuzzifier, product inference and Gaussian membership functions is obvious. The output y of DNFARX model is sum of outputs y_i of fuzzy subsystems:

$$y = \frac{1}{n_u + n_y} \sum_{i=1}^{n_u+n_y} y_i \qquad y_i = \sum_{l=1}^{M} \frac{\bar{y}_{il} \cdot e^{-\frac{1}{2}\frac{(x_i - \bar{x}_{il})^2}{\sigma_{Ail}^2 + \varepsilon}}}{\sigma_{Bil}^2 + \varepsilon} \Bigg/ \sum_{l=1}^{M} \frac{e^{-\frac{1}{2}\frac{(x_i - \bar{x}_{il})^2}{\sigma_{Ail}^2 + \varepsilon}}}{\sigma_{Bil}^2 + \varepsilon}, \tag{5}$$

where M is number of bases fuzzy sets, $i = 1,2, \dots n_u+n_y$ is number of fuzzy subsystems, $x_i \in \{ u(k-\tau-1), \dots, u(k-\tau-n_u), y(k-1), \dots, y(k-n_y) \}$ are particular crisp input variables of fuzzy subsystems and $\bar{x}_{il}, \sigma_{Ail}, \bar{y}_{il}, \sigma_{Bil}$ are $4(n_u+n_y)$ free parameters of all subsystems. The purpose of adding the small $\varepsilon > 0$ to the fuzzy membership functions is that even the σ's = 0 the fuzzy membership functions are still well defined. This modification will make the learning procedure simpler because we do not require the σ's ≠ 0. All free parameters are collected in vector $\theta_i \equiv [\bar{x}_{i1}, \dots, \bar{x}_{iM}, \sigma_{A_{i1}}, \dots, \sigma_{A_{iM}}, \bar{y}_{i1}, \dots, \bar{y}_{iM}, \sigma_{B_{i1}}, \dots, \sigma_{B_{iM}}]^T$. Thanks to supervised learning methods, it is possible to optimise both the antecedent and consequent parts of a linguistic rule-based fuzzy systems. A many of learning methods are derived from back-propagation algorithm. We used the back-propagation training algorithm proposed by Wang [14] to adjust the parameters in vector θ_i.

3 Numerical Example

With the aim of analysing the behaviour of the proposed modelling method based on decomposed neuro-fuzzy ARX models we have chosen a popular example of the input-output dynamic system modelling, namely, Box and Jenkins's gas furnace data set [1]. The data set consists of 296 input-output observations, where the input $u(k)$ is the rate of the gas flow into the furnace and the output $y(k)$ is the CO_2 concentration in the outlet gases. The sampling interval is 9 s.

In order to compare the performance quality, the mean square error criterion

$$J_{MSE} = \frac{1}{296-\tau-1} \sum_{k=\tau+2}^{296} \left[y(k) - \hat{y}(k) \right]^2 \tag{6}$$

is used. A structure of the input-output identification method using DNFARX with two neuro-fuzzy based FIS f_u and f_y. is shown on Fig. 1. The parameter τ was varied between 0 and 4. The best performance was achieved with choice of $\tau = 3$.

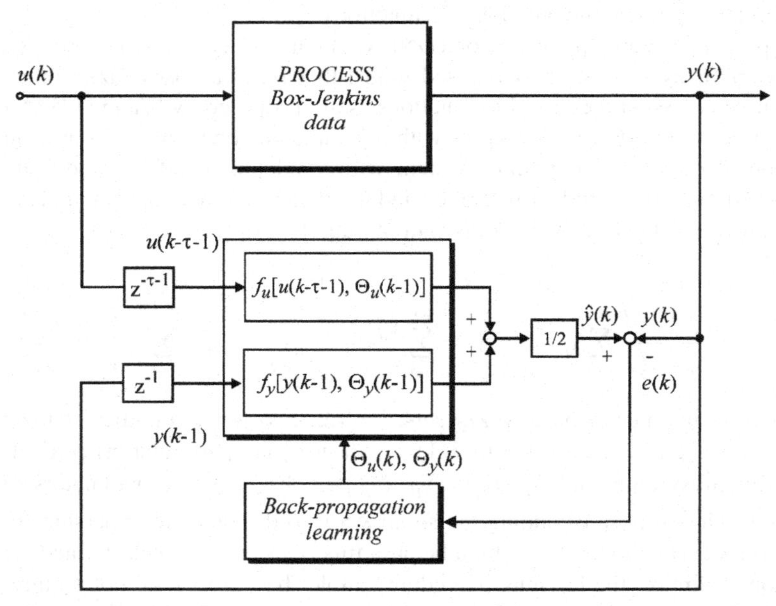

Fig. 1. A linguistic input-output identification method using decomposed ARX model with two neuro-fuzzy based FIS f_u and f_y.

We simulated two cases: all free parameters in vectors θ_u and θ_y were chosen randomly; and free parameters were chosen according to the corresponding IF-THEN rules, obtained as result of our previous research [5]. Adaptation speed and accuracy were greatly improved by incorporating the linguistic rules. These initial fuzzy membership functions were distributed over the input-output space so that the space was completely covered. Verification of this work resulted in $J_{MSE} = 0.196$.

We shall compare the accuracy of the linguistic models generated from our method with the ones designed by means of other methods with different characteristics (four methods are typical non-optimize methods based on fuzzy relational matrix identification [4], [9], [3], [5] and two methods based on Sugeno type of fuzzy inference [12], [13]). From all of mentioned methods only two [5] are designed with decomposed fuzzy structures.

In Table 1 the performance index achieved by other authors are compared with the proposed.

Table 1. Identifications results for Box-Jenkins gas furnace

Reference	No. of input variables	No. and type of rules	J_{MSE}
Pedrycz [4], [9]	$y(k\text{-}1)$, $u(k\text{-}4)$	81, complex	0.320
Costa Branco [3]	$y(k\text{-}1)$, $u(k\text{-}4)$	49, complex	0.312
Sugeno-Yasukawa [12]	$y(k\text{-}1),y(k\text{-}2),y(k\text{-}3)$ $u(k\text{-}1),u(k\text{-}2),u(k\text{-}3$	6, complex	0.190
Takagi Sugeno [13]	$y(k\text{-}1)$, $u(k\text{-}3),u(k\text{-}4)$	2, complex	0.068
Golob ARX min-max [5]	$y(k\text{-}1)$, $u(k\text{-}4)$	14, simple	0.73
Golob ARX sum-prod [5]	$y(k\text{-}1)$, $u(k\text{-}4)$	14, simple	0.57
Proposed method	$y(k\text{-}1)$, $u(k\text{-}4)$	14, simple	0.196

Verification result $J_{MSE} = 0.196$ indicate that methods based on DNFARX, although they use simple structures, results in better identification performances.

4 Conclusion

In this paper a method based on decomposed neuro-fuzzy ARX model has been proposed as a powerful new approach to design linguistic models. Decomposed neuro-fuzzy ARX model is well suited for input-output modelling of non-linear dynamical process. The main idea is based on the decomposed model structure, which is supported with the neural network learning algorithm. Advantages of the proposed model are; the reduction of number of rules in rule base, the simple composition of a FIS; a two-dimensional fuzzy relations in case of relational matrix based FIS or simple single input single output neural network structures in case of the neuro-fuzzy FIS. Decomposed neuro-fuzzy ARX models are appropriate to program code optimisation or hardware inference realisations. Close relationship between structure of the proposed model and the discrete linear ARX model is evident.

References

1. Box, G.E.P., Jenkins, G.M.: Time Series Analysis, Forecasting and Control. Holden Day, San Francisco (1970)
2. Chung, F.-L., Duan, J.-C.: On Multistage Fuzzy Neural Network Modeling. IEEE Trans. Fuzzy Syst. 8 (2000) 125-142
3. Costa Branco, P. J., Dante, J. A.: A New Algorithm for On-line Relational Identification of Nonlinear Dynamic Systems. Proceedings of the Second IEEE Int. Conf. Fuzzy Syst. (FUZZ-IEEE'93) (1993) 1173-1178
4. Czogala, E., Pedrycz, W.: On Identification in Fuzzy Systems and its Applications in Control Problems. Fuzzy Sets and Systems 6 (1981) 73-83
5. Golob, M., Tovornik, B.: Identification of Non-linear Dynamic Systems with Decomposed Fuzzy Models. In Proceedings of the 2000 IEEE Int. Conf. on Systems, Man & Cybernetics (SMC-IEEE'2000), October 8-11, 2000, Nashville (2000) 3520-3525.

6. Jin, Y.: Fuzzy Modeling of High-Dimensional Systems: Complexity Reduction and Interpretability Improvement. IEEE Trans. Fuzzy Syst. 8 (2000) 212-221
7. Lee, C.C.: Fuzzy Logic in Control Systems: Fuzzy Logic Controller – Part I and II. IEEE Trans. System Man Cybernet. SMC-20 (1990) 404-435
8. Ljung, L., Soderstrom, T.: Theory and Practice of Recursive Identification. The MIT Press, London (1983)
9. Pedrycz, W.: Identification Algorithm in Fuzzy Relational Systems. Fuzzy Sets and Systems 13 (1984) 153-167
10. Jang, J.-S.: ANFIS: Adaptive-Network-Based Fuzzy Inference Systems. IEEE Trans. Systems Man Cybernet. 23 (1993) 665-684
11. Jang, J.-S.: Input Selection for ANFIS Learning. In Proc. IEEE Int. Conf. Fuzzy Syst. (FUZZ-IEEE'97) (1996) 1493-1499
12. Sugeno, M., Yasukawa, T.: A Fuzzy-Logic-Based Approach to Qualitative Modeling. *IEEE Trans. on Fuzzy Systems*, 1(1) (1993) 7-31
13. Takagi, T., Sugeno, M.: Fuzzy Identification of Systems and its Applications to Modelling and Control. IEEE trans. System Man Cybernet. SMC-15 (1985) 116-132
14. Wang, L.-X.: Design and Analysis of Fuzzy Identifiers of Nonlinear Dynamic Systems, IEEE Trans. on Automatic Control, 40(1) (1995) 11-23
15. Yam, Y., Baranyi, P., Yang, C.-T.: Reduction of Fuzzy Rule Base Via Singular Value Decomposition. IEEE Trans. Fuzzy Syst. 7 (1999) 120-132

Weather Forecasting System Based on Satellite Imageries Using Neuro-fuzzy Techniques

Chien-Wan Tham, Sion-Hui Tian, and Liya Ding

The Institute of Systems Science, National University of Singapore,
25 Heng Mui Keng Terrace, Singapore 119615
cwtham@singnet.com.sg

Abstract. We have built an automated Satellite Images Forecasting System with *Neuro-Fuzzy techniques*. Firstly, *Subtractive Clustering* is applied on to a satellite image to extract the locations of the clouds. This is followed by *Fuzzy C-Means Clustering* which operates on the next satellite image, seeded with the cloud clusters of the previous image. With the matching of cloud clusters across successive images, cloud cluster velocities are deduced. Using a *Generalized Regression Neural Network*, we interpolate the cloud cluster velocities over the whole area of interest. Finally, the linear forecasting scheme then moves each cloud pixel in that satellite image according to the velocities of the past hour.

1 Introduction

There are two basic neural net approaches in forecasting radar or satellite imageries. The first is by *brute force method* where the last image is used as direct input into an error back-propagation feed forward neural net, trained with the hour later image as target. However, it has been shown that such black box usage of neural net gives only a slight improvement over traditional statistics [6].

The second approach, which draws upon domain knowledge, is by *intelligent feature extraction.* In this method, it is assumed that the linear propagation of cloud clusters and rain cells are neither growing nor decaying within the time frame of interest. By extracting the cloud velocities, we can linearly extrapolate them to get a forecast.

One instance of this feature extraction approach is based on the approximation capabilities of the neural network. Lee [7], and Denœux and Rizand [3] have developed a modified radial basis function neural net in which a hidden node is created for every significant rain cells on the neural net. The weights of each modified radial basis neural net are trained to represent the sample image itself. With the linear extrapolation of the neural net weights to give a set of new weights, a forecast radar image is hence produced.

A second instance is where the characteristics of the rain cells such as size, shape, and mass are extracted. Einfalt et al. [5] derive the velocity of these cells by matching the adjacent radar images where the neural net is employed as classifier. In this way, each possible pair between successive radar images are input into the classifier that are pre-trained with labeled examples.

N.R. Pal and M. Sugeno (Eds.): AFSS 2002, LNAI 2275, pp. 267–273, 2002.

In this paper, we present a Weather Satellite Forecasting system in which *Fuzzy Clustering* techniques are used to match the cloud clusters between satellite images. This is augmented with a *Generalized Regression Neural Network* that interpolates a velocity field to forecast the movement of clouds on a satellite image. We found that there is a significant reduction of error by 20-40% against the persistence forecast. The whole system is implemented with Matlab and its Neural Net Toolbox.

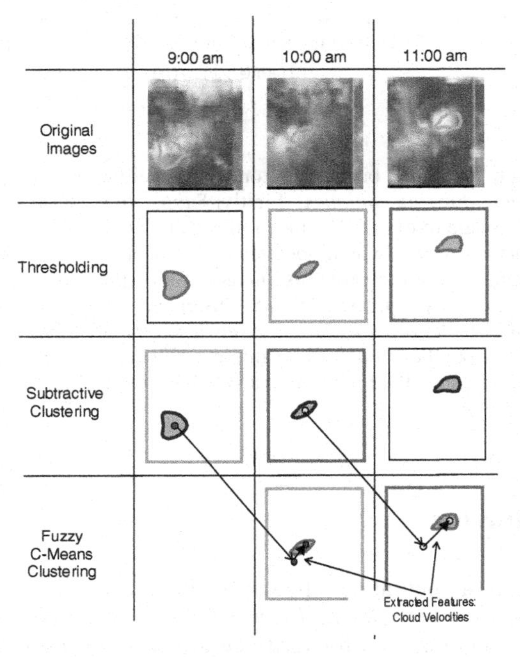

Fig. 1. The schematic matrix for the features extraction procedure, characterized by the processing steps (*vertical axis*) and time series (*horizontal axis*).

2 Fuzzy Clusterings for Features Extraction

The aim of the features extraction step is to identify the cloud clusters and their velocities, from a given set of the infrared satellite images. The steps are illustrated, as rows, in Figure 1.

2.1 Preprocessing of Satellite Images

For every image, a simple crisp thresholding is first used to detect pixels that belong to deep convective clouds. These cloud pixels are then through a clustering algorithm grouped into clusters and identified as cloud clusters objects.

The objective of the thresholding procedure is to transform a given rectangular matrix of satellite image into cloud pixels objects. This somewhat trivial procedure is a necessary step in our feature extraction step, because clustering (both hard and fuzzy) algorithms expect objects, not images nor matrices, as input in order to cluster them. It is this important conceptalization that allows us to deploy clustering algorithms on to satellite images.

2.2 Subtractive Clustering

The most prevalent and elementary unsupervised clustering method is the K-Means Clustering method [4]. A fuzzified version of it, the Fuzzy C-Means Clustering is proposed by Bezdek [1]. The difficulty of using either of these two methods is that they both require as initial parameters, *a pre-determined number of clusters* and the initial position of these clusters.

In our application, however, there is no simple way of knowing the number of clusters in advanced. As such, we employed Subtractive Clustering, proposed by Chiu [2], which does not require any initial number of clusters.

In essence, the method starts off with the premise that each data point is a potential cluster centre and then calculates a measure of the likelihood that each data point will define the cluster centre, based on the density of the surrounding data points.

2.3 Fuzzy C-Means Clustering

After the subtractive clustering procedure, the extracted cloud clusters in the current image are used as the *initial seeds for Fuzzy C-Means Clustering* of the subsequent (an hour later) image. The application of Fuzzy C-Means on the Satellite Images requires that we define a distance measure, and we choose the Euclidean distance.

For the initial seeds, we use the cluster centres of the previous image, obtained from applying subtractive clustering on it. This has dual purposes. Firstly, noting the difference between each of the seed initial and final values (at the start of the Fuzzy C-Means clustering and at the end), we are able to extract the velocities of the cloud clusters. Secondly, by using seeds that are close to the final position of the cloud cluster, there is a higher chance that the iterative algorithm will converge, and within a shorter time, albeit not guaranteed.

2.4 Extracting Cloud Velocities

With the pairing of cloud clusters across images, we extract the cloud clusters velocities by subtracting the difference in position between the cloud centers across adjacent time frames. This technique of extracting cloud velocity is the *main original idea* of this work. An instance of the cloud centers and velocities is given in figure 2.

3 Neural Net Interpolation for Forecasting

For simplicity of implementation, we target a *one-hour ahead deep convective cloud cluster movement* forecast. Meteorologically, for such short duration forecast, probably the most effective and accurate forecasting scheme is the *linear extrapolation of cloud movements.* In essence, we use the past hour cloud clusters movements, extrapolate them linearly into the next hour and use that as the prediction.

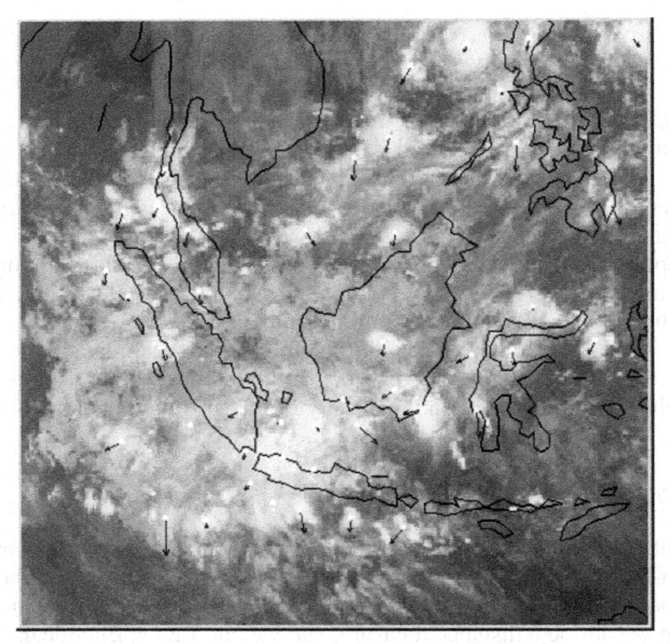

Fig. 2. Derived Cloud Clusters (*white dots*) and their Velocities (*black arrows*) for a particular satellite image.

3.1 Generalized Regression Neural Network

The function served by the Generalised Regression Neural Network (GRNN) is to interpolate a 2-dimension vector field of velocities.

The GRNN is basically similar to interploation/extrapolation calculation. It is depolyed because it is both a suitable way of representing the knowledge captured from clustering, as well as an appropiate way of implementing the inference structure for advecting the cloud pixels.

We have used Matlab's implementation of the GRNN, which has 2 layers of neurons: a radial basis layer and a special linear layer. The radial basis layer stores the cloud clusters locations, while the linear layer holds the cloud clusters velocities. For a given input location, the GRNN outputs a vector sum of the existing cloud cluster velocities weighted on the physical proximity of the input to its neighbouring clusters.

3.2 Cloud Cluster Forecasting

We use the current cloud pixels and move them in accordance to the velocity at that position. The velocity at the various cloud pixels position is obtained by inputting the required x-y coordinates into the GRNN which will return the velocity at that point. An example of the cloud cluster advection is given in Figure 3.

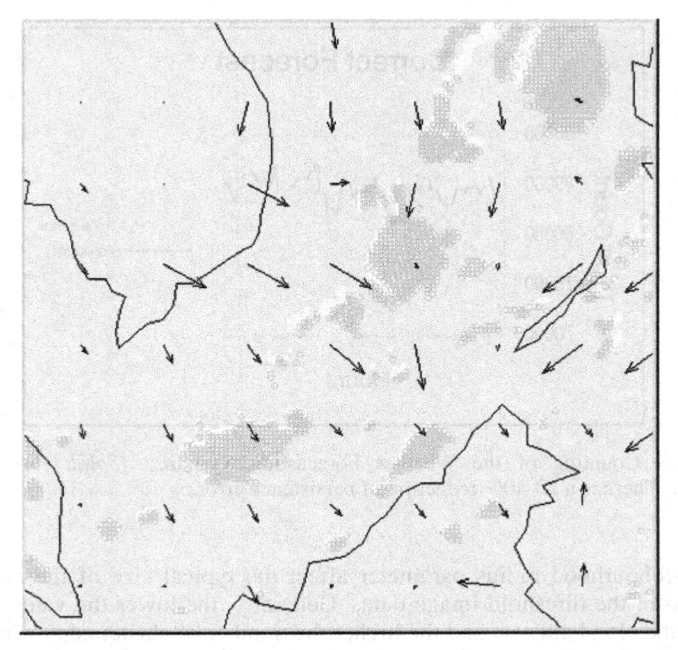

Fig. 3. Advections of cloud cluster pixels. (*White: current; Gray: forecast*)

4 Numerical Result

In the context of our weather forecasting system, the weather situation is encapsulated as pixel values in the form of satellite images. Therefore, we introduce the method of pixel counting, as a way of determining the accuracy of our forecast results.

The method works by comparing the value of the same pixel, between the image of the actual weather situation, against our system's image prediction. Our system scores a point, for every pixel value correctly predicted (i.e. same value as the pixel in the actual image), for each corresponding pixel location. Conversely, the system can also compute the pixel error count, for every pixel value predicted incorrectly.

As part of the sensitivity analysis, we have two parameters in out system. The first parameter is the threshold pixel value, in which pixels whose values do not meet this threshold would not be considered in the subsequent part of the image processing. The second is a neighbourhood radius parameter used in the Subtractive Clustering process to determine the critical radius from the data point in which neighbouring points will have a high contribution to that data point.

Numerically, Figure 4 holds, even when these two parameters are varied around its neighbourhood region. The threshold pixel value parameter influences the amount of data considered as clouds, for subsequent feature extraction steps. In general, the higher the threshold value, less pixels will be able to meet the high pixel cut-off value, and less pixel data would be considered in the next image processing step.

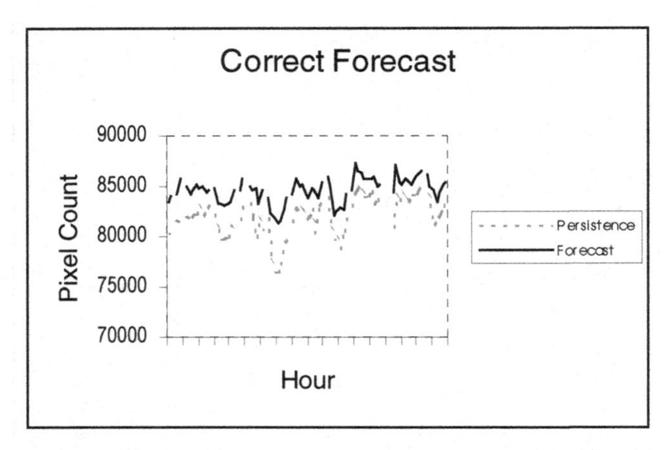

Fig. 4. Pixel Counting of the Weather Forecasting System. *(Solid: forecast; Dotted: persistence)*. There is a 20-40% reduction of persistence error.

The neighbourhood radius parameter affect the typical size of the cloud clusters extracted from the threshold image data. Generally, the lower the value, the smaller the size of the cloud clusters and the higher the number of cluster centres identified.

5 Conclusion

We have build an automated Weather Forecasting System based on satellite images using a neuro-fuzzy approach. The *validation and verification* procedures compare our model against the persistence forecast, finding a 20-40% reduction in error.

There are two parameters in our system. They are the *threshold value* used to defined cloud pixels; and the *neighbourhood radius* used in Subtractive Clustering to define the circle of contribution of a data value to the density of the cloud cluster in consideration. The result of our numerical experiments shows that *our model is insensitive to both model parameters.* Thus we have built a robust satellite image forecasting system capable of giving 1-hour ahead forecasts.

There are two possible future extensions. The first is to *fuzzify the crisp threshold used to determine cloud pixels.* To do this, we need to reinterpret the pixels values as data density points in the Subtractive and Fuzzy C-Means Clustering algorithms. The second is to use some non-linear statistics, such as the multilayer feed forward neural network with error back-propagation, to train and forecast the extracted cloud cluster velocities to give longer lead-time forecast.

References

1. Bezdek, J.C., *Fuzzy Mathematics in Pattern Classification.* PhD thesis, Applied Math. Center, Cornell University, Ithaca, 1973.
2. Chiu, S.L., Fuzzy Model Identification based on Cluster Estimation. *Journal of Intelligent and Fuzzy Systems*, Vol. 2, No. 3, pp. 267–278, 1994.
3. Denœux, T., and Rizand P., Analysis of Radar Images for Rainfall Forecasting using Neural Networks. *Neural Computing and Applications* (3), 1, pp. 50–61, 1995.
4. Duda, R.O., R.E. Hart, and D.G. Stork, *Pattern Classification*, second edition, John Wiley & Sons, 2001.
5. Einfalt, T., T. Denœux, and G. Jacquet, A Radar Rainfall Forecasting Method Designed for Hydrological Purposes. *Journal of Hydrology*, 114, pp.229–244, 1990.
6. French, M. N., W. F. Krajewski, and R. R. Cuykendall, Rainfall Forecasting in Space and Time using a Neural Network. *J. Hydrology,* 137, pp.1–31, 1992.
7. Lee, S., Supervised Learning with Gaussian Potentials. In *Neural Networks for Signal Processing*, edited by B. Kosko, Prentice-Hall, Englewood Cliffs, NJ, pp.189–227, 1992.

Evolutionary Subsethood Product Fuzzy Neural Network

C. Shunmuga Velayutham[1], Sandeep Paul[2], and Satish Kumar[1]

[1] Dept. of Physics and Computer Science
Faculty of Science
Dayalbagh Educational Institute
Dayalbagh, Agra 282005, India
[2] Dept. of Electrical Engineering
Technical College
Dayalbagh Educational Institute
Dayalbagh, Agra 282005, India

Abstract. This paper employs a simple genetic algorithm (GA) to search for an optimal set of parameters for a novel subsethood product fuzzy neural network introduced elsewhere, and to demonstrate the pattern classification capabilities of the network. The search problem has been formulated as an optimization problem with an objective to maximize the number of correctly classified patterns. The performance of the network, with GA evolved parameters, is evaluated by computer simulations on Ripley's synthetic two class data. The network performed excellently by being at par with the Bayes optimal classifier, giving the best possible error rate of 8%. The evolutionary subsethood product network outperformed all other models with just two rules.

1 Introduction

Synergistic fuzzy neural models combine the merits of connectionist and fuzzy approaches. Many models realize fuzzy knowledge representation and inference in a connectionist way [1,2,3,4,5]. By virtue of the ability to refine initial domain knowledge, operate and adapt in both numeric as well as linguistic environments, these models are being extensively used in a wide range of application domains. An important problem in the formulation of fuzzy neural network models is searching a correct set of parameters of the network. This has an important bearing on the performance of such models. Commonly used training procedures employ either gradient descent [6,7] or genetic algorithms [8,9] or a combination of these two [10,11,12], to cite a few examples.

In this paper we concentrate on the GA-based training of a novel fuzzy neural network model introduced in [13] and refined subsequently in [14] and [15]. The model admits numeric as well as fuzzy inputs. Numeric inputs are first fuzzified at input nodes by treating the numeric value as the center of a Gaussian fuzzy set with a certain spread. All signals transmitted forward from the input layer are uniformly fuzzy. Although in [14] the spread for numeric inputs was held

N.R. Pal and M. Sugeno (Eds.): AFSS 2002, LNAI 2275, pp. 274–280, 2002.

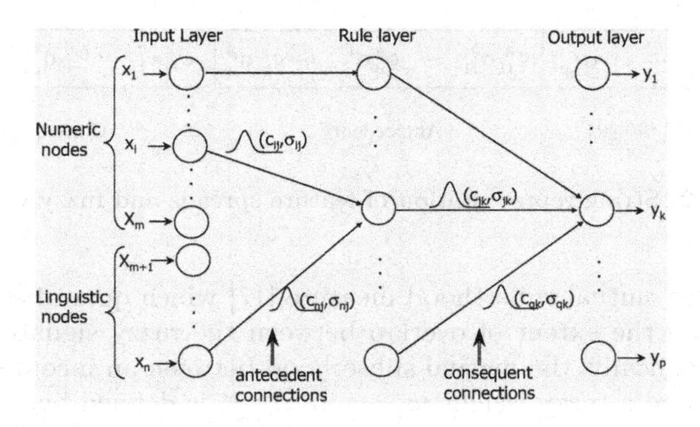

Fig. 1. Architecture of the subsethood product fuzzy neural network

fixed, here we allow these feature spreads to evolve over a pre-defined range of values. The flow of fuzzy information through the fuzzy weights is based on a mutual subsethood measure [17]. To compute the strength of firing of a rule, the conventional *min* operator commonly used in fuzzy systems (such as [6,10,11]) has been replaced by the *product* operator. A volume defuzzification scheme generates numeric outputs, in contrast with traditional center of gravity/area methods or weighted sum/mean methods. The tunability of feature spreads, novel composition mechanism, product operator and volume defuzzification collectively distinguish the model from conventional fuzzy neural networks. The classification capabilities of the GA evolved network is evaluated on Ripley's synthetic two class data classification problem [16].

The organization of this paper is as follows: Section 2 describes the subsethood product fuzzy neural network, Section 3 details the simple genetic algorithm used, Section 4 presents the simulation results and Section 5 concludes the paper.

2 Subsethood Product Fuzzy Neural Network

The subsethood-product fuzzy neural network architecture [14] shown in Fig. 1 embeds fuzzy rules of the form "If X is A_1 AND Y is A_2 then Z is B_1" in the standard way.

Input nodes in the network represent features, output nodes represent pattern classes. Since a hidden node represents a rule, each input-hidden connection represents a fuzzy rule antecedent, and each hidden-output connection represents a fuzzy rule consequent. A connection from node i to node j is modeled by a center c_{ij} and a spread σ_{ij} of a symmetric Gaussian fuzzy set. Thus the connection weights are represented as the center-spread pairs, $w_{ij} = (c_{ij}, \sigma_{ij})$. A numeric input is fuzzified by treating it as the center c_i of a Gaussian fuzzy set with spread σ_i.[1] Signal transmission along the input-hidden connections is computed

[1] Note the use of the single subscript for the signal and the double subscript for the weight.

σ^f_1 ... σ^f_n	$c^a_{11}\sigma^a_{11}$ ... $c^a_{ij}\sigma^a_{ij}$ ... $c^a_{nj}\sigma^a_{nj}$	$c^c_{11}\sigma^c_{11}$ ... $c^c_{qk}\sigma^c_{qk}$ ... $c^c_{qp}\sigma^c_{qp}$
Feature spreads	Antecedents	Consequents

Fig. 2. String representation of feature spreads and fuzzy weights

using the fuzzy mutual subsethood measure [17] which quantifies the net signal transmission as the extent of overlap between the fuzzy signals and the fuzzy weights. Symbolically, the mutual subsethood between an incoming fuzzy signal $s_i = (c_i, \sigma_i)$ and a fuzzy weight $w_{ij} = (c_{ij}, \sigma_{ij})$ is defined by:

$$\mathcal{E}_{ij} = \mathcal{E}(s_i, w_{ij}) = \text{Degree}(s_i \subseteq w_{ij} \text{ and } w_{ij} \subseteq s_i) \tag{1}$$

$$= \frac{\mathcal{C}(s_i \cap w_{ij})}{\mathcal{C}(s_i) + \mathcal{C}(w_{ij}) - \mathcal{C}(s_i \cap w_{ij})} . \tag{2}$$

Here, $\mathcal{C}(\cdot)$ denotes the cardinality of a fuzzy set. This is defined for a Gaussian fuzzy set A with center c and spread σ as $\mathcal{C}(A) = \int_{-\infty}^{\infty} e^{-\left(\frac{x-c}{\sigma}\right)^2} dx$. $\mathcal{E}_{ij}$ can have values in the interval [0, 1] depending upon the relative values of centers and spreads of s_i and w_{ij}.

A subsethood based product aggregation operator at the rule nodes aggregates all the transmitted signals, to compute the strength of firing of a rule, z_j as a *fuzzy inner product*:

$$z_j = \prod_{i=1}^{n} \mathcal{E}_{ij} . \tag{3}$$

Volume defuzzification [17] is employed to generate the output:

$$y_k = \frac{\sum_{j=1}^{q} z_j c_{jk} \sigma_{jk}}{\sum_{j=1}^{q} z_j \sigma_{jk}} \tag{4}$$

where the activation (and signal) of the output node k is y_k and q is the number of rule nodes, and (c_{jk}, σ_{jk}) are respectively centers and spreads of consequent fuzzy weights. Equation (4) essentially computes a convex sum of consequent set centers. In [14] and [15] the subsethood product network learns antecedent and consequent fuzzy weights using gradient descent.

3 GA Representation and Operators

In this paper we use a genetic algorithm to evolve all the trainable parameters (feature spreads, antecedent and consequent fuzzy weights) of the network. The centers and the spreads to be evolved are represented as bits in a binary string. The fuzzy weights of all the connections that fan-in to each rule node are concatenated with those of the output nodes.

Fig. 2 shows the string representation of the parameters of the network in which σ^f_i denotes the spread for the ith feature, c^a_{ij} σ^a_{ij} represents the antecedent fuzzy weight between input node i and rule node j, and c^c_{qk} σ^c_{qk} represents the consequent fuzzy weight between rule node q and output node k. With the centers and the spreads respectively represented by sub-strings of length l and m bits, the length of each chromosome becomes $(m \times n) + (l + m)[(n \times q) + (q \times p)]$, where n, q, and p respectively denote the number of input, rule and output nodes in the network. The tournament selection method with tournament size 2 was used to select the chromosomes for subsequent generations. Two copies of the best chromosome were always preserved in the population (using an *elitist* strategy [18]). In the experiments presented single-point crossover was employed with crossover probability $p_c = 1.0$ and mutation probability was $p_m = 0.01$. The disruptive nature of the selected crossover operator is balanced by the elitist selection used in this paper.

4 Simulation Results: Ripley's Synthetic Two Class Data

4.1 Description of Training and Testing Data

We trained and tested the network on Ripley's synthetic two class data available from `http://markov.stats.ox.ac.uk/pub/PRNN`. This data set comprises patterns having two features divided into two classes. Each class has a bimodal distribution generated from equal mixtures of Gaussian distributions with identical covariance matrices [16]. The class distributions have been chosen to allow a Bayesian classifier error rate of 8%. The training set consists of 250 patterns with 125 patterns in each class. The test set consists of 1000 patterns with 500 patterns in each class. To demonstrate the high level of parameter economy and the classification capabilities of the model, a minimal network with two input nodes, two rule nodes, and two class nodes was considered for the experiments presented.

Experiment I: Evolution of fuzzy weights with fixed feature spread. The centers and the spreads were represented by 16 and 12 bits respectively, and with corresponding ranges $(-1.5, 1.5)$ and $(0.05, 0.9)$. A fixed spread of 0.5 was used to fuzzify each input feature. The GA was applied with a population of 500 individuals each of length 224 bits. The network gave an 8.2% test error rate.

Experiment II: Evolution of fuzzy weights with variable feature spreads. In this experiment, the feature spreads were also allowed to evolve in the same range as other spreads, i.e., in the range $(0.05, 0.9)$. Using a population size of 800, the network gave an 8% test error rate, at par with the error rate of the Bayes optimal classifier. The final GA evolved input feature spreads were 0.3524 and 0.1887 respectively. The class separating boundaries learnt by the network for the experiment is shown in Fig. 3 alongwith the Bayesian decision boundary.

Table 1 shows the evolved weights of the network for both the experiments. Connections are depicted from source nodes (rows) to destination nodes (columns). For the 2-2-2 network architecture input nodes are denoted I1, I2; rule nodes R1, R2; and output nodes O1, O2.

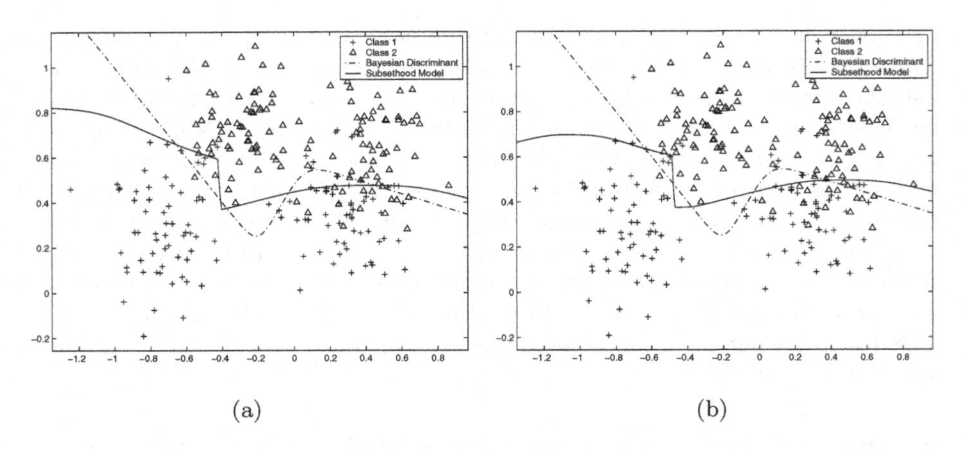

Fig. 3. Class separating boundary learnt by the evolutionary subsethood network along with the Bayesian decision boundary for Ripley's data (a) 8.2% error boundary for fixed feature spread, (b) 8.0% error boundary for evolvable feature spreads

Table 1. The centers and the spreads of the network for Experiments I and II where I, R, O denote input, rule and output nodes

Experiment I: Fixed feature spread–Error rate 8.2%

	R1	**R2**		**O1**	**O2**
I1	(−1.2824, 0.1337)	(0.4104, 0.8618)	**R1**	(−1.3102, 0.5594)	(0.5102, 0.7950)
I2	(1.3956, 0.2742)	(−0.1136, 0.3454)	**R2**	(1.0322, 0.6021)	(0.4989, 0.7960)

Experiment II: GA evolved feature spreads–Error rate 8.0%

	R1	**R2**		**O1**	**O2**
I1	(−1.0232, 0.0982)	(0.4696, 0.6075)	**R1**	(−0.3911, 0.8653)	(1.0956, 0.1029)
I2	(1.3332, 0.3379)	(−0.3759, 0.4363)	**R2**	(0.2398, 0.6416)	(−0.5588, 0.1519)

A comparison with the results of other classification methods (adapted from [19]) is presented in Table 2. The subsethood product model [14] which employs gradient descent learning gave test error rates of 9.3% and 8.1% for 2 and 6 rules respectively. In contrast, the evolutionary version of the model with two rules yields an error rate of 8.0%—outperforming all other models, and at par with the Bayes optimal classifier. This demonstrates the success of evolutionary methods in searching a set of optimal parameters for the fuzzy neural network.

5 Discussion and Conclusions

Ripley's synthetic two class data is designed to have an error rate of 8% with the Bayes optimal classifier. We have shown that this error rate is also achievable with our subsethood product network which outperforms all other models as reported in Table 2, including an earlier gradient descent version of the model [14,15]. As shown in this paper, GA based learning can improve solutions fur-

Table 2. Comparison of performance of various classification methods

Classification Methods	% Test error
Linear discriminant	10.8
Logistic discriminant	11.4
Quadratic discriminant	10.2
1-nearest neighbor	15.0
3-nearest neighbor	13.4
5-nearest neighbor	13.0
MLP with 3 hidden nodes	11.1
MLP with 3 hidden nodes (weight decay)	9.4
MLP with 6 hidden nodes (weight decay)	9.5
LVQ(12 centers)	9.5
CTM(4 units)	8.1
Subsethood Product Model (fixed feature spreads, 2 rules)	9.3
Subsethood Product Model (fixed feature spreads, 6 rules)	8.1
Evolutionary Subsethood Model (fixed feature spreads, 2 rules)	8.2
Evolutionary Subsethood Model (variable feature spreads, 2 rules)	8.0

ther. In addition, a high level of parameter economy is demonstrated by the fact that this low error rate is achievable with only two fuzzy rules. This kind of economy has been observed in other applications as well [15]. We feel that the economy of subsethood-product network can be exploited even further by embedding an effective connection and node pruning strategy into the learning procedure, coupled with GA based optimization of feature spreads and connections. This should help improve the performance of the network and thus realize the full capabilites of the model. Future work will involve this.

References

1. Mitra, S., Hayashi, Y.: Neuro-fuzzy rule generation: survey in soft computing framework. IEEE Transactions on Neural Networks **11** (May 2000) 748–768
2. Kasabov, N.: Neuro-fuzzy techniques for intelligent processing. Physica Verlag (1999)
3. Fuller, R.: Introduction to neuro-fuzzy systems. Heidelberg New York: Physica-Verlag (2000)
4. Lin, C., Lee, C.S.G.: Neural fuzzy systems: A neuro-fuzzy synergism to intelligent systems. Upper Saddle River NJ: Prentice Hall P T R (1996)
5. Bezdek, J.C., Keller, K., Krishnapuram, R., Pal, N.R.: Fuzzy models and algorithms for pattern recognition and image processing. Kluwer Academic Press (1999)
6. Kim, J., Kasabov, N.: HyFIS: Adaptive neuro-fuzzy inference systems and their application to nonlinear dynamical systems. IEEE Transactions on Neural Networks **12**(9) (1999) 1301–1321
7. Mitra, S., Pal, S.: Fuzzy multi-layer perceptron, inferencing and rule generation. IEEE Transactions on Neural Networks (1995)
8. Belarbi, K., Titel, F.: Genetic algorithm for the design of a class of fuzzy controllers: An alternative approach. IEEE Transactions on Fuzzy Systems **8** (August 2000) 398–405

9. Pujol, J.C.F., Poli, R.: Evolving the topology and the weights of neural networks using a dual representation. Applied Intelligence **8**(1) (1998) 73–84
10. Russo, M.: FuGeNeSys - A fuzzy genetic neural system for fuzzy modeling. IEEE Transactions on Fuzzy Systems **6** (August 1998) 373–388
11. Russo, M.: Genetic fuzzy learning. IEEE Transactions on Evolutionary Computation **4** (September 2000) 259–273
12. Russo, M.: Distributed fuzzy learning using the MULTISOFT machine. IEEE Transactions on Neural Networks **12**(3) (May 2001) 475–484
13. Paul, S., Kumar, S.: Adaptive rule-based linguistic networks for function approximation. 4th International Conference on Advances in Pattern Recognition and Digital Techniques N. R. Pal, A. K. De, and J. Das, eds., Indian Statistical Institute Calcutta Narosa Publishing House (December 1999) 246–250
14. Paul, S., Kumar, S.: Subsethood based adaptive linguistic networks for pattern classification. IEEE Transactions on Systems, Man and Cybernetics, to appear.
15. Paul, S., Kumar, S.: Fuzzy neural inference system using mutual subsethood products with applications in medical diagnosis and control. Proceedings IEEE 10th IEEE International Conference on Fuzzy Systems (FUZZ-IEEE) Melbourne, Australia, December 2-5 2001
16. Ripley, B.: Pattern recognition and neural networks. Cambridge University Press (1996)
17. Kosko, B.: Fuzzy engineering Englewood Cliffs: Prentice Hall (1997)
18. Goldberg, D. E.: Genetic algorithms in search, optimization and machine learning Reading, MA: Addison Wesley (1989)
19. Cherkassky, V., Mulier, F.: Learning from data: concepts, theory, and methods. John Wiley and Sons (1998)

VSS Learning Based Intelligent Control of a Bioreactor System

Ugur Yildiran[1] and Okyay Kaynak[1,2]

[1]UNESCO Chair on Mechatronics,
Bogaziçi University, 80815, Bebek, Istanbul, Turkey
[2]Aeronautical and Space Technologies Institute, Air Force Academy,
Yesilkoy, Istanbul, Turkey
{yildirau,kaynak}@boun.edu.tr

Abstract. Intelligent control of a class of biochemical reactor systems is discussed. For this purpose, an Adaptive Neuro-fuzzy Inference System (ANFIS) architecture, whose parameters are updated using a VSS based learning algorithm, is utilized as controller. Incorporation of this new update mechanism guarantees stability and robustness of the learning dynamics under the existence of parametric and nonparametric uncertainties. Furthermore, the proposed approach does not require measurement of cell concentration that is very hard to do in practical applications. Simulation results presented demonstrate the efficacy of the control architecture.

1 Introduction

In analysis and control of real world problems, classical mathematical tools with their precise nature often result in complicated solutions. This problem is somewhat alleviated by the recent developments in technology. The substantial increase in the available computing power allows the implementation of complex algorithms. Furthermore, a novel approach to computation, named soft computing, can provide more efficient solutions without the necessity of using complicated methods of analysis. Soft computing methodologies are especially promising with their capabilities such as learning from available samples, adapting to changes in the environment and dealing with imprecision. Numerous studies have appeared in the literature that utilize the possibilities offered by soft computing methodologies in a wide range of control applications. However, most of them suffer from a disadvantage, being based on techniques like Steepest Descent and Levenberg-Marquardt for training. These techniques cannot guarantee stability for learning dynamics. The presence of noise and uncertainties in real control applications reduces stability further. A solution to this problem, which is first proposed by Sira-Ramirez et al. [1], is the use of a robust control method known as Sliding Mode Control (SMC) to stabilize the learning dynamics. In this work, the zero error level at the output of an Adaptive Linear Element (ADALINE) network is utilized as the sliding line and the output of the system is maintained on this line after a finite reaching phase. Yu et al. [2], develops results of [1] further by introducing adaptive uncertainty bound dynamics. Recently Efe [3] has introduced another method to robustify gradient-based

N.R. Pal and M. Sugeno (Eds.): AFSS 2002, LNAI 2275, pp. 281–287, 2002.

learning algorithms and shown the equivalence between his approach and the one proposed in [1].

Soft computing methodologies with the capabilities discussed above are good candidates for process control applications, which inherently entail strong nonlinearities and uncertainties. The capability of learning from past input-output data samples of a system makes them especially useful for applications in a wide spectrum of industrial processes. This paper discusses such an application, namely the direct adaptation of the parameters of the control system of a biochemical reactor.

Organization of this paper is as follows. The second section describes the control problem and gives the governing equation of the bioreactor process. ANFIS structure and the learning algorithm are introduced in the third and fourth sections respectively. In the fifth section, control of bioreactor system is investigated and simulation results are evaluated. A discussion on the result is given in the last section.

2 Bioreactor System

The control objective in a bioreactor process is to maintain the substrate and the cell concentrations at some desired levels. Any transient stage must be kept as short as possible because undesirable by-product formation can occur while the substrate concentration is above a certain limiting value. There are several methods proposed in the literature in this direction [4], but most of them suffer from the disadvantages such as nonrealistic modeling assumptions and the need for both substrate and cell concentration measurements. The approach proposed in this paper achieves control objectives while requiring the measurement of only the substrate concentration.

The mathematical model of the process based on Haldane growth kinetics may be given as below [4].

$$\dot{c} = \mu(s,c)c - c\tau, \quad c(0) > 0 \tag{1}$$

$$\dot{s} = -\frac{\mu(s,c)}{Y} + (S_F - s)\tau, \quad s(0) > 0 \tag{2}$$

$$\mu(s,c) = \mu_0 s / (K_S + s + \frac{s^2}{K_I}) \tag{3}$$

Here c and s denote cell and substrate concentrations respectively. μ is the growth model of the process, τ denotes dilution rate, Y denotes yield coefficient and S_F denotes influent substrate concentration. Parameters of the growth model are, μ_0, K_s and K_I.

3 Adaptive Neuro-fuzzy Inference System

Today, it is well known that there is a strong relationship between fuzzy inference systems and artificial neural networks. This relationship allows designers to represent a fuzzy inference system as a multi-layered network, and thus, exploit benefits of both structures. While NN representation enables the use of gradient-based optimization techniques, fuzzy reasoning mechanism can utilize available expert knowledge to reduce the training time. Adaptive Neuro-fuzzy Inference System (ANFIS) is one of the proposed structures in this direction. The rule base of an ANFIS with first order Sugeno model has the from of

$$\text{If } x \text{ is } A_i \text{ then } f_i = \Phi_{i1} x + \Phi_{i2}$$

for a single input architecture. This structure is shown in Figure 2 in graphical form and its mathematical representation is given below.

$$\tau = \underline{f}^T(x)\underline{u}^n(x) \tag{4}$$

$$\underline{f}_i(x) = \phi_{i1}x + \phi_{i2} \tag{5}$$

$$u_i^n(x) = u_i(x) / (\sum_{j=1}^{R} u_j(x)) \tag{6}$$

In above, u^n represents normalized firing strength vector which is calculated using firing strengths given by

$$u_i(x) = \left\{1 + \{(x - \mu_i)/\rho_i\}^{2b_i}\right\}^{-1} \tag{7}$$

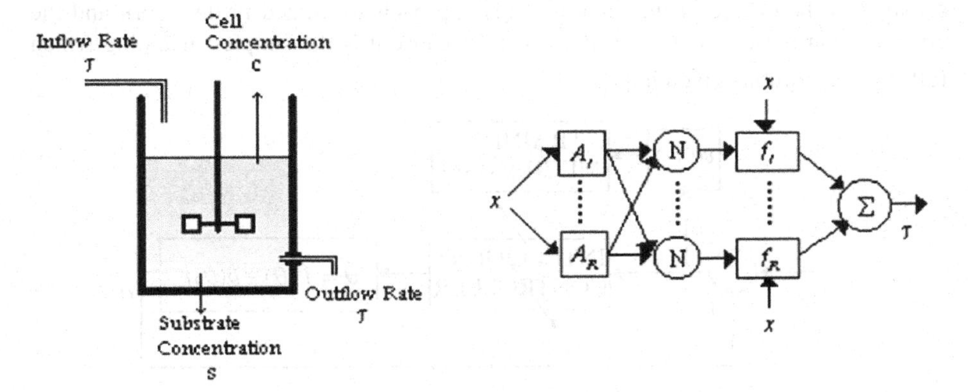

Fig. 1. Bioreactor system

Fig. 2. Basic ANFIS Structure

4 Learning Algorithm

The convergence of the learning algorithm is one of the most critical issues in intelligent control applications. It is well known that gradient-based parameter update methods may result in unstable learning dynamics. Recently some novel techniques, based on Variable Structure Systems (VSS) theory, have been proposed to overcome this difficulty. The following parameter adaptation mechanism, which utilizes SMC approach for training of flexible structures, is used in this work.

Consider the flexible structure, described by

$$\tau = \underline{\phi}^T \underline{u} \tag{8}$$

where $\underline{\phi}$ denotes the parameter vector and $\underline{u}$ denotes the input vector of the output layer of the structure. The parameter update mechanism, described as

$$\dot{\underline{\phi}} = -\frac{\underline{u}}{\underline{u}^T \underline{u}} K \operatorname{sgn}(s_c) \tag{9}$$

enforces the flexible structure to the sliding surface (s_c), which is defined as the learning-error level at the output of the controller, provided that K, uncertainty bound parameter, is chosen sufficiently large. In mathematical terms, the sliding surface s_c can be given as

$$s_c = \tau - \tau_d \tag{10}$$

where τ_d is the desired control sequence. Mathematical proof and the criterion for choice of the constant K can be found in [1].

Although, the adaptation mechanism given by (5) achieves parametric convergence, it requires a measure of s_c which is not available in control applications because τ_d is not available. To alleviate this problem, Efe [1] has assumed a mathematical relation between the sliding line at the output of the plant (s_p) and the learning-error level of the flexible controller and investigated conditions that must be satisfied in the choice of the relation. This approach is utilized in this work and the error relation is chosen as $s_c = \Psi(s_p) = s_p$ because it is the simplest mapping which fulfill the conditions given in [1].

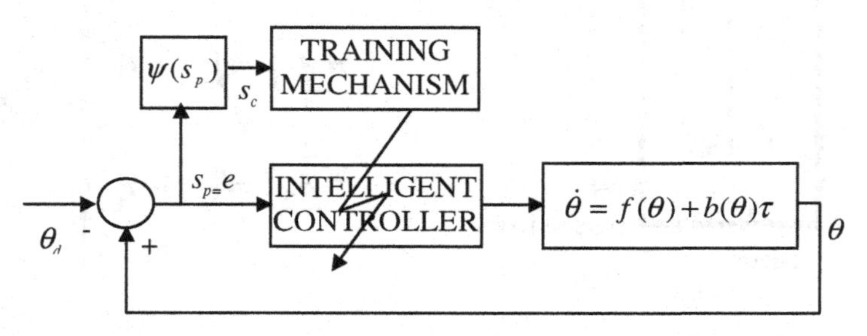

Fig. 3. Intelligent control sceheme

The overall structure of a control scheme, which uses the adaptation mechanism given in (5)-(6) for a first order nonlinear system, is given in Figure 3. If ANFIS structure, desribed by (7)-(10), is used in this system, one can come up with the following weight update rules.

$$\dot{\phi}_{i1} = -\frac{e u_i^n(e)}{(e^2+1)(\underline{u}^n(e))^T \underline{u}^n(e)} K \operatorname{sgn}(s_c) = -\frac{u_i^n(e)}{(e^2+1)(\underline{u}^n(e))^T \underline{u}^n(e)} K|e| \tag{11}$$

$$\dot{\phi}_{i2} = -\frac{u_i^n(e)}{(e^2+1)(\underline{u}^n(e))^T \underline{u}^n(e)} K \operatorname{sgn}(s_c) = -\frac{u_i^n(e)}{(e^2+1)(\underline{u}^n(e))^T \underline{u}^n(e)} K \operatorname{sgn}(e) \tag{12}$$

Close investigation of (7) reveals the fact that this term is negative and decreases continuously as long as $e \neq 0$. In order to eliminate this parameter drift problem, the part of the adaptation rule given in (7) is switched off when the absolute value of error between the desired output and actual output is below a certain value, which can be determined by the system designer.

5 Control of Bioreactor System

The control strategy shown in Figure 3 is used for control of bioreactor process. The dilution rate is utilized as the control input and substrate concentration is chosen as the state to be controlled. The target output is set to s=0.2 [g/l], and initial states of the plant are chosen as c_0=10 [g/l], s_0=10 [g/l]. The initial parameters of the controller are set using a priori knowledge which is acquired following a trail and error procedure. Simulations are performed in MATLAB 5.3 environment and step size is chosen as 0.001 hours. During the simulations, it is assumed that the parameters of the growth model possesses time varying behavior described by the following equations while the yield coefficient and influent substrate concentration are kept at constant levels given as Y=0.5 [g cells/g substrate], S_F=200 [g/l].

$$\mu_0 = 0.35 + 0.15\cos(2\pi t/10)\ [1/h] \tag{13}$$

$$K_S = 0.105 + 0.095\sin(2\pi t/15 + 3\pi/2)[g/l] \tag{14}$$

$$K_I = 5 + 4.975\cos(2\pi t/25)[g/l] \tag{15}$$

In order to not to be in conflict with practical reality, the output measurement is subjected to Gaussian noise with zero mean and variance equal to 1.64e-6. To smooth out chattering, the sgn function is replaced by the following approximating function

$$\text{sgn}(s_c) \approx \frac{s_c}{|s_c| + \delta} \tag{16}$$

where δ=0.005. Furthermore, the uncertainty bound in the parameter update algorithm is chosen as K=3 and the update term given in (7) is switched off when the absolute value of the substrate concentration error is below 0.1 g/l. Lastly, output of the controller is passed through a saturation function because dilution rate cannot take negative values.

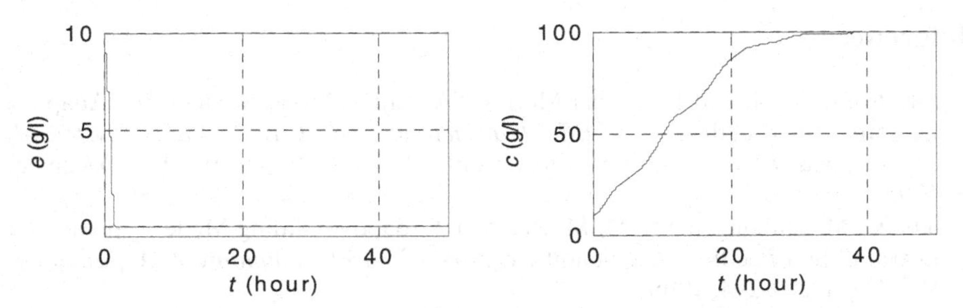

Fig. 4. Substrate concentration error and cell concentration

As can be seen from Figure 4, the substrate concentration error reaches zero level very rapidly. This is especially important because undesirable by-product formation may occur while the substrate concentration is above a limiting value. Moreover, the cell concentration enters the steady state regime in 40 hours as depicted in the same figure. By investigating Figure 5, it can be seen that there is no chattering on applied control signal because of sign function smoothing in the vicinity of the decision boundary characterized by e=0. Furthermore, as Figure 6 suggests adjustable parameters of the controller remains bounded.

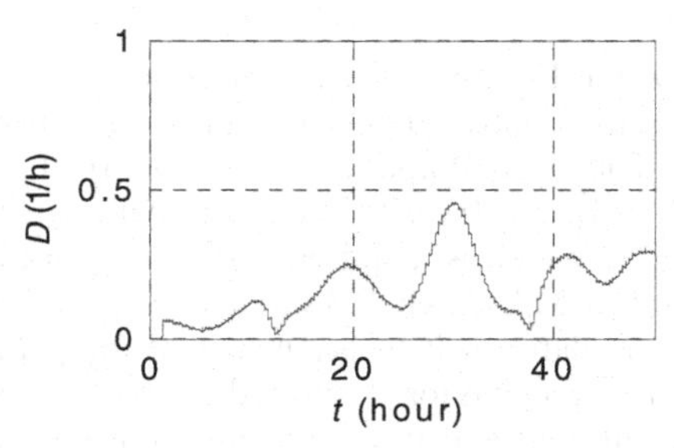

Fig. 5. Dilution rate

6 Conclusions

In this paper, the control of a class of bioreactor systems is investigated. An ANFIS structure having five rules is utilized as intelligent controller. Different from the classical gradient based methods, adaptation algorithm of the controller utilizes SMC strategy to guarantee bounded evolution of the controller parameters. Simulation results show that transient period is short and system output is maintained at the desired level in steady state. Another advantage of the proposed method is that it requires only the measurement of the substrate concentration.

Acknowledgement

This work is supported by TUBITAK (Project no: 100E042).

References

1. Sira-Ramirez, H. and E. Colina-Morles, “A sliding Mode Strategy for Adaptive Learning in Adalines”, *IEEE Transactions on Circuits and Systems-I: Fundamental Theory and Applications*, Vol. 42, No. 12, pp. 1001-1012, December 1995.
2. Yu, X., M. Zhihong and S. M. M. Rahman, “Adaptive Sliding Mode Approach for Learning in a Feedforward Neural Network”, *Neural Computing & Applications*, Vol. 7, pp. 289-294, 1998.
3. M.Ö. Efe, “*Variable Structure Systems Theory Based Training Strategy for Computationally Intelligent Systems*”, Ph.D. Thesis, Boğaziçi University, 2000.
4. Boskovic, J.D., “*Novel Feeding Strategy for Adaptive Control of Bioreactor Processes*”, J. Proc. Cont., Vol. 7, No. 3, pp. 209-217, 1996.

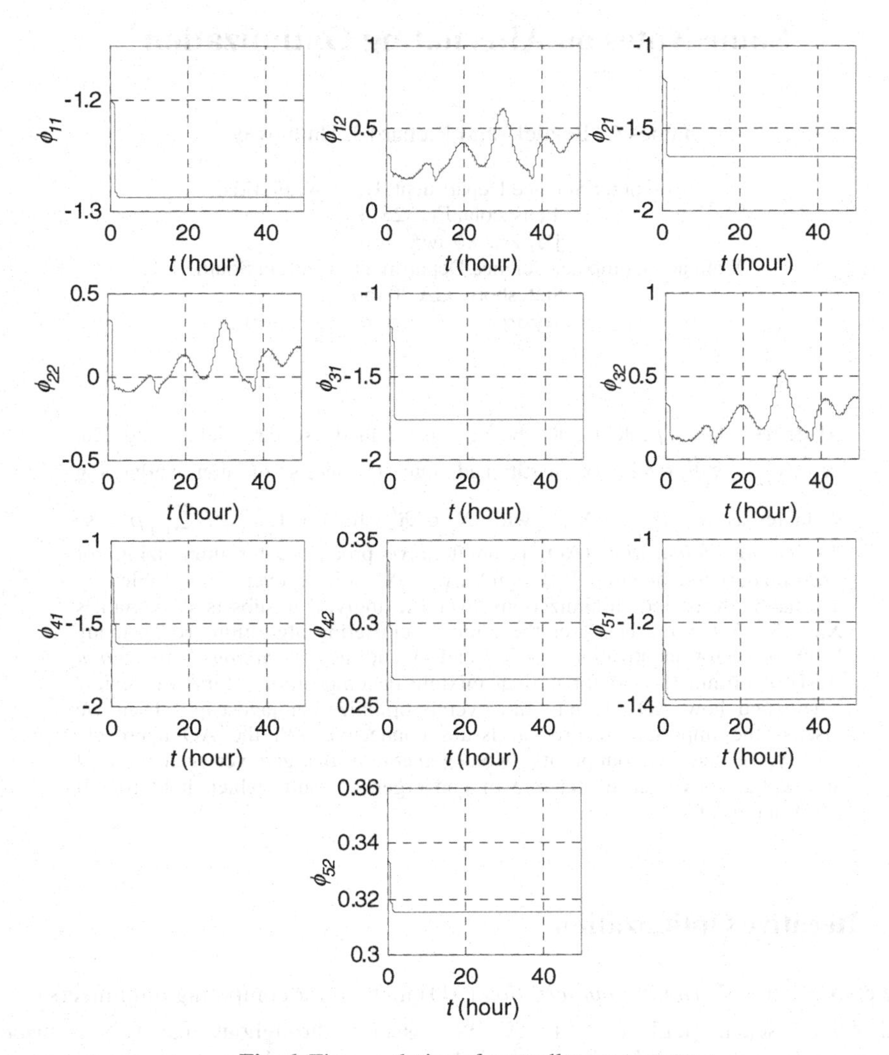

Fig. 6. Time evolution of controller parameters

Some Notes on Alternating Optimization*

James C. Bezdek[1] and Richard J. Hathaway[2]

[1] Computer Science Department, U. of W. Florida,
Pensacola, FL 32514
jbezdek@uwf.edu
[2] Math and Computer Science Department , Georgia Southern U.,
Statesboro, GA 30460
hathaway@gsaix2.cc.gasou.edu

Abstract. Let $f:\Re^s \mapsto \Re$ be a real-valued scalar field, and let $x=(x_1,\ldots,x_s)^T \in \Re^s$ be partitioned into t subsets of non-overlapping variables as $x=(X_1,\ldots,X_t)^T$, with $X_i \in \Re^{p_i}$, for i = 1, …, t, $\sum_{i=1}^{t} p_i = s$. *Alternating optimization* (AO) is an iterative procedure for minimizing (or maximizing) the function $f(x) = f(X_1,X_2,\ldots,X_t)$ jointly over all variables by alternating restricted minimizations over the individual subsets of variables $X_1,\ldots,X_t$. AO is the basis for the c-means clustering algorithms (t=2), many forms of vector quantization (t = 2, 3 and 4), and the *expectation-maximization* (EM) algorithm (t = 4) for normal mixture decomposition. First we review where and how AO fits into the overall optimization landscape. Then we discuss the important theoretical issues connected with the AO approach. Finally, we state (without proofs) two new theorems that give very general local and global convergence and rate of convergence results which hold for all partitionings of x.

1. Iterative Optimization

We consider the *alternating optimization* (AO) method for computing minimizers of a real valued scalar field $f:\Re^s \mapsto \Re$. We assume throughout that f is a *twice continuously differentiable* function of the vector variable x. Our discussion is restricted to minimization, but the theory is equally applicable to the maximization problem. Our presentation begins with a definition of the *nonlinearly constrained optimization problem* (NCOP):

$$\min_{x \in \Re^s} \{f(x)\}, \text{ subject to} \tag{NCOP.1}$$

$$c_i(x) = 0, \quad i = 1,\ldots,k; \text{ and} \tag{NCOP.2}$$

$$c_i(x) \geq 0, \quad i = k+1,\ldots,m. \tag{NCOP.3}$$

* Research supported by ONR Grant 00014-96-1-0642.

N.R. Pal and M. Sugeno (Eds.): AFSS 2002, LNAI 2275, pp. 288–300, 2002.

The $\{c_i(x)\}$ are *constraint functions* and f is the *objective function* of NCOP; together, these are called the *problem functions*. The set of points $\Psi \subseteq \Re^s$ that satisfy all of the constraints at (NCOP.2) and (NCOP.3) is the *feasible region* for NCOP. With Ψ specified, we can replace (NCOP.1-NCOP.3) with the more compact form

$$\min_{x \in \Psi \subseteq \Re^s} \{f(x)\}. \qquad \text{(NCOP)}$$

Only feasible points can be optimal, and the optimality of a *solution* x^* of NCOP depends on its relationship to its neighbors. The point x^* is called a *strong local minimizer* when $f(x^*) < f(x)$ for all $x \neq x^*$ in a neighborhood $N(x^*, \delta) \subset \Psi$. The local minimizer is *weak* when inequality is not strict. A strong or weak local minimizer is a *global* minimizer when the inequality holds for all $x \in \Psi$. Unless qualified otherwise, a "solution" of (NCOP) means a global solution. Generally it is very difficult to find a global minimizer of f. However, local solutions of (NCOP) are usually satisfactory; and as we shall see, are often introduced as an artifact of building iterative optimization algorithms to solve (NCOP) from its *first order necessary conditions*, or FONCs.

The simple idea underlying AO is to replace the sometimes difficult joint optimization of f over all s variables with a sequence of easier optimizations involving grouped subsets of the variables. The A0 approach begins with a partitioning of $x = (x_1, \ldots, x_s)^T \in \Re^s$ into t subsets of non-overlapping variables as $x = (X_1, \ldots, X_t)^T$, with $X_i \in \Re^{p_i}$ for i = 1, …, t, $\sum_{i=1}^t p_i = s$. For example, suppose s = 8. The vector variable x might be partitioned into two subsets as $X_1 = (x_1, x_4, x_7)^T$; $X_2 = (x_2, x_3, x_5, x_6, x_8)^T$; or four subsets as $X_1 = (x_4, x_6)^T$; $X_2 = (x_1, x_2)^T$, $X_3 = (x_3, x_7)^T$; $X_4 = (x_5, x_8)^T$; and so on. For any reasonable value of s, there are a lot of ways to partition x, so the interesting question of how *best* to choose these subsets is an important issue.

We list a very few examples of alternating optimization that appear in the clustering and mixture decomposition literature: *hard c-means* [1]; the *expectation-maximization* algorithm for estimating the parameters in a mixture of normal distributions [2]; *fuzzy c-means* [3]; *fuzzy c- varieties* [4]; *assignment-prototype* (AP) clustering [5]; *fuzzy c-shells* [6]; *fuzzy c-regression models* [7]; *possibilistic c-means* [8]; and *optimal completion strategy* fuzzy c-means [9].

After a partitioning of x is chosen, AO attempts to minimize the function f(x) = $f(X_1, X_2, \ldots, X_t)$ jointly over all variables by alternating restricted minimizations over the individual sets of vector variables $X_1, \ldots, X_t$. Specifically, AO defines an iterate sequence $\{(X_1^{(r)}, X_2^{(r)}, \ldots, X_t^{(r)})$: r = 0, 1,…} that begins at an initial iterate $(X_1^{(0)}, X_2^{(0)}, \ldots, X_t^{(0)})$ via a sequence of restricted minimizations of the form

$$\min_{X_i \in \Psi_i \subset R^{p_i}} \left\{ f(\not{X}_1^{(r+1)}, \ldots, \not{X}_{i-1}^{(r+1)}, X_i, \not{X}_{i+1}^{(r)}, \ldots, \not{X}_t^{(r)}) \right\}, \; i = 1, \ldots, t, \qquad \text{(AO)}$$

where $\{\Psi_i \subset \Re^{p_i}\}$ are the sets over which the (global) restricted optimizations are done[1]. The *strikethrough notation* ($\not{X}$) in (AO) indicates that these vectors are *fixed*

[1] We use NCOP and AO without parentheses as acronyms; (AO) and (NCOP) with parentheses stand for the equations shown here.

with respect to the current *subproblem* at index i. The values of the variables $X_1,\ldots,X_t$ are successively updated via (AO) for each r as i runs from 1 to t until either a maximum number of iterations on r is reached, or there is sufficiently little change between successive iterates. Notice that when t = 1, (AO) reduces to (NCOP), and there are no subproblems as in (AO). Hence, NCOP is the special case of AO with t = 1 where the optimization is done jointly (or *directly*) in the original variables.

If everything goes just right, a termination point $(X_1^*, X_2^*, \ldots, X_t^*) \in \Re^s$ of AO will be a solution of NCOP. Unfortunately, things can go wrong. To see this, let

$$S_{NCOP} = \{x^* \in \Re^s : x^* \text{ solves (NCOP)}\} \quad \text{, and}$$

$$S_{AO} = \{x^* = (X_1^*, X_k^*, \ldots, X_t^*) \in \Re^s : x^* \text{ solves (AO)}\}$$

If we put $\Psi = \Psi_1 \times \ldots \times \Psi_t$, iteratively solving (for r = 0,1,...) the t subproblems in (AO) is sometimes simpler and equivalent in practice to solving (NCOP) directly. Our hope is, of course, that a termination point of AO solves (NCOP), i.e., that $x^* = (X_1^*, \ldots, X_t^*)^T \in S_{NCOP}$. However, we make two sacrifices when we solve (AO) instead of (NCOP): first, we may find purely local solutions to (NCOP); and second, we may sacrifice joint optimality for a weaker result given by the global convergence theory of Section 6.

Theorem 1 shows that S_{NCOP} *and* S_{AO} are *not* necessarily equal. Any x in S_{NCOP} must also be in S_{AO}; but there could be points in S_{AO} that are not in S_{NCOP}. We illustrate this in Example 1, and then give a simple argument that establishes the general case.

Example 1. Consider the unconstrained minimization of $f(x,y) = x^2 - 3xy + y^2$ over $\Re^2$ by AO using the partitioning $X_1 = x$; $X_2 = y$. For this function, S_{NCOP} is empty. To see this, consider the restriction of f(x,y) to the line y = x, which gives g(x) = $f(x,x) = -x^2$. Clearly g has no minimizer, so neither does f. The function f does have a saddle point at $(0,0)^T$, but this point is NOT in S_{NCOP}. On the other hand, $(0,0)^T \in S_{AO}$, since the restricted (global) minimizer (over x) of $f(x,0) = x^2 - 3x0 + 0^2 = x^2$ is x = 0, and the restricted (global) minimizer (over y) of $f(0,y) = 0^2 - 3(0)y + y^2 = y^2$ is also y = 0. Since S_{NCOP} is empty and S_{AO} is not, we have $S_{NCOP} \subset S_{AO}$ for this example. ∎

To see that S_{NCOP} is *always* a subset of S_{AO}, we prove an even more general result, which answers the question: how does the partitioning of x affect the solution set for (AO)? Stated roughly, if partitioning $\hat{P}$ is obtained from partitioning P by splitting one or more (P partitioning) vector variables, then all solutions for partitioning P are inherited by partitioning $\hat{P}$, and $\hat{P}$ may have more solutions of its own.

Theorem 1. Let t satisfy $1 \le t \le s$; let P denote a partitioning of $x \in R^s$ into $(X_1,\ldots,X_t)^T$; let $\hat{P}$ denote the refined partitioning of x into $(\hat{X}_1, \ldots, (\hat{X}_i, \hat{X}_k), \ldots, \hat{X}_t)$, obtained by splitting one vector variable, say X_i in P, into two parts, $X_i = (\hat{X}_i, \hat{X}_k)$. Now let S_{AO} and $\hat{S}_{AO}$ denote the sets of points in $\Re^s$ solving (AO) for the partitionings P and $\hat{P}$. Then

$$S_{AO} \subseteq \hat{S}_{AO}. \tag{T1}$$

Proof: Let $x^* = (X_1^*, \ldots, X_{i-1}^*, X_i^*, X_{i+1}^*, \ldots, X_t^*) \in S_{AO}$, Let $X_i^* = (\hat{X}_i^*, \hat{X}_k^*)$, where k is an index inserted between i and i+1 in the original partitioning of x. Since $X_i^* = \underset{X_i \in \Psi_i \subset R^{p_i}}{\arg\min} \left\{ f(X_1^*, \ldots, X_{i-1}^*, (\hat{X}_i^*, \hat{X}_k^*), X_{i+1}^*, \ldots, X_t^*) \right\}$, it follows that

$$\text{(i) } \hat{X}_i^* = \underset{\hat{X}_i \in \hat{\Psi}_i \subset R^{\hat{p}_i}}{\arg\min} \left\{ f(\hat{X}_1^*, \ldots, \hat{X}_{i-1}^*, \hat{X}_i, \hat{X}_k^*, \hat{X}_{i+1}^*, \ldots, \hat{X}_t^*) \right\}; \text{ and}$$

$$\text{(ii) } \hat{X}_k^* = \underset{\hat{X}_k \in \hat{\Psi}_k \subset R^{p_k}}{\arg\min} \left\{ f(\hat{X}_1^*, \ldots, \hat{X}_i^*, \hat{X}_k, \hat{X}_{i+1}^*, \ldots, \hat{X}_t^*) \right\},$$

for otherwise X_i could be altered from X_i^* to a different location so that one or both of conditions (i) or (ii) yielded a smaller function value, contradicting the assumption that X_i^* is a global minimizer of the restriction $f(X_1^*, \ldots, X_{i-1}^*, X_i, X_{i+1}^*, \ldots, X_t^*)$ over $X_i \in \Psi_i$. This is true for all i, which means that (AO) is satisfied, and hence, that x* is also in $\hat{S}_{AO}$. ■

Comparing (NCOP) at t = 1 to (AO) with t>1, (T1) immediately yields $S_{NCOP} \subseteq S_{AO}$. Since (AO) reduces to (NCOP) at t = 1, equality holds in (C1) at least in this case. It might be that the solution sets for (NCOP) and (AO) are equal when t > 1, but this must be verified for a specific instance. Next we turn to the questions; when are these sets empty? When are they singletons? Or when do they possess many solutions?

2. Existence and Uniqueness

Perhaps the most important mathematical questions associated with (NCOP) and (AO) are: *existence* (does any x^* exist?) and *uniqueness* (when x^* exists, is it unique?). It is impossible to resolve these two issues for a general instance of NCOP. Instead, our analysis will rely on an existence and uniqueness *assumption* for the subproblems comprising the general AO method. Specifically, the notation "=" used in (AO) implies a property of f, X_i and Ψ_i for i = 1,…,t, that we *assume* to be true in order to secure proofs (given in [10]) of the theorems stated in Sections 6 and 7.

Existence and Uniqueness (EU) Assumption. Let $\Psi_i \subseteq \Re^{p_i}$, i = 1,…,t; and let $\Psi = \Psi_1 \times \ldots \times \Psi_t$. Partition $x = (X_1, \ldots, X_t)^T$, $X_i \in \Re^{p_i}$, and let $g(X_i) = f(X_1, \ldots, X_{i-1}, X_i, X_{i+1}, \ldots, X_t)$, i = 1,…,t, If $x \in \Psi$, then

$$g(X_i) \text{ has a unique (global) minimizer for } X_i \in \Psi_i. \tag{EU}$$

The *EU assumption* is pretty strong. On the other hand, most instances of (NCOP) that have been analyzed do possess it. In any case, without these two properties, there can be no convergence theory. Because of the difficulty of establishing existence and uniqueness, many practitioners skip these two questions and proceed directly to the main issue for NCOP, which is – how to find (any) x^*? Our specification of how to search for a solution using AO follows.

3. The Alternating Optimization Algorithm

Here is a specification of the iteration procedure implied by (AO).

AO-1 Let $\Psi_i \subseteq \Re^{p_i}$, for i = 1,...,t, and let $\Psi = \Psi_1 \times \ldots \times \Psi_t$. Partition $x \in \Re^s$ as $x = (X_1, X_2, \ldots, X_t)^T$, with $X_i \in \Re^{p_i}$ for i = 1,...,t, $\bigcup_{i=1}^{t} X_i = X;\ X_i \cap X_j = \varnothing$ for $i \neq j$; and $s = \sum_{i=1}^{t} p_i$. Pick an initial iterate $x^{(0)} = (X_1^{(0)}, X_2^{(0)}, \ldots, X_t^{(0)})^T \in \Psi = \Psi_1 \times \ldots \times \Psi_t$, a vector norm $\|\cdot\|$, termination threshold ε, and iteration limit L. Set r = 0.

AO-2 For i = 1,..., t, compute the restricted minimizer

$$X_i^{(r+1)} = \underset{X_i \in \Psi_i \subset R^{p_i}}{\arg\min} \left\{ f(X_1^{(r+1)}, \ldots, X_{i-1}^{(r+1)}, X_i, X_{i+1}^{(r)}, \ldots, X_t^{(r)}) \right\} \tag{1}$$

AO-3 If $\| x^{(r+1)} - x^{(r)} \| \le \varepsilon$ or r > L, then quit; otherwise, set r = r+1 and go to AO-2.

The AO sequence is well defined if (EU) holds, which is typically true (but difficult to verify) in practice. When minimizers exist but are not unique, a simple tie-breaking strategy can be incorporated into (1) so that the AO sequence is still well defined. The heart of AO lies in finding the restricted minimizers at (1). We will discuss some strategies for solving these equations soon, but first, we give an example to illustrate the general idea.

Example 2. We illustrate the AO approach using a simple quadratic objective function. Let $f: \Re^4 \mapsto \Re$ be defined as $f(x) = x^T A x$,

$$f(x) = \left(x_1, x_2, x_3, x_4\right)^T \begin{pmatrix} 100 & 80 & 5 & 1 \\ 80 & 90 & 2 & 1 \\ 5 & 2 & 70 & 40 \\ 1 & 1 & 40 & 80 \end{pmatrix} \begin{pmatrix} x_1 \\ x_2 \\ x_3 \\ x_4 \end{pmatrix}. \tag{2}$$

We want to minimize f over all of $\Psi = \Re^4$, so this is a case of unconstrained optimization. Suppose we choose $X_1 = (x_1, x_2)^T$ and $X_2 = (x_3, x_4)^T$ as the non-overlapping partitioning of the variables, take $\Psi_1 = \Psi_2 = \Re^2$, and guess $x^{(0)} = (1,1,1,1)^T$ as the point of initialization. It is easily verified that the symmetric matrix in (2) is positive definite, which implies that f is strictly convex everywhere in the plane and has

$(0,0,0,0)^T$ as its sole minimizer. The restricted minimizers in (1) can be found using calculus to find FONCs for each *subset* of variables. Setting the gradient $\nabla_{X_1} f(X_1, X_2^{(r)})$ to zero ($\in \Re^2$) and solving for X_1 :

$$X_1^{(r+1)} = -\begin{pmatrix} 100 & 80 \\ 80 & 90 \end{pmatrix}^{-1} \begin{pmatrix} 5 & 1 \\ 2 & 1 \end{pmatrix} X_2^{(r)} . \quad (3)$$

Similarly, rearranging $\nabla_{X_2} f(X_1^{(r+1)}, X_2) = 0 \ (\in \Re^2)$ gives:

$$X_2^{(r+1)} = -\begin{pmatrix} 70 & 40 \\ 40 & 80 \end{pmatrix}^{-1} \begin{pmatrix} 5 & 2 \\ 1 & 1 \end{pmatrix} X_1^{(r+1)} . \quad (4)$$

The results of iteratively minimizing f(x) at (2) by alternately satisfying the FONCs for each subset of partitioned variables are given in Table 1, where the numerical results are truncated to 4 decimal places.

Table 1. AO of f(x) in (2) using $x^{(0)} = (1,1,1,1)^T$, L = 100, Euclidean Norm $\|\cdot\|$, and $\varepsilon = 0.001$.

r	$X_1^{(r+1)}$	$X_2^{(r+1)}$	$\| x^{(r+1)} - x^{(r)} \|$
0	$(-0.1154, 0.0692)^T$	$(0.0083, -0.0036)^T$	2.0251
1	$(-0.0009, 0.0006)^T$	$(0.0000, -0.0000)^T$	0.1337
2	$(-0.0000, 0.0000)^T$	$(0.0000, -0.0000)^T$	0.0011
3	$(-0.0000, 0.0000)^T$	$(0.0000, -0.0000)^T$	0.0000

The AO scheme converges very quickly, requiring only 4 iterations to satisfy the stopping criterion $\| x^{(r+1)} - x^{(r)} \| < 0.001$. The *speed* of convergence depends on the chosen partitioning of the variables. In fact, *every* partitioning of the variables in (2) *except* the one used here yields an AO iterate sequence that terminates more slowly (by about an order of magnitude) than the termination seen in Table 1 [10]. ■

The quadratic function in Example 2 is easily minimized directly in closed form (or in a single iteration of AO, where the partitioning groups the four variables together as a single subset, t=1). So, Example 2 is *not* a situation where AO is a good alternative to the direct approach. What Example 2 provides us with is a first look at several important ingredients of the AO approach.

4. When Is Alternating Optimization a Good Choice?

Alternating optimization is but one competitor in the race to solve (NCOP). There are many, many other ways to look for x^*, Rarely, closed form optimization via the calculus of scalar fields admits a direct solution [11]. More typically, we resort to computational methods such as gradient-based descent in all variables [12-14]; and gradient-based alternating descent (AO) on subsets of the variables [15-17], A relatively new set of interesting techniques that eschew the use of optimality conditions from calculus are based on evolutionary computation [18, 19]. And finally, there are many data driven approaches to solving NCOP such as neural networks [20] and rule-based fuzzy systems [21] that require input-output pairs of *training data* (as opposed to derivative information) to search for solutions to NCOP. In a nutshell, the problems for which AO *is* worth considering are those for which the simultaneous optimization over all variables is much more difficult than the restricted optimizations of (1).

We offer five reasons why AO *may be* preferable to its competitors when considering a specific instance of (NCOP). Some of our assertions are based on computational evidence, while others rest solely on our intuition about differences between (NCOP) and (AO). We believe that AO should be considered:

1. When there is a natural division of variables into t subsets for which *explicit* partial minimizer formulae exist. This strong reason is exemplified many times in the pattern recognition literature.
2. When there is a division of variables for which explicit partial minimizer formulas exist for *MOST* of the variables. This potentially greatly reduces the number of variables which require application of a numerical solution such as Newton's method.
3. Because AO can be *faster* in some cases. Hu and Hathaway [22] found that for the FCM functional, AO in some examples is computationally cheaper (in time) than the best standard optimization routines applied to a reformulated version of the FCM function.
4. Because of savings in *development time.* When explicit formulas exist for AO, it may be easier to program the AO approach than to try to get a general approach written or adapted to the problem. Remember, AO comes with good convergence properties, so no time is wasted trying to pick a good global convergence strategy.
5. Because AO may be more adept at *bypassing local minimizers* than other approaches. Each restricted iteration of AO is typically global, and is therefore able to "hop" great distances through the reduced variable space in order to find an optimal iterate. On the other hand, Newton's method builds a model based on the current iterate, and in this sense is more trapped by local information about the function. For example, if the axes in Example 1 are rotated 45 degrees in the x-y plane, then $(0,0)^T$ would still be a saddle point of f, but it would no longer satisfy the (A0) conditions and would therefore not be "findable" using AO!

5. How Do We Solve (1), the Heart of the AO Algorithm?

The search for solutions to (NCOP) or (AO) is almost always made by an iterative algorithm. An *iteration function* $T:\Re^s \mapsto \Re^s$ generates an infinite sequence $\{x^{(r+1)} = T(x^{(r)}) : r = 0,1,\ldots,\}$ of approximations to x^*, The *choice* of T can be guided to some extent by classifying the problem functions and their derivatives. Gill et al. [13] itemize the following factors: number of variables, constraints on the variables, smoothness of the problem functions and their derivatives, highest level of derivatives that can be efficiently coded, sparseness of the Hessian and Jacobian matrices, number of general linear constraints compared to the number of variables, number of constraints likely to be active at a solution, and whether the problem functions can be evaluated outside the feasible region.

Most of the problems we see in practice can be usefully subdivided into five categories in a slightly different way, according as the *constraints* are: missing or not, linear or not, and equality or not. The easiest case is the unconstrained problem ($\Psi = \Re^s$). A pair of cases of intermediate difficulty occurs when there are (only) one or more equality constraints (linear or not); and the most difficult cases occur when there are one or more (linear or not) inequality constraints, as at (NCOP.3). The type of constraints will lead us to algorithms based on necessary conditions for the unconstrained theory, or to the Kuhn-Tucker conditions for any of the four constrained cases.

In all of these cases iterative algorithms are most often defined via *optimality conditions* associated with the first two derivatives of f. The first and second derivatives of f at a point $x \in \Re^s$ are represented by, respectively, the *gradient vector* ($\nabla f(x)$) and *Hessian matrix* ($\nabla^2 f(x)$). Zeroing the gradient of f (or its LaGrangian) at x^* provides FONCs for x^*. The necessary conditions provide equations on which to base iterative search algorithms that seek an $x^* \in \Psi$ to solve either (NCOP) or (AO). The Hessian of f at x^* may also play a useful role in algorithmic development, and its eigenstructure is useful in determining the type of point found by an algorithm.

We used solutions of the FONCs for minimizing each of the restricted functions in Example 2 to define the steps of the AO illustrated by the iterate sequence in Table 1. Equations (3) and (4) capture the basic structure of almost all AO algorithms. Equation (3) gives $X_1^{(r+1)} = F_1(X_2^{(r)})$, and conversely, equation (4) gives $X_2^{(r+1)} = F_2(X_1^{(r+1)})$. The iteration function $T: \Re^s \mapsto \Re^s$ in this example is the vector field $T = (F_1, F_2)$.

The functions F_1 and F_2 are gotten by solving the FONCs for equations that are satisfied by any *stationary point* (point at which the gradient vanishes) of each restricted function. In Example 2 this strategy leads us to $X_1^{(r)}$ and $X_2^{(r)}$ as explicit functions of each other. Thus, given either estimate, we can compute the "other half" directly.

To illustrate this point, we exhibit the general situation for the easiest case, viz., unconstrained optimization, $\Psi = \Re^s$. We want to compute minimizers $X_i^{(r+1)} = \underset{X_i \in \Psi_i \subset R^{p_i}}{arg\,min} \left\{ f(X_1^{(r+1)}, \ldots, X_{i-1}^{(r+1)}, X_i, X_{i+1}^{(r)}, \ldots, X_t^{(r)}) \right\}$ for i = 1, 2, ..., t. The necessity for the gradients of these t restricted functions to vanish at any candidate solution leads to a set of t coupled equations:

$$\nabla f_{X_1}(X_1, X_2^{(r)}, \ldots, X_i^{(r)}, \ldots, X_t^{(r)}) = 0 \ (\in \Re^{p_1}); \tag{5.1}$$

$$\vdots$$

$$\nabla f_{X_i}(X_1^{(r+1)}, \ldots, X_{i-1}^{(r+1)}, X_i, X_{i+1}^{(r)}, \ldots, X_t^{(r)}) = 0 \ (\in \Re^{p_i}); \tag{5.i}$$

$$\vdots$$

$$\nabla f_{X_t}(X_1^{(r+1)}, X_2^{(r+1)}, \ldots, X_{t-1}^{(r+1)}, X_t) = 0 \ (\in \Re^{p_t}). \tag{5.t}$$

There are two possibilities for the system of equations at (5). When we can solve each equation explicitly for the active variables as functions of the remaining variables (this happens (more often than you might suspect!), system (5) leads to:

$$X_1^{(r+1)} = F_1(X_2^{(r)}, \ldots, X_i^{(r)}, \ldots, X_t^{(r)}); \tag{6.1}$$

$$\vdots$$

$$X_i^{(r+1)} = F_i(X_1^{(r+1)}, \ldots, X_{i-1}^{(r+1)}, X_{i+1}^{(r)}, \ldots, X_t^{(r)}); \tag{6.i}$$

$$\vdots$$

$$X_t^{(r+1)} = F_t(X_1^{(r+1)}, \ldots, X_i^{(r+1)}, \ldots, X_{t-1}^{(r+1)}). \tag{6.t}$$

System (6) is the basis of *Explicit* AO. The equations at (6) allow us to immediately define the iteration function $T: \Re^s \mapsto \Re^s$, so that $x^{(r+1)} = T(x^{(r)})$, r = 0, 1, ...

Less cooperative problem functions yield a system at (5) which can *not* be (completely) solved explicitly. We call this situation *Implicit* AO. In this case system (6) can look pretty intimidating, since one or more subsets of variables will appear on both sides of their subproblem equations. We suspect that some readers have arrived at a system like (5) in their research, but found it to be implicit, leading them to conclude that AO could not be used to solve their problem. AO in this harder case is still possible, but requires an additional level of effort- namely, *numerical solution* of each implicit necessary condition *at each iteration*. This sounds very bad – but sometimes it looks worse than it is. This harder type of alternating optimization is encountered, for example, in the *fuzzy c-shells* clustering algorithm [6]. But, each alternation through the implicit FONCs for this model requires but one iteration of Newton's method at each half step for maximum attainable accuracy [23], so things may not be as bad as they seem.

Many, if not most, readers are automatically conditioned to think of AO in terms of iterative methods based on searching through the FONCs for the problem. When we zero gradients as at (5) to arrive at this stage, we get optimality conditions that identify *all* the stationary points of f, comprising its extrema (local minima and maxima), as well as its saddle points. While most examples in the literature define AO equations by zeroing gradients, the restricted AO minimizations are being done globally. This does not mean that AO will terminate at a global minimizer of f, but it does mean that points of convergence will *look like* global minimizers when viewed (with the blinders we put on when we partition x into t subsets) along the $X_1,\ldots,X_t$ coordinate directions. This is exactly how $(0,0)^T$ looks (like a global minimizer) for f(x,y) in Example 1 if we look from $(0,0)^T$ in the $X_1 = x$ coordinate direction (by itself); and then look along the $X_2 = y$ coordinate direction (by itself).

Are the stationary points of f in (AO) and (NCOP) the same? To answer this question, let

$$E_{NCOP} = \{x^* \in \Re^s : \nabla f_x(x^*) = 0\}, \text{ and}$$

$$E_{AO} = \{x^* = (X_1^*, X_k^*, \ldots, X_t^*) \in \Re^s : \nabla f_{X_k}(x^*) = 0; k = 1, 2, \ldots, t\}.$$

Points in E_{AO} are determined by choosing to break the bigger gradient given by the condition for membership in E_{NCOP} into smaller pieces. In both cases the sets consist of points in $\Re^s$ for which all partials of f with respect to $X_1, X_2,\ldots,X_t$ are 0. So these two sets are equal, $E_{NCOP} = E_{AO}$. In particular, any saddle point of f found by AO is certainly a saddle point of f. Moreover, it is not a saddle point or minimizer for AO other than a point that satisfies the conditions in (AO). A tricky point in this regard is that f can have saddle points which can never be *found by* AO, because they don't look like a minimizer when viewed along the coordinate axis restrictions defined by the partitioning $X_1,\ldots,X_t$

The problem of local solutions and saddle points for (NCOP) and (AO) is in some sense *introduced* by users who base their methods on FONCs, and hence, search for stationary points of f. This is not to say that, for example, a genetic algorithm approach to the solution of (NCOP) or (AO), which eschews the FONCs in favor of a search based on other conditions, always avoids unwanted stationary points of f. But on the face of it, it is more likely that optimization of f using a method that is not guided by necessary conditions for a stationary point will not be so trapped.

6. To What Does AO Converge?

Since $\{x^{(r+1)} = T(x^{(r)})\}$ is a sequence in $\Re^s$, convergence to a limit point is well defined, $\{x^{(r)}\} \to x^* \Leftrightarrow \{x_i^{(r)}\} \to x_i^*$, i = 1, 2,,..., s. We hope that $\{x^{(r+1)}\}$ converges, and that it converges to the right thing (a global solution of (NCOP)). In this section we give a *global* (numerical) *convergence* result, which is one that holds for all possible initializations.

Global results have been obtained for particular instances of A0, such as the FCM algorithm [24]. Our new convergence result (Theorem 2) is obtained by application of convergence theorem A [17, p.91). We need iterates that lie in a compact set, and force this to happen by choosing $\Psi_1,\ldots, \Psi_t$ to be compact. While the assumption of compactness is strong, it is usually enforced during implementation; e.g., there are often bounds on the possible values of variables in an optimization routine. Here is the general global result which holds for all partitionings of x and initializations $x^{(0)}$.

Theorem 2 [10]. Suppose that (EU) holds for $f:\Re^s \mapsto \Re$. Let $x = (X_1,\ldots,X_t)^T$, and $\Psi = \Psi_1 \times \ldots \times \Psi_t$, where Ψ_i is a compact subset of $\Re^{p_i}$, i = 1,...,t. Let $\{x^{(r+1)} = T(x^{(r)})\}$ denote the AO iterate sequence begun at $x^{(0)} \in \Psi$, and denote the fixed points of T as $\Omega = \{x \in \Psi : x = T(x)\}$. Then:

(i) if $x^* \in \Omega$, then $x^* = (X_1^*,\ldots,X_t^*)^T$ satisfies, for i = 1,...,t,

$$X_i^* = \underset{X_i \in \Psi_i \subset R^{p_i}}{\arg\min} \left\{f(X_1^*,\ldots,X_{i-1}^*,X_i,X_{i+1}^*,\ldots,X_t^*)\right\};$$

(ii) $f(x^{(r+1)}) \le f(x^{(r)})$, equality if and only if $x^{(r)} \in \Omega$;

(iii) either: (a) $\exists\, x^* \in \Omega$ and $r_o \in \Re$ so that $x^{(r)} = x^*$ for all $r \ge r_o$;

or (b) the limit of every convergence subsequence of $\{x^{(r)}\}$ is in Ω.

Part (i) of Theorem 2 describes the set of points Ω to which convergence can occur. This set includes global minimizers of f, but it can also contain some purely local minimizers and a certain type of saddle point; viz., a point which behaves like a global minimizer when looking only along the various grouped coordinate (X_1,X_2, etc.) directions (as in Example 1). While it is extremely difficult to find examples where convergence occurs to a saddle point rather than a minimizer, they do exist [24]. This potential for convergence to a saddle point is one "price" alluded to in Section 1 for swapping a difficult joint optimization for a sequence of easier ones involving subsets of the variables.

7. When and How Fast Does Local Convergence Occur?

A *local convergence result* is one that gives the properties of $\{x^{(r)}\}$ if $x^{(0)}$ is chosen sufficiently near a solution. Often, local convergence theory addresses the "speed" with which $\{x^{(r)}\}$ converges. A sequence *converges q-linearly* $\Leftrightarrow$ $\exists n_o \geq 0; \exists \rho \in [0,1); \ni \forall k \geq n_o$,

$$\| x_{k+1} - x^* \| \leq \rho \| x_k - x^* \| \tag{7}$$

The "q" stands for "quotient", and is used to distinguish this type of convergence from the weaker notion of root (or "r") orders of convergence. If an iterative algorithm converges to the correct answer at a certain rate when started close enough to the answer, the algorithm is *locally convergent* at that rate. Algorithms may be faster than q-linear; for example, q-superlinear, q-quadratic, q-cubic, etc. Certainly "bragging rights" accrue to the iterative method with the fastest rate. We are tempted to place more confidence in a method with a faster rate, on the presumption that we can get closer to a solution in a reasonable time, than when a slower method is used. However, the theoretical conditions whereby an iterative method actually achieves its convergence rate may be very rare, and in almost all cases, the theory specifies "starting close enough" to a solution. This places a heavy burden on the initial guess - one which cannot often be verified in practice. It may be that most of the value in having convergence rates lies with the psychological reassurance they provide users (and, of course, the papers and grants they generate for the authors that secure the theory).

AO is locally, q-linearly convergent for the special case $t = 2$ [25]. Local q-linear convergence is maintained even if the restricted minimizations for one of the vector variables is only done approximately, using a single iteration of Newton's method [26]. Recently, the local theory was extended to the case $t = 3$ [27]. Our new result completes the local theory, by giving a proof of the local convergence result for all values $2 \leq t \leq s$.

Theorem 3 [10]. Let x^* be a local minimizer of $f: \Re^s \mapsto \Re$ for which $\nabla^2 f(x^*)$ is positive definite, and let f be C^2 in a neighborhood $N(x^*, \delta)$. Let $0 < \varepsilon \leq \delta$ be chosen so that f is strictly convex on $N(x^*, \varepsilon)$. Finally, assume that if $y = (\mathbf{X}_1, \ldots, \mathbf{X}_{i-1}, Y_i, \mathbf{X}_{i+1}, \ldots, \mathbf{X}_t)^T \in N(x^*, \varepsilon)$, and Y_i^* locally minimizes

$g_i(Y_i) = f(\mathbf{X}_1,\ldots,\mathbf{X}_{i-1}, Y_i, \mathbf{X}_{i+1},\ldots,\mathbf{X}_t)$, then Y_i^* is also the unique global minimizer of g_i:

Then for any $x^{(0)} \in N(x^*,\varepsilon)$, the corresponding AO iterate sequence $\{x^{(r+1)} = T(x^{(r)})\} \to x^*$ q-linearly.

This result is not specific to a partiticular choice of $2 \le t \le s$ and partitioning $(X_1,\ldots,X_t)$. The actual speed of convergence depends on how much the error is reduced at each iteration, and this in turn does depend on which partitioning is used [10]. In other words, Theorem 3 holds for all partitionings of x , but the particular value of ρ that can be used in (7) is dependent on the partitioning.

8. Conclusions

Alternating optimization is intuitive and works well in a surprising number of cases. Theorems 2 and 3 capture the fundamental global and local convergence properties of the *general case* of alternating optimization. Additional assumptions would enable cleaner results. For example, if f(x) is a strictly convex function on $\Re^s$ that has a minimizer x* at which $\nabla^2 f(x^*)$ is continuous and positive definite, then AO (with $\Psi = \Re^s$) will converge q-linearly to the minimizer using *any* initialization.

Theorems 2 and 3 can also be applied in some important cases when variable constraints are present. For example, the constraints of subsets of variables summing to 1, which are commonly found in clustering methods, can be handled by substituting out the last membership in each constraint. Then the local theory given here can be applied to the reduced function and variables. The global result is much more flexible regarding constraints than the local; as long as the constraints are handled internal to the calculation of the partial minimizers so that the assumptions hold, then the global convergence result can be applied.

References

1. Lloyd, S.P. (1957). Least squares quantization of PCM, [originally an unpublished Bell Labs technical note], reprinted in *IEEE Trans. IT*, 28, March, 1982, 129-137.
2. Wolfe, J.H. (1970). Pattern clustering by multivariate mixture analysis, *Multivariate Behavioral Research*, v. 5, 329-350.
3. Bezdek, J.C. (1973). *Fuzzy Mathematics for Pattern Classification*, PhD Thesis, Cornell University, Ithaca, NY.
4. Bezdek, J.C., Coray, C., Gunderson, R. and Watson, J. (1981). Detection and Characterization of Cluster Substructure II. Fuzzy c-Varieties and Convex Combinations Thereof , *SIAM J. Appl. Math.*, 40(2), 358-372.
5. Windham, M.P. (1985). Numerical classification of proximity data with assignment measures, *Journal of Classification*, v. 2, 157-172.
6. Dave, R.N. (1990). Fuzzy shell-clustering and applications to circle detection in digital images, *International Journal of General Systems*, v. 16, 343-355.

7. Hathaway, R.J. and Bezdek, J. C. (1993). Switching Regression Models and Fuzzy Clustering, *IEEE Trans. Fuzzy Systems*, 1(3), 195-204.
8. Krishnapuram, R. and J.M. Keller (1993). A possibilistic approach to clustering, *IEEE Trans. on Fuzzy Systems*, v. 1, 98-110.
9. Hathaway, R.J., and J.C. Bezdek (2001). Fuzzy c-means clustering of incomplete data, to appear, *IEEE Trans. on Systems, Man and Cybernetics B.*
10. Bezdek, J.C. and R.J. Hathaway (2002). Alternating optimization, parts 1 and 2, in preparation for *IEEE Trans. on Systs., Man and Cybernetics.*
11. Apostol, T.M. (1969). *Calculus*, Ginn Blaisdell, Waltham, MA.
12. Ortega, J.M. and W.C. Rheinboldt (1970). *Iterative Solution of Nonlinear Equations in Several Variables.* New York: Academic Press.
13. Gill, P.E., Murray, W. and Wright, M. H. (1981). *Practical Optimization*, Academic Press, NY.
14. Dennis, J.E., Jr., and R.B. Schnabel (1983). *Numerical Methods for Unconstrained Optimization and Nonlinear Equations.* Englewood Cliffs, NJ: Prentice-Hall.
15. Luenberger, D.G. (1969). *Optimization by Vector Space Methods*, Wiley, NY.
16. Luenberger, D.G. (1984). *Linear and Nonlinear Programming.* Second Edition, Reading, MA: Addison-Wesley.
17. Zangwill, W. (1969). Nonlinear Programming: A Unified Approach. Englewood Cliffs, NJ: Prentice-Hall.
18. Goldberg, D.E. (1989). *Genetic algorithms in Search, Optimization and Machine Learning*, Addison-Wesley, Reading, MA.
19. Fogle, D.B. (1995). *Evolutionary Computation*, IEEE Press, Piscataway, NJ.
20. Golden, R.M. (1996). *Mathematical methods for Neural Network Analysis and Design*, Bradford Books, MIT Press, Cambridge, MA.
21. Nguyen, H.T. and Sugeno, M. (eds.) (1997). *Fuzzy Systems: Modeling and Control*, Kluwer, Norwell, MA.
22. Hu, Y., and R.J. Hathaway (2001). On efficiency of optimization in fuzzy c-means, preprint.
23. Bezdek, J.C. and R.J. Hathaway (1992). Numerical convergence and interpretation of the fuzzy c-shells clustering algorithms, *IEEE Trans. on Neural Networks*, v. 3, 787-793.
24. Bezdek, J.C., Hathaway, R.J., Sabin, M.J., and W.T. Tucker (1987a). Convergence theory for fuzzy c-means: counterexamples and repairs, *IEEE Trans. on Systems, Man and Cybernetics*, SMC-17, 873-877.
25. Bezdek, J.C., Hathaway, R.J., Howard, R.E., Wilson, C.A., and M.P. Windham (1987b). Local convergence analysis of a grouped variable version of coordinate descent, *Journal of Optimization Theory and Applications*, v. 54, 471-477.
26. Hathaway, R.J. and J.C. Bezdek (1991). Grouped coordinate minimization using Newton's method for inexact minimization in one vector coordinate, *Journal of Optimization Theory and Applications*, v. 71, 503-516.
27. Hathaway, R.J., Hu. Y.K., and J. C. Bezdek (2001). Local convergence analysis of tri-level alternating optimization, *Neural, Parallel, and Scientific Computation*, 9, 19-28.

Noisy Speech Segmentation/Enhancement with Multiband Analysis and Neural Fuzzy Networks

Chin-Teng Lin, Der-Jenq Liu, Rui-Cheng Wu, and Gin-Der Wu

Department of Electrical and Control Engineering, National Chiao-Tung University
Hsinchu, Taiwan, R.O.C.
ctlin@fnn.cn.nctu.edu.tw
http://falcon1.cn.nctu.edu.tw/

Abstract. Background noise added to speech can decrease the performance of speech segmentation and enhancement. To solve this problem, new methods have been developed in this thesis. First, a new speech segmentation method (ATF-based SONFIN algorithm) is proposed in fixed noise-level environment. This method contains the multiband analysis and a neural fuzzy network, and it achieves higher recognition rate than the TF-based robust algorithm by 5%. In addition, a new speech segmentation method called RTF-based RSONFIN algorithm is proposed for variable noise-level environment. The RTF-based RSONFIN algorithm contains a recurrent neural fuzzy network. This method contains the multiband analysis and achieve higher recognition rate than the TF-based robust algorithm by 12%.

1. Introduction

An important problem in speech processing is to detect the presence of speech in noisy environment. The energy (in time domain), zero-crossing rate, duration, LPC, linear prediction error energy, and pitch information parameters have been usually used to find the boundary between the word signal and background noise [1,2,3,4]. It has been found that these parameters are not sufficient to get reliable word boundaries in noisy environment, even if more complex decision strategies are used [4]. In the connection, Junqua et al. [4] proposed the time-frequency (TF) parameter and a robust algorithm to get the word boundary in noisy environment (TF-based robust algorithm). However, it needs to empirically determine thresholds and ambiguous rules which are not easily determined by human. Some researchers [2,5] used the neural network's learning ability to solve this problem. However, the proper structure of the network (including numbers of hidden layers and nodes) is not easy to decide.

To solve the problems of the above approaches, we proposed the adaptive time-frequency (ATF) parameter and refined time-frequency (RTF) parameter to make the distinction between speech signal and noise clear. Based on the ATF parameter, we proposed the ATF-based SONRIN algorithm in fixed noise-level environment. This SONFIN is a self-constructing neural fuzzy inference network that we proposed

N.R. Pal and M. Sugeno (Eds.): AFSS 2002, LNAI 2275, pp. 301–309, 2002.

previously in [6]. Based on the RTF parameter, we proposed the RTF-based RSONFIN algorithm in variable noise-level environment. This RSONFIN is a recurrent self-organizing neural fuzzy inference network that we proposed previously in [7]. Due to the self-learning ability of SONFIN and RSONFIN, the ATF-based SONFIN algorithm and RTF-based RSONFIN algorithm avoid the need of empirically determining thresholds and ambiguous rules in normal word boundary detection algorithms. Also, since the SONFIN and RSONFIN house the human-like IF-THEN rules in its network structure, expert knowledge can be put into these networks as a priori knowledge, which can usually increase its learning speed and detection accuracy [8,9].

This paper is organized as follows. In Section 2, we proposed ATF-based SONFIN algorithm for speech segmentation in fixed noise-level environment. We further proposed RTF-based RSONFIN algorithm for speech segmentation in variable noise-level environment in Section 3. Finally, conclusions are made in Section 4.

2. ATF-Based SONFIN Algorithm

In this section, we proposed ATF-based SONFIN algorithm for speech segmentation in fixed noise-level environment.

2.1 Adaptive Time-Frequency (ATF) Parameter

The relation between mel-scale frequency and frequency (Hz) is described by the following equation:

$$\text{mel} = 2595\ (1 + \text{f} / 700). \tag{1}$$

A filter bank is then designed according to the mel scale, where the filters of 20 bands are approximated by simulating 20 triangular band-pass filers, $f(i,k)$ $(1 \le i \le 20,$ $0 \le k \le 63)$, over a frequency range of 0 ~ 4000 Hz. Hence, each filter band has a triangular bandpass frequency response, and the spacing as well as the bandwidth is determined by a constant mel frequency interval by Eq (1). We first find the spectrum, $x_{freq}(m,k)$, of this signal by Discrete Fourier Transform (128-point DFT): $x_{freq}(m,k) = \sum_{n=0}^{N-1} x_{time}(m,n) W_N^{kn}$, $0 \le k \le N-1$, $0 \le m \le M-1$, $W_N = \exp(-j2\pi / N)$, where $x_{freq}(m,k)$ is the magnitude of the kth point of the spectrum of the mth frame, N is 128 in our system, and M is the number of frames of the speech signal for analysis. We then multiply the spectrum $x_{freq}(m,k)$ by the weighting factors f(i,k): $x(m,i) = \sum_{k=0}^{N-1} x_{freq}(m,k) f(i,k)$, $0 \le m \le M-1$, $1 \le i \le 20$, where i is the filter band index, k is the spectrum index, m is the frame number, and M is the number of frames for analysis. We further smooth it by using a three-point median filter : $\hat{x}(m,i) = (x(m-1,i) + x(m,i) + x(m+1,i))/3$, $1 \le m \le M - 2$.

Finally, the smoothed energy is normalized by removing the frequency energy of the beginning interval, Noise_f, to get X(m,i) where the energy of background noise is estimated by averaging the frequency energy of the first five frames of the recording: $X(m,i) = \hat{x}(m,i) - Noise\_f = \hat{x}(m,i) - \sum_{m=0}^{4} \hat{x}(m,i)/5$. We can calculate the total energy of the almost pure speech signal at ith band as E(i): $E(i)\sum_{m=0}^{M-1} | X(m,i) |$. Since the band with higher E(i) contains more pure speech signal, we should sort the twenty mel-scale frequency bands according to their E(i) value. Let S be the set of all E(i): $S = \{E(i) \mid i = 1,2,3,......,20\}$. The sorting is performed as follows:

$$\begin{aligned}
&P(1) = Max\{S\} \\
&P(2) = Max\{S - P(1)\} \\
&P(1) = Max\{S - \{P(1), P(2)\}\} \\
&\cdots \\
&\cdots \\
&P(1) = Max\{S - \{P(1), P(2), \cdots, P(19)\}\}.
\end{aligned} \tag{2}$$

We denote the number of bands useful for producing reliable frequency energy as Na:

$$(Na - 18)/(Noise\_f - 18) = (18 - 3)/(83 - 93), \tag{3}$$

$$Na = -1.5 \times Noise\_freq + 142.5, \tag{4}$$

$$\begin{aligned}
&Na = \lfloor A \times Noise_freq + B \rfloor \\
&3 \le Na \le 18, A = -1.5 \text{ and } B = 142.5,
\end{aligned} \tag{5}$$

where Na is an integer. With the number of useful bands Na, we then sum the total energies of the first Na bands (after ordering) in (2) to get the final frequency energy, F(m), of frame m:

$$F(m) = \sum_{i=1}^{Na} X(m, I(i)). \tag{6}$$

The proposed adaptive time-frequency (ATF) parameter is derived as follows:

$$x_{rms} = \log\sqrt{\sum_{n=0}^{L-1} x_{time}^{2}(m,n)/L}, \tag{7}$$

$$\hat{x}_{rms}(m) = (x_{rms}(m-1) + x_{rms}(m) + x_{rms}(m+1))/3, \tag{8}$$

$$\begin{aligned}
T(m) &= \hat{x}_{rms}(m) - Noise_time, \\
&= \hat{x}_{rms}(m) - \sum_{m=0}^{4} \hat{x}_{rms}(m)/5,
\end{aligned} \tag{9}$$

$$ATF\ (m) = SMOOTHING\ \ (T(m) + cF\ (m)), \tag{10}$$

where SMOOTHING is performed by a three-point median filter as in Eq. (8) and constant c is a proper weighting factor. The procedure to calculate the ATF parameter is illustrated in Fig. 1.

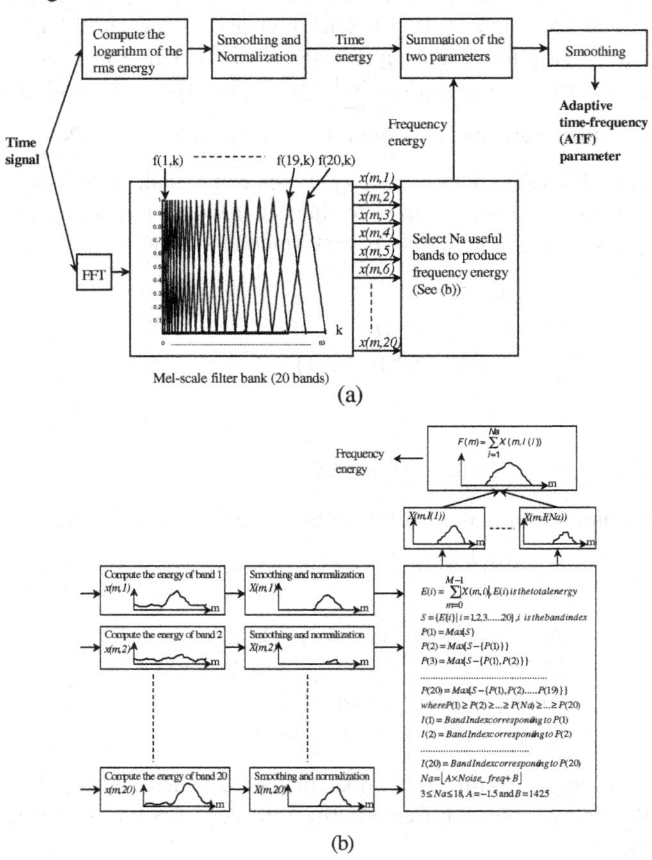

Fig. 1. Flowchart for computing the ATF parameter.

2.2 Self-Constructing Neural Fuzzy Inference Network (SONFIN)

The neural fuzzy network that we used for word boundary detection is called self-constructing neural fuzzy inference network (SONFIN) that we proposed previously in [6]. The SONFIN is a general connectionist model of a fuzzy logic system, which can find its optimal structure and parameters automatically. It can always find itself an economic network size, and the learning speed as well as the modeling ability are all superior to normal neural networks. The structure is shown in Fig. 2 and the details of function of each layer can be found in [6].

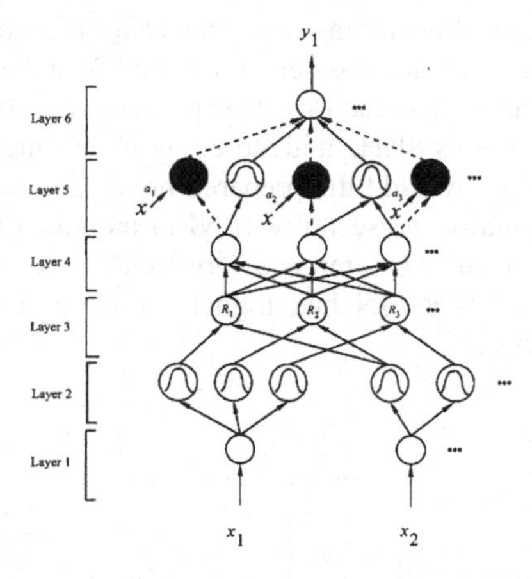

Fig. 2. Network structure of the SONFIN

2.3 SONFIN for Word Boundary Detection

The procedure of using the SONFIN for word boundary detection is illustrated in Fig. 3. The input feature vector of the SONFIN consists of the average of the logarithmic root-mean-square (rms) energy on the first five frames of recording interval (Noise_time), adaptive time-frequency (ATF) parameter, and zero-crossing rate (ZCR). The three parameters in an input feature vector are obtained by analyzing a frame of signal. Hence there are three (input) nodes in layer 1 of SONFIN. Before entering the SONFIN, the three input parameters are normalized to be in [0, 1]. For each input vector (corresponding to a frame), the output of SONFIN indicates whether the corresponding frame is a word signal or noise. For this purpose, we used two (output) nodes in layer 6 of SONFIN, where the output vector of (1, 0) standing for word signal, and (0, 1) for noise.The SONFIN is trained by a set of 80 training patterns, which were randomly selected from four noise conditions with different SNRs. 40 patterns are from word sound category, and the other 40 patterns are from noise category. After training, there were only 14 rules generated in the SONFIN.

2.4 Experiments

We shall test the performance of the proposed ATF-based SONFIN algorithm, and compare it to the robust algorithm proposed by Junqua et al. [4]. The detected word signal is then sent into a speech recognizer. Since inaccurate detection of word boundary is harmful to recognition, the performance of the word boundary detection process is examined by the recognition rate of speech recognizer. The speech recognition system used in this paper is a robust isolated word recognition system consisting of two parts, feature extractor and classifier. In the feature extractor, the modified two-dimensional cepstrum (Modified TDC – MTDC) [10] is used as the speech feature. In the classifier, a Gaussian clustering algorithm is used.The speech data used for our experiment are the set of isolated Mandarin digits. They are ten digits spoken by 10 speakers and each speaker pronounced 20 times of the ten digits. The recording sampling rate is 8KHz and stored as 16-bit integer. To set up the noisy speech database for testing, we add the prepared noisy signals to the recorded speech signals with different signal-to-noise-ratios (SNRs) including 0dB, 5dB, 10dB, 15dB, 20dB and ∞ dB. A total of 600 utterances are used in our experiment. The results show that the ATF-based SONFIN had higher recognition rate than the TF-based robust algorithm [4] by 5%.

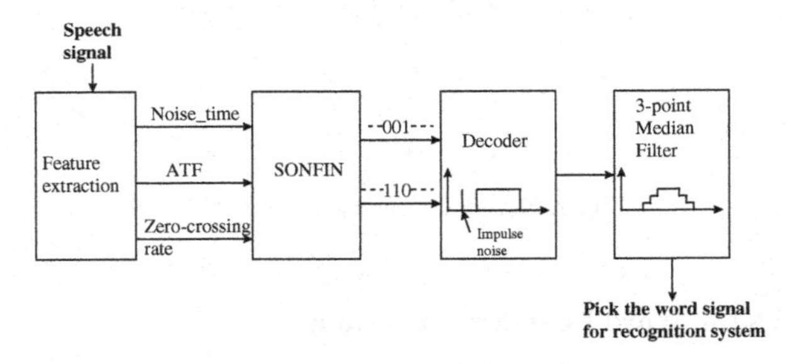

Fig. 3. Flowchart of the SONFIN-based word boundary detection algorithm

3. RTF-Based RSONFIN Algorithm

In this section, we proposed RTF-based RSONFIN algorithm for speech segmentation in variable noise-level environment.

3.1 Refined Time-Frequency (RTF) Parameter

We adopt the maximum X(m,i) defined in section 2 to get the final frequency energy: $F(m) = \max[X(m,i)]_{i=1,2,\ldots,20.}$.

The RTF parameter can be get as follow:

$$x_{rms} = \log\sqrt{\sum_{n=0}^{L-1} x_{time}^{2}(m,n)/L}, \tag{11}$$

$$\hat{x}_{rms}(m) = (x_{rms}(m-1) + x_{rms}(m) + x_{rms}(m+1))/3, \tag{12}$$

$$T(m) = \hat{x}_{rms}(m) - Noise\_time = \hat{x}_{rms}(m) - \sum_{m=0}^{4} \hat{x}_{rms}(m)/5, \tag{13}$$

$$RTF\ (m) = SMOOTH\ \ (T(m) + cF(m)), \tag{14}$$

where c is constant and set to be 0.8.

3.2 TF-Based RSONFIN Algorithm

To solve the problem of variable background noise-level environment, we modified the structure of the proposed SOFIN as a Recurrent SONFIN, called RSONFIN. The functions of its learning algorithm can be found in [7]. With the learning ability of temporal relations, a procedure of using the RSONFIN for word boundary detection in variable background noise level condition is illustrated in Fig. 4. The input feature vector of the RSONFIN consists of the average of the logarithmic root-mean-square (rms) energy on the first five frames of recording interval (Noise_time), refined time-frequency (RTF) parameter, and zero-crossing rate (ZCR). The three parameters in an input feature vector are obtained by analyzing a frame of signal. Hence there are three (input) nodes in layer 1 of RSONFIN. Before entering the RSONFIN, the three input parameters are normalized to be in [0, 1]. For each input vector (corresponding to a frame), the output of RSONFIN indicates whether the corresponding frame is a word signal or noise. For this purpose, we used two (output) nodes in layer 5 of RSONFIN, where the output vector of (1, 0) standing for word signal, and (0, 1) for noise.

In the training process, the noisy speech waveform is sampled, and each frame is transformed into the desired input feature vector of the RSONFIN (Noise_time, RTF parameter, and zero-crossing rate). These training vectors are classified as word signal or noise by using waveform, spectrum displays and audio output. Among these training vectors, some are from word sound category with the desired RSONFIN output vector being (1, 0), and the others are from noise category with the desired RSONFIN output vector being (0, 1).

The RSONFIN after training is ready for word boundary detection. As shown in Fig. 4, the outputs of RSONFIN are processed by a decoder. The decoder decodes the RSONFIN's output vector (1, 0) as value 100 standing for word signal, and (0, 1) as value 0 standing for noise. We observed that the decoding waveform (i.e., the outputs of the decoder) contains impulse noise sometimes. Hence, we let the output waveform of the decoder pass through a three-point median filter to eliminate the isolated ‘‘impulse" noise. Finally, we recognize the word-signal island as the part of the filtered waveform whose magnitude is greater than 30, and duration is long enough (by setting a threshold value). We then regard the parts of the original signal corresponding to the allocated word-signal island as the word signal, and the other ones as the background noise.

3.3 Experiments

The speech data used for our experiment are the set of isolated Mandarin digits. They are ten digits spoken by 10 speakers and each speaker pronounced 20 times of the ten digits. The recording sampling rate is 8KHz and stored as 16-bit integer. To set up the noisy speech database for testing, we add the prepared noisy signals to the recorded speech signals with different signal-to-noise-ratios (SNRs) including 5dB, 10dB, 15dB, 20dB and ∞ dB. A total of 600 utterances are used in our experiment; 300 utterances are in the condition of increasing background noise level, and 300 utterances are in the condition of decreasing background noise level. Experimental results show that our new algorithm achieves higher recognition rate than the TF-based algorithm which has been shown to outperform several commonly used word boundary detection algorithms by about 12% in variable background noise level condition.

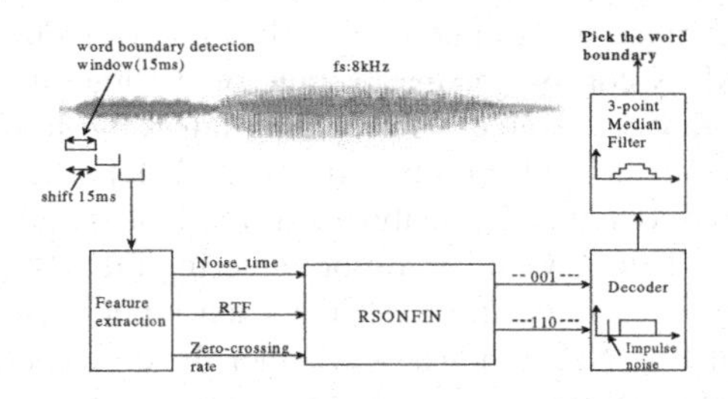

Fig. 4. The RTF-based RSONFIN algorithm for automatic word boundary detection.

4. Conclusions

A new speech segmentation method (ATF-based SONFIN algorithm) is proposed in fixed noise-level environment. This method contains the multiband analysis and a neural fuzzy network, and it achieves higher recognition rate than the TF-based robust algorithm by 5%. In addition, a new speech segmentation method RTF-based RSONFIN algorithm is proposed in variable noise-level environment. The RTF-based RSONFIN algorithm contains a recurrent neural fuzzy network. This method contains the multiband analysis and achieve higher recognition rate than TF-based robust algorithm by 12%.

References

1. L.F. Lamel, L.R. Rabiner, A.E. Rosenberg, and J.G. Wilson, "An improved endpoint detector for isolated word recognition," *IEEE ASSP Mag.*, vol.29, pp.777-785, August, 1981.
2. Y. Qi and B.R. Hunt, "Voiced-unvoiced-silence classification of speech using hybrid features and a network classifier," *IEEE Tran. Speech Audio Processing*, vol.1, pp. 250-255, April, 1993.
3. B. Reaves, "Comments on an improved endpoint detector for isolated word recognition," *IEEE Trans. Signal Processing*, vol.39, pp.526-527, February, 1991.
4. J.C. Junqua, B. Mak, and B. Reaves, "A robust algorithm for word boundary detection in the presence of noise," *IEEE Trans. Speech Audio Processing*, vol.2, pp.406-412, July, 1994.
5. T. Ghiselli-Crippa and A. El-Jaroudi, "A fast neural net training algorithm and its application to voiced- unvoiced-silence classification of speech," *ICASSP91*, vol.1, pp.441-444, 1991.

6. C.F. Juang and C.T. Lin, "An on-line self-constructing neural fuzzy inference network and its application," *IEEE Trans. Fuzzy System*, vol. 6, pp. 12-32, February 1998.
7. C.F. Juang and C.T. Lin, "A Recurrent Self-Organizing Neural Fuzzy Inference Network", *IEEE Trans. Neural Networks*, vol. 10, no. 4, pp. 828-845, July, 1999.
8. C.T. Lin, *Neural Fuzzy Control Systems with Structure and Parameter Learning*, World Scientific, 1994.
9. C.T. Lin and C.S.G. Lee, *Neural Fuzzy Systems: A Neural-Fuzzy Synergism to Intelligent Systems*, Englewood Cliffs, NJ: Prentice-Hall, May, 1996.
10. C.T. Lin, H.W. Nein, and J.Y. Hwu, "GA-based Noisy Speech Recognition using Two Dimensional Cepstrum," accepted to appear in *IEEE Trans. Speech and Audio Processing* .

Towards Optimal Feature and Classifier for Gene Expression Classification of Cancer*

Jungwon Ryu and Sung-Bae Cho

Dept. of Computer Science, Yonsei University,
134 Shinchon-dong, Sudaemoon-ku, Seoul 120-749, Korea
rjungwon@candy.yonsei.ac.kr, sbcho@csai.yonsei.ac.kr

Abstract. Recently, demand on the tools to efficiently analyze biological genomic information has been on the rise. In this paper, we attempt to explore the optimal features and classifiers through a comparative study with the most promising feature selection methods and machine learning classifiers. In order to predict the cancer class, the gene information from patient's marrow expressed by DNA microarray, who has either the acute myeloid leukemia or acute lymphoblastic leukemia. Pearson and Spearman's correlation, Euclidean distance, cosine coefficient, information gain, mutual information and signal to noise ratio have been used for feature selection. Backpropagation neural network, self-organizing map, structure adaptive self-organizing map, support vector machine, inductive decision tree and *k*-nearest neighbor have been used for classification. Experimental results indicate that backpropagation neural network with Pearson's correlation coefficients is the best method, obtaining 97.1% of recognition rate on the test data.

1 Introduction

Although cancer detection and class discovery have been seriously investigated over the past 3 decades, there has been no general and perfect way to work out this problem. It is because there can be so many pathways causing cancer, and there exist tremendous numbers of varieties. Recently, array technologies have made it straightforward to monitor the expression patterns of thousands of genes during cellular differentiation and response [1, 2]. These gene expression numbers are just simple sequence of numbers, which are meaningless, and the necessity of tools analyzing these to get useful information has risen sharply.

In this paper, we attempt to explore the optimal feature and classifier that efficiently detect the class of the disease. To find out the genes that has cancer-related function, we have applied seven feature selection methods, which are commonly used in the field of data mining and pattern recognition: Pearson's correlation and Spearman's correlation based on statistical approach, information gain and mutual information based on information theoretic approach, and Euclidean distance and cosine coefficient is based on similarity distance measure. As classifiers, we have

* This paper was supported by Brain Science and Engineering Research Program sponsored by Korean Ministry of Science and Technology.

N.R. Pal and M. Sugeno (Eds.): AFSS 2002, LNAI 2275, pp. 310–317, 2002.

adopted multilayer perceptron, self-organizing map, decision tree and *k*-nearest neighbor.

2 DNA Microarray

DNA arrays consist of large numbers of DNA molecules spotted in a systemic order on a solid substrate. Depending on the size of each DNA spot on the array, DNA arrays can be categorized as microarrays, when the diameter of DNA spot is less than 250 microns, and macroarrays ,when the diameter is bigger than 300 microns. When the solid substrate used is small in size, arrays are also referred to as DNA chips. It is so powerful that we can investigate the gene information in short time, because at least hundreds of genes are put on the DNA microarray to be analyzed.

DNA microarrays consist of thousands of individual DNA sequences printed in a high density array on a glass microscope slide using a robotic arrayer. The relative abundance of these spotted DNA sequences in two DNA or RNA samples may be assessed by monitoring the differential hybridization of the two samples to the sequences on the array. For mRNA samples, the two samples are reverse-transcribed into cDNA, labeled using different fluorescent dyes mixed (red-fluorescent dye Cy5 and green-fluorescent dye Cy3). After the hybridization of these samples with the arrayed DNA probes, the slides are imaged using scanner that makes fluorescence measurements for each dye. The log ratio between the two intensities of each dye is used as the gene expression data.

$$gene\_expression = \log_2 \frac{\text{Int(Cy5)}}{\text{Int(Cy3)}} \tag{1}$$

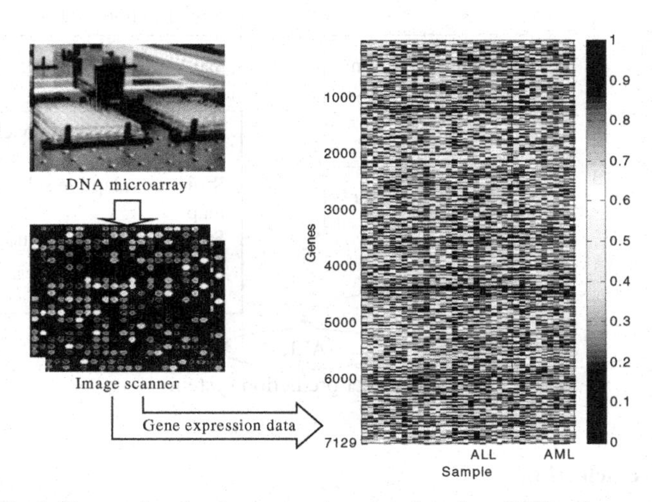

Fig. 1. Preparation for the gene expression data from DNA microarray

3 Cancer Prediction

We have used the data set that is comprised of DNA microarray from the myeloid samples of the patients who have either the acute lymphoblastic leukemia (ALL) or the acute myeloid leukemia (AML). There are 7,129 known genes in human genome, so that each sample has 7,129 numbers of gene expression. The system developed in this paper to predict the cancer class of patients is as shown in Fig. 2. After acquiring the gene expression data calculated from the DNA microarray, our prediction system has 2 stages in general, which are the feature selection and pattern classification stage.

The feature selection can be thought of as the gene selection, which is to get the list of genes that might be informative for the prediction by statistical, or information theoretical methods, etc. Since it is known that only a very few genes out of the 7,129 have the information related to the cancer and using every genes results in too big dimensionalities, it is necessary to explore the efficient way to get the best feature. We have extracted 25 genes using seven methods described in Section 3.1, and the cancer predictor classifies the category only with these genes.

Given the gene list, a classifier makes decision where the gene pattern may belong to at prediction stage. We have adopted six most widely used classification methods as shown in Fig. 2.

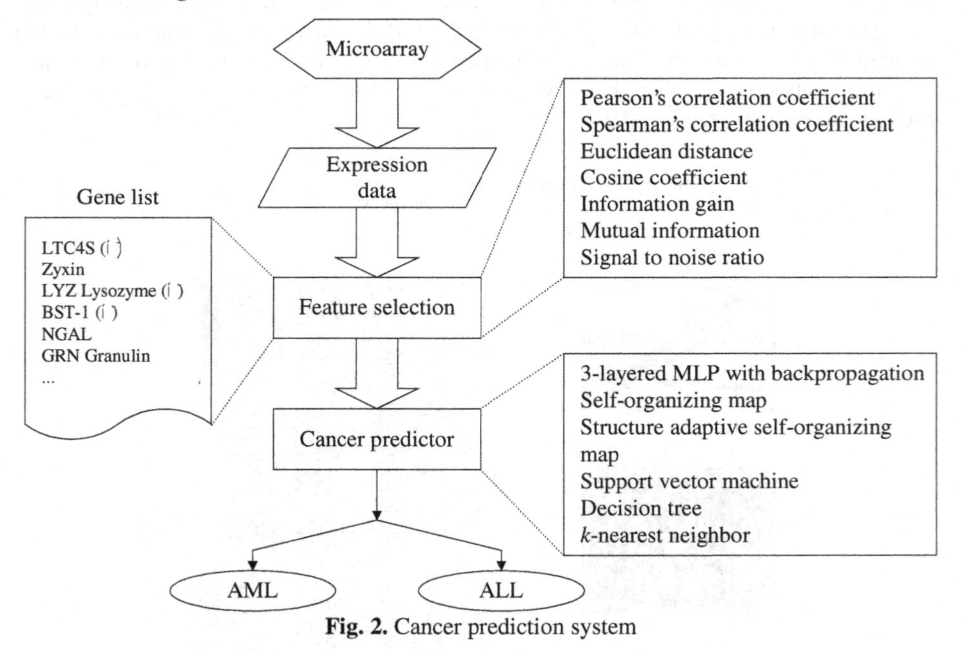

Fig. 2. Cancer prediction system

3.1 Feature Selection

Using the statistical correlation analysis, we can see the linear relationship and the direction of relation between two variables. Correlation coefficient r varies from –1 to +1, so that the data distributed near the line biased to (+) direction will have positive coefficients, and the data near the line biased to (-) direction will have negative coefficients.

Suppose that we have a gene expression pattern g_i (i = 1 ~ 7,129). Each g_i has a gene expression number from 38 samples, $g_i = (e_1, e_2, e_3, \ldots, e_{38})$. The first 27 numbers $(e_1, e_2, \ldots, e_{27})$ are examples of ALL, and the other 11 $(e_{28}, e_{29}, \ldots, e_{38})$ are those from AML. Let us define an ideal gene pattern which belongs to ALL class, called g_{ideal_ALL} = (1, 1, …, 1, 0, …, 0), so that all numbers from the ALL samples are 1, and the others are 0. In this paper, we have calculated the correlation coefficient between this g_{ideal} and each gene's expression pattern. When we have two vectors X and Y, which have - N elements, $r_{Pearson}$ and $r_{Spearman}$ are calculated as follows:

$$r_{Pearson} = \frac{\sum XY - \frac{\sum X \sum Y}{N}}{\sqrt{\left(\sum X^2 - \frac{\left(\sum X\right)^2}{N}\right)\left(\sum Y^2 - \frac{\left(\sum Y\right)^2}{N}\right)}} \tag{2}$$

$$r_{Spearman} = 1 - \frac{6\sum\left(D_x - D_y\right)^2}{N\left(N^2 - 1\right)} \tag{3}$$

where, D_x and D_y are the rank matrices of X and Y, respectively.

The similarity between two input vectors X and Y can be thought of as the distance of those. Distance is a measure on how far the two vectors are located, and the distance between g_{ideal_ALL} and g_i tells us how much the g_i is likely to the ALL class. Calculating the distance between them, if it is bigger than certain threshold, the gene g_i would belong to ALL, otherwise g_i belongs to AML. In this paper, we have adopted Euclidean distance ($r_{Eclidean}$) and cosine coefficient (r_{Cosine}) expressed by following equations:

$$r_{Eclidean} = \sqrt{\sum (X - Y)^2} \tag{4}$$

$$r_{Cosine} = \frac{\sum XY}{\sqrt{\sum X^2 \sum Y^2}} \tag{5}$$

We have utilized the information gain and mutual information that are widely used in many fields such as text categorization and data mining. When we count the number of genes excited or not excited in category c_j, the coefficient of the information gain and mutual information is as follows:

$$IG(g_i, c_j) = P(g_i \mid c_j)\log\frac{P(g_i \mid c_j)}{P(c_j)\cdot P(g_i)} + P(\bar{g}_i \mid c_i)\log\frac{P(\bar{g}_i \mid c_j)}{P(c_j)\cdot P(\bar{g}_i)} \tag{6}$$

$$MI(g_i, c_j) = \log\frac{P(g_i \wedge c_j)}{P(g_i)\times P(c_j)} = \log\frac{A\times N}{(A+C)\times(A+B)} \quad (7)$$

For the each genes g_i, some are from ALL, and some are from AML samples. When we calculate the mean μ and standard deviation σ from the distribution of gene expressions within their classes, the signal to noise ratio of gene g_i, $SN(g_i)$, is defined by:

$$SN(g_i) = \frac{\mu_{ALL}(g_i) - \mu_{AML}(g_i)}{\sigma_{ALL}(g_i) - \sigma_{AML}(g_i)} \quad (8)$$

3.2 Classification

Error backpropagation neural network is a feed-forward multilayer perceptron (MLP) that is applied in many ways due to its powerful and stable learning algorithm. The neural network learns the training examples by adjusting the synaptic weight of neurons according to the error occurred on the output layer. The power of the backpropagation algorithm lies in two main aspects: local for updating the synaptic weights and biases and efficient for computing all the partial derivatives of the cost function with respect to these free parameters.

Self-organizing map (SOM) defines a mapping from the input space onto an output layer by unsupervised learning algorithm. SOM has an output layer consisting of N nodes, each of which represents a vector that has the same dimension as the input pattern. For a given input vector X, the winner node m_c is chosen using Euclidean distance between X and its neighbors, m_i.

$$\|x - m_c\| = \min_i \|x - m_i\| \quad (9)$$

$$m_i(t+1) = m_i(t) + \alpha(t)\times n_{ci}(t)\times\{x(t) - m_i(t)\} \quad (10)$$

Even though SOM is well known for its good performance of topology preserving, it is difficult to apply it to practical problems such as classification since it should have the topology fixed from the beginning of training. A structure adaptive self-organizing map (SASOM) is proposed to overcome this shortcoming [3]. SASOM starts with 4×4 map, and dynamically splits the output nodes of the map, where the data from different classes are mixed, trained with the LVQ learning algorithm.

Support vector machine (SVM) estimates the function classifying the data into two classes. SVM builds up a hyperplane as the decision surface in such a way to maximize the margin of separation between positive and negative examples. SVM achieves this by the structural risk minimization principle that the error rate of a learning machine on the test data is bounded by the sum of the training-error rate and a term that depends on the Vapnik-Chervonenkis (VC) dimension. Given a labeled set of M training samples (X_i, Y_i), where $X_i \in R^N$ and Y_i is the associated label, $Y_i \in \{-1, 1\}$, the discriminant hyperplane is defined by:

$$f(X) = \sum_{i=1}^{M} Y_i \alpha_i k(X, X_i) + b \tag{11}$$

where $k(\ .)$ is a kernel function and the sign of $f(X)$ determines the membership of X. Constructing an optimal hyperplane is equivalent to finding all the nonzero α_i (support vectors) and a bias b. This paper has used SVM^{light} module.

The concept-learning induction method such as decision tree (DT) aims to construct rules for the classification from the set of objects of which class labels are known. Quinlan's C4.5 uses an information-theoretical approach based on the energy entropy. C4.5 builds the decision tree as follows: select an attribute, divide the training set into subsets characterized by the possible values of the attribute, and follow the same partitioning procedure recursively with each subset until no subset contains objects from more than one class. The single class subsets correspond them to the leaves. The entropy-based criterion that has been used for the selection of the attribute is called the gain ratio criterion [4]. The gain ratio criterion selects that test X such that the gain ratio (X) is maximized.

k-nearest neighbor(KNN) is one of the most common methods among memory based induction. Given an input vector, KNN extracts *k* number of most close vectors in the reference set based on similarity measures, and makes decision the label of input vector using distribution of labels *k* neighbors have and similarity.

4 Experimental Results

4.1 Environments

72 people have participated to construct the data set. A sample of DNA microarray is submitted by each person, so that the database consists of 72 samples. 38 samples are for training and the other 34 are for test of the classifiers. The training data has 27 ALL and 11 AML samples, whereas the test data has 20 ALL and 14 AML samples. ALL class is encodes as 1, whereas AML class as 0. Each sample is comprised of 7,129 gene expression numbers.

We used 3-layered MLP, with 5~15 hidden nodes, 2 output nodes, 0.03~0.5 of learning rate and 0.1~0.9 of momentum. SOM used 2x2~5x5 map with rectangular topology and 0.05 of learning rate. and KNN used k =1~38. The best parameters have been chosen after several trial-and-errors. The final number is averaged by 10 times of repetition.

4.2 Result Analysis

Fig. 3 shows the Ids and names of 11 out of 25 genes chosen by Pearson's method. These are those selected by one up to three other feature selection methods. 8 of them are appeared in the result of cosine method, 4 appeared in Spearman and 3 in information gain. g_{2288} has been 4 times appeared overlapped, g_{6200} has been top-ranked both in information gain and Spearman, which implies very informative. There were no genes appeared in every methods at the same time.

```
ID   Name
====+==============
3320 Leukotriene C4 synthase (LTC4S) gene
2020 FAH Fumarylacetoacetate
1745 LYN V-yes-1 Yamaguchi sarcoma viral related oncogene homolog
5039 LEPR Leptin receptor
4196 "PRG1 Proteoglycan 1, secretory granule"
2288 DF D component of complement (adipsin)
6201 INTERLEUKIN-8 PRECURSOR
1882 CST3 Cystatin C (amyloid angiopathy and cerebral hemorrhage)
2121 CTSD Cathepsin D (lysosomal aspartyl protease)
6200 Interleukin 8 (IL8) gene
2043 "LGALS3 Lectin, galactoside-binding, soluble, 3 (galectin 3) (NOTE: redefinition of symbol)"
```

Fig. 3. Genes chosen by Pearson's correlation

The results of recognition rate on the test data are as shown in Table 1. The MLP seems to have the best recognition rate among the classifiers on the average. SASOM performs better than SOM, DT has good results in some cases, but it seems to be very dependent on the feature. KNN doesn't seem to do the classification at all. In the meanwhile, Pearson's and cosine coefficient have good numbers among the feature methods, obtaining 87% on the average (except KNN results). Information gain has 83.5% and Spearman 79.4% of recognition rate. This fact agrees that several genes are chosen overlapped in those four feature methods, which means they may be correlated somehow.

Table 1. Recognition rate by features and classifiers [%]

Classifier Feature	MLP	SOM	SASOM	SVM	DT	KNN (k=10)
Pearson coefficient	97.1	73.5	88.2	79.4	97.1	29.4
Spearman coefficient	70.6	73.5	82.4	88.2	82.4	32.4
Euclidean distance	97.1	70.6	70.6	58.5	73.5	32.4
Cosine coefficient	79.4	97.1	74.1	94.1	94.1	23.5
Information gain	91.2	73.5	82.4	88.2	82.4	58.8
Mutual information	67.6	67.6	64.7	58.5	47.1	58.8
S/N ratio	94.1	52.9	64.7	58.5	55.9	8.8

Fig. 4 shows the examples of misclassification made be MLP, SASOM and DT. Sample 28 is the only one misclassified by MLP with Pearson feature. Many of other classifiers also fail to answer this sample, but MLP with Euclidean, information gain and mutual information, SASOM with cosine and decision tree with Pearson and mutual information feature have classified it correctly. On the other hand, sample 29 and 11, which are misclassified by MLP with signal-to-noise ratio and decision tree with Pearson respectively, have been correctly classified by most of other classifiers.

Fig. 5 is the expression level of genes chosen by Pearson' correlation.

5 Concluding Remarks

We have done a thorough quantitative comparison among the 42 combinations of features and classifiers. Pearson, Spearman, cosine and information gain were the top four feature selection methods and MLP, SASOM and decision tree were the best classifiers. They also have shown some correlations between features and classifiers.

Feature	Multilayer perceptron	Structure-adaptive SOM	Decision tree
Pearson	28	21 23 27 28	11
Spearman	15 16 20 22 23 24 25 27 28 29	14 17 21 24 25 28	5 6 9 17 21 28
Euclidean	22	11 14 16 19 22 23 24 25 26 28	16 19 20 22 23 24 25 26 28
Cosine	11 16 22 24 25 26 28	23 29	23 28
Ig	16 22 26	5 6 9 17 21 28	14 17 21 24 25 28
Mi	4 9 12 16 18 19 22 24 25 26 31	2 13 16 19 20 22 23 24 26 28 29 33	1 2 3 5 6 9 13 14 16 17 18 19 21 22 23 24 26 27
Sn	28 29	9 11 12 14 16 21 22 23 24 27 28 32	1 4 5 6 9 10 12 17 18 19 21 28 29 30 33

Fig. 4. Misclassified samples

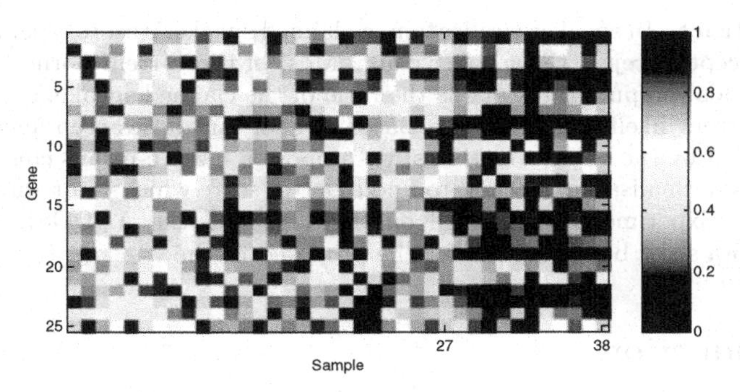

Fig. 5. Expression level of genes chosen by $r_{Pearson}$

References

1. Golub, T.R. *et al.*: Molecular classification of cancer: Class discovery and class prediction by gene-expression monitoring. Science, Vol. 286. (1999) 531-537
2. Tamayo, P.: Interpreting patterns of gene expression with self-organizing map: Methods and application to hematopoietic differentiation. Proceedings of the National Academy of Sciences of the United States of America. Vol. 96. (1999) 2907-2912
3. Kim, H.D. and Cho, S.B.: Genetic optimization of structure-adaptive self-organizing map for efficient classification. Proceedings of International Conference on Soft Computing, pp. 34-39, (October 2000)
4. Quinlan, J.R.: The effect of noise on concept learning. Machine Learning: an Artificial Intelligence Approach. R. S. Michalski, J. G. Carbonell and T. M. Mitchell. Eds. San Mateo, CA: Morgan Kauffmann, Vol. 2. (1986) 149-166

Fuzzy C-Means Clustering-Based Speaker Verification

Dat Tran and Michael Wagner

University of Canberra, School of Computing, ACT 2601, Australia
{Dat.Tran,Michael.Wagner}@canberra.edu.au
http://www.ise.canberra.edu.au/DatT

Abstract. In speaker verification, a claimed speaker's score is computed to accept or reject the speaker claim. Most of the current normalisation methods compute the score as the ratio of the claimed speaker's and the impostors' likelihood functions. Based on analysing false acceptance error occured by the current methods, we propose a fuzzy c-means clustering-based normalisation method to find a better score which can reduce that error. Experiments performed on the TI46 and the ANDOSL speech corpora show better results for the proposed method.

1 Introduction

Fuzzy c-means (FCM) clustering approach to speaker verification is proposed in this paper. For an input utterance and a claimed identity, most of the current speaker verification methods compute a claimed speaker's score, which is the ratio of the claimed speaker's and the impostors' likelihood functions, and compare this score with a given threshold to accept or reject this speaker. In all verification paradigms, there are two classes of errors: false rejections and false acceptances. An equal error rate (EER) condition is often used to adjust system parameters so that the two types of errors are equal. Speaker verification performance is strongly affected by variations in signal characteristics, therefore normalisation methods have been applied to compensate for these variations [4]-[8].

For practical implementations, the use of a background speaker set that is representative of the impostors' population close to the claimed speaker has been proposed [11]. However, false rejections of the claimed speaker can arise because the likelihood values of the background speakers are assumed to be of equal weight. This assumption is often not true as the similarity measures between each background speaker and the claimed speaker might be different. To overcome this drawback, the speaker verification problem is considered in fuzzy set theory, the claimed speaker's score is viewed as the fuzzy membership function of the input utterance in the claimed speaker's fuzzy set of utterances. The FCM membership function is proposed as a fuzzy membership score, which is the ratio of functions of the claimed speaker's and impostors' likelihood functions. Distances in FCM clustering are also redefined.

N.R. Pal and M. Sugeno (Eds.): AFSS 2002, LNAI 2275, pp. 318–324, 2002.

Speaker verification experiments performed on the TI46 corpus including 16 speakers and the ANDOSL corpus including 108 speakers show better results for the proposed method.

2 Current Normalisation Methods

Let λ_0 be the claimed speaker model and λ be a model representing all other possible speakers, i.e. impostors. $P(X|\lambda_0)$ and $P(X|\lambda)$ are the likelihood functions of the claimed speaker and impostors, respectively. For a given input utterance X and a claimed identity, a claimed speaker's score $S(X)$ is used as follows

$$S(X) \begin{cases} > \theta & accept \\ \leq \theta & reject \end{cases} \tag{1}$$

where θ is the decision threshold. The simplest likelihood ratio score in the log domain is as follows

$$S_1(X) = \log P(X|\lambda_0) - \log P(X|\lambda) \tag{2}$$

The term $\log P(X|\lambda)$ is called the normalisation term and requires calculation of all impostors' likelihood functions. However when the size of the population increases, a subset of the impostor models consisting of B "background" speaker models λ_i, $i = 1, \ldots, B$ is used [5] and is representative of the population close to the claimed speaker. Current normalisation methods are as follows

$$S_2(X) = \log P(X|\lambda_0) - \log \left\{ \frac{1}{B} \sum_{i=1}^{B} P(X|\lambda_i) \right\} \tag{3}$$

$$S_3(X) = \log P(X|\lambda_0) - \log \sum_{i=0}^{B} P(X|\lambda_i) \tag{4}$$

$$S_4(X) = \log P(X|\lambda_0) - \frac{1}{B} \sum_{i=1}^{B} \log P(X|\lambda_i) \tag{5}$$

$$S_5(X) = \sum_{t=1}^{T} \left[\log P(x_t|\lambda_0) - \log \sum_{i=1}^{B} P(x_t|\lambda_i) \right] \tag{6}$$

where $S_2(X)$ was proposed by Reynolds [10], $S_3(X)$ by Matsui and Furui [8], $S_4(X)$ by Liu [6], and $S_5(X)$ by Markov and Nakagawa [7].

3 Fuzzy C-Means Clustering

The most successful technique in fuzzy cluster analysis is fuzzy C-means (FCM) clustering [1] , it is widely used in both theory and practical applications of fuzzy clustering techniques to unsupervised classification [3].

Let $\mathbf{X} = \{X_1, \ldots, X_T\}$ be a set of feature vectors, $U = [u_{it}]$ be a matrix whose elements are memberships of X_t, $t = 1, \ldots, T$ in class i, $i = 1, \ldots, C$. Fuzzy C-partition space [1] for $\mathbf{X}$ is the set of matrices U such that

$$0 \leq u_{it} \leq 1, \qquad \sum_{i=1}^{C} u_{it} = 1 \quad \forall t, \qquad 0 < \sum_{t=1}^{T} u_{it} < T \quad \forall i \tag{7}$$

and the fuzzy objective function is defined as follows

$$J_m(U, \theta; \mathbf{X}) = \sum_{i=1}^{C} \sum_{t=1}^{T} u_{it}^m d^2(X_t, \theta_i) \tag{8}$$

where $m > 1$ is called the degree of fuzziness, $\theta = \{\theta_1, \ldots, \theta_C\}$ is a C tuple of prototypes, each of which characterises one of the C clusters, and $d(X_t, \theta_i)$ is the distance from X_t to prototype θ_i, known as a measure of dissimilarity. Minimising the fuzzy objective function $J_m(U, \theta; \mathbf{X})$ in (8) gives

$$u_{it} = \left\{ \sum_{j=1}^{C} \left[d(X_t, \theta_i)/d(X_t, \theta_j) \right]^{2/(m-1)} \right\}^{-1}, \qquad m > 1 \tag{9}$$

4 FCM-Based Normalisation Method

Consider the speaker verification problem in fuzzy set theory. To accept or reject the claimed speaker, the task is to make a decision whether the input utterance X is either from the claimed speaker λ_0 or from the set of impostors λ, based on comparing the score for X and a decision threshold θ. Since input utterances belong to either the claimed speaker or impostors, the set of input utterances can be divided into two subsets for the claimed speaker and impostors. Computing the claimed speaker's score for an input utterance can be viewed as a fuzzification process, where two above subsets are considered as *fuzzy* subsets and the score means the fuzzy membership function, which denotes the degree of belonging of the input utterance to the claimed speaker. Accepting (or rejecting) the claimed speaker is thus viewed as a defuzzification process, where the input utterance is (or is not) in the claimed speaker's fuzzy subset if the fuzzy membership value is (or is not) greater than the given threshold.

According to this fuzzy set theory-based viewpoint, some currently used scores are viewed as fuzzy membership scores and inversely, other fuzzy memberships can be selected to use as the claimed speaker's scores. In theory, there are many ways to define the fuzzy membership function, therefore it can be said that the fuzzy approach proposes more general scores than the current likelihood ratio scores for speaker verification. These are termed fuzzy membership scores, which can denote the belonging of X to the claimed speaker.

The next task is to find effective fuzzy membership scores. As introduced in Section 1, false rejections of the claimed speaker can arise because the likelihood values of the background speakers are assumed to be of equal weight. However, this assumption is often not true as the similarity measures between each background speaker and the claimed speaker might be different. To overcome this drawback, the likelihood values of the background speakers are transformed into certain values by applying the idea of the fuzzy C-means membership function. The FCM membership score is proposed as follows

$$S_6(X) = \left\{ \sum_{i=0}^{B} \left[\log P(X|\lambda_0) / \log P(X|\lambda_i) \right]^{1/(m-1)} \right\}^{-1}, \qquad m > 1 \quad (10)$$

where B was defined in Section 2, the distance in (9) is redefined as $d^2(X_t, \lambda_i) = -\log P(X_t|\lambda_i)$ and the impostors' fuzzy set is approximately represented by B background speakers' fuzzy subsets. Note that $S_6(X)$ is not written in the log domain.

To illustrate the effectiveness of the FCM membership score, for simplicity, we compare $S_2(X)$ with $S_6(X)$ in a numerical example.

Table 1. The likelihood values for 4 input utterances $X_1 - X_4$ against the claimed speaker λ_0 and 3 impostors $\lambda_1 - \lambda_3$, where X_1^c, X_2^c are from the claimed speaker and X_3^i, X_4^i are from impostors

Utterance	$P(X\|\lambda_0)$	$P(X\|\lambda_1)$	$P(X\|\lambda_2)$	$P(X\|\lambda_3)$
X_1^c	3.0×10^{-4}	2.2×10^{-5}	4.2×10^{-5}	1.5×10^{-6}
X_2^c	1.1×10^{-3}	1.4×10^{-4}	1.4×10^{-4}	2.0×10^{-5}
X_3^i	7.9×10^{-6}	1.2×10^{-7}	1.2×10^{-6}	7.2×10^{-8}
X_4^i	1.2×10^{-4}	7.2×10^{-6}	6.6×10^{-8}	6.5×10^{-8}

Table 1 presents the likelihood values for 4 input utterances $X_1 - X_4$ against the claimed speaker λ_0 and 3 impostors $\lambda_1 - \lambda_3$, where X_1^c, X_2^c are from the claimed speaker and X_3^i, X_4^i are from impostors (these are real values from experiments on the TI46 database). Given a score $S(X)$, the EER = 0 in the case that all the scores for X_1^c and X_2^c are greater than all those for X_3^i and X_4^i.

Table 2. Scores of 4 utterances using $S_2(X)$ and $S_6(X)$

Score	X_1^c	X_2^c	X_3^i	X_4^i
$S_2(X)$	2.628	2.399	2.810	3.899
$S_6(X)$	0.316	0.316	0.302	0.350

Table 2 shows scores in (3) and (10) computed using these likelihood values where $m = 2$ is applied to $S_6(X)$ in (10). It can be seen that with the score $S_2(X)$, we always have the EER $\neq 0$ since scores for X_3^i and X_4^i are higher than those for X_1^c and X_2^c. However using $S_6(X)$, the EER is reduced since the score for X_3^i is lower than those for X_1^c and X_2^c.

5 Experimental Results

The Gaussian mixture model (GMM) [10] speaker verification experiments were performed on the TI46 and the ANDOSL corpora in text-independent mode.

5.1 The TI46 Corpus

The TI46 corpus was designed and collected at Texas Instruments (TI). The speech was produced by 16 speakers, 8 female and 8 male, labelled f1-f8 and m1-m8 respectively, consisting of two vocabularies—TI-20 and TI-alphabet. The TI-20 vocabulary contains the ten digits from 0 to 9 and ten command words: *enter, erase, go, help, no, rubout, repeat, stop, start,* and *yes*. The TI-alphabet vocabulary contains the names of the 26 letters of the alphabet from a to z. For each vocabulary item, each speaker produced 10 tokens in a single training session and another two tokens in each of 8 testing sessions. The corpus was sampled at 12500 samples per second and 12 bits per sample.

The vocabulary used in this experiment was the 10-command set. In the training phase, 100 training tokens (10 utterances x 1 training session x 10 repetitions) of each speaker were used to train GMMs of 16, 32, 64, and 128 mixtures. The degree of fuzziness was set to $m = 1.06$.

Since the TI46 corpus has a small speaker set consisting of only 8 female and 8 male speakers, each speaker was used as a claimed speaker with the remaining speakers (including 7 same-gender background speakers) acting as impostors and rotating through all speakers. Therefore the total numbers of claimed test utterances and impostor test utterances are 2560 (16 claimed speakers x 10 test utterances x 8 sessions x 2 repetitions) and 38,400 ((16 x 15) impostors x 10 test utterances x 8 sessions x 2 repetitions), respectively. Experimental results are shown in Table 3. The lowest EER is 3.3% is obtained for 128-mixture GMMs.

Table 3. EER results (%) for the TI46 corpus using GMMs with 7 closest speakers.

Similarity Score	Number of mixtures			
	16	32	64	128
$S_2(X)$	6.6	5.8	4.2	3.6
$S_3(X)$	6.6	5.8	4.2	3.6
$S_4(X)$	6.6	5.8	4.2	3.6
$S_5(X)$	6.8	5.9	4.2	3.6
$\mathbf{S_6(X)}$	**6.4**	**5.5**	**3.9**	**3.3**

5.2 The ANDOSL Corpus

The Australian National Database of Spoken Language (ANDOSL) [9] corpus comprises carefully balanced material for Australian speakers, both Australian-born and overseas-born migrants. Current holdings are divided into those from native speakers of Australian English (born and fully educated in Australia) and those from non-native speakers of Australian English (first generation migrants having a non-English native language). A subset used in this paper consists of 108 native speakers, divided into 36 speakers of General Australian English, 36 speakers of Broad Australian English, and 36 speakers of Cultivated Australian English comprising 6 speakers of each gender in each of three age ranges (18-30, 31-45 and 46+). So there are total of 18 groups of 6 speakers labelled "ijk", where i denotes f (female) or m (male), j denotes y (young) or m (medium) or e (elder), and k denotes g (general) or b (broad) or c (cultivated). For example, the group fyg contains 6 female young general Australian English speakers. Each speaker contributed in a single session, 200 phonetically rich sentences. The averaged duration for each sentence is approximately 4 sec.

The speech was recorded in an anechoic chamber using a B&K 4155 microphone and a DSC-2230 VU-meter used as a preamplifier and was digitised direct to computer disk using a DSP32C analog-to-digital converter mounted in a PC. All waveforms were sampled at 20 kHz and 16 bits per sample and converted to 8 kHz bandwidth. Low and high pass cut-offs were set to 300 Hz and 3400 Hz. The speech data were processed in 32 ms frames at a frame rate of 10 ms. Frames were Hamming windowed and preemphasised. The basic feature set consisted of 12th-order mel-frequency cepstral coefficients (MFCCs) and the normalised short-time energy, augmented by the corresponding delta MFCCs and delta energy to form a 26-dimensional vector for each individual frame.

Table 4. EER results (%) for the ANDOSL corpus using GMMs with 5 closest speakers in set 1 and 5 same-group speakers in set 2.

Similarity	Set 1		Set 2	
Score	16-mixture GMM	32-mixture GMM	16-mixture GMM	32-mixture GMM
$S_2(X)$	2.2	1.8	2.7	2.1
$S_3(X)$	2.2	1.8	2.7	2.1
$S_4(X)$	3.2	3.0	3.5	3.3
$S_5(X)$	2.3	2.0	2.9	2.3
$\mathbf{S_6(X)}$	**1.9**	**1.5**	**2.3**	**1.7**

The 16 and 32-mixture GMMs were trained for each speaker using the first 10 sentences numbered from 001 to 010. Sentences numbered from 011 to 200 were used for verification. Experiments were performed on 108 speakers using each speaker as a claimed speaker with 5 closest background speakers or 5 same-group background speakers as indicated above and 102 mixed-gender impostors (excluding the 5 background speakers) and rotating through all speakers. The total number of claimed test utterances and impostor test utterances are 20,520 (108 claimed speakers x 190 test utterances) and 2,093,040 ((108 x 102) impostors x 190 test utterances), respectively. Results are shown in Table 4. The degree of fuzziness is set to $m = 1.05$. The table shows that the proposed method performs better than the current methods.

6 Conclusion

The fuzzy C-means clustering-based normalisation method has been proposed to speaker verification. Based on considering the false rejections and the idea of the fuzzy membership function, the fuzzy C-means membership has been redefined and used as a claimed speaker's fuzzy score. Experiments show better results for the proposed method.

References

1. J. C. Bezdek, *Pattern Recognition with Fuzzy Objective Function Algorithms*, Plenum Press, New York and London, 1981.
2. F. Chen, B. Millar and M. Wagner, "Hybrid-threshold approach in text-independent speaker verification", in *Proc.* ICSLP'94, Yokohama, pp. 1855-1858, 1994.
3. J. Dunn, "A Fuzzy Relative of the ISODATA Process and Its Use in Detecting Compact Well-Separated Cluster", *J. Cybern.*, vol. 3, pp. 32-57, 1974.
4. S. Furui, "An Overview of Speaker Recognition Technology", in *Proc. ESCA Workshop on Automatic Speaker Recognition, Identification & Verification*, pp. 1-9, 1994.
5. A. L. Higgins, L. Bahler and J. Porter, "Speaker Verification using Randomized Phrase Prompting", *Digital Signal Processing*, vol. 1, pp. 89-106, 1991.
6. C. S. Liu, H. C. Wang and C.-H. Lee, "Speaker Verification using Normalization Log-Likelihood Score", *IEEE Trans. Speech and Audio Processing*, vol. 4, pp. 56-60, 1980.
7. K. P. Markov and S. Nakagawa, "Text-independent speaker recognition using non-linear frame likelihood transformation", *Speech Communication*, vol. 24, pp. 193-209, 1998.
8. T. Matsui and S. Furui, "Concatenated Phoneme Models for Text Variable Speaker Recognition", in *Proc.* ICASSP'93, pp. 391-394, 1993.
9. J. B. Millar, J. P. Vonwiller, J. M. Harrington and P. J. Dermody, "The Australian National Database of Spoken Language", in *Proc.* ICASSP'94, vol. 1, pp. 97-100, 1994.
10. D. A. Reynolds, "Speaker identification and verification using Gaussian mixture speaker models", *Speech Communication*, vol. 17, pp. 91-108, 1995.
11. A. E. Rosenberg, J. Delong, C.-H. Lee, B.-H. Juang and F. K. Soong, "The use of cohort normalised scores for speaker verification", in *Proc.* ICSLP'92, pp. 599-602, 1992.
12. D. Tran and M. Wagner, "Fuzzy Normalisation Methods for Speaker Verification", in *Proc.* ICSLP2000, vol. 1, pp. 446-449, Beijing, China, 2000.
13. D. Tran and M. Wagner, "A Proposed Likelihood Transformation for Speaker Verification", in *Proc.* ICASSP2000, vol. 2, pp. 1069-1072, Turkey, 2000.

Noise Clustering-Based Speaker Verification

Dat Tran and Michael Wagner

University of Canberra, School of Computing, ACT 2601, Australia
{Dat.Tran,Michael.Wagner}@canberra.edu.au
http://www.ise.canberra.edu.au/DatT

Abstract. The normalisation method for speaker verification proposed in this paper is based on the idea of the noise clustering method in fuzzy clustering. The proposed method can reduce false acceptance errors and apply to all current normalisation scores. Experiments performed on the ANDOSL and YOHO speech corpora show better results for the proposed method.

1 Introduction

For an input utterance and a claimed identity, most of the current speaker verification methods compute a claimed speaker's score, which is the ratio of the claimed speaker's and the impostors' likelihood functions, and compare this score with a given threshold to accept or reject this speaker. There are two classes of errors in all speaker verification paradigms: false rejections and false acceptances. An equal error rate (EER) condition is often used to adjust system parameters so that the two types of errors are equal. The fuzzy C-means (FCM) clustering-based normalisation method proposed in the previous paper [14] can reduce false rejection errors. Interestingly, the noise clustering-based normalisation method proposed in this paper can reduce false acceptance errors. The use of the background speaker set in the current and FCM scores can cause the false acceptances of impostors because of the relativity of the ratio-based values. This problem can be overcome by applying the idea of the noise clustering (NC) method [3] in fuzzy clustering, where impostor's utterances are considered as noisy data and thus should have arbitrarily small fuzzy membership scores in the claimed speaker's fuzzy subset. This is implemented by simply adding to the normalisation term a constant membership value, which denotes belonging of all input utterances to impostors' fuzzy subset.

Speaker verification experiments are performed on the ANDOSL corpus including 108 speakers and on the YOHO corpus including 138 speakers. All noise clustering-based methods show better results than the other methods.

2 Current and FCM Normalisation Methods

Let λ_0 be the claimed speaker model and λ be a model representing all other possible speakers, i.e. impostors. $P(X|\lambda_0)$ and $P(X|\lambda)$ are the likelihood functions

N.R. Pal and M. Sugeno (Eds.): AFSS 2002, LNAI 2275, pp. 325–331, 2002.

of the claimed speaker and impostors, respectively. For a given input utterance X and a claimed identity, a claimed speaker's score $S(X)$ is used as follows

$$S(X) \begin{cases} > \theta & accept \\ \leq \theta & reject \end{cases} \tag{1}$$

where θ is the decision threshold. Current scores $S_i(X)$, $i = 1, \ldots, 5$ and the FCM score $S_6(X)$ are defined as follows [14]:

$$S_1(X) = \log P(X|\lambda_0) - \log P(X|\lambda) \tag{2}$$

$$S_2(X) = \log P(X|\lambda_0) - \log \left\{ \frac{1}{B} \sum_{i=1}^{B} P(X|\lambda_i) \right\} \tag{3}$$

$$S_3(X) = \log P(X|\lambda_0) - \log \sum_{i=0}^{B} P(X|\lambda_i) \tag{4}$$

$$S_4(X) = \log P(X|\lambda_0) - \frac{1}{B} \sum_{i=1}^{B} \log P(X|\lambda_i) \tag{5}$$

$$S_5(X) = \sum_{t=1}^{T} \left[\log P(x_t|\lambda_0) - \log \sum_{i=1}^{B} P(x_t|\lambda_i) \right] \tag{6}$$

$$S_6(X) = \left\{ \sum_{i=0}^{B} \left[\log P(X|\lambda_0) / \log P(X|\lambda_i) \right]^{1/(m-1)} \right\}^{-1}, \qquad m > 1 \tag{7}$$

where a subset of the impostor models consisting of B "background" speaker models λ_i, $i = 1, \ldots, B$ is used [5] and is representative of the population close to the claimed speaker.

3 Noise Clustering

FCM clustering has a disadvantage in the problem of sensitivity to outliers. An idea of a noise cluster was proposed [3] to deal with noisy data or outliers. The noise is considered to be a separate class and is represented by a prototype that has a constant distance $\delta > 0$ from all feature vectors. Therefore the sum of memberships for the good clusters is smaller than one: $\sum_{i=1}^{C} u_{it} < 1$, $t = 1, \ldots, T$. This allows noisy data and outliers to have arbitrarily small membership values in good clusters. The NC objective function is defined as follows

$$J_m(U, \theta; \mathbf{X}) = \sum_{i=1}^{C} \sum_{t=1}^{T} u_{it}^m d^2(X_t, \theta_i) + \sum_{t=1}^{T} \delta^2 \left(1 - \sum_{i=1}^{C} u_{it} \right)^m \tag{8}$$

where $m > 1$, $U = [u_{it}]$, u_{it} is the membership of X_t in class i, and $d(X_t, \theta_i)$ is the distance from X_t to prototype θ_i. Minimising (8) gives

$$u_{it} = \frac{1}{\sum_{k=1}^{C} \left[d(X_t, \theta_i) / d(X_t, \theta_k) \right]^{2/(m-1)} + \left[d(X_t, \theta_i) / \delta \right]^{2/(m-1)}} \tag{9}$$

The second term in the denominator of (9) becomes quite large for outliers, resulting in small membership values in all the good clusters for outliers.

4 NC-Based Normalisation Method

The use of the background speaker set in the current and proposed scores can cause the false acceptances of impostors because of the relativity of the ratio-based values. Indeed, assuming that we have

$$S_2(X_1) = \log\left(\frac{0.07}{0.03}\right) \qquad S_2(X_2) = \log\left(\frac{0.0000007}{0.0000003}\right) \tag{10}$$

where X_1 is and X_2 is not from the claimed speaker. It is clear that $S_2(X_1) = S_2(X_2)$, thus if the given threshold is 2 to obtain a true acceptance for X_1 then we cause a false acceptance for X_2 although the likelihood value for X_2 is very small.

Interestingly, the false acceptances of impostors because of the relativity of the ratio-based values can be overcome by applying the idea of the NC method, where impostor's utterances are considered as noisy data and thus should have arbitrarily small fuzzy membership scores in the claimed speaker's fuzzy subset. This fuzzy approach can reduce the false acceptance error by forcing the membership value of the input utterance X to become as small as possible if X is really from impostors, not from the claimed speaker or background speakers. According to the fuzzy set theory-based viewpoint considered in [14], the above current scores $S_i(X)$, $i = 1, \ldots, 5$, are viewed as fuzzy membership scores and hence we can apply the NC method to all of these scores. From (9) we can see that the NC approach to the normalisation scores is implemented by simply adding to the normalisation term a constant membership value $\epsilon > 0$, which denotes the belonging of all input utterances to the impostors' cluster. The NC-based versions of $S_i(X)$, $i = 1, \ldots, 6$ are as follows

$$S_{1nc}(X) = \log P(X|\lambda_0) - \log\left[P(X|\lambda) + \epsilon_1\right] \tag{11}$$

$$S_{2nc}(X) = \log P(X|\lambda_0) - \log\left\{\frac{1}{B+1}\left[\sum_{i=1}^{B} P(X|\lambda_i) + \epsilon_2\right]\right\} \tag{12}$$

$$S_{3nc}(X) = \log P(X|\lambda_0) - \log\left[\sum_{i=0}^{B} P(X|\lambda_i) + \epsilon_3\right] \tag{13}$$

$$S_{4nc}(X) = \log P(X|\lambda_0) - \frac{1}{B+1}\left[\sum_{i=1}^{B} \log P(X|\lambda_i) + \log \epsilon_4\right] \tag{14}$$

$$S_{5nc}(X) = \sum_{t=1}^{T}\left\{\log P(x_t|\lambda_0) - \log\left[\sum_{i=1}^{B} P(x_t|\lambda_i) + \epsilon_5\right]\right\} \tag{15}$$

$$S_{6nc}(X) = \frac{1}{\sum_{i=1}^{C} \left[\log P(X|\lambda_0)/\log P(X|\lambda_i)\right]^{\frac{1}{m-1}} + \left[\log P(X|\lambda_0)/\log \epsilon_6\right]^{\frac{1}{m-1}}} \tag{16}$$

Applying to the above example, with $\epsilon_3 = 0.01$ we obtain

$$S_{2nc}(X_1) = \log \frac{0.07}{0.03 + 0.01} = 0.6 \; > \; S_{2nc}(X_2) = \log \frac{0.0000007}{0.0000003 + 0.01} = -9.6 \tag{17}$$

Therefore the false acceptance cannot occur for X_2 if the threshold is 0.

Table 1. The likelihood values for 4 input utterances $X_1 - X_4$ against the claimed speaker λ_0 and 3 impostors $\lambda_1 - \lambda_3$, where X_1^c, X_2^c are from the claimed speaker and X_3^i, X_4^i are from impostors

Utterance	$P(X\|\lambda_0)$	$P(X\|\lambda_1)$	$P(X\|\lambda_2)$	$P(X\|\lambda_3)$
X_1^c	3.0×10^{-4}	2.2×10^{-5}	4.2×10^{-5}	1.5×10^{-6}
X_2^c	1.1×10^{-3}	1.4×10^{-4}	1.4×10^{-4}	2.0×10^{-5}
X_3^i	7.9×10^{-6}	1.2×10^{-7}	1.2×10^{-6}	7.2×10^{-8}
X_4^i	1.2×10^{-4}	7.2×10^{-6}	6.6×10^{-8}	6.5×10^{-8}

To illustrate the effectiveness of the NC-based score, we consider the scores $S_2(X)$ and $S_6(X)$ in a numerical example using real values from experiments on the TI46 speech corpus.

Table 1 presents the likelihood values for 4 input utterances $X_1 - X_4$ against the claimed speaker λ_0 and 3 impostors $\lambda_1 - \lambda_3$, where X_1^c, X_2^c are from the claimed speaker and X_3^i, X_4^i are from impostors (these are real values from experiments on the TI46 database). Given a score $S(X)$, the EER $= 0$ in the case that all the scores for X_1^c and X_2^c are greater than all those for X_3^i and X_4^i.

Table 2. Scores of 4 utterances using $S_2(X)$ and $S_6(X)$

Score	X_1^c	X_2^c	X_3^i	X_4^i
$S_2(X)$	2.628	2.399	2.810	3.899
$S_6(X)$	0.316	0.316	0.302	0.350

Table 2 shows the two scores $S_2(X)$ and $S_6(X)$ in (2) computed using these likelihood values where $m = 2$ is applied. It can be seen that with the score $S_2(X)$, we always have the EER $\neq 0$ since scores for X_3^i and X_4^i are higher than those for X_1^c and X_2^c. Using $S_6(X)$, the EER is reduced since the score for X_3^i is lower than those for X_1^c and X_2^c.

Table 3. Scores of 4 utterances using $S_{2nc}(X)$ and $S_{6nc}(X)$

Score	X_1^c	X_2^c	X_3^i	X_4^i
$S_{2nc}(X)$	2.251	1.710	-2.547	0.158
$S_{6nc}(X)$	0.139	0.152	0.109	0.136

However, with $\epsilon_3 = 10^{-4}$ for S_{2nc} and $\epsilon_6 = 0.5$ for S_{6nc}, from Table 3 we can see that with thresholds $\theta_3 = 1.0$ and $\theta_6 = 0.137$, the EER is 0 for both of these NC-based scores.

5 Experimental Results

The Gaussian mixture model (GMM) and Vector Quantisation (VQ) speaker verification experiments were performed on the ANDOSL and YOHO corpora in text-independent mode [10].

The YOHO corpus was collected by ITT under a US government contract and was designed for speaker verification systems in office environments with limited vocabulary. There are 138 speakers, 108 males and 30 females. The vocabulary consists of 56 two-digit numbers ranging from 21 to 97 pronounced as "twenty-one", "ninety-seven", and spoken continuously in sets of three, for example "36-45-89", in each utterance. There are four enrolment sessions per speaker, numbered 1 through 4, and each session contains 24 utterances. There are also ten verification sessions, numbered 1 through 10, and each session contains 4 utterances. All waveforms are low-pass filtered at 3.8 kHz, sampled at 8 kHz, and were processed as for the ANDOSL corpus [14] to form a 26-dimensional vector for each individual frame.

Table 4. EER results (%) for the YOHO corpus using GMMs and VQ codebooks.

Similarity	GMM			VQ		
Score	16-mixture	32-mixture	64-mixture	16-vector	32-vector	64-vector
$S_2(X)$	4.4	3.1	2.4	5.9	4.6	3.6
$\mathbf{S_{2nc}(X)}$	**4.2**	**2.9**	**2.0**	**4.9**	**4.2**	**3.5**
$S_3(X)$	4.4	3.1	2.4	5.9	4.6	3.6
$\mathbf{S_{3nc}(X)}$	**4.2**	**2.9**	**2.0**	**4.9**	**4.2**	**3.5**
$S_4(X)$	4.5	3.3	2.4	6.8	5.1	4.2
$\mathbf{S_{4nc}(X)}$	**3.9**	**2.7**	**1.9**	**4.5**	**3.4**	**2.9**
$S_5(X)$	5.0	3.4	2.4	6.9	5.1	4.2
$\mathbf{S_{5nc}(X)}$	**4.5**	**2.9**	**2.1**	**2.1**	**4.9**	**4.0**
$S_6(X)$	4.2	3.0	2.0	5.3	4.0	3.1
$\mathbf{S_{6nc}(X)}$	**4.2**	**2.9**	**2.0**	**4.5**	**3.4**	**3.0**

GMMs with 16, 32, and 64 mixtures and VQ codebooks with 16, 32 and 64 codevectors were trained for each speaker using 48 training tokens in enrolment sessions 1 and 2 only. Using all four enrolment sessions, as done for example by Reynolds [10], resulted in error rates that were too low to allow meaningful comparisons between the different normalisation methods. Experiments were performed on 138 speakers using each speaker as a claimed speaker with 5 closest background speakers and 132 mixed-gender impostors (excluding 5 background speakers) and rotating through all speakers. The total number of claimed test utterances and impostor test utterances are 5,520 (138 claimed speakers x 40 test utterances) and 728,640 ((138 x 132) impostors x 40 test utterances), respectively. Results for the YOHO database are shown in Table 4, where $m = 1.05$ and $\epsilon = 10^{-9}$ for all NC-based methods using GMMs and $\epsilon = 10^{-22}$ for all NC-based methods using VQ codebooks. The current normalisation method $S_5(X)$ produced the highest EER of 6.9% for 16-codevector VQ codebooks and the proposed methods $S_6(X)$ and $S_{6nc}(X)$ produced the lowest EER of 2.0% for 64-mixture GMMs. Results for the ANDOSL are shown in Table 5, where $m = 1.05$ and $\epsilon = 10^{-7}$ for all NC-based methods. The current normalisation method $S_4(X)$ produced the highest EER of 3.5% for 16-mixture GMMs and the proposed method $S_{6nc}(X)$ produced the lowest EER of 1.3% for 32-mixture GMMs. These two tables show that all of the proposed methods perform better than the current and FCM methods.

Table 5. EER results (%) for the ANDOSL corpus using GMMs

Similarity	Set 1: 5 closest speakers		Set 2: 5 same-group speakers	
Score	16-mixture GMM	32-mixture GMM	16-mixture GMM	32-mixture GMM
$S_2(X)$	2.2	1.8	2.7	2.1
$\mathbf{S_{2nc}(X)}$	**1.8**	**1.4**	**2.0**	**1.4**
$S_3(X)$	2.2	1.8	2.7	2.1
$\mathbf{S_{3nc}(X)}$	**1.8**	**1.4**	**2.0**	**1.4**
$S_4(X)$	3.2	3.0	3.5	3.3
$\mathbf{S_{4nc}(X)}$	**2.0**	**1.9**	**2.3**	**2.1**
$S_{5nc}(X)$	2.3	2.0	2.9	2.3
$\mathbf{S_{5nc}(X)}$	**2.0**	**1.7**	**2.2**	**1.7**
$S_6(X)$	1.9	1.5	2.3	1.7
$\mathbf{S_{6nc}(X)}$	**1.7**	**1.3**	**1.9**	**1.3**

6 Conclusion

The noise clustering-based normalisation method for speaker verification has been considered. This is an interesting fuzzy approach to speaker verification. This approach is simple but robust and has very good experimental evaluations. Based on considering the false acceptances, six noise clustering-based scores have been proposed and experimentally evaluated. Experiments show better results for the proposed methods. With 2,093,040 test utterances for each ANDOSL result and 728,640 test utterances for each YOHO result, these evaluation experiments are sufficiently reliable.

References

1. J. C. Bezdek, *Pattern Recognition with Fuzzy Objective Function Algorithms*, Plenum Press, New York and London, 1981.
2. F. Chen, B. Millar and M. Wagner, "Hybrid-threshold approach in text-independent speaker verification", in *Proc.* ICSLP'94, Yokohama, pp. 1855-1858, 1994.
3. R. N. Davé, "Characterization and detection of noise in clustering", *Pattern Recognition Lett.*, vol. 12, no. 11, pp. 657-664, 1991.
4. S. Furui, "An Overview of Speaker Recognition Technology", in *Proc. ESCA Workshop on Automatic Speaker Recognition, Identification & Verification*, pp. 1-9, 1994.
5. A. L. Higgins, L. Bahler and J. Porter, "Speaker Verification using Randomized Phrase Prompting", *Digital Signal Processing*, vol. 1, pp. 89-106, 1991.
6. C. S. Liu, H. C. Wang and C.-H. Lee, "Speaker Verification using Normalization Log-Likelihood Score", *IEEE Trans. Speech and Audio Processing*, vol. 4, pp. 56-60, 1980.
7. K. P. Markov and S. Nakagawa, "Text-independent speaker recognition using non-linear frame likelihood transformation", *Speech Communication*, vol. 24, pp. 193-209, 1998.
8. T. Matsui and S. Furui, "Concatenated Phoneme Models for Text Variable Speaker Recognition", in *Proc.* ICASSP'93, pp. 391-394, 1993.
9. J. B. Millar, J. P. Vonwiller, J. M. Harrington and P. J. Dermody, "The Australian National Database of Spoken Language", in *Proc.* ICASSP'94, vol. 1, pp. 97-100, 1994.
10. D. A. Reynolds, "Speaker identification and verification using Gaussian mixture speaker models", *Speech Communication*, vol. 17, pp. 91-108, 1995.
11. A. E. Rosenberg, J. Delong, C.-H. Lee, B.-H. Juang and F. K. Soong, "The use of cohort normalised scores for speaker verification", in *Proc.* ICSLP'92, pp. 599-602, 1992.
12. D. Tran and M. Wagner, "A Proposed Likelihood Transformation for Speaker Verification", in *Proc.* ICASSP2000, vol. 2, pp. 1069-1072, Turkey, 2000.
13. D. Tran and M. Wagner, "Fuzzy Normalisation Methods for Speaker Verification", in *Proc.* ICSLP2000, vol. 1, pp. 446-449, Beijing, China, 2000.
14. D. Tran and M. Wagner, "Fuzzy *C*-means clustering-based speaker verification", in *Proc.* AFSS2002, Calcutta, India, Feb., 2002.

A Combination Scheme for Fuzzy Clustering

Evgenia Dimitriadou, Andreas Weingessel, and Kurt Hornik

Institut für Statistik, Wahrscheinlichkeitstheorie
Technische Universität Wien
Wiedner Hauptstraße 8–10/1071
A-1040 Wien, Austria
{dimi,weingessel,hornik}@ci.tuwien.ac.at
http://www.ci.tuwien.ac.at

Abstract. In this paper we present a voting scheme for cluster algorithms. This voting method allows us to combine several runs of cluster algorithms resulting in a common partition. This helps us to tackle the problem of choosing the appropriate clustering method for a data set where we have no a priori information about it, and to overcome the problems of choosing an optimal result between different repetitions of the same method. Further on, we can improve the ability of a cluster algorithm to find structures in a data set and to validate the resulting partition.[1]

1 Introduction

Combination strategies in classification are a popular way of improving the recognition rate in classification tasks [12,1,6,13]. The combination can be implemented using a variety of strategies, such as majority vote of the individual classifiers, voting networks, and several combinations of these schemes [17,7,8]. A direct application of ideas such as "voting" to cluster analysis problems is not possible, as no a priori class information for the patterns is available. We present a methodology for combining ensembles of partitions obtained by clustering, discuss the properties of such combination strategies and relate them to the task of assessing partition "agreement".

In data mining, the true structure, especially the number and shapes of the clusters, is unknown. Different cluster algorithms and even multiple replications of the same algorithm result in different solutions due to random initializations and stochastic learning methods. Moreover, there is no clear indication which of the different solutions of the replications of an algorithm is the best one [10,15,16,14,2]. Our "greedy" algorithm combines partitions on a sequential basis, thus overcoming the computationally infeasible simultaneous combination of all partitions, and improving also the ability of a cluster algorithm to find structures in a data set.

[1] This piece of research was supported by the Austrian Science Foundation (FWF) under grant SFB#010 ('Adaptive Information Systems and Modeling in Economics and Management Science').

N.R. Pal and M. Sugeno (Eds.): AFSS 2002, LNAI 2275, pp. 332–338, 2002.

The paper is organized as follows. In Section 2 we present our combination scheme. Section 3 demonstrates our experimental results and some comments on them. A conclusion of the paper is given in Section 4.

2 Voting Scheme

2.1 Description of the Voting Algorithm

We developed the following algorithm. Our input data x is clustered several times. Let $U_j^{(m)}$ denote the jth cluster in the mth run and $P_j^{(m)}$ denote the jth cluster in the combination of the first m runs. The first two runs are combined in the following way. First a mapping between the clusters of the two runs is defined. To do this we compute for each cluster $U_j^{(2)}$ how many percent of its points have been assigned to which cluster $U_k^{(1)}$. Then, the two clusters with the highest percentage of common points are assigned towards each other. Of the remaining clusters, again the two with the highest similarity are matched and so on. After renumbering the clusters of the second run so that $U_j^{(2)}$ corresponds to $U_j^{(1)}, \forall j$, we assign the points to the common clusters $P_j^{(2)}$ in the following way. If a data point x has been assigned to both $U_j^{(2)}$ and $U_j^{(2)}$ it will be assigned to $P_j^{(2)}$. If x has been assigned to $U_j^{(1)}$ in the first run and to $U_k^{(2)}$ with $j \neq k$ in the second run then it will be assigned to both, $P_j^{(2)}$ and $P_k^{(2)}$, with weight 0.5.

If we have already combined the first n runs in a common clustering $P_j^{(n)}$ we add an additional run $U_j^{(n+1)}$ by combining it with $P_j^{(n)}$ in the same way as for two runs, but give weight $n/(n+1)$ to the common clusters of the first n runs and weight $1/(n+1)$ to the new cluster. Note, that the algorithm is described for the case where the data are clustered only to the same prespecified number of clusters in all simple clustering runs.

After voting of M runs we get for every data point x and every cluster k a value $p_k^{(M)}$ which gives the fraction of times this data point has been assigned to this cluster. For interpreting the final result we can either accept this fuzzy decision or assign every data point to that cluster $k = \text{argmax}_j(P_j^{(M)})$ where it has been assigned most often. We define the *sureness* of a data point as the percentage of times it has been assigned to its "winning" cluster k, that is $sureness(x) = \max_j(P_j^{(M)})$. Then we can not only see how strong a certain point belongs to a cluster but we can also compute the average sureness of a cluster (*Avesure*) as the average sureness of all the points of a cluster that belong to it. Thus, we can notice which clusters have a clear data structure and which not.

2.2 Error Function

The final partition P is encoded as an $N \times k$ (k classes) matrix with elements p_{ij}. The task is to find P which optimally represents a given set of M partitions of the same set in such a way that

$$l(U^{(1)}, \ldots, U^{(M)}; P) := \frac{1}{M} \sum_{m=1}^{M} h(U^{(m)}, P) \tag{1}$$

is minimized over all P, where $h(U^{(m)}, P)$ is an appropriate distance function between the partitions $U^{(m)}$ and P.

As distance $h(U^{(m)}, P)$ we choose the average square distance between the membership functions [3,5] $u_i^{(m)}$ and p_i for all x_i. That is,

$$h(U^{(m)}, P) := \frac{1}{N} \sum_{i=1}^{N} ||u_i^{(m)} - p_i||^2 \tag{2}$$

Inserting (2) into (1) we get the minimization problem

$$\min_P l(U^{(1)}, \ldots, U^{(M)}; P) = \min_{p_1, \ldots, p_N} \left(\frac{1}{M} \sum_{m=1}^{M} \frac{1}{N} \sum_{i=1}^{N} ||u_i^{(m)} - p_i||^2 \right) \tag{3}$$

By interchanging the sums and using the property of the arithmetic mean to minimize the mean square error we get the solution

$$p_i = \frac{1}{M} \sum_{m=1}^{M} u_i^{(m)}, \forall i \tag{4}$$

Since it is not clear which class number (label) is assigned to which class in each run. That is, any relabeling of the classes, which can be written as a column permutation $\Pi_m(U^{(m)})$, are to be considered the same partition. Thus, partitions $U^{(m)}$ and $\Pi_m(U^{(m)})$ which only differ by a permutation of the class labels, are to be considered the same and thus we demand that

$$h(U^{(m)}, P) = h(\Pi_m(U^{(m)}), P), \quad \forall P \tag{5}$$

where $\Pi_m(U^{(m)})$ is any permutation of the columns of $U^{(m)}$.

Thus, the task of finding an optimal partition P is given by the minimization problem

$$\min_P l(U^{(1)}, \ldots, U^{(M)}; P) = \min_{p_1, \ldots, p_N} \min_{\pi_1, \ldots, \pi_M} \left(\frac{1}{M} \sum_{m=1}^{M} \frac{1}{N} \sum_{i=1}^{N} ||\pi_m(u_i^{(m)}) - p_i||^2 \right) \tag{6}$$

Note that, for finding an optimal P, we have to minimize p_i and π_m simultaneously, because the choice of the permutations π_m depends on the values of p_i. If we assume that the π_m are fixed, the optimal choice for the p_i is given by

$$p_i = \frac{1}{M} \sum_{m=1}^{M} \pi_m(u_i^{(m)}), \tag{7}$$

Note that the solution of our minimization problem is not unique, since the numbers of the classes (labels) in the final result are arbitrary. We can set for example $\pi_1 \equiv id$, thus avoiding a global maximum which is trivially non-unique.

A direct solution of the minimization problem 6 is infeasible, since it requires an enumeration of all possible column permutations.

For establishing an approximation algorithm we note that another view on this problem can be obtained by writing

$$\frac{1}{M}\sum_{m=1}^{M}||\pi_m(u_i^{(m)}) - p_i||^2 = \frac{1}{M}\sum_{m=1}^{M}||\pi_m(u_i^{(m)})||^2 - ||p_i||^2,$$

Since $||\pi_m(u_i^{(m)})||^2$ is permutation invariant, under the constraint 7, the problem 6 reduces to maximizing the *information*

$$\max_{\pi_1,\dots,\pi_M} \frac{1}{N}\sum_{i=1}^{N}||p_i||^2 \tag{8}$$

Using matrix notation, equation 7 becomes

$$P = \frac{1}{M}\sum_{m=1}^{M}\Pi_m(U^{(m)}) \tag{9}$$

Then, we can rewrite problem (8) as

$$\max_{\Pi_1,\Pi_2,\dots,\Pi_M} \frac{1}{N}\operatorname{tr}(P'P) \tag{10}$$

since

$$\operatorname{tr}(P'P) = \operatorname{tr}(PP') = \sum_{i=1}^{N} p_i'p_i = \sum_{i=1}^{N}||p_i||^2. \tag{11}$$

Combining Equations 9 & 10 and skipping constants we have to maximize

$$\max_{\Pi_1,\Pi_2,\dots,\Pi_M} \operatorname{tr}\left((\sum_{m=1}^{M}\Pi_m(U^{(m)}))'(\sum_{n=1}^{M}\Pi_n(U^{(n)})\right) \tag{12}$$

The following lemma tells us that we can skip the non-mixed terms in the above maximization.

Lemma: Let X be an arbitrary $n \times k$ matrix and $\Pi(X)$ be a column permutation of X, then

$$\operatorname{tr}(\Pi(X)'\Pi(X)) = \operatorname{tr}(X'X). \tag{13}$$

Proof: *Let $\pi(i)$ be the value of the permutation of the i-th column of X; then*

$$\operatorname{tr}(X'X) = \sum_{i=1}^{k} x_i'x_i = \sum_{i=1}^{k} x_{\pi(i)}'x_{\pi(i)}.$$

Considering the symmetry and linearity of the trace-operator we finally get

$$\begin{aligned}\max_{\Pi_m} \operatorname{tr}(P'P) &\hat{=} \max_{\Pi_m} \operatorname{tr}\left(\sum_{m=1}^{M-1}\sum_{n=m+1}^{M} \Pi_m(U^{(m)})'\Pi_n(U^{(n)})\right)\\ &= \max_{\Pi_m} \sum_{m=1}^{M-1}\sum_{n=m+1}^{M} \operatorname{tr} \Pi_m(U^{(m)})'\Pi_n(U^{(n)}) \end{aligned} \tag{14}$$

which can be written as maximization over $\Pi_1, \ldots, \Pi_M$ of

$$\begin{aligned}&\left(\operatorname{tr}\left(\Pi_1(U^{(1)})'\Pi_2(U^{(2)})\right) + \operatorname{tr}\left((\Pi_1(U^{(1)}) + \Pi_2(U^{(2)}))'\Pi_3(U^{(3)})\right) + \cdots\right.\\ &\left. + \operatorname{tr}\left((\sum_{m=1}^{M-1} \Pi_m(U^{(m)}))'\Pi_M(U^{(M)})\right)\right) \end{aligned} \tag{15}$$

This equation gives an iterative algorithm to find an approximate solution to the minimization problem 6. If we have only 2 partitions $U^{(1)}$ and $U^{(2)}$ we have to maximize

$$\max_{\Pi_1,\Pi_2} \operatorname{tr}(\Pi_1(U^{(1)})'\Pi_2(U^{(2)})) \tag{16}$$

which is exactly the first term of the sum (15). Let $\hat{\Pi}_1$ and $\hat{\Pi}_2$ denote the solution of (16). Inserting into (9) yields the resulting voting partition $P^{(2)}$

$$P^{(2)} = \frac{1}{2}\left(\hat{\Pi}_1(U^{(1)}) + \hat{\Pi}_2(U^{(2)})\right)$$

If we add now a third partition $U^{(3)}$, we can approximate the maximization of

$$\max_{\Pi_1,\Pi_2,\Pi_3} \left(\operatorname{tr}(\Pi_1(U^{(1)})'\Pi_2(U^{(2)})) + \operatorname{tr}((\Pi_1(U^{(1)}) + \Pi_2(U^{(2)}))'\Pi_3(U^{(3)}))\right)$$

by maximizing first (16) to get $P^{(2)}$ and then maximizing

$$\max_{\Pi_3} \operatorname{tr}((\hat{\Pi}_1(U^{(1)}) + \hat{\Pi}_2(U^{(2)}))'\Pi_3(U^{(3)})) \hat{=} \max_{\Pi_3} \operatorname{tr}\left((P^{(2)})'\Pi_3(U^{(3)})\right).$$

Thus, the simultaneous maximization over $\{\Pi_1, \Pi_2, \Pi_3\}$ is split into the maximization over $\{\Pi_1, \Pi_2\}$ and then maximization over Π_3 . Additional partitions $U^{(m)}, m = 4, \ldots$ are added accordingly.

3 Experimental Results

For our experimental results we use two of the most popular fuzzy clustering algorithms in the literature: fuzzy c-means (also known as FCM) as an example of an off-line algorithm, and UFCL (unsupervised fuzzy competitive learning) as an online version of c-means (see for example [11]). For the comparison with the partition resulting from Voting the following validity indexes are used: Fuzzy hypervolume (FHV) and Average partition density (APD) [4], Xie-Beni index (XB) [16], Partition coefficient (PC) [2], Partition entropy (PE) [2], Separation index (known as CS Index) [2], Classification rate (CLRATE), Avesure (see Section 2). The experiments presented here consist of 100 voting runs on every 100 normal clustering ones. Note, that the 100 runs consist of both the FCM and UFCL algorithms results and the partitions used for voting are the crisp ones (using maximum membership). It is possible to use any clustering results and to proceed to voting on them as long as the crisp partition matrix is built for each of them. The results given in Table 1 are the mean value and standard deviation for the mentioned validity measures. Due to space constraints the results of only one data set are presented.

3.1 Iris Data Set

We apply a cluster algorithm to the well known Iris data set (see for example [9]). Voting does not improve significantly the performance (from 87.8% to 89.3%) but it renders absolute stable results opposite to c-means (sd of 7.75). The "setosa cluster", which is linearly separable from the others, is the surest (0.990 *Avesure*). Since the two clusters are not linearly separable the separation index yields a value smaller than 1. In the case of the iris data voting also concludes to less fuzzy partitions than c-means (PC, PE measures). Note that this data set is often benchmarked because of the two existing overlapping clusters. For this reason c-means does not yield a better result for the partition density criteria as well as for the Xie-Beni indexes. However, voting still yields more stable results.

Table 1. Comparative Performance

	FHV		APD		XB		PC		PE		CLRATE	
	mean	sd	mean	sd	mean	sd	mean	sd	mean	sd	mean	sd
Cl.alg.	0.048	0.003	586.0	42.82	0.219	0.362	0.780	0.015	0.401	0.024	87.85	7.750
Voting	0.030	0.007	1933	21.40	0.161	0.001	0.995	0.018	0.092	0.003	89.33	0.000

4 Conclusion

In this paper we present a method to combine the results of several independent cluster runs by voting between their results. This sequential scheme helps us to overcome the computationally infeasible simultaneous combination of all partitions. It allows us to deal with the problem of local minima of cluster algorithms and to find a partition of the data which is supported by repeated applications of the cluster algorithm and not influenced by the randomness of initialization or the cluster process itself.

References

1. Eric Bauer and Ron Kohavi. An empirical comparison of voting classification algortihms: Bagging, boosting, and variants. *Machine Learning*, 36:105–139, 1999.
2. James C. Bezdek. *Pattern Recognition with Fuzzy Objective Function Algorithms*, chapter Cluster Validity, pages 100–120. Plenum Press, N.Y., 1981.
3. Didier Dubois and Henri Prade. *Fuzzy Sets and Systems: Theory and Applications*, chapter Fuzzy Sets, page 24. Academic Press, Harcourt Brace Jovanovich, 1980.
4. I. Gath and A. B. Geva. Unsupervised optimal fuzzy clustering. *IEEE Transactions on Pattern Analysis and Machine Intelligence*, 11(7):773–781, 1989.
5. Janusz Kacprzyk. Fuzzy-set-theoretic approach to the optimal assignment of work places. In G. Guardabassi and A. Locatelli, editors, *Large Scale Systems: Theory and Applications, Proceedings of the IFAC Symposium*, pages 123–131, Udine, Italy, June 1976.
6. L. Lam and C. Y. Suen. Application of majority voting to pattern recognition: An analysis of its behavior and performance. *IEEE Transactions on Systems, Man, and Cybernetics*, 22(5):553–568, 1997.
7. D. S. Lee and S. N. Srihari. Handprinted digit recognition: a comparison of algorithms. In *Proc. 3rd Int. Workshop Frontiers Handwriting Recognition*, pages 153–162, Buffalo, NY, May 1993.
8. P. R. Lorczak, A. K. Caglayan, and D. E. Eckhardt. A theoretical investigation of generalized voters for redundant systems. In *Proc. Int. Symp. Fault Tolerant Computing*, pages 444–451, Chicago, June 1989.
9. Hans-Joachim Mucha. *Clusteranalyse mit Mikrocomputern*. Akademie Verlag, 1992.
10. Nikhil R. Pal and James C. Bezdek. On cluster validity for the fuzzy c-means model. *IEEE Transactions on Fuzzy Systems*, 3(3):370–379, 1995.
11. N. R. Pal, J. C. Bezdek, and R. J. Hathaway. Sequential competitive learning and the fuzzy c-means clustering algorithm. *Neural Networks*, 9(5):787–796, 1996.
12. Behrooz Parhami. Voting algorithms. *IEEE Transactions on Reliability*, 43(4):617–629, 1994.
13. C. Y. Suen, C. Nadal, T. A. Mai, R. Legault, and L. Lam. Computer recognition of unconstrained handwritten numerals. In *Proc. IEEE*, volume 80, pages 1162–1180, July 1992.
14. M. Sugeno and T. Yasakawa. A fuzzy-logic-based approach to qualitative modelling. *IEEE Transactions on Fuzzy Systems*, 1(1):7–31, 1993.
15. M. P. Windham. Cluster validity for the fuzzy c-means clustering algorithm. *IEEE Transactions on Pattern Analysis and Machine Intelligence*, 4(4):357–363, 1982.
16. L. X. Xie and G. Beni. Validity measure for fuzzy clustering. *IEEE Transactions on Pattern Analysis and Machine Intelligence*, 3(8):841–847, 1991.
17. L. Xu, A. Krzyzak, and C. Y. Suen. Methods of combining multiple classifiers and their applications to handwriting recognition. *IEEE Transactions on Syst., Man, and Cybern.*, 22:418–435, 1992.

Clustering of Symbolic Data and Its Validation

Kalyani Mali[1] and Sushmita Mitra[2]

[1] Dept. of Computer Science, Kalyani University, Kalyani 741 235
[2] Machine Intelligence Unit, Indian Statistical Institute, Kolkata 700035
sushmita@isical.ac.in

Abstract. Categorical clustering of symbolic data and its validation has been studied. Symbolic objects include linguistic, nominal, boolean, and interval-type data. Clustering in this domain involves the use of symbolic similarity and dissimilarity between the objects. The optimal number of meaningful clusters are determined in the process. The effectiveness of the symbolic clustering is demonstrated on a real life benchmark dataset.

Keywords: Categorical clustering, data mining, validation, symbolic processing.

1 Introduction

The digital revolution has made digitized information easy to capture and fairly inexpensive to store. With the development of computer hardware and software and the rapid computerization of business, huge amount of data have been collected and stored in databases. The rate at which such data is stored is growing at a phenomenal rate. However, raw data is rarely of direct benefit. Its true value is predicated on the ability to extract information useful for decision support or exploration, and understanding the phenomenon governing the data source. All these have prompted the need for intelligent data analysis methodologies, which could discover useful knowledge from data. The term KDD refers to the overall process of *knowledge discovery in databases. Data mining* is a particular step in this process, involving the application of specific algorithms for extracting patterns (models) from data. One of the major challenges to data mining [1] is handling of *mixed media data.* This implies learning from data that is represented by a combination of various media, like (say) numeric, symbolic, images and text.

Clustering is a useful technique for the discovery of some knowledge from a dataset. It maps a data item into one of several clusters, where clusters are natural groupings of data items based on similarity metrics or probability density models. Clustering of data is broadly based on two approaches, *viz.*, hierarchical and partitive [2]. Hierarchical methods can again be categorized as *agglomerative* and *divisive* algorithms, corresponding to bottom-up and top-down strategies, to build a hierarchical clustering tree (*dendogram*). The optimal number of clusters is usually determined based on a validation index. There exist several clustering algorithms and validation indices in literature [2,3], but they mostly deal with numerical data.

N.R. Pal and M. Sugeno (Eds.): AFSS 2002, LNAI 2275, pp. 339–344, 2002.

Categorical clustering refers to the clustering of symbolic/categorical data. This is important from the point of view of data mining, where one has to mine for information from a set of symbolic objects. Symbolic objects are defined by attributes that can be quantitative (numeric or intervals) as well as qualitative. The similarity and dissimilarity measures between symbolic objects are determined based on their position, span and content [4,5]. Validation of symbolic clusters, however, is an issue hitherto untouched in literature.

In the present article we have studied the effectiveness of agglomerative symbolic clustering and its validation on real life categorical data. The symbolic benchmark dataset *Zoo* [6] is used for the purpose. An optimal number of meaningful clusters are obtained in the process. Section 2 provides a brief overview of symbolic objects, followed by a description of the symbolic/categorical clustering algorithm. Section 3 concentrates on the validation aspect of the clustering, using two indices in the symbolic framework. The implementation on the benchmark dataset is provided in Section 4. Section 5 concludes the article.

2 Categorical Clustering

In this section we describe symbolic objects and an agglomerative symbolic clustering algorithm.

2.1 Symbolic Objects

A symbolic object A can be expressed as the Cartesian product of specific values of its features A_i as

$$A = A_1 * A_2 * \ldots * A_n.$$

The dissimilarity between two symbolic objects A and B is defined as [4]

$$D(A, B) = \sum_{i=1}^{n} D(A_i, B_i), \tag{1}$$

where

$$D(A_i, B_i) = D_p(A_i, B_i) + D_s(A_i, B_i) + D_c(A_i, B_i)$$

with D_p, D_s and D_c (normalized to [0,1]) indicating the components due to position, span and content, respectively.

We have

$$D_p(A_i, B_i) = \frac{|\text{lower limit of } A_i - \text{lower limit of } B_i|}{\text{length of maximum interval along feature } i}, \tag{2}$$

where the denominator indicates the difference between the highest and lowest values of the ith feature over all the objects. This measure holds for quantitative attributes only. The remaining two measures are defined for both quantitative and qualitative attributes.

$$D_s(A_i, B_i) = \frac{|\text{length of } A_i - \text{length of } B_i|}{\text{span length of } A_i \text{ and } B_i}, \tag{3}$$

where span length denotes the length of the minimum interval containing both A_i and B_i for quantitative values. The length of a qualitative feature value is the number of its elements, and the span length of two such feature values is the number of elements in their union.

$$D_c(A_i, B_i) = \frac{|\text{length of } A_i + \text{length of } B_i - 2 * \text{length of intersection of } A_i \text{ and } B_i|}{\text{span length of } A_i \text{ and } B_i}. \tag{4}$$

2.2 Agglomerative Clustering

Agglomerative algorithms typically involve a repetition of the steps (i) assignment of pattern vectors X to clusters, (ii) inter-cluster distance computation, and (iii) merging of closest clusters, until an "optimum" number of clusters is obtained by minimizing the *within-cluster* distances and maximizing the *between-cluster* distances. Like typical hierarchical methods, the partitioning at any stage depends on the previously found clusters.

Merging of symbolic objects/clusters generally results in a composite representative object or their centroid. When the ith feature is quantitative it is $[\min(A_{il}, B_{il}), \max(A_{iu}, B_{iu})]$, where A_{il} and A_{iu} stand for the lower and upper limits respectively of A_i. In case of qualitative features, we have $A_i \cup B_i$. Let us explain this by an example. Consider the ith attribute to represent the *display* feature of a set of computer data. Let A_i be *black-and-white* and B_i be *color* monitor. Then $A_i \cup B_i$ will be {*black-and-white, color*}.

The symbolic clustering algorithm used is given below. It tends to favor the merging of singleton clusters, or of small clusters with large ones, as compared to the merging of medium sized clusters [4,7].

1. Let $\{X_1, X_2, \ldots, X_N\}$ be a set of N symbolic objects. Let the initial number of clusters be N, with each cluster having a weight (number of objects) of 1, *i.e.*, $c_i = 1$ for $i = 1, \ldots, N$.
2. Compute the weighted dissimilarities between all pairs of symbolic objects in the data set as
$$D_w(X_i, X_j) = D(X_i, X_j)\left(\frac{|c_i|.|c_j|}{|c_i| + |c_j|}\right)^{0.5}, \tag{5}$$
where $|c_i|$, $|c_j|$ are the weights of clusters/objects X_i, X_j respectively, and $D(X_i, X_j)$ is the dissimilarity value given by eqn. (1).
3. Determine the mutual pair having the lowest weighted dissimilarity $D_{w_{\min}}$ by eqn. (5). Form a composite object $X_{(i,j)_{cen}}$ by merging the individuals of this pair, such that
$$X_{(i,j)_{cen}} = \frac{|c_i|X_i + |c_j|X_j}{|c_i| + |c_j|} \tag{6}$$
and $|c_{(i,j)_{cen}}| = |c_i| + |c_j|$. Reduce the number of clusters by 1.
4. Compute the cluster validity index at the tth iteration as V_t.

5. Repeat steps 2 to 4 until the number of clusters equals 1.
6. Determine the stage t_o, with clusters $c = c_0$, $c = 2, \ldots, \sqrt{N}$, at which V_t is optimum. This indicates the optimal number of clusters. The composite objects at this stage denote the symbolic objects representing the clusters.

3 Cluster Validity Indices

To select the best among different partitioning, each of these can be evaluated using some validity index. The procedure is repeated for $c = 2, \ldots, \sqrt{N}$ number of clusters, where N is the size of the data set. Several validation methods have been proposed in literature [3] for quantitative data. In this section we modify the Davies-Bouldin index in the symbolic framework. We use the weighted dissimilarity D_w of eqn. (5) between object pairs, both for the average distance S_a within a cluster and the average linkage d_a between clusters, in the computations. The performance is compared with that of the Cluster indicator [4].

3.1 Davies-Bouldin Index

The index is a function of the ratio of the sum of within-cluster scatter to between-cluster separation. The best clustering, for $c = c_0$, minimizes

$$\frac{1}{c}\sum_{k=1}^{c} \max_{l \neq k} \left\{ \frac{S_a(U_k) + S_a(U_l)}{d_a(U_k, U_l)} \right\}. \tag{7}$$

Here the within-cluster scatter is minimized and the between-cluster separation is maximized. This gives good results for spherical clusters. Note that the novelty of our method lies in extending the Davies-Bouldin index from the existing quantitative domain to the symbolic framework.

3.2 Cluster Indicator

The cluster indicator value at the tth iteration is defined as

$$CI_t = \frac{R_t}{R_{t+1}}, \tag{8}$$

where

$$R_t = \frac{\min_{k \neq l} d_a^t(U_k, U_l)}{\min_{k' \neq l'} d_a^{t+1}(U_{k'}, U_{l'}) + \min_{k'' \neq l''} d_a^{t-1}(U_{k''}, U_{l''})}. \tag{9}$$

This is maximized over different iterations for $t = 2, \ldots, N-2$.

4 Results

The symbolic clustering algorithm was implemented on a benchmark dataset, *viz.*, *Zoo* [6]. Since the objective was to do clustering, the class information was eliminated from the data.

The *Zoo* data set consists of 100 instances of animals with 17 features and 7 output classes. The name of the animal constitutes the first attribute. There are 15 boolean features corresponding to the presence of hair, feathers, eggs, milk, backbone, fins, tail; and whether airborne, aquatic, predator, toothed, breathes, venomous, domestic, catsize. The numeric attribute corresponds to the number of legs lying in the set {0,2,4,5,6,8}.

The clustering algorithm provided four clusters for the *Zoo* data with both validity indices. These are enumerated in Table 1, with the first and second columns indicating the validity index used and the number of elements in each cluster (in parentheses) respectively. The last column provides the individual elements in the sequential order of their entry in the corresponding cluster, while optimizing the two indices. It is observed that the resulting clusters are semantically meaningful, and also very similar to those obtained by Kohonen's self-organizing feature map [8].

Table 1. Symbolic clustering of *Zoo* data

Index	Cluster No.	Animals
CI & Davies-Bouldin	1 (41)	aardvark, bear, girl, boar, cheetah, leopard, lion, raccoon, wolf, lynx, mongoose, polecat, puma, mink, platypus, dolphin, porpoise, seal, sealion, antelope, buffalo, deer, elephant, giraffe, oryx, gorilla, wallaby, calf, goat, pony, reindeer, pussycat, cavy, hamster, fruitbat, vampire, squirrel, hare, vole, mole, opossum.
	2 (22)	bass, catfish, piranha, chub, herring, carp, haddock, seahorse, sole, dogfish, pike, tuna, stingray, frog, toad, newt, tuatara, pitviper, slowworm, scorpion, seasnake.
	3 (21)	chicken, dove, parakeet, lark, pheasant, sparrow, wren, flamingo, ostrich, tortoise, crow, hawk, vulture, kiwi, rhea, penguin, duck, swan, gull, skimmer, skua.
	4 (17)	clam, seawasp, crab, starfish, crayfish, lobster, octopus, flea, termite, slug, worm, gnat, ladybird, housefly, moth, honeybee, wasp.

5 Conclusions and Discussion

Real life data is not necessarily restricted to the numeric domain. Hence the need for symbolic processing to efficiently handle data like linguistic, nominal, boolean, interval, etc. Partitioning of such data demands the use of symbolic measures for determining the similarity and dissimilarity between objects. In this article we have studied the effectiveness of categorical clustering and its validation on a benchmark symbolic dataset. The resultant optimal clusters are found to be *naturally* meaningful.

References

1. S. Mitra, S. K. Pal, and P. Mitra, "Data mining in soft computing framework: A survey," *IEEE Transactions on Neural Networks* (to appear).
2. A. K. Jain and R. C. Dubes, *Algorithms for Clustering Data.* NJ: Prentice Hall, 1988.
3. J. C. Bezdek and N. R. Pal, "Some new indexes for cluster validity," *IEEE Transactions on Systems, Man, and Cybernetics, Part-B*, vol. 28, pp. 301–315, 1998.
4. K. Chidananda Gowda and E. Diday, "Symbolic clustering using a new dissimilarity measure," *Pattern Recognition*, vol. 24, no. 6, pp. 567–578, 1991.
5. K. Chidananda Gowda and T. V. Ravi, "Divisive clustering of symbolic objects using the concepts of both similarity and dissimilarity," *Pattern Recognition*, vol. 28, no. 8, pp. 1277–1282, 1995.
6. C. Blake and C. Merz, "UCI repository of machine learning databases," 1998. University of California, Irvine, Dept. of Information and Computer Sciences, *http://www.ics.uci.edu/~mlearn/MLRepository.html*.
7. K. Chidananda Gowda and E. Diday, "Unsupervised learning through symbolic clustering," *Pattern Recognition Letters*, vol. 12, pp. 259–264, 1991.
8. D. Alahakoon, S. K. Halgamuge, and B. Srinivasan, "Dynamic self-organizing maps with controlled growth for knowledge discovery," *IEEE Transactions on Neural Networks*, vol. 11, no. 3, pp. 601–614, 2000.

Generalised Fuzzy Hidden Markov Models for Speech Recognition

Dat Tran and Michael Wagner

University of Canberra, School of Computing, ACT 2601, Australia
{Dat.Tran,Michael.Wagner}@canberra.edu.au
http://www.ise.canberra.edu.au/DatT

Abstract. A generalised fuzzy approach to statistical modelling techniques for speech recognition is proposed in this paper. Fuzzy C-means (FCM) and fuzzy entropy (FE) techniques are combined into a generalised fuzzy technique and applied to hidden Markov models (HMMs). A more robust version of the above fuzzy technique based on the noise clustering (NC) method is also proposed. Experimental results were performed on the TI46 speech data corpus. A significant result for isolated-word recognition performed on a highly confusable vocabulary consisting of the nine English E-set words is that, a 33.8% recognition error rate for the HMM-based system was reduced to 30.5%, 29.9%, 29.8% and 27.8%, respectively, by using the FCM-HMM, the FE-HMM, the NC-FE-HMM, and the NC-FCM-HMM-based systems.

1 Introduction

The pattern recognition approach is the method of choice for speech recognition because of its simplicity of use, proven high performance, and relative robustness and invariance to different speech vocabularies, users, algorithms and decision rules. This method has two steps: training of speech patterns, and recognition of patterns via pattern comparison. The most successful approach is to treat the speech signal as a stochastic pattern and to use hidden Markov modelling of the speech signal. The hidden Markov model [12] is a doubly stochastic process with an underlying Markov process which is not directly observable (hidden) but which can be observed through another set of stochastic processes that produce observable events in each of the states.

An alternative successful approach in pattern recognition and cluster analysis is the fuzzy set theory-based approach [13], where fuzzy C-means (FCM) and noise clustering (NC) are widely used methods [4]. An early application based on fuzzy set theory in decision making for speech and speaker recognition was proposed in 1977 [7]. Recent applications are based on the use of the FCM algorithm instead of the hard C-means (HCM) algorithm in coding feature vector sequences for the discrete HMM [2]. Hybrid neuro-fuzzy systems are also recent approaches to speech and speaker recognition [6]. It can be said that, fuzzy set theory-based applications have been developed intensively for speech recognition. Therefore, this paper proposes an alternative approach of fuzzy set theory to

N.R. Pal and M. Sugeno (Eds.): AFSS 2002, LNAI 2275, pp. 345–351, 2002.

statistical modelling techniques. To obtain a general fuzzy approach to HMMs in speech recognition, we combine three methods in fuzzy clustering— FCM clustering [1], NC [3], and fuzzy entropy (FE) clustering [9]—and propose a fuzzy reestimation algorithm for training fuzzy HMMs.

Speech recognition results for evaluation of proposed fuzzy modelling techniques are reported using the TI46 (Texas Instruments) and the ANDOSL (Australian National Database Of Spoken Language) speech data corpora. Fuzzy HMMs and HMMs are used in isolated word recognition experiments. Fuzzy models, especially NC-based models are more effective than conventional models in isolated word recognition performed on the highly confusable vocabulary consisting of the nine English E-set words, the 10-digit set, and the 10-command words.

2 Conventional HMMs

Let $O = (o_1\, o_2 \dots o_T)$ be the observation sequence, $S = (s_1\, s_2 \dots s_T)$ the unobservable state sequence, $V = \{v_1, v_2, \dots, v_K\}$ the discrete symbol set, and N the number of states. A compact notation $\lambda = \{\pi, A, B\}$ is proposed to indicate the complete parameter set of the HMM, where

- $\pi = \{\pi_i\}, \quad \pi_i = P(s_1 = i|\lambda), \quad 1 \le i \le N$: the initial state distribution;
- $A = \{a_{ij}\}, \quad a_{ij} = P(s_{t+1} = j|s_t = i, \lambda), \quad 1 \le i, j \le N$, and $1 \le t \le T-1$: the state transition probability distribution, denoting the transition probability from state i at time t to state j at time $t+1$; and
- $B = \{b_j(o_t)\}, \quad b_j(o_t) = P(o_t|s_t = j, \lambda), \quad 1 \le j \le N$, and $\quad 1 \le t \le T$: the observation probability distribution, denoting the probability of generating an observation o_t in state j at time t with probability $b_j(o_t)$.

The Baum-Welch algorithm yields an iterative procedure to reestimate the model parameters λ. The Q-function for the HMM is as follows [5]

$$Q(\lambda, \overline{\lambda}) = \sum_{t=0}^{T-1} \sum_{i=1}^{N} \sum_{j=1}^{N} P(s_t = i, s_{t+1} = j|O, \lambda) \log \left[\overline{a}_{ij} \overline{b}_j(o_{t+1})\right] \tag{1}$$

Regrouping (1) into three terms for the π, A, B coefficients and applying the Lagrange multipliers subject to the following constraints:

$$\sum_{i=1}^{N} \pi_i = 1, \qquad \sum_{j=1}^{N} a_{ij} = 1, \qquad \text{and} \qquad \sum_{k=1}^{K} b_j(k) = 1 \tag{2}$$

we obtain the HMM parameter estimation equations

$$\overline{\pi}_j = \gamma_1(i) \qquad \overline{a}_{ij} = \frac{\sum_{t=1}^{T-1} \xi_t(i,j)}{\sum_{t=1}^{T-1} \gamma_t(i)} \qquad \overline{b}_j(k) = \frac{\sum_{\substack{t=1 \\ s.t.\ o_t = v_k}}^{T} \gamma_t(j)}{\sum_{t=1}^{T} \gamma_t(j)} \tag{3}$$

where

$$\gamma_t(i) = \sum_{j=1}^{N} \xi_t(i,j), \qquad \xi_t(i,j) = \frac{\alpha_t(i) a_{ij} b_j(o_{t+1}) \beta_{t+1}(j)}{\sum_{i=1}^{N}\sum_{j=1}^{N} \alpha_t(i) a_{ij} b_j(o_{t+1}) \beta_{t+1}(j)} \tag{4}$$

and

$$\alpha_t(i) = P(o_1 o_2 \ldots o_t, s_t = i|\lambda), \qquad \beta_t(i) = P(o_{t+1} o_{t+2} \ldots o_T | s_t = i, \lambda) \tag{5}$$

3 Generalised Fuzzy HMMs

3.1 FCM-FE-HMMs

The generalised fuzzy objective function for the fuzzy HMM (FHMM) is proposed as follows

$$G_{mn}(U,\lambda) = \sum_{t=0}^{T-1} \sum_{s_t} \sum_{s_{t+1}} u^m_{s_t s_{t+1}} d^2_{s_t s_{t+1}} + n\,\delta_{m1} \sum_{t=0}^{T-1} \sum_{s_t} \sum_{s_{t+1}} u_{s_t s_{t+1}} \log u_{s_t s_{t+1}} \tag{6}$$

where $m \geq 1$ is the degree of fuzziness, $n > 0$ is the degree of fuzzy entropy, $u_{s_t s_{t+1}} = u_{s_t s_{t+1}}(O)$ is the fuzzy membership function denoting the degree to which the observation sequence O belongs to state sequence being state s_t at time t and state s_{t+1} at time $t+1$, satisfying

$$0 \leq u_{s_t s_{t+1}} \leq 1, \qquad \sum_{s_t=1}^{N} \sum_{s_{t+1}=1}^{N} u_{s_t s_{t+1}} = 1, \qquad 0 < \sum_{t=1}^{T} u_{s_t s_t} < T \qquad \forall s_t, s_{t+1} \tag{7}$$

δ_{m1} is the Kronecker delta function ($\delta_{m1} = 1$, if $m = 1$; $= 0$, otherwise), and $d^2_{s_t s_{t+1}} = -\log P(O, s_t, s_{t+1}|\lambda)$. Note that π_{s_1} is denoted by $a_{s_0 s_1}$ in (6) for simplicity. The FCM-HMM and the FE-HMM functions are derived from the FHMM function as follows:

- As $m > 1$, the FHMM objective function reduces to the following FCM-HMM function

$$G_{(m>1)n}(U,\lambda) = \sum_{t=0}^{T-1} \sum_{s_t} \sum_{s_{t+1}} u^m_{s_t s_{t+1}} d^2_{s_t s_{t+1}} \tag{8}$$

- As $m = 1, n > 0$, the FHMM objective function reduces to the following FE-HMM function

$$G_{(m=1)(n>0)}(U,\lambda) = \sum_{t=0}^{T-1} \sum_{s_t} \sum_{s_{t+1}} u_{s_t s_{t+1}} d^2_{s_t s_{t+1}} + n \sum_{t=0}^{T-1} \sum_{s_t} \sum_{s_{t+1}} u_{s_t s_{t+1}} \log u_{s_t s_{t+1}} \tag{9}$$

The task is to minimise the generalised function on variables U and λ, e.g., finding a pair of $(\overline{U}, \overline{\lambda})$ such that $G_{mn}(\overline{U}, \overline{\lambda}) \leq G_{mn}(U, \lambda)$. This task is implemented by two alternative steps: 1) Finding $\overline{U}$ such that $G_{mn}(\overline{U}, \lambda) \leq G_{mn}(U, \lambda)$, and then 2) Finding $\overline{\lambda}$ such that $G_{mn}(\overline{U}, \overline{\lambda}) \leq G_{mn}(\overline{U}, \lambda)$. Assuming that we are in state i at time t and state j at time $t+1$. Minimising the G-function on U is based on minimising the FCM and FE objective functions, therefore

$$\overline{u}_{ijt} = \begin{cases} \dfrac{1}{\sum_{k=1}^{N}\sum_{l=1}^{N} \left(d_{ijt}^2 / d_{klt}^2\right)^{1/(m-1)}} & m > 1 \\ \dfrac{1}{\sum_{k=1}^{N}\sum_{l=1}^{N} \left(e^{d_{ijt}^2} / e^{d_{klt}^2}\right)^{1/n}} & m = 1 \quad n > 0 \end{cases} \tag{10}$$

where $u_{ijt} = u_{ijt}(O)$ is the fuzzy membership function denoting the degree to which the observation sequence O belongs to state sequence being state i at time t and state j at time $t+1$, satisfying

$$0 \leq u_{ijt} \leq 1 \quad \forall i, j, t, \qquad \sum_{i=1}^{N}\sum_{j=1}^{N} u_{ijt} = 1 \quad \forall t, \qquad 0 < \sum_{t=1}^{T} u_{iit} < T \quad \forall i \tag{11}$$

and

$$d_{ijt}^2 = -\log P(O, s_t = i, s_{t+1} = j|\lambda) = -\log\left[\alpha_t(i) a_{ij} b_j(o_{t+1}) \beta_{t+1}(j)\right] \tag{12}$$

For minimising the G-function on λ, since $P(O, s_t = i, s_{t+1} = j|\lambda) = \alpha_t(i) a_{ij} b_j(o_{t+1}) \beta_{t+1}(j)$, regrouping the first term in (6) into five terms, where two terms for $\alpha_t(i)$ and $\beta_{t+1}(j)$ are neglected, and applying Lagrange multipliers for three terms of the π, A, B coefficients, we obtain model parameter estimation equations for the fuzzy discrete HMM as follows [11]

$$\overline{\pi}_j = \sum_{i=1}^{N} \overline{u}_{ij1}^m \qquad \overline{a}_{ij} = \frac{\sum_{t=1}^{T-1} \overline{u}_{ijt}^m}{\sum_{t=1}^{T-1}\sum_{j=1}^{N} \overline{u}_{ijt}^m} \qquad \overline{b}_j(k) = \frac{\sum_{\substack{t=1 \\ s.t.\, o_t = v_k}}^{T} \sum_{i=1}^{N} \overline{u}_{ijt}^m}{\sum_{t=1}^{T}\sum_{i=1}^{N} \overline{u}_{ijt}^m} \tag{13}$$

Note that in practical implementation, a scaling procedure is required [12]. From (10) and (13) as $m = n = 1$ we obtain $\overline{u}_{ijt} = \xi_t(i, j)$, therefore it can be said that the conventional HMM is a special case of the FHMM.

3.2 NC-FCM-FE-HMMs

For the fuzzy HMM in the NC approach, a separate state is used to represent outliers and can be named the *garbage state* [10]. This state has a constant distance δ from all observation sequences. The membership $u_{\bullet t}$ of an observation sequence O at time t in the garbage state is defined to be

$$u_{\bullet t} = 1 - \sum_{i=1}^{N}\sum_{j=1}^{N} u_{ijt} \qquad 1 \le t \le T \tag{14}$$

Therefore, the membership constraint for the good states is relaxed to

$$\sum_{i=1}^{N}\sum_{j=1}^{N} u_{ijt} < 1 \qquad 1 \le t \le T \tag{15}$$

This allows noisy data and outliers to have arbitrarily small membership values in good states. The fuzzy objective function for the NC-FHMM is

$$G_{mn}(U,\lambda) = \sum_{t=0}^{T-1}\sum_{i=1}^{N}\sum_{j=1}^{N} u_{ijt}^{m} d_{ijt}^{2} + \sum_{t=0}^{T-1} u_{\bullet t}^{m}\delta^{2}$$
$$+ n\,\delta_{m1}\sum_{t=0}^{T-1}\sum_{i=1}^{N}\sum_{j=1}^{N} u_{ijt}\log u_{ijt} + n\,\delta_{m1}\sum_{t=0}^{T-1} u_{\bullet t}\log u_{\bullet t} \tag{16}$$

The fuzzy membership function for the NC-FHMM can be computed as follows

$$\overline{u}_{ijt} = \begin{cases} \dfrac{1}{\sum_{k=1}^{N}\sum_{l=1}^{N}\left(d_{ijt}^{2}/d_{klt}^{2}\right)^{1/(m-1)} + \left(d_{ijt}^{2}/\delta^{2}\right)^{1/(m-1)}} & m > 1 \\ \dfrac{1}{\sum_{k=1}^{N}\sum_{l=1}^{N}\left(e^{d_{ijt}^{2}}/e^{d_{klt}^{2}}\right)^{1/n} + \left(e^{d_{ijt}^{2}}/e^{\delta^{2}}\right)^{1/n}} & m = 1 \quad n > 0 \end{cases} \tag{17}$$

4 Experimental Results

The TI46 database was used for speech recognition experiments. The data were processed as in [11]. The 6-state left-to-right HMMs in speaker-dependent mode were used for all experiments. In the training phase, 10 training tokens of each word were used to train conventional HMMs, FE-HMMs ($m = 1, n = 2.5, \delta = 0$), NC-FE-HMMs ($m = 1, n = 2.5, \delta = 1.5$), FCM-HMMs ($m = 1.2, \forall n, \delta = 0$) and NC-FCM-HMMs ($m = 1.2, \forall n, \delta = 1.5$) using vetor quantisation (VQ) codebook sizes of 16, 32, 64, and 128. In the recognition phase, isolated word recognition was carried out by testing all 160 test tokens against conventional HMMs, FE-HMMs, NC-FE-HMMs, FCM-HMMs and NC-FCM-HMMs of each of 16 speakers.

Table 1 presents the experimental results for the recognition of the E set consisting of 9 letters *b,c,d,e,g,p,t,v,z*. The results show that FHMMs are consistently better than HMMs. For example, using a VQ codebook size of 64,

Table 1. Isolated word recognition error rates (%) for the E set

Code book size	HMM	FE-HMM	NC-FE-HMM	FCM-HMM	NC-FCM-HMM
16	54.5	41.5	41.4	51.9	42.7
32	39.4	34.3	34.3	37.4	33.4
64	33.8	29.9	29.8	30.5	27.8
128	33.9	31.2	31.2	32.2	31.8

Table 2. Isolated word recognition error rates (%) for the 10-digit set

Code book size	HMM	FE-HMM	NC-FE-HMM	FCM-HMM	NC-FCM-HMM
16	6.2	4.8	4.3	5.8	4.7
32	2.3	1.8	1.7	2.2	2.1
64	0.4	0.4	0.3	0.4	0.4
128	0.4	0.3	0.3	0.3	0.3

Table 3. Isolated word recognition error rates (%) for the 10-command set

Code book size	HMM	FE-HMM	NC-FE-HMM	FCM-HMM	NC-FCM-HMM
16	15.7	6.4	6.3	13.8	13.7
32	4.3	3.6	3.5	3.8	3.7
64	2.4	2.2	2.2	2.2	2.2
128	1.6	1.4	1.4	1.6	1.5

the average recognition error reductions by the FE-HMMs, the NC-FE-HMMs, the FCM-HMMs, and the NC-FCM-FHMMs in comparison with the HMMs are $(54.5 - 41.5)\% = 13.0\%$, $(54.5 - 41.4)\% = 13.1\%$, $(54.5 - 51.9)\% = 2.6\%$, and $(54.5 - 42.7)\% = 11.8\%$, respectively. Since limited training data was used (10 training tokens for each word), the errors for the codebook size of 128 are larger than those for a codebook size of 64. Table 2 is for the 10-digit set consisting of 10 digits from 0 to 9. Table 3 is for the 10-command set consisting of 10 commands: *enter, erase, go, help, no, rubout, repeat, stop, start, yes.* In most of the experiments, FHMMs show better results than conventional HMMs.

5 Conclusion

A generalised fuzzy technique including fuzzy C-means (FCM), fuzzy entropy (FE), and noise clustering (NC) has been applied to hidden Markov models (HMMs) in this paper. Experimental results in speech recognition indicate that fuzzy modelling techniques are very flexible since the degree of fuzziness m, the degree of fuzzy entropy n, and the constant distance δ can be chosen flexibly in each modelling problem to minimise recognition errors.

References

1. J. C. Bezdek, *Pattern recognition with fuzzy objective function algorithms*, Plenum Press, New York and London, 1981.
2. H. J. Choi and Y. H. Oh, "Speech recognition using an enhanced FVQ based on codeword dependent distribution normalization and codeword weighting by fuzzy objective function", in *Proceedings of the International Conference on Spoken Language Processing* (ICSLP), vol. 1, pp. 354-357, 1996.
3. R. N. Davé, "Characterization and detection of noise in clustering", *Pattern Recognition Lett.*, vol. 12, no. 11, pp. 657-664, 1991.
4. R. N. Davé and R. Krishnapuram, "Robust clustering methods: a unified view", *IEEE Trans. Fuzzy Syst.*, vol. 5, no.2, pp. 270-293.
5. X. D. Huang, Y. Ariki and M. A. Jack, *Hidden Markov models for speech recognition*, Edinburgh University Press, 1990.
6. N. Kasabov, R. Kozma, R. Kilgour, M. Laws, J. Taylor, M. Watts and A. Gray, "Hybrid connectionist-based methods and systems for speech data analysis and phoneme-based speech recognition" in *Neuro-Fuzzy Techniques for Intelligent Information Processing*, N. Kasabov & R.Kozma, eds., Physica Verlag, 1999.
7. S. K. Pal and D. D. Majumder, "Fuzzy sets and decision making approaches in vowel and speaker recognition", *IEEE Trans. Syst. Man Cybern.*, pp. 625-629, 1977.
8. E. Tsuboka and J. Nakahashi, "On the fuzzy vector quantisation based hidden Markov model", in *Proc. Inter. Conf. on Acoustics, Speech & Signal Processing* (ICASSP'94), vol. 1, pp. 537-640, 1994.
9. Dat Tran and Michael Wagner, "Fuzzy entropy clustering", the FUZZ-IEEE'2000 Conf., vol. 1, pp. 152-158, USA.
10. Dat Tran and Michael Wagner, "Hidden Markov models using fuzzy estimation", in *Proc. EUROSPEECH'99 Conf.*, vol. 6, pp. 2749-2752, Hungary, 1999.
11. Dat Tran and Michael Wagner, "Fuzzy hidden Markov models for speech and speaker recognition", in *Proc.* NAFIPS'99, pp. 426-430, USA, 1999.
12. L. R. Rabiner and B. H. Juang, *Fundamentals of speech recognition*, Prentice Hall PTR, USA, 1993.
13. L. A. Zadeh, "Fuzzy sets", *Inf. Control.*, vol. 8, no. 1, pp. 338-353, 1965.

On Generalization and K-Fold Cross Validation Performance of MLP Trained with EBPDT

Pinaki Roy Chowdhury[1] and K.K. Shukla[2]

[1] Scientist, Defence Terrain Research Laboratory, Metcalfe House, Delhi-110054, India
prc@ieee.org
[2] Reader, Department of Computer Engineering, Institute of Technology, BHU, Varanasi-221005, India
shukla@ieee.org

Abstract. This paper presents the generalization capability of multilayer perceptrons (MLP). The learning algorithm is based on mixing the concepts of dynamic tunneling along with error backpropagation (EBPDT), which enables detrapping of the local minimum point. In this study, the generalization capability is presented on three standard datasets, and the k-fold cross validation results is presented for two of the datasets. A comparative study of the performance of the proposed method with EBP clearly demonstrates the power of tunneling applied in conjunction with EBP type of learning.

1 Introduction

MLP– a class of artificial neural networks (ANN) has found applications in wide variety of tasks like pattern recognition, classification, function approximation etc. The advantage of ANN lies in its massive interconnections among nodes enabling high degree of parallelism. Besides this MLP has also been observed to posses a very strong generalization capability [1]. This facilitates processing of large data sets in an efficient manner.

Typically the performance measure of the network are two folds. The first is its training performance, i.e., how efficiently it can recognise data used for training, after the training is accomplished. Secondly, it's operational performance where it is supposed to perform on unknown/future data, taken from the same distribution as that of the training data. This is generally called the testing phase and this measures the generalization performance of the trained network. It is widely accepted fact that, EBP type of learning has very poor generalization ability, and thereby it requires the addition of regularization term along with the objective function to enhance the performance beyond the given training set.

EBPDT type of learning [2], which is recently reported in the literature, avoids the local trap by employing dynamic tunneling technique. This type of scheme is accomplished by modelling the system in such a manner that it violates the Lipschitz condition [3] at the equilibrium point, thereby enabling the search process to continue and possibly lead to global minimum in error surface [4]. In this work the focus is on the generalization capability of EBPDT learning algorithm, also to validate the

N.R. Pal and M. Sugeno (Eds.): AFSS 2002, LNAI 2275, pp. 352–359, 2002.

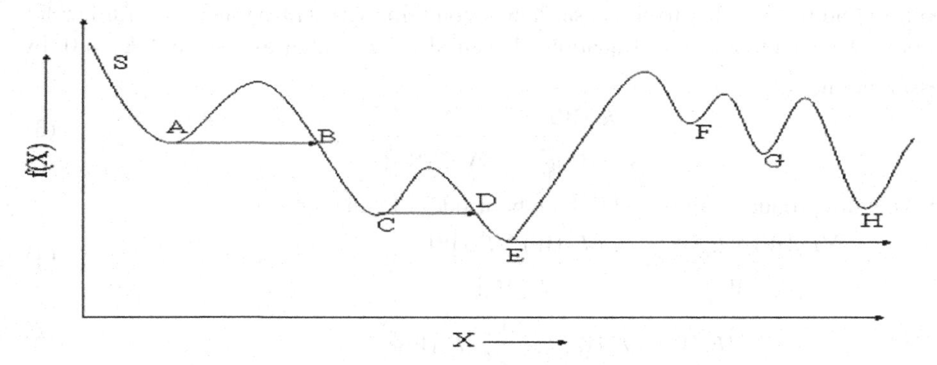

Fig. 1. A schematic view of tunneling operation in 1-D

performance of EBPDT with cross validation schemes, and to assess the results statistically by performing *student's t-test* [5] on the various training/testing errors obtained from multiple folds.

A schematic representation of EBPDT in 1-D is depicted above in Fig. 1. To explain the operation of tunneling, a simple objective function is taken for depiction. In Fig. 1, S is the randomly selected start point, A is the first local minimum obtained, C and E are subsequent minimum points obtained, where E is the point of global minimum. Clearly, in this scheme of operation, traversal of state spaces like F, G and H are eliminated.

2 Mathematical Formulation

This section will present the mathematical formulation necessary to show the similarity between MLP training with EBPDT and training MLP with a regularized objective function using EBP.

It is well known that the form of the regularised objective function that minimises the total risk is given as [1]:

$$R(W) = E_s(W) + \lambda E_c(W) \tag{1}$$

Where $E_s(W)$ is the well known mean of the sum squared error criteria and called as MSE, λ is the regularization parameter and $E_c(W)$ is the complexity penalty term, whose forms are given in [1]. Here W is the weight vector of the network under consideration. From equation (1), it is obvious that at the points of local or global optimal solutions, the gradient of $R(W)$ is zero, i.e., $\nabla R(W) = 0$. This implies the following:

(i) both $\nabla E_s(W)$ and $\nabla E_c(W)$ are zero, or

(ii) both $\nabla E_s(W)$ and $\nabla E_c(W)$ are non zero such that

$$\nabla E_s(W) + \lambda \nabla E_c(W) = 0 \tag{2}$$

Since condition (i) is a trivial case, hence condition (ii) will be analysed further for more relevant information. Equation (2) can also be written as, (since $\nabla E_S \neq 0$ by assumption)

$$\frac{\nabla E_s(W)}{\|\nabla E_s(W)\|^2} + \lambda \frac{\nabla E_c(W)}{\|\nabla E_s(W)\|^2} = 0 \cdot \tag{3}$$

Taking dot product with $\nabla E_s(W)$ on both sides of equation (3)

$$\frac{\nabla E_s(W).\nabla E_s(W)}{\|\nabla E_s(W)\|^2} + \lambda \frac{\nabla E_c(W).\nabla E_s(W)}{\|\nabla E_s(W)\|^2} = 0 \cdot \tag{4}$$

Or
$$\nabla E_c(W).\nabla E_s(W) = -\frac{1}{\lambda}\|\nabla E_s(W)\|^2 \,. \tag{5}$$

Which is the condition at equilibrium point.

Now the weight updation rule can be given as follows:

$$W(t+1) = W(t) - \eta \frac{\partial R}{\partial W} \tag{6}$$

From equation (6) it can be shown that

$$\Delta W(t) = -\eta \nabla R \tag{7}$$

Using equation (1)

$$\Delta W(t) = -\eta \nabla E_s(W) - \eta\lambda \nabla E_c(W) \,. \tag{8}$$

Or
$$\nabla E_c(W) = -\left[\frac{\Delta W(t) + \eta \nabla E_s(W)}{\eta\lambda}\right] . \tag{9}$$

Taking dot product with $\nabla E_s(W)$ on both sides of equation (9)

$$\nabla E_c(W).\nabla E_s(W) = -\left[\frac{\Delta W(t) + \eta \nabla E_s(W)}{\eta\lambda}\right] . \nabla E_s(W) \cdot \tag{10}$$

Using equation (5) along with (10) gives,

$$\Delta W(t).\nabla E_s(W) = 0 \,.$$

Or
$$\frac{dW}{dt} . \nabla E_s(W) = 0 \,. \tag{11}$$

For further details in interpreting equation (11) the readers are referred to [6].

Finally, before concluding this section, the weight updation scheme for the above two phases are presented below for the sake of clarity [6],[2];

$$\Delta w^l_{k,j} = \begin{cases} -\eta \partial R / \partial w^l_{k,j} & EBP \\ \int_t^{t+\Delta t} \rho\left(w^l_{k,j} - w^{l*}_{k,j}\right)^{1/3} dt & DTT \end{cases} \tag{12}$$

In equation (12), η is the learning rate, ρ is the strength of learning, w^l_{kj} is the weight connecting kth neuron in layer l to jth neuron in layer $l-1$. In equation (12), the weight updation for the EBP phase is quite simple and it follows from equation (6). However, the updation rule for DTT phase needs some explanation that is presented below.

Generally, the repeller system that is used for modelling the equilibrium point, which violates the Lipschitz condition, is typically given by [3],[4]:

$$\frac{du}{dt} = u^{1/3} \,. \tag{13}$$

The above system has a repelling unstable equilibrium point at u = 0, which violates the Lipschitz condition at the equilibrium point. Any initial condition which is infinitesimally close to the repelling point u = 0 will escape the repeller, to reach point $u_0 \neq 0$ in a finite time given by

$$t_1 = \int_0^{u_0} u^{-1/3} du = \frac{3}{2} u_0^{\;2/3} \,. \tag{14}$$

This concept of terminal repelling, thereby implementing tunneling along with error backpropagation forms the basis of the EBPDT learning algorithm. The equation given below, is a dynamical system which is modelled from the above discussion, where, $w^l_{j,k}$ is the perturbed weight vector and $w^{l*}_{j,k}$ is the corresponding weight vector at the equilibrium point.

$$\frac{dw^l_{j,k}}{dt} = \rho \int_t^{t+\Delta t} \left(w^l_{j,k} - w^{l*}_{j,k}\right)^{1/3} \,. \tag{15}$$

From equation (15) it is easy to obtain the DTT phase of equation (12).

Remark 1. Equation (11) is derived using regularized criteria function. For non-regularized, i.e., MSE criteria function, $\nabla E_s(W)=0$ at the equilibrium point. Hence, under the tunneling framework, equation (11) is valid.

3 Simulation Experiments

The three data sets used for testing the generalization performance are Iris data, British town data [7], and Pima Indians diabetes data [8]. Before presenting the generalization results, a brief description about the datasets are presented.

A. IRIS data (IRIS): The Iris dataset consists of 150 patterns. Each pattern is 4-dimensional, having the size measurements on various Iris species as given below:

x_1 : Sepal length, x_2 : Sepal width, x_3 : Petal length, x_4 : Petal width

Each of the 150 samples has been grouped into 3 classes as follows:

Group I: Iris setosa, Group II: Iris versicolor, Group III: Iris virginica

Amongst these 150 patterns, training/testing combination was formed and generalization ability of MLP trained with EBPDT is investigated. The combination, studied is as follows:

Size of Training Set	Size of Testing Set
90	60

The training/testing performance of the above mentioned pair is detailed in Table 1.

B. British town data (BTD): Original British town data consists of 155 samples of social-economic data collected from 155 British towns with each sample described by

57 variables. In this paper four principal components of the first 50 samples have been used because many of the original variables have been found to be highly correlated. These 50 samples form four cluster groups as follows:

Group I: Samples 1-10, Group II: Samples 11-20, Group III: Samples 21-36, and Group IV: Samples 37-50.

Amongst these 50 samples, the following training/testing combination is investigated.

Size of Training Set	Size of Testing Set
30	20

Generalization ability of MLP on the above mentioned data set is presented in Table 1.

C. Pima Indians Diabetes data (PIMA): The diagnostic variable investigated is whether the patient shows signs of diabetes according to WHO criteria. In this, there are 768 instances, which are 8 dimensional. The attributes in their chronological orders are:

1. No. of times pregnant, 2. Plasma glucose concentration, 3. Diastolic BP, 4. Triceps skin fold thickness, 5. 2-hour serum insulin, 6. Body mass index, 7. Diabetes pedigree function, 8. Age.

Amongst these 768 samples, the following training/testing combination is investigated

Size of Training Set	Size of Testing Set
400	368

In all the above mentioned training/testing pairs, sufficient care has been taken to generate the proportionate population from each class randomly. Table 1 given below, presents the results in terms of MLP architecture, training error, testing error, % classification in the testing case.

Table 1. Table presenting the percentage classification in test cases for the three datasets along with MLP architecture and training/testing errors

	MLP architecture	Training error	Testing error	% Classification in test cases
IRIS	4-10-2	0.0277	0.0183	96.66
BTD	4-12-2	0.0069	0.0401	95.00
PIMA	8-8-1	0.0634	0.0881	81.00

4 Cross Validation Performance

In this section the k-fold cross validation results will be presented, only for IRIS and BTD datasets. In particular, experiments were performed for 10-fold cross validation for IRIS data and 4-fold cross validation for BTD. The generation of folds will be presented subsequently. Student's *t-test* was performed on both the datasets, to validate the following hypothesis, called as null hypothesis in text [5].

Hypothesis, H_0: The training/Testing pattern combination does not have much effect on the generalization performance of EBPDT algorithm.

4.1 Fold Generation: IRIS Data

IRIS: As 10-fold cross validation is performed, so in each fold, number of training patterns are 135, whereas number of testing patterns are 15. Particularly it is emphasised to take 5 patterns for testing from each class. While generating the test patterns for fold, k, where $1 \leq k \leq 10$, it is carefully observed that any test pattern appearing in a particular fold does not appear in any other fold as a test pattern.

4.2 Fold Generation: BTD Data

BTD: Since the dataset is small so 4-fold cross validation is performed. In each fold, the number of training patterns is 38, whereas the number of testing patterns is 12. The selection of training/testing pairs is made cautiously as described earlier.

Table 2 and Table 3 given below presents the results of the 10-fold cross validation for Iris data and 4-fold cross validation for British town data respectively.

4.3 Fold Results and Computation of *t*

Table 2a. Cross validation results for Iris data indicating 5 out of 10 folds

Fold #	1	2	3	4	5
X_1(Training error)	0.025	0.023	0.021	0.020	0.020
X_2(Testing error)	0.007	0.005	0.003	0.033	0.056

Table 2b. Cross validation results for Iris data indicating the remaining 5 folds

Fold #	6	7	8	9	10
X_1(Training error)	0.022	0.017	0.020	0.022	0.022
X_2(Testing error)	0.019	0.048	0.034	0.004	0.006

Table 3. Cross validation results for British town data

FOLD #	1	2	3	4
X_1(Training error)	0.0079	0.0077	0.0110	0.0067
X_2(Testing error)	0.0061	0.0057	0.0418	0.0053

The Tables 2a, 2b and 3 above actually present mean squared error obtained in training and testing phases for various folds. Put in other words, Iris dataset went through 10 runs on different training/testing pairs, whereas the British town dataset went through 4 runs again on different training/testing pairs. In *student's t test,* t is computed by the formulae given below [5]

$$t = \frac{\overline{X_1} - \overline{X_2}}{S}\sqrt{\frac{n_1 n_2}{n_1 + n_2}} \; . \tag{16}$$

Where $S = \sqrt{\frac{n_1 \sum \left(X_1 - \overline{X}_1\right)^2 + n_2 \sum \left(X_2 - \overline{X}_2\right)^2}{n_1 + n_2 - 2}}$. (17)

and $\overline{X_1}$, $\overline{X_2}$ are the respective means.

Based on the data presented in Table 2a, 2b and Table 3, the value of S for IRIS and BTD are 0.045483, 0.025665 respectively. The respective mean values for IRIS and BTD are (0.0216, 0.02202) and (0.008325, 0.014725) for $\overline{X}_1$ and $\overline{X_2}$ taken together. Two sample convergence curve for IRIS and BTD are presented in Fig.2 and Fig. 3 respectively.

Hence the computed t value for IRIS is 0.020, and that of BTD is 0.35265. Tabulated value of t at 0.5% confidence level for 18 and 6 degrees of freedom are 2.88 and 3.71 respectively. In both the cases, it is observed that $t_{computed} < t_{tabulated}$. Thus, H_0 cannot be rejected at 0.005 level of significance.

Lastly, before concluding this Section, a comparative performance of the proposed method vis-à-vis EBP is included in Table 4 given below. It is noteworthy to mention that the classification result of Iris dataset using EBP is directly taken from literature [9].

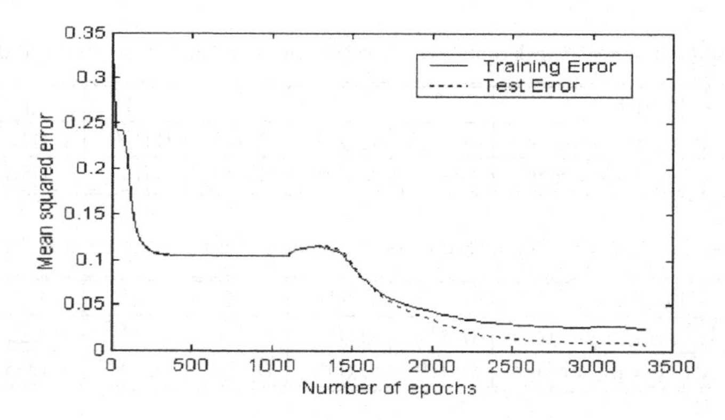

Fig. 2. Convergence curve for IRIS dataset

Table 4. Comparison of performance of EBPDT algorithm with EBP. First figure within the bracket depicts the % accuracy in classification for test cases, whereas the second figure represents the number of patterns used for testing the classification performance. The standard deviation for PIMA dataset is 4.09 and 1.76 for EBP and EBPDT respectively for 10 runs

Problem	Test case classification accuracy of EBP	Test case classification accuracy of EBPDT
IRIS	(78,53%, 50)	(96.66%, 60)
BTD	(75.0%, 20)	(95.0%, 20)
PIMA	(68.20%, 368)	(81.0%, 368)

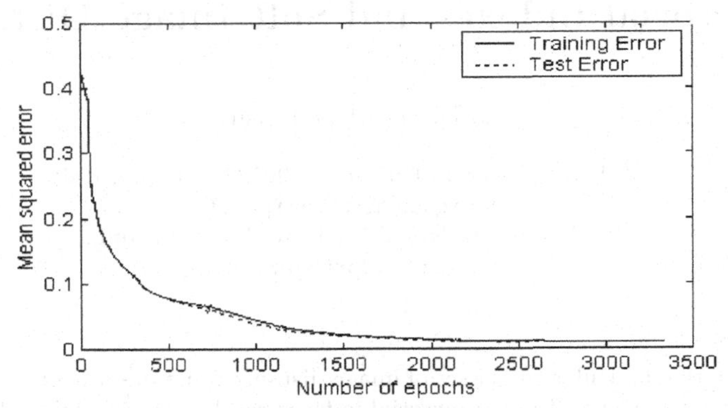

Fig. 3. Convergence curve for BTD dataset

5 Conclusion

The ability of EBPDT to correctly generalize beyond the training set is established by giving sufficient statistical and analytical reasons. This in a way puts emphasis on the power of dynamic tunneling applied in conjunction with EBP as a robust learning framework for MLP. It is proved and also experimentally validated that EBPDT can be treated as an automatic regularizer. The performance comparison with EBP indicates significant enhancement in classification in test case. This clearly establishes the effectiveness of the proposed method.

References

1. Haykin, S.: Neural Networks: A Comprehensive Foundation. 2nd edn. Prentice Hall, New Jersey (1999)
2. RoyChowdhury, P., Singh, Y.P., Chansarkar, R.A.: Dynamic Tunneling Technique for Efficient Training of Multilayer Perceptrons. IEEE Trans. On Neural Networks. **10** (1999) 48-55
3. Barhen, J., Protopopescu, V., Reister, D.: TRUST: A Deterministic Algorithm for Global Optimization. Science. **276** (1997) 1094-1097
4. RoyChowdhury, P., Singh, Y.P., Chansarkar, R.A.: Hybridization of Gradient Descent Algorithms with Dynamic Tunneling Methods for Global Optimization. IEEE Trans. On Systems, Man, and Cybernetics, Part A. **30** (2000) 386-392
5. Spiegel, M.R.: Theory and Problems of Statistics. 1st edn. McGraw–Hill, Singapore (1981)
6. Singh, Y.P., RoyChowdhury, P.: Dynamic tunneling based regularization in feedforward neural networks. Artifical Intelligence. **131** (2001) 55-71
7. Chien, Y.–t.: Interactive Pattern Recognition (Electrical Engineering and Electronics; v.3). Marcel Dekker Inc., New York (1978)
8. Murphy, P., Aha, D.: Repository of Machine Learning Databases. Dept. Inform. Comput. Sci.,
9. Univ. Calif., Irvine, CA, http://www.ics.uci.edu/AI/ML/MLDBRepository.html
10. Joshi, A., Ramakrishnan, N., Houstis, E.N., Rice, J.R.: On neurobiological, neuro-fuzzy, machine learning, and statistical pattern recognition techniques. IEEE Trans. On Neural Networks. **8** (1997) 18-31

Pictorial Indexes and Soft Image Distances

Vito Di Gesú[1] and Sisir Roy[2]

[1] DMA, Palermo University, 90123 Palermo, Italy
digesu@math.unipa.it
[2] Indian Statistical Institute, Calcutta, India
sisir@Springer.de

Abstract. Different classes of image-distance functions are often used in computer vision. Robust pictorial indexes can be also constructed based on distance criteria for image retrieval purposes. This paper introduces two new classes of entropic distances that are based on the concept of convex transformations. Their formal properties are studied and tested on real images. Experiments on the comparison of images and the matching of objects are presented. A comparison of image distances, here, is proposed and carried out with measures of closeness.

1 Introduction

Visual tasks are often based on the evaluation of distances or similarities between objects that are represented in an appropriate feature space. For example, an image querying system, grounded on the image content, processes the query on the basis of a matching procedure that assigns the unknown to the closest prototype. The performance of the whole query system depends on the definition of suitable Image Distance Functions (IDF's).

In the past, several definitions of distance between images have been considered [1,2]. The choice of the feature space (type and number of the parameters describing the image) is one of the crucial aspect of the problem. *Global* features, directly derived from gray levels (e.g. first and second order statistics, colour), can be used to define measures of similarity. However, global features may produce wrong results. *Structural* features (e.g. skeleton, medial axis, convex hull, symmetry) are more sensitive to shapes and they could give us complementary information. Distance functions, based on the combination of global and contextual information, seem to characterize more adequately the difference between images.

Soft approaches can be more effective whenever both probability and geometry fault. Soft distances are recommended whenever classification and matching are not determined by sharp conditions. Examples of fuzzy distances have been considered in image classification and segmentation [3].

Menger [4,5] first proposed a probabilistic generalization of the metric spaces as developed by Frechet and Hausdorff. A metric space is a pair (S, d) where S is a non-empty set and d is a metric on S. Menger replaced the number $d(p, q)$ by a real function F_{pq} whose value $F_{pq}(x)$, for any real x, is interpreted as the

N.R. Pal and M. Sugeno (Eds.): AFSS 2002, LNAI 2275, pp. 360–366, 2002.

probability that the distance between p and q is less than x. Since probabilities can neither be negative nor be greater than 1, we have $0 \leq F_{pq}(x) \leq 1$

The concept of entropy is closely related to those of *order* and *information*; for example it is well known that Boltzmann related the entropy with the very concept of probability. Moreover, the uncertainty of a system can be measured as a function of its entropy. Again, non probabilistic entropy of a fuzzy set has been discussed by several authors [6].

These considerations suggested us the use of the entropy function for the definition of fuzzy IDF's. For this purpose, a convex transformation, T, will be defined on a normalized feature space; where values range in the interval $[0, 1]$. Two new classes of entropic IDF, named E_1 and E_2, are then built by using T. They are based on fuzzy-entropy transformations that must be convex in the interval $[0, 1/2]$.

The main purpose of the paper is to evaluate the performance of these entropic IDS's, by comparing them with standard ones. For this purpose, matching experiments have been done on real data, that are in the JACOB image database [7]. Preliminary results indicate that a better discrimination performance can be reached by using the new entropic distance functions.

Sect. 2 describes the theoretical foundation of the new IDF's. Experimental results are shown in Sect. 3. Final remarks are given in Sect. 4.

2 IDF Based on Fuzzy-Entropy

In the following, elements are represented in a d-dimensional hypercube $H = \{\underline{x} \equiv (x_1, x_2, ..., x_d) | x_i \in [0, 1]\}$. In the case of images, the elements of H are normalized features that are assigned to each pixel. The normalization is done in the interval $[0, 1]$ and it can be obtained by using classical *image-fuzzifiers* and/or scaling functions.

Let us consider a vector space generated by a distance function $f : H \times H \rightarrow H$ that associates to each pair of elements $\underline{x}, \underline{y} \in H$ a new element: $\underline{a} \equiv (a_1, a_2, ..., a_i, ..., a_d)$; where $a_i = f(x_i, y_i)$.

Proposition 1 For each $\underline{x}, \underline{y}, \underline{z} \in H$, let us denote $\underline{a} = f(\underline{x}, \underline{y})$, $\underline{b} = f(\underline{y}, \underline{z})$, and $\underline{c} = f(\underline{x}, \underline{z})$, then:

1) $\underline{a} \geq 0$, $\underline{a} = 0 \Leftrightarrow \underline{x} \equiv \underline{y}$
2) $\underline{a} = f(\underline{x}, \underline{y}) = f(\underline{y}, \underline{x})$
3) $\underline{a} + \underline{b} \geq \underline{c}$

The statements 1),2) and 3) follow from the definition of f that is a distance function. The vector space generated by f will be denoted by DS (Distance Space). DS is in general a subspace of H. The most natural example of f function is given by $|x_i - y_i|$.

Now, we consider a class of convex transformations, $T : DS \longrightarrow [0, 1]$, such that $\forall \underline{h}, \underline{r}, \underline{s} \in DS$:

a) $T(\underline{h}) + T(\underline{r}) \geq T(\underline{h} + \underline{r})$ *convexity*
b) $T(\underline{h}) \geq T(\underline{r})$ if $\underline{h} \geq \underline{r}$ *not decreasing*
c) $T(\underline{O}) = 0$ where $\underline{O} \equiv (0, 0, ..., 0)$

From the definition of DS, it follows that T can be also considered as an application from $H \times H$ to $[0, 1]$.

Proposition 2: $T : H \times H \longrightarrow [0, 1]$ is a fuzzy-distance function.

Proof: In fact, it satisfies the following conditions:

$T(\underline{x}, \underline{y}) \geq 0$ (positive);
$T(\underline{x}, \underline{y}) = T(\underline{y}, \underline{x})$ (reflexivity), it follows from 3) of Proposition 1;
$T(\underline{x}, \underline{y}) = 0$ if $\underline{x} = \underline{y}$ (idempotent);
$T(\underline{x}, \underline{y}) + T(\underline{y}, \underline{z}) \geq T(\underline{x}, \underline{z})$ (triangular inequality) it follows directly from a) and b).

A second class of IDF's, $S \subset T$, can be introduced on the space that is generated by a normalized distance function, $0 \leq \delta(\underline{x}, \underline{y}) \leq 1$.

Lemma 1:Let $S : [0, 1] \longrightarrow [0, 1]$ be a convex and not decreasing function, such that $S(0) = 0$, then $S(\delta(\underline{x}, \underline{y}))$ is also a distance function in H.

The proof can be performed as in Proposition 2.

2.1 Examples of Entropic Distances

Entropy-distances of type E_1. In the following $\underline{h} = f(\underline{x}, \underline{y}) \in DS$, with $\underline{x}, \underline{y} \in H$. Four IDF's have been considered; they are based on the result shown in Propositions 1,2.

- $D_0(\underline{h}) = -\frac{2}{d}\sum_{i=1}^{d} \frac{h_i}{2} \times log(\frac{h_i}{2})$
  This function has been directly inspired by Shannon's definition of entropy.
- $D_1(\underline{h}) = \frac{1}{d}\sum_{i=1}^{d} h_i \times e^{(1-h_i)}$
  This has been introduced by [8].
- $D_2(\underline{h}) = \frac{1}{d}\sum_{i=1}^{d} 2 \times h_i \times (1 - \frac{h_i}{2})$

These distances are all entropy based; in some cases $D_3(\underline{h}) = \frac{1}{d}\sum_{i=1}^{d} \sqrt{h_i}$ can be considered and it has been used for comparison purposes.

It is easy to see that D_0, D_1, D_2, and D_3 satisfy conditions a), b), and c); in fact they are convex, not decreasing, and $D_n(\underline{h}) = 0$ for $\underline{h} = \underline{O}$ (n=0,1,2,3). Fig. 1 shows their plot for $d = 1$.

Therefore, they are distance function in $H \times H$; moreover, they take values in the interval $[0, 1]$. It is easy to see that their maximum is obtained for $\underline{h} \equiv (1, 1, ...1) \equiv \underline{1}$ (the unitary features vector).

Note that near $\underline{1} \in H$ the distances are compressed, while, on the contrary, they are expanded around the origin ($\underline{0}$). This property can be useful in classification problems, because a better separation can be obtained between similar (but separated) classes.

Entropic distances of type E_2. Examples of fuzzy IDF's, named E_2, can be given by using the result shown in the Lemma 1. Three entropic IDF's have been considered.

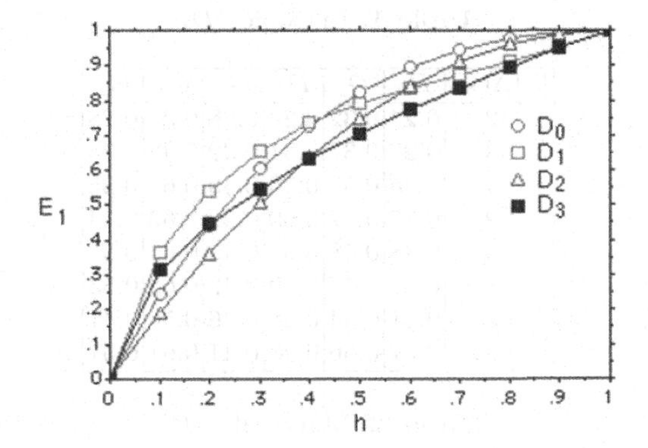

Fig. 1. The E_1 distances for d=1.

$G_0(\underline{h}) = -\frac{\delta(\underline{x},\underline{y})}{2} \times log\frac{\delta(\underline{x},\underline{y})}{2} - \frac{1-\delta(\underline{x},\underline{y})}{2} \times log\frac{1-\delta(\underline{x},\underline{y})}{2}$

This is a variant of the entropy function introduced in [9].

$G_1(\underline{h}) = \delta(\underline{x}, \underline{y}) \times \exp(1 - \delta(\underline{x}, \underline{y}))$

$G_2(\underline{h}) = 2 \times \delta(\underline{x}, \underline{y}) \times (1 - \frac{\delta(\underline{x},\underline{y})}{2})$

It can be easily shown that G_n $(n = 0, 1, 2)$ is a distance function and it is limited in the interval $[0, 1]$.

The distance function: $G_3(\underline{h}) = \sqrt{\delta(\underline{x}, \underline{y})}$ can be introduced too.

For example, for $\delta(\underline{x}, \underline{y}) = \sum_i^d |x_i - y_i|/d$:

$$G_0(\delta(\underline{x}, \underline{y})) = -\frac{\sum_i^d |x_i-y_i|/d}{2} \times log\frac{\sum_i^d |x_i-y_i|/d}{2} - \frac{1-\sum_i^d |x_i-y_i|/d}{2} \times log\frac{1-\sum_i^d |x_i-y_i|/d}{2}$$

Intuitively, distances of type E_2 modulate the δ-distance. Note that, as in the case of E_1 distances, near $\underline{1} \in H$ they are compressed, while they are expanded around the origin ($\underline{0}$).

3 Experimental Results

In this section, the performance of the entropic distances is evaluated. For this purpose, standard distances (the correlation, CO, and the normalized Euclidean, NE) have been also considered for comparison. The comparison has been also performed using three fuzzy-measures, recently introduced in [10,11] (the averaged distance AD, the global feature distance GD, and the symmetry distance SD.

The testing set was composed of 200 digital photos from the pictorial database JACOB [7]. They represent faces and people with slightly changing positions and in different conditions of illumination.

Table 1. Entropic IDF

IMAGE	D_0	D_1	D_2	G_0	G_1	G_2
a2	0.24	0.32	0.26	0.28	0.35	0.28
a3	0.27	0.37	0.31	0.32	0.42	0.34
b1	0.38	0.57	0.50	0.43	0.64	0.55
b2	0.37	0.57	0.50	0.42	0.63	0.54
b3	0.38	0.58	0.51	0.43	0.65	0.55
c1	0.38	0.57	0.50	0.42	0.62	0.53
c2	0.31	0.44	0.38	0.36	0.50	0.41
c3	0.37	0.56	0.48	0.41	0.61	0.51

Table 2. Standard IDF

IMAGE	AD	GD	SD	CO	NE
a2	0.09	0.15	0.10	0.05	0.27
a3	0.09	0.19	0.20	0.06	0.39
b1	0.20	0.31	0.36	0.06	0.43
b2	0.18	0.23	0.36	0.11	0.43
b3	0.21	0.30	0.38	0.11	0.49
c1	0.26	0.40	0.43	0.14	0.75
c2	0.26	0.32	0.42	0.13	0.75
c3	0.30	0.33	0.43	0.14	0.72

The first experiment regards the comparison of images, considered as a whole. The metric NE has been used to compute the entropic distances G_n. The experiments show that highest distances are assigned to images representing different persons.

Table 1 reports the evaluation of the new distances between the prototype a1, which is the leftmost image in Figure 2a and other images in the figure. The metric NE has been used to compute the entropic distances G_n. From Table 1 it is evident that highest distances are assigned to images representing different persons.

Table 2 reports the distance values obtained for IDF's which have been chosen for the comparison. Also in this case a better matching is obtained among elements of the same class.

Fig.2 shows the *fuzzy-classes* generated by: AD, GD, D_0, and G_0. The *degree of belonging* of an image x to a "fuzzy"-class, A, is defined as: $\chi_{A_x} = 1 - \delta_{A_x}$, where δ_{A_x} is the distance of x from the prototype representing the class A. Four IDF's are shown to be in agreement. The separation among classes is wider in the case of the measures G_0 and D_0. The comparison has been performed by using the prototype $A1$ as an ideal element of the fuzzy-class, A. Note that the first 20 images belong to class A.

A second experiment refers the matching of objects. In this case, the problem consists in retrieving the presence of an object in a sequence of image data grabbed from TV programs. The test database contains about two hundreds of short (9 frames each one) sequences, with images representing people and objects

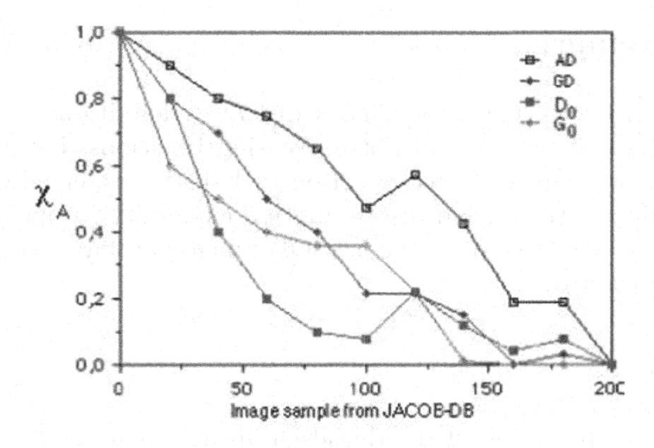

Fig. 2. The fuzzy-sets generated by distance function AD, GD, E_1, and E_2.

Fig. 3. The prototype: a woman face in the square (size $Z = 24 \times 24$).

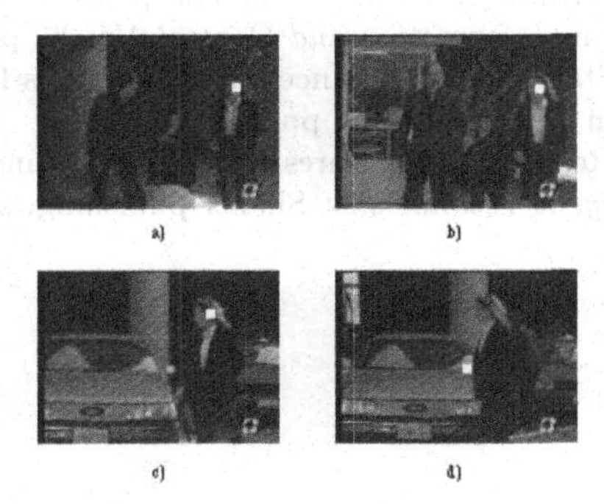

Fig. 4. Examples of correct (a,b,c) and missed (d) matches.

of various kind. The percentage of correct matches was: 81.5 for E_1, 92.6 for E_2, 80 for CO, and 75.5 for AD. The run time of the shape based query was about 0.3sec./frame. A prototype picture is shown in Fig.3 the results of the query in different shots are shown in Fig.4.

4 Final Remarks

This paper introduces two new classes of IDF, definition of which is based on the properties of convex and not decreasing functions. Examples, based on fuzzy-entropy, are proposed. A comparison with distances, usually referred in the literature, indicates their good discrimination capability in matching problems; this feature has been tested on the JACOB image database, and used in query by content procedures.

References

1. J.C.Russ, J.C.Russ, Uses of the Euclidean distance map for the measurement of features in images, in *J.Comput.Assist.Microsc.*, Vol.1, N.4, pp.343-455, 1989.
2. P.E.Danielson, Euclidean Distance Mapping, in *Comput.Graphics Image Proc.*, Vol.14, pp.227-248, 1980.
3. V.Di Gesú: Integrated Fuzzy Clustering. in *Int.Journal of Fuzzy Sets and Systems*, Vol.68, pp.293-308, 1994.
4. K.Manger : Probabilistic Geometry ,*Procd. nat.Acad.Sc* Vol.37, pp.226-229,1951
5. K.Menger :Probailistic Theories of relations,*Procd. Nat.Acad.Sc*, Vol.37, pp.178-180,1951.
6. N.R.Pal, Pattern Recognition in Soft Computing Paradigm, Fuzzy Logic Systems Institute (FLSI) Soft Computing Series - Vol. 2, 2001.
7. E. Ardizzone, M. La Cascia, D. Molinell, Motion and colour based video indexing and retrieval. in *IEEE Proc. of ICPR'96*, 1996.
8. N.R. Pal and S.K. Pal, Object-background segmentation using new definitions of entropy, IEE Proceedings-E, Vol. 136, No. 4, pp 284-295, 1989.
9. A. De Luca and S.Termini, A definition of a non-probabilistic entropy in the setting of fuzzy set theory, in *Information and Control*, Vol.20, pp.301-322, 1972.
10. V.Di Gesú and V.Starovoitov, Distance-based functions for image comparison, in Pattern Recognition Letters, Vol.20, pp.207-214, 1999.
11. V. Di Gesú and S.Roy, Fuzzy Measures for Image Distance, in *Advanced in Fuzzy Systems and Intelligent Technologies*, Shaker pubishing, pp.156-164, 1999.

Stereo Correspondence Using a Fuzzy Approach

S. Srinivas Kumar[1] and B.N. Chatterji[2]

[1]Research Scholar, E&ECE Department, Indian Institute of Technology,
Kharagpur-721302, India
samay_ssk@yahoo.com
[2]Professor , E&ECE Department, Indian Institute of Technology,
Kharagpur-721302, India
bnc@ece.iitkgp.ernet.in

Abstract. Normalized Cross Correlation (NCC) and Sum of Squared Differences (SSD) are the measures generally used in area-based techniques for stereo correspondence. They fail to establish correspondence in the presence of specular reflection and in occluded regions. Two algorithms for stereo correspondence based on fuzzy relations are presented. A novel idea of finding the correspondence based on Weighted Normalized Cross Correlation (WNCC) is proposed. Experiments with various real stereo images suggest the superiority of these algorithms over Normalized Cross Correlation under non-ideal conditions.

1 Introduction

Stereo Correspondence is an important and difficult step in Stereovision. The algorithms available in the current literature are broadly classified into three classes: Area based [3,4], Pixel based [2], and Feature based algorithms [1]. Feature based algorithms provide sparse disparity i.e., only at the positions of features. The accuracy of the results at intermediate positions depends on the interpolation scheme used. Hence, these techniques are less preferred. Area based (window based) and Pixel based algorithms are advantageous, as they provide dense disparity. These algorithms perform matching at each pixel, using absolute intensity value of pixels. Window-based algorithms are based on the assumption that the intensity values of pixels within the windows of the left image and the right image have identical intensity distributions. However, this assumption is violated due to number of physical phenomena such as specular reflection, depth discontinuity, occlusion, and projective distortion. Pixel intensity values at corresponding points may vary due to varying camera parameters such as camera gain, bias factor etc. Pixels in the left window may not be visible in the right window and also vice versa due to occlusion. In summary, area based algorithms that can establish correspondences in such regions of images are to be developed. In this paper, we present two algorithms based on fuzzy relation [5] of fuzzy sets. This paper is organized as follows. Algorithm-I proposed to solve the stereo correspondence problem in the presence of specular reflection is presented in section-2. Algorithm-II proposed to solve the stereo

N.R. Pal and M. Sugeno (Eds.): AFSS 2002, LNAI 2275, pp. 367–374, 2002.

correspondence problem in the presence of specular reflection and occlusion is dealt in section 3.Experimental results are discussed in section 4. This paper concludes in section 5.

2 Stereo Matching Algorithm-I

Linear correlation methods based on Sum of Squared Differences (SSD) and Normalized Cross Correlation (NCC) are generally used in area-based techniques for stereo correspondence. Noise in the images tends to corrupt the image agreement measure. Specular reflection is one of the problems in the area-based techniques to establish the correspondence. The object is projected in the left camera and the right camera with different intensity values due to different illumination, and different camera parameters. This affects the intensity values in the left image and the right image to differ. In this paper, Algorithm-I is proposed using fuzzy relations of fuzzy data to solve stereo correspondence problem in the presence of specular reflection. The intensity data in the left and right windows are fuzzified using Gaussian membership function. The fuzzy relation of this fuzzified data is used to find the strength of relationship of the left and the right windows. The steps of Algorithm-I are as follows.

2.1 Algorithm-I

Step 1: The matching is restricted to

$$(x',y')| \; y' \leq y \pm d_{max}$$
$$x - \xi \leq x' \leq x + \xi$$

Where $\xi = 3$ is scan lines, d_{max} is the maximum disparity estimated. (x,y) are the co-ordinates of the pixel in left image. (x',y') are the co-ordinates of pixel in right image.

Step 2: For each pixel at (x,y) in the left image, the corresponding pixel at (x',y') in the right image is to be determined. Define a stationary window of size 11X11 around (x,y) and a moving window of size 11X11 around the estimated corresponding pixel (x',y'). [A] and [B] are the matrices of size 11X11 with intensity values around (x,y) and (x',y') respectively. A_{ij} and B_{ij} represent the elements in A and B.

Step 3: Fuzzify the matrices [A] and [B] by considering a Gaussian membership function with mean and variance equal to mean and variance of intensity values of corresponding matrices. Let $[\mu(A)]$ and $[\mu(B)]$ are the fuzzified matrices of [A] and [B] with elements μ_{ij} (A) and μ_{ij} (B) respectively.

Step 4: Determine the matrix [R] of size 11X11 with values r_{ij} as

$$r_{ij} = \frac{\mathrm{Min}\ (\mu_{ij}(A), \mu_{ij}(B))}{\mathrm{Max}\ (\mu_{ij}(A), \mu_{ij}(B))}$$

The value of r_{ij} represents the strength of relationship between A_{ij} and B_{ij}.

Step 5: The strength of relationship (S) between [A] and [B] can be computed as

$$S = \frac{\sum_i \sum_j r_{ij}}{N}$$

Where N is number of elements in the matrix [R].
Repeat steps 2 to 5 when the window in right image moves within the estimated disparity. The matching pixel (x',y') in the right image corresponding to (x,y) in left image is the center pixel of the moving window for which the strength of relationship (S) is the maximum.

2.2 Justification

Due to the specular reflection, the pixel intensities within the stationary window in the left image and the corresponding window in the right image may differ. Hence, the mean and the variance of the intensity data in two windows also differ. The intensity values within the windows of the left image and the right image are fuzzified using Gaussian membership function with the mean and the variance of intensity data of the corresponding windows. This results in fuzzified data that is not affected by specular reflection. Thus, the step 3 of this algorithm eliminates the effects of specular reflection.

The strength of relationship (r_{ij}) of A_{ij} and B_{ij} depends on the fuzzy intersection (t - norm) that allows the lowest degree of membership to dictate the degree of membership in their fuzzy intersection. This value is normalized by its highest degree of membership using fuzzy union (s - norm). Hence, the value of r_{ij} lies in between 0 and 1. This idea is presented in the step 4 of the algorithm.

The strength of relationship (S) of intensity values of the windows in the left image and the right image depends on every value r_{ij} of matrix [R]. Hence, the average value of all elements decides the strength of relationship (S) between [A] and [B].

3 Stereo Matching Algorithm –II

In the presence of depth discontinuities and occlusion, only part of the data in the windows is valid for correlation. The rest of the data is insignificant and can be regarded as outliers. The similarity measure should either be less sensitive to these outliers or these outliers should be discarded in finding the corresponding pixel. In this paper, Algorithm-II is proposed based on the latter approach. This algorithm is based on the idea of giving relative importance to the outliers in finding the strength of relationship of the left and right windows. The strength of relationship of pixels at the positions of this insignificant data in the left image and the right image is very less. These pixels should be made insignificant in establishing the correspondence. The Algorithm-II proposed in this paper is based on the strength of relationship of pixels in windows of the left image and the right image. The fuzzy relations of this intensity data in both the windows are considered in the proposed Algorithm-II.

3.1 Algorithm –II

The steps 1 to 4 of the Algorithm II are the same as those of the Algorithm-I. Remaining steps are as follows:

Step 5: Let $[\alpha_1, \alpha_2, \alpha_3 \ldots \alpha_n]$ be the distinct elements of matrix [R].

Step 6: $A'_{\alpha i}$ and $B'_{\alpha i}$ are intensity values of [A] and [B] which have the strength of relationship greater than α_i. Find the Normalized Cross Correlation ($C_{\alpha i}$) of $A'_{\alpha i}$ and $B'_{\alpha i}$.

Repeat the step 6 for different α_i (i = 1,2,3…n).

Step 7: Find the strength of relationship (S) between [A] and [B] as

$$S = \frac{\sum_{i} \alpha_i C_{\alpha i}}{\sum_{i} \alpha_i}$$

Repeat steps 2 to 7 when the window in right image moves within the estimated disparity. The matching pixel (x',y') in the right image corresponding to (x,y) in left image is the center pixel of the moving window for which the strength of relationship (S) is the maximum.

3.2 Justification

Specular reflection problem is solved by fuzzifing the intensity data in the windows of the left image and the right image with Gaussian membership function as discussed in Algorithm-I. However, the pixels around (x,y) within the window of the left image may not be visible in the window of the right image around its corresponding pixel (x',y') due to occlusion. Algorithm-I fails to establish correspondence in such regions as it establishes a correspondence based on the strength of relationship of all pixels within windows of the left image and the right image. Algorithm-II can provide correspondence in such regions. In this algorithm, the correlation co-efficient ($C_{\alpha i}$) of the pixel intensity values of the windows in the left image and the right image in the absence of occluded pixels is given relative importance by multiplying with a weight factor α_i. The weight factor α_i at the positions of occluded pixels is less. The occluded pixels are made insignificant by multiplying $C_{\alpha i}$ with weight factor (α_i).. Hence, the occluded pixels are made insignificant in finding the strength of relationship (S). The value of strength of relationship lies between 0 and 1.

4 Experimental Results

Experiments are conducted on real stereo images satisfying epi-polar constraint, such as SHR_RUB [1] and TELEPHONE available on the Web [6] with each image of size 256X256 pixels. These stereo images are shown in Fig.1 and Fig.2.The disparity is

[1] Available in the image database of Computer vision laboratory, E&ECE Department of Indian Institute of Technology, Kharagpur

estimated to be ± 20 pixels on either side of the pixel in the left image. The results of stereo correspondence by NCC, Algorithm-I and Algorithm-II are reported in Table 1 and Table 2. The results are compared with those obtained by linear correlation measures such as NCC with window size 11x11 same as that of Algorithms-I and II. False matching pixels by NCC and Algorithm-I are more in occluded regions. Algorithm-II is successful in such regions. However, the computational cost for implementing Algorithm -II is more compared to Algorithm-I. The success of Algorithm-II is illustrated in Fig. 3 and Fig. 4 near occluded regions. The left top corner pixels of the introduced patch in stereo images are the matching pixels obtained by Algorithm-II.

(a)

(b)

Fig. 1. Shr_rub stereo Images with epipolar constraint (a) Shr_rub left image (b) Shr_rub right image

(a)

(b)

Fig. 2. Telephone Stereo Images with epipolar constraint (a) Telephone left image (b) Telephone right image

5 Conclusions

Stereo correspondence algorithms based on fuzzy relations of fuzzy sets are presented in this paper. The Algorithms- I and II are successful, even if the scale of intensity values within the windows of left and right images differ due to specular reflection. The Algorithm-II is successful in establishing correspondence even in occluded regions. The performance of the Algorithm-II can be improved significantly in the presence of noise, if NCC ($C_{\alpha i}$) is replaced by Kendall's rank correlation coefficient[7]. The computational cost increases substantially with such changes in algorithm, a bottleneck for real-time applications.

Table 1. Comparison of Stereo Correspondence results for Shr_rub Stereo images using NCC, Algorithm-I and Algorithm-II for pixels in both non-occluded and occluded regions. (* - indicates false matching pixel)

S.No	Left Image Coordinates		Right Image Coordinates using NCC		Right Image Coordinates using Algorithm – I		Right Image Coordinates using Algorithm - II	
	x_l	y_l	x_r	y_r	x_r	y_r	x_r	y_r
1	189	140	189	161*	189	160	189	160
2	100	83	100	84*	100	86	100	86
3	181	95	181	101	181	101	181	101
4	168	164	168	162*	168	161	168	161
5	125	125	125	148*	125	105	125	105
6	174	100	174	106	174	106	174	106
7	225	30	225	29	225	29	225	29
8	185	192	185	188*	185	187	185	187
9	161	196	161	190	161	190	161	190
10	160	75	160	100*	160	81	160	81
11	185	110	185	130*	185	130*	185	190
12	243	120	243	142*	243	120*	243	140
13	210	93	210	98*	210	73*	210	97
14	161	112	161	133*	161	132*	161	112
15	126	197	126	182*	126	181*	126	177
16	95	96	95	101*	95	76*	95	87
17	140	112	140	115*	140	127*	140	112
18	181	148	181	166	181	165*	181	166
19	185	129	185	150*	185	146*	185	148
20	200	52	200	56*	200	54*	200	58

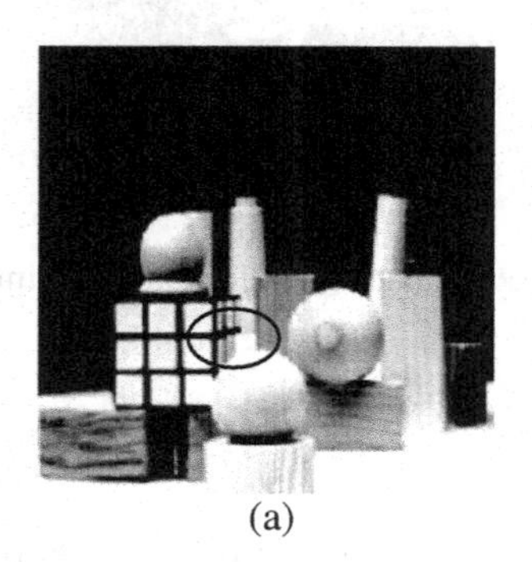

(a)

(b)

Fig. 3. Shr_rub stereo images with introduced patches to illustrate the success of Algorithm-II. The left top corner pixels in the introduced patches in stereo images are matching pixels. (a) Shr_rub left image with an introduced patch at pixel (161,112) (b) Shr_rub right image with an introduced patch at pixel (161,112).

Table 2. Comparison of Stereo Correspondence results for Telephone Stereo images using NCC, Algorithm-I and Algorithm-II for pixels in both non-occluded and occluded regions. (*-indicates false matching pixel)

S.No	Left Image Coordinates		Right Image Coordinates using NCC		Right Image Coordinates using Algorithm – I		Right Image Coordinates using Algorithm – II	
	x_l	y_l	x_r	y_r	X_r	y_r	x_r	y_r
1	138	124	138	126*	138	125	138	125
2	125	135	125	144	125	144	125	144
3	125	185	125	186	125	186	125	186
4	130	210	130	214	130	214	130	214
5	166	88	166	95	166	95	166	95
6	122	200	122	205*	122	206	122	206
7	118	224	118	228*	118	230	118	230
8	160	220	160	230	160	230	160	230
9	177	42	177	44	177	44	177	44
10	132	181	132	182	132	182	132	182
11	141	187	141	180*	141	184*	141	179
12	145	89	145	96*	145	96*	145	97
13	134	97	134	107*	134	108*	134	109
14	132	105	132	127*	132	113*	132	114
15	134	164	134	175*	134	175*	134	174
16	123	165	123	177*	123	176	123	176
17	132	150	132	167*	132	163*	132	143
18	136	103	136	113*	136	114*	136	112
19	123	183	123	184*	123	185*	123	183
20	150	123	150	122*	150	137*	150	121

(a)

(b)

Fig. 4. Telephone stereo images with introduced patches to illustrate the success of Algorithm-II. The left top corner pixels in the introduced patches in stereo images are matching pixels. (a) Telephone left image with an introduced patch at pixel (132,150). (b) Telephone right image with an introduced patch at pixel (132,143).

Acknowledgements

We are grateful to Prof. N.R. Pal, ECSU, Indian Statistical Institute, Kolkata, India for very helpful discussions and encouragement.

References

1. Stephane Mallat,: Zero-Crossings of a Wavelet Transform. IEEE Transactions on Information Technology, Vol. 37, No. 4, (1991).
2. Guo-Qing Wei, Wilfried Brauer, and Gerd Hirzinger, : Intensity – and Gradient – Based Stereo Matching Using Hierarchial Gaussian Basis Functions. IEEE Transactions on Pattern Analysis and Machine Intelligence, Vol.20, No.11, (1998).
3. Takeo Kanade, and Masatoshi Okutomi, : A stereo Matching Algorithm with an Adaptive Window: Theory and Experiment. IEEE Transactions on Pattern Analysis and Machine Intelligence, Vol.16,
4. No.9, (1994).
5. Dinakar N. Bhat and Shree K. Nayer,: Ordinal Measures for Image Correspondence. IEEE Transactions on Pattern Analysis and Machine Intelligence, Vol.20, No.4, (1998).
6. H.J. Zimmermann, Fuzzy Set Theory – And Its Applications. 2nd ed., Kluwer Academic Publishers, Published and Reprinted by Allied Publishers Limited, New-Delhi.
7. http://www.vasc.ri.cmu.edu/idb/html/stereo/index.html
8. Sidney Siegel: Non-parametric Statistics for the behavioral sciences. McGraw Hill Book Company, Inc. International Student Edition (1956)

Implementation of BTTC Image Compression Algorithm Using Fuzzy Technique

Munaga. V.N.K. Prasad, K.K. Shukla, and R.N. Mukherjee

Department of Computer Engineering, Institute of Technology, Banaras Hindu University, Varanasi - 221 005, India.
munagasayee@yahoo.com, shukla@ieee.org

Abstract. This paper presents a new algorithm for image compression based on fuzzy domain decomposition, which is an improvement of the recently published Binary Tree Triangular Coding (BTTC) algorithm. The algorithm is based on recursive decomposition of the image domain into right-angled triangles arranged in a binary tree and uses a fuzzy measure of image compactness. The algorithm executes in O(nlogn) time for encoding and $\theta(n)$ time for decoding, where n is the number of pixels in the image. Simulation results on standard test images show that the new algorithm produces significantly less triangles as compared with conventional BTTC while providing the same quality of reconstructed image as good as BTTC. Further, the fuzzy algorithm is more robust with respect to noise. Both these algorithms have faster execution time than JPEG.

1 Introduction

Image Compression is the art/science of efficiently coding digital images to reduce the number of bits required in representing an image. The purpose of doing so is to reduce the storage and transmission costs while maintaining good quality. A 512 x 512 pixel, 8-bit gray-level image requires 2,097,152 bits for storage and 3.64 minutes for transmission using 9600-buad (bits/s) modem.

Image data compression algorithms fall into two major categories[4]: one is lossless compression: where no loss of information takes place while compressing and the reconstructed data is same as original data. The second one is lossy compression: where the algorithm allows loss of data to facilitate higher compression. BTTC using fuzzy sets discussed in this paper belong to the second category.

The coding scheme described in this paper is based on fuzzy domain decomposition. The decomposition scheme involves triangulating the domain recursively. Linear interpolation has been chosen since, while yielding an acceptable quality in the decoding image, it has a quite convenient computational cost if compared to other methods[9]: the interpolation performed by BTTC using fuzzy coding requires only four floating-point multiplications and one floating-point division per pixel. As a result, it runs much faster than the standard techniques based on transforms[1-3].

N.R. Pal and M. Sugeno (Eds.): AFSS 2002, LNAI 2275, pp. 375–381, 2002.

The paper is organized as follows. Section 2 introduces the Fuzzy domain decomposition, section 3 explains coding scheme of BTTC using Fuzzy technique, the experimental results are discussed in section 4, and concluding remarks are given in section 5.

2 Fuzzy Domain Decomposition

In natural images, many regions are associated with ambiguity or fuzziness. Several authors have used the fuzzy set concept for image processing[5-7]. Pal and Rosenfeld[6] have introduced several fuzzy geometric properties of images. They have used a fuzzy measure of image compactness for enhancement and thresholding. Nobuhara[8] et al. have recently introduced the application of fuzzy relational equation to lossy image compression. Sinha[9] et al. introduced several fuzzy morphological algorithms for image processing tasks like shape detection, edge detection, and cutter removal. In the present work, we report the novel idea of applying fuzzy compactness measure to triangular domain decomposition for efficiently encoding gray scale images.

Let X be an image of size MxN having L gray levels. This image can be interpreted as an array of fuzzy singletons. Each pixel has a membership value depending upon its brightness relative to some level l, l=0,1,2...L-1 i.e. $X = \{\mu_x(X_{mn}) = \mu_{mn}/X_{mn};$ $m=1,2\ldots M,\ n=1,2\ldots N\}$ where $\mu_x(X_{mn})$ or μ_{mn}/X_{mn} $(0 \leq \mu_{mn} \leq 1)$ denotes the grade of possessing some brightness property μ_{mn} by the (m, n) th pixel intensity X_{mn}.

Rosenfeld[10-13] extended the concepts of digital picture geometry to fuzzy subsets and generalized some of the standard geometric properties of and relationships among regions to fuzzy sets. Among the extensions of the various properties, we only discuss here the area, perimeter and compactness of fuzzy image subset, characterized by $\mu_x(X_{mn})$, which will be used in the following section for developing fuzzy BTTC algorithm. In defining the above mentioned parameters we replace $\mu_x(X_{mn})$ by μ for simplicity.

The area of μ is defined as

$$a(\mu) = \int \mu$$

Where the integral is taken over any region outside which $\mu=0$. For μ being piecewise constant (in case of a digital image X of dimension MxN) the area is

$$a(\mu) = \sum \mu = \sum_x \sum_y \mu(x, y) \quad \text{with } x = 1,2\ldots M,\ y = 1,2\ldots N.$$

For the piecewise constant case the perimeter of μ is defined as

$$P(\mu) = \sum_{i,j} \sum_k |\mu_i - \mu_j| \, |A_{ijk} \qquad i, j = 1,2\ldots r,\ i<j;\ k = 1,2\ldots rij$$

This is just weighted sum of the lengths of the arcs A_{ijk} along which the i-th and j-th regions having constant μ values μ_i and μ_j respectively meet, weighted by the absolute difference of the values. In the case of an image, if we consider the pixels as the piecewise constant regions, then the perimeter of an image is defined by

$$P(\mu)=\sum_{i,j} |\mu(x_{i,j})-\mu(x_{i,j+1})| + |\mu(x_{i,j} - \mu(x_{i+1,j})|$$

Where $\mu(x_{i,j})$,$\mu(x_{i,j+1})$ and $\mu(x_{i+1,j})$ are the membership values of adjacent pixels. As shown in Fig.1 only the east and south neighbors are considered and i, j $\in$ k^{th} triangle.

The compactness of μ is defined as

$$\text{Comp}(\mu) = a(\mu)/P^2(\mu) \quad (1)$$

For the purpose of image domain decomposition, we assign memberships to pixels belonging to triangular domain of the image as follows

For 8 bits/pixel image number of intensity levels L=256 and the intensity varies in [0,255]

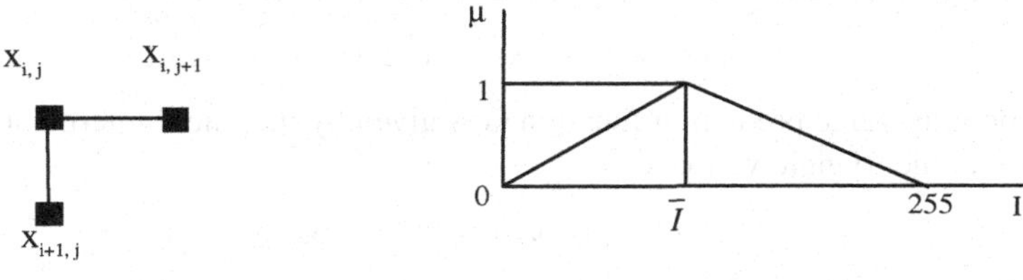

Fig. 1. **Fig. 2.**

If $I = \bar{I}$ then $\mu = 1$

Else If $I < \bar{I}$ then $\mu = I/\bar{I}$

Else $\mu = (255 - I)/(255 - \bar{I})$

Where $\bar{I} = \frac{1}{n_k} \sum_{m=1}^{k} I_m$, $m \in k^{th}$ triangle is the mean intensity of the kth triangle and n_k is the total number of pixels in k^{th} triangle.

These membership values denote the similarities of the pixel brightness to the brightness of a typical prototype pixel in the region. The typical brightness value is taken as the arithmetic mean of all the pixels belonging to that region.

After the memberships are assigned, the fuzzy compactness of the region is computed using (1). The decision to further sub divide a triangle region into smaller triangles is taken based on the value of computed compactness against a user specified quality parameter. The details of the algorithm are given in next section. The use of compactness rather than intensity deviation at a single pixel gives us an image compression algorithm which is robust and requires less number of subdivisions while giving the reconstructed image quality as good as BTTC algorithm.

3 Coding Scheme

The image to be encoded can be regarded as a discrete surface i.e. a finite set of points in three dimensional (3-D) space, by considering a nonnegative discrete function of two variables F (x, y) and establishing the correspondence between the image and the surface A={(x, y, z) | c = F (x, y)}, so that each point in A corresponds to a pixel in the image. The couple (x, y) gives the pixel's position in the XY plane, while c (the point's height) is the pixel's gray value.

Our goal is to approximate A by a discrete surface B={(x, y, d) | d= G (x, y)}, defined by means of a finite set of points. Let T be a generic triangle on the XY plane of vertices:

$$P_1=(x_1, y_1),\ P_2=(x_2, y_2),\ P_3=(x_3, y_3) \tag{2}$$

And let

$$c_1 = F(x_1, y_1),\ c_2 = F(x_2, y_2),\ c_3 = F(x_3, y_3)$$

Represent the gray values at P_1, P_2, and P_3 respectively so that

$$(x_1, y_1, c_1),\ (x_2, y_2, c_2),\ (x_3, y_3, c_3) \in A \tag{3}$$

The gray value prediction function G is given by the linear interpolation of the gray values at the triangle vertices:

$$G(x, y) = c_1 + \alpha (c_2 - c_1) + \beta (c_3 - c_1) \tag{4}$$

Where α and β are defined by the two relations

$$\alpha = \frac{(x - x_1)(y_3 - y_1) - (y - y_1)(x_3 - x_1)}{(x_2 - x_1)(y_3 - y_1) - (y_2 - y_1)(x_3 - x_1)} \tag{5}$$

$$\beta = \frac{(x_2 - x_1)(y - y_1) - (y_2 - y_1)(x - x_1)}{(x_2 - x_1)(y_3 - y_1) - (y_2 - y_1)(x_3 - x_1)} \tag{6}$$

From this definition it can be seen that the values of F and G coincide at the vertices of T:

$$F(P_1) = G(P_1);\ F(P_2) = G(P_2);\ F(P_3) = G(P_3) \tag{7}$$

$$\text{Compactness} \leq e \tag{8}$$

Where $e > 0$ is an adjustable quality factor.

If compactness exceeds e, the triangle T is divided into two triangles by dividing along its height relative to the hypotenuse as shown in Fig.3 (a). The relevant information known, as topological information is stored in a hierarchical structure – a B tree. Fig. 3(b) shows how the partition process works: each time a triangle is subdivided the resulting triangles become its children in the tree. Hence, each node has either zero or two children.

A triangle of vertices P_1, P_2, P_3 is represented by notation $<P_1P_2P_3>$. The first vertex corresponds to the right angle: the other two vertices follow clockwise. The structure of the resulting binary tree is stored in a binary string s obtained from a breadth-first visit of the tree: the value 0 represents a leaf node, the value 1 an intermediate node; the root is not stored at all. We need one bit for each node (with exception of root). Indeed, as illustrated in Fig. 4, it is possible to store only a part of s. Decoding is analogous to encoding, with the only difference being that the quality of the approximation need not be tested and the pseudo code of fuzzy BTTC is shown in Fig. 5.

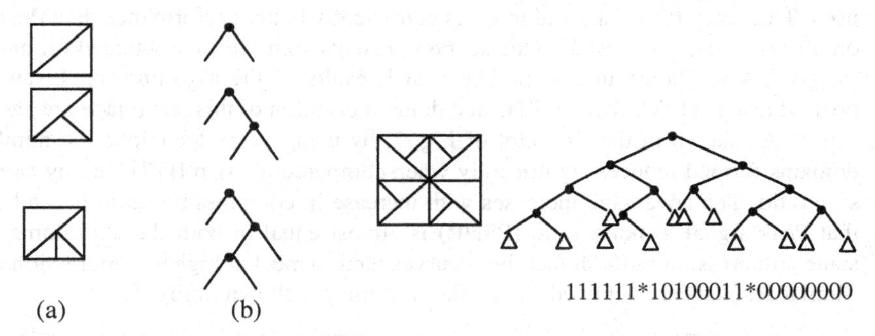

Fig. 3. (a). Example of domain partition process, (b). Tree representation of the decomposition.

Fig. 4. Possible domain partition and the resulting binary string s

// L is list of leaves with no processing

1. Set L = NULL;

   Set T_1 = <(1,1)(1,m)(m, 1)>, T_2 = <(m, m)(m, 1)(1,m)>.

2. Push T_1 and T_2 into stack.

3. Pop the Triangle T from stack.

   Let P_1 = (x_1, y_1), P_2 = (x_2, y_2), P_3=(x_3, y_3) be it's vertices.

   Set C_1 =F (x_1, y1), C_2 =F (x_2, y2), C_3 =F (x_3, y3).

4. Calculate the compactness using (1) of the triangle T.

5. If condition (8) is satisfied then go to step 7.

6. Set P_{max} = (P_2+P_3)/2;Set T_1= <P_{max} P_3 P_1>; T_2 = <P_{max} P_1 P_2>

   Go to step 2.

7. Insert T into L.

8. If the stack is empty then stop, otherwise go to step 3.

Fig. 5. Fuzzy BTTC Algorithm

4 Experimental Results

To evaluate the performance of fuzzy BTTC method several 8- bit test images were used. The fuzzy BTTC algorithm gives consistently better performance than the BTTC on all test images we used. This section presents some results obtained on one such image, Lisaw, shown in Fig. 6. The visual results of the algorithm on Lisaw compressed using BTTC, fuzzy BTTC and domain partition of this test image are shown in Fig. 6. As shown in the first plot of Fig. 7, by using fuzzy technique the number of domains formed reduces significantly when compared to crisp BTTC at any compression ratio. The advantage increases with increase in compression ratio. It can be seen that Peak signal to noise ratio (PSNR) is almost equal in both the algorithms at the same compression ratio, in fact, both curves tend to meet at higher compression ratios. Similar results were obtained using other test images shown in Fig. 8.

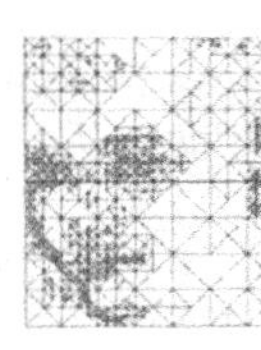

Fig. 6. Original lisaw image, domain triangulation for lisaw, reconstructed images using BTTC and fuzzy BTTC from left to right

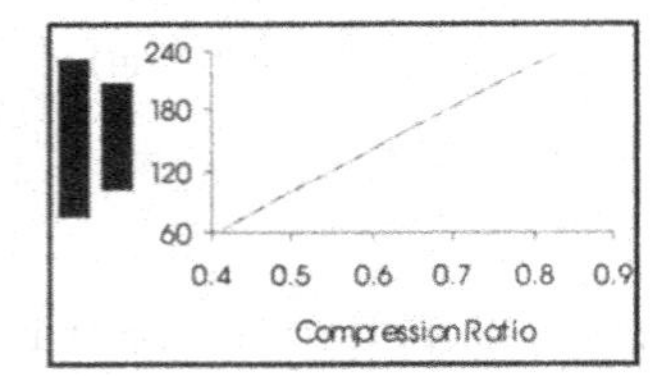

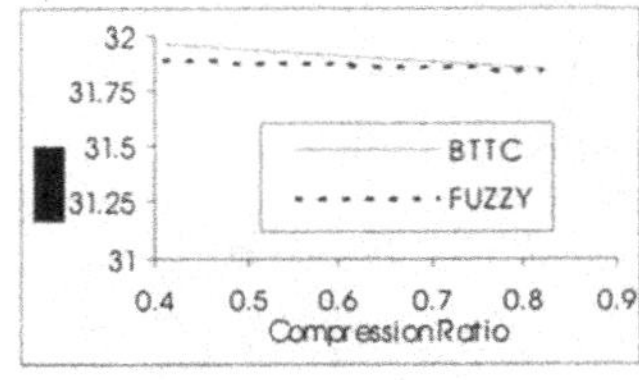

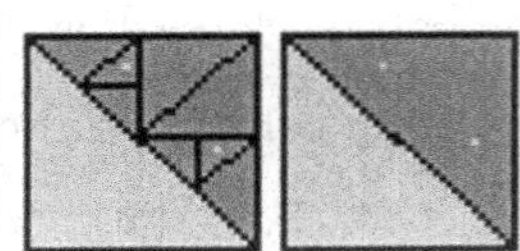

Fig. 7. From left to right- Variation of difference in number of triangles formed using BTTC and fuzzy BTTC, variation in peak-signal-to-noise-ratio against compression ratio, zoomed view of a portion of noisy image domain decomposition using BTTC and Fuzzy BTTC, respectively – illustrates how unnecessary triangulations are avoided by fuzzification.

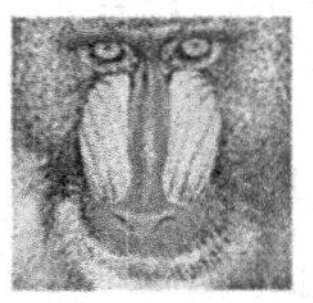
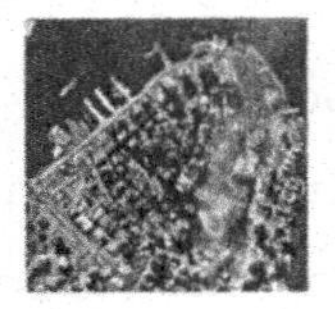

(Barbara) (Cameraman) (Baboon) (World Trade Centre)

Fig. 8. Other test images used for performance evaluation.

5 Conclusion

In this paper we have introduced the idea of using the fuzzy measure of compactness as a basis of image domain decomposition. This idea was incorporated into recently reported BTTC algorithm. The modified algorithm reported in this paper avoids the disadvantages of hard decisions during decomposition. The fuzzy BTTC offers the advantages of reduced triangulation and better noise immunity as shown in Fig.7. Since BTTC take a hard decision to subdivide based on the maximum error at a single pixel, noisy pixels force many unnecessary subdivisions. On the other hand, the fuzzy BTTC takes a subdivision decision based on fuzzy compactness of the entire parent triangle, thereby avoiding unnecessary subdivisions due to outliers. The new algorithm has same computational complexity as the original version of BTTC, and gives comparable quality of reconstructed image at similar compression ratios.

References

1. Riccardo Distasi, Michele Nappi and Sergio Vitulanco, "Image Compression by B-Tree Triangular coding", IEEE Transaction communications, vol. 45, (1997) 1095-1100.
2. G.K. Wallance, "The JPEG Still Picture Compression Standard", Commun. ACM, Vol. 34, (1991) 31-44.
3. B.A.D. Vore, B. Jawerth and B.J. Lucien, "Image Compression Through Wavelet Transform Coding", IEEE transaction Inform. theory, Vol. 38, (1992) 719-746.
4. khalid Sayood, "Introduction to data compression," Margan Kaufmann Publishers, (1996).
5. Rosenfield. A. "The fuzzy geometry of image subsets", Pattern Recognition Letters 2, (1984) 311-317.
6. Pal. S.K. and A. Rosenfeld, "Image enhancement and thresholding by optimization of fuzzy compactness", Pattern Recognition Letters 7, (1988) 77-86.
7. Pal. S.K, "Fuzzy skeletanization of an image", Pattern Recognition Letters 10, (1989), 17-23.
8. Hajime Nobuhara, Witold Pedrycz and Kaoru Harota, "Fast solving method of fuzzy relational equation and its application to lossy image compression/reconstruction", IEEE transaction Image Processing, Vol. 1, (2000) 325-317.
9. Divyendu Sinha, Purnendu Sinha, Edward R. Dougherty, and Sinan Batman, "Design and analysis of fuzzy morphological algorithms for image processing", IEEE transaction Fuzzy Systems, Vol. 5, (1997) 570-584.
10. Rosenfeld. A., "Fuzzy digital topology", Inform. and Control 40, (1979) 76-87.
11. Rosenfeld. A., "On connectivity properties of gray scale pictures", Patt. Recog. 16, (1983) 47-50.
12. Rosenfeld, A. and S.Haber, "The perimeter of fuzzy set", Patt. Recog. 18, (1985) 125-130.
13. Rosenfeld, A., "The diameter of fuzzy set", Fuzzy Sets and Systems 13, (1984) 241-246.

Applications of the ILF Paradigm in Image Processing

Aureli Soria-Frisch

Fraunhofer IPK, Dept. Pattern Recognition, Pascalstr 8-9, 10587 Berlin, Germany
aureli.soria_frisch@ipk.fhg.de

Abstract. This paper describes some applications of the recently introduced Intelligent Localized Fusion (ILF) paradigm in multisensorial computer vision systems. The paradigm is based on a new interpretation of the fuzzy integral as fusion operator in image processing and thus related to Soft-Computing methodologies. The application of the paradigm for color edge detection in outdoor scenes and for preprocessing in automated visual inspection of high reflective materials is described. While in the first application the usage of the paradigm allows the avoidance of false edges caused by shadows, it succeeds suppressing the annoying highlights in the second one.

1 Introduction

The simultaneous advances in computer technology, in terms of computing capability and memory capacity, and in sensor technology, in terms of cost and reliability of the acquisition devices, allow the successful implementation of multisensorial computer vision systems. These are characterized by the inclusion of more than one image information source and the processing of the image signals delivered from them. Some examples of multisensorial computer vision systems can be found in color, multiband remote-sensing, multifocal microscopy image, or range-visual analysis systems.

In traditional multisensorial systems the fusion operator reduces the n-dimensionality introduced in the system by the usage of n sensors, being n bigger than one. The literature on multisensorial fusion maintains that a quantitative and qualitative gain should be derived from the usage of a fusion operator [1][8]. Nevertheless the fusion operator is practically consider to be a mean of combining the data coming from different sensors into one representational form [8]. Specially fusion paradigms used in the context of computer vision [2][7][9][10] are limited to the consideration of a very narrow group of them, disregarding recent developments in operator research related specially to fuzzy logic. The semantic interpretation of the images processed is ignored in the fusion procedure. Moreover the characterization of the information sources is made in a module separated from the fusion operator and mostly based on gaussian and independence suppositions on them [7].

Fuzzy Aggregation operators overcome this lack of flexibility. A large number of operators have been developed and successful employed in different application

N.R. Pal and M. Sugeno (Eds.): AFSS 2002, LNAI 2275, pp. 382–387, 2002.

fields: Uninorms, OWAs, weighted ranking operators, fuzzy integrals, symmetric sums. Such operators are seldom used for data fusion in image processing applications.

A fuzzy integral based paradigm already presented [13] is used in this paper to implement different image processing applications. The Intelligent Localized Fusion (ILF) paradigm uses a local parameterization of the fuzzy integral. The paradigm is presented in section 2. Its applications for edge detection on outdoor scenes and in the preprocessing stage of a system for the automated visual inspection of high reflective materials are described in sections 3 and 4 respectively.

2 Intelligent Localized Fusion for Multisensorial Image Processing

The introduction of fuzzy aggregation strategies allows overcoming the shortages of classical fusion operators. Fuzzy aggregation operators present some positive features: adaptability, reinforcement capability [15], inclusion of meta-knowledge [15], characterization of the interaction between information sources [5], and tractability of fuzzy information.

Specially fuzzy integrals present all mentioned properties [5], i.e. other fuzzy aggregation operators are not capable of reflecting the interaction between information sources. Moreover, fuzzy integrals generalized other fuzzy aggregation operators, e.g. OWAs, weighted ranking operators, beyond most aggregation operators usually employed in computer vision.

The information fusion can be undertaken in different stages of a multisensorial image processing system. The position of the fusion operator in the processing chain leads to different types of fusion, e.g. signal, pixel, feature or segment fusion [11]. In the here presented paradigm the operator acts in the local level, i.e. the fusion undertaken is a pixel fusion. The employment of the fuzzy integral as fusion operator and its parameterization through fuzzy measures at the pixel level were taken as starting points for the development of the Intelligent Localized Fusion (ILF) paradigm.

The Theory of Fuzzy Measures was built on the conclusions made in Sugeno's pioneering work [14], where the fuzzy integral was introduced. There are basically two types of fuzzy integral known as Sugeno's and Choquet's Fuzzy Integral. A deeper description on different aspects of Fuzzy Integrals can be found in [5].

One important property of fuzzy integrals for the development of the here presented paradigm is related to their limits. The value of the fuzzy integral has its lower and upper bound in the minimum ($\wedge$) and maximum ($\vee$) operator, respectively:

$$\bigwedge[x_1, \ldots, x_n] \leq \mathcal{S}_\mu(x_1, \ldots, x_n) \leq \bigvee[x_1, \ldots, x_n] \tag{1}$$

where this possible interval of output values for the fusion operation is induced by the value of the fuzzy measures coefficients.

2.1 "Localized" Fuzzy Measures

The fuzzy measure coefficients are used in the fuzzy integral for multisensorial fusion to characterize the importance of the information sources. These coefficients are determined based on *a priori* knowledge of the contribution of each channel in the fulfillment of the hypothesis being proved. This characterization refers to the individual information sources, but also to their possible coalitions or combinations. Usually the coefficients of the fuzzy measures are determined globally in the image, i.e. the fuzzy measure coefficients have a value characterizing the importance of each image channel and that of their possible coalition taking the image as a unit.

A novelty concerning this last aspect is introduced in the ILF paradigm. Here the used fuzzy measure is not unique, but depends on the region where the fusion is undertaken, i.e. the coefficients of the fuzzy measure change over the image. It can be talked then from "localized" fuzzy measures. Taking into consideration the lower and upper bounds of the fuzzy integral's output and its dependence on the value of the fuzzy measures coefficients, the utilization of these "localized" fuzzy measures makes possible operating in the same image with different fusion strategies. The localization of these different fuzzy measures is given in form of a mask. Such a mask is obtained through the application of standard image processing algorithms on the input images. Here the goal is the different characterization of the regions of interest from the other ones through a label. These mask images contain then an indication of the fuzzy measure to be used in each region and not the coefficients itself. The "localized" fuzzy measures exploit the flexibility of the fuzzy integral formerly mentioned and constitutes the main new feature of the ILF paradigm.

3 Color Edge Detection on Outdoor Color Images

Edge detection is an important previous stage in several image processing algorithms. In the case of edge detection on color images different mathematical approaches have been chosen in order to cope with the vectorial nature of color information [6].

A system with an ILFO for color edge detection following the paradigm formerly detailed and based on the Sugeno's Fuzzy Integral was implemented. The system was applied for detecting edges on color images. A Haar edge detector is used in each channel. Implementation details of the framework for color edge detection can be found in [12].

One important problem when evaluating edge maps of images taken under natural illumination conditions is the appearance of false edges due to the presence of shadows. Such artifacts can for instance bring a camera guided autonomous vehicle to detect a non-existing terrain change. Such a problem is a current term of research in image processing [4]. The here presented paradigm can be used to avoid the detection of shadow edges. "Localized" fuzzy measures are defined for the outdoor scenes, where the regions on shadowed areas present a coefficient configuration different from that of the other areas. Such a localization allows the fuzzy integral to behave minimum-like in the shadow regions and maximum-like in the other ones.

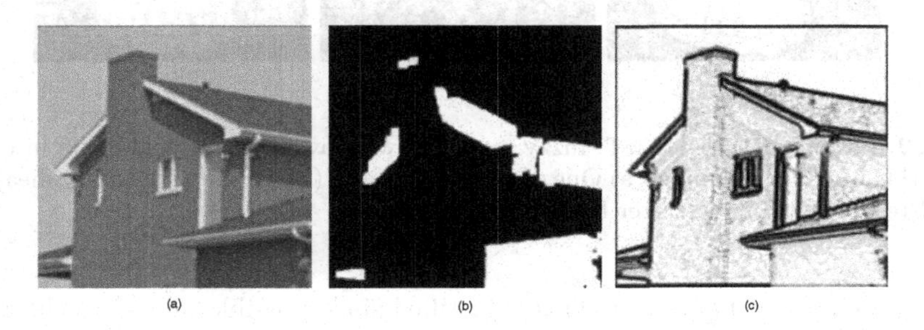

Fig. 1. Fusion with "localized" fuzzy measures with avoidance of shadow false edges. (a) Original color image. (b) Mask for the fuzzy measures. (c) Color edge maps after fusion.

The framework for color edge detection was tested on outdoor color images. The result can be found in figure 1. The avoidance of false edges caused by shadows through the the usage of "localized" fuzzy measures is demonstrated (see figure 1). The mask with the region distribution for the different fuzzy measures, which can be observed in figure 1b, was obtained through following operations: a transformation of the image into the HSI space, the addition of the S and I channels, a threshold and a morphological dilation with a cross mask. The fuzzy measure coefficients for the shadowed regions were set in order for the result of the fuzzy integral to approximate that of a minimum operator. As a consequence the edge due to the roof shadow (see figure 1a) could be avoided (see figure 1c). This result is a direct consequence of the localized parameterization of the fusion operators through the fuzzy measures.

4 Avoidance of Highlights for the Visual Inspection of High Reflective Materials

The inspection of high reflective materials is a long term of research in image processing. A new solution based on the ILF paradigm is presented in the following. In this case the paradigm was implemented with a Choquet's Fuzzy Integral. Moreover the purpose was not the total suppression of all the highlights, but the detection of possible faults in the lamp body. Allowing possible faults to appear highlighted can facilitate its detection in the inspection system.

The images constituting the multisensorial set are in this case different images of the object to be inspected taken under different illumination conditions. For the synthesis of the mask following operations were used: all the images were added, a threshold was applied, a morphological flooding [3] and finally a different label for each connected region was given depending on the number of pixels

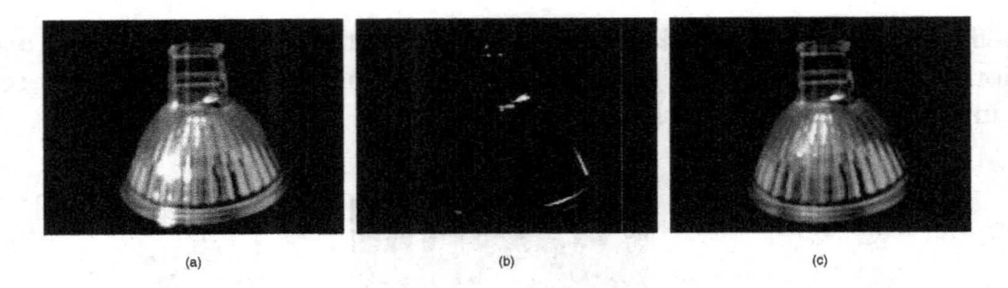

Fig. 2. Fusion with "localized" fuzzy measures with avoidance of highlights in the inspection of hallogen lamps. (a) One of original images. (b) Mask for the fuzzy measures. (c) Pre-processed image after fusion.

of the region (see figure 2b). This is justified since possible faults are characterized by areas smaller than just annoying highlights. As a consequence possible fault areas received different coefficients from those of "normal" highlights areas (such a fault candidate is the lighter triangular area in the center of the image, in figure 2b). Although the final result (see figure 2c) is not perfect, it can be considered as encouraging for the further development of the paradigm.

5 Conclusions and Prospective Work

The utility of the recently introduced ILF paradigm has been shown in the here presented paper. Two different possible applications were developed and successful implemented following this paradigm.

In case of the application for color edge detection a generalization of its capabilities on hand of a more extended image data set is necessary. This would be the goal of future research works. One of the problems to be solved in this case is the utilization of alternative morphological operators for the completion of the fuzzy measures mask. The morphological dilation currently used could difficult the detection of edges of interest.

The application for the suppression of highlights in automated visual inspection is in a more advanced stage of development. The problem of the automated determination of the fuzzy measure coefficients will be treated in future communications. For that purpose preliminary positive results are already obtained through the employment of genetic algorithms.

The exploitation of the ILF paradigm capabilities into other application fields is also considered.

References

1. M.A. Abidi, R.C. Gonzalez, eds. (1992). *Data Fusion in Robotics and Machine Intelligence*. San Diego: Academic Press.
2. F. Alkoot and J.Kittler (2000). Improving the performance of the Product Fusion Strategy. *Proc. 15th International Conference on Pattern Recognition*, . ICPR 2000, Barcelona, Catalonia.

3. S. Beucher (1982). Watersheds of functions and picture segmentation. *IEEE Int. Conf. on Acoustics, Speech and Signal Processing*, Paris, 1928-1931.
4. E.D. Dickmanns (1997). Improvements in Visual Autonomous Road Vehicle Guidance 1987-94,in *Visual Navigation: From Biological Systems to Unmanned Ground Vehicles*, Y. Aloimonos ed., Mahwah (New Jersey): Lawrence Erlbaum Associates, Pubs.
5. M. Grabisch, H.T. Nguyen and E.A. Walker (1995). *Fundamentals of Uncertainty Calculi with Applications to Fuzzy Inference*, Kluwer Ac. Pub.
6. M. Koeppen, C. Nowack and G. Rsel (1999). Pareto-Morphology for Color Image Processing. *Proc. of the 11th Scandinavian Conference in Image Analysis*, Greenland, Denmark.
7. H. Li, B.S. Manjunath and S.K. Mitra (1995). Multisensor Image Fusion Using the Wavelet Transform. *Graphical Models and Image Processing*, 57 (3): 235-245.
8. R.C. Luo and M.G. Kay eds. (1995). *Multisensor Integration and Fusion for Intelligent machines and systems.* Norwood, NJ: Ablex Publishing Corporation.
9. N. Nandhakumar (1994). Robust physics-based analysis of thermal and visual imagery. *Journal of the Opt. Soc. Am. A*, 1994: 2981-2989.
10. R.A. Salinas, C. Richardson, M.A. Abidi and R.C. Gonzalez (1996). Data Fusion: Color Edge Detection and Surface Reconstruction Through Regularization. *IEEE Trans. on Industrial Electronics*, 43(3): 355-363.
11. A. Soria-Frisch and J. Ruiz-del-Solar (1999). Towards a Biological-based Fusion of Color and Infrared Textural image Information. *Proc. of the IEEE W. on Intelligent Signal Processing'99*, Budapest, Hungary.
12. A. Soria-Frisch (2000). Intelligent Localized Fusion Operators for Color Edge Detection. *Proc. 12th Scandinavian Conference on Image Analysis, SCIA 2001*, Bergen, Norway.
13. A. Soria-Frisch (2001). A New Paradigm for Fuzzy Aggregation in Multisensorial Image Processing. *Proc. 7th Fuzzy Days in Dortmund, 2001*, Dortmund, Germany.
14. M. Sugeno (1974). *Theory of Fuzzy Integral and its applications.* Ph.D. thesis, Tokyo Institute of Technology.
15. R.R. Yager and A. Kelman (1996). Fusion of Fuzzy Information With Considerations for Compatibility, Partial Aggregation, and Reinforcement. *Int. J. of Approximate Reasoning*, 15:93-122.

On OCR of Degraded Documents Using Fuzzy Multifactorial Analysis

U. Garain and B.B. Chaudhuri

Computer Vision & Pattern Recognition Unit
Indian Statistical Institute
203, B. T. Road, Kolkata 700 035, India
{utpal,bbc}@isical.ac.in

Abstract. *Optical Character Recognition* (OCR) systems show poor performance while processing documents like old books or newspapers, Xerox materials, faxed documents, etc. Such documents are considered as degraded documents. One of the important reasons for poor recognition rate for degraded documents is existence of touching or connected characters, which create a major problem for designing an effective character segmentation procedure. In this paper, a new technique is proposed for segmentation of touching characters. The technique is based on fuzzy multifactorial analysis. A predictive algorithm is developed for effectively selecting cut-points to segment touching characters. Initially, our proposed method has been applied for segmenting touching characters that appear in Devnagari (Hindi) and Bangla, two major scripts in Indian sub-continent. The results obtained from a test-set of considerable size show that a high recognition rate can be achieved with a reasonable amount of computations.

1 Introduction

In the field of OCR technology, researchers have noted that a large number of recognition errors in OCR systems are due to character segmentation errors [1, 2, 3, 4]. In documents like old books or newspapers, photocopied or faxed materials, very often, adjacent characters touch each other and separation of such touching characters is a major problem.

A number of strategies for segmenting touching characters are available in the literature [1, 3, 5-8]. But the techniques mainly deal with Roman Scripts. To the best of our knowledge no similar study is available for Indian language scripts. On the other hand, development of OCR systems for the Indian language scripts are gaining importance because of their enormous market potential (*more than 600 million people in India, Bangladesh and Nepal use Devnagari and Bangla scripts to write their language*) and as a result, a Bangla OCR system [10] and later on, a bilingual (Devnagari & Bangla) OCR system [11, 12] is developed.

The work described in this paper is a part of our OCR project. Our original system yields recognition accuracy of 98% for documents of good quality. But it shows poor performance for degraded documents. For such documents, the recognition accuracy comes down to 85% - 90% due to the presence of a large number of touching characters.

N.R. Pal and M. Sugeno (Eds.): AFSS 2002, LNAI 2275, pp. 388–394, 2002.

Fig. 1. Some touching characters in (a) Devnagari (b) Bangla documents.

This paper deals with touching characters in Devnagari and Bangla documents. We propose a new fuzzy decision making approach to solve this problem. The concept of using this approach is motivated after examining the complex ways by which characters touch each other in the Devnagari and Bangla scripts (see fig. 1). Sometimes, it is really difficult even for humans to isolate the characters from certain touching patterns. Hence, we visualize that the task of finding the cut points for segmenting touching characters is itself a fuzzy phenomenon. Moreover, the selection of appropriate cut points depends on many factors making it a multi-factorial decision making problem. Rest of the paper describes details of the proposed fuzzy multifactorial analysis and its application to solve problem related to touching characters.

2 Our Proposed Method

2.1 Fuzzy Multifactorial Analysis

P.-Z. Wang [13] first defined the concept of factor spaces and applied it to the study of artificial intelligence. In our study, we deal with a special type of factors called *degree* or *fuzzy* factors (see [14] for discussion on different types of factors). State spaces for such factors are represented by the interval [0, 1]. We use another concept called *objects* identified based on the objectives to be satisfied in a decision-making problem. Objects are related to a number of factors and distinguished by the states of these factors. Basically, an object like a point in Cartesian space can be viewed as residing in a space constructed by the factors.

Let $f = \{f_1, f_2 ... f_n\}$ be the set of n-factors and $o = \{o_1, o_2 ... o_m\}$ be the set of m objects. Let $E = \{e_1, e_2 ... e_m\}$ be the set of m-evaluations for m objects and for each e_i there are n-values $\{v_{i1}, v_{i2} ... v_{in}\}$ for n-factors. We represent these values in a matrix called evaluation matrix V (n and m are the number of factors and evaluations, respectively and each column in V represents one evaluation):

$$V = \begin{bmatrix} v_{11} & v_{12} & \cdots & v_{1m} \\ v_{21} & v_{22} & \cdots & v_{2m} \\ \vdots & \vdots & \cdots & \vdots \\ v_{n1} & v_{n2} & \cdots & v_{nm} \end{bmatrix} \quad (1)$$

Now from this matrix it is very difficult to choose any evaluation as the solution because each of them consists of n-different values for n-different factors. Hence, a function M_n is used to map the n-dimensional vector $f = \{f1, f_2 ... f_n\}$ into a one-dimensional scalar as: $M_n(f) = M_n(f_{1,}f_2 \cdots, f_n)$.

M_n being a function of factors is called a multifactorial function. Design of M_n is restricted in a way such that the synthesized value should never be greater than the largest of the component values and should never be less than the smallest of the component values, i.e.

$$\bigwedge_{i=1}^{m} f_i \leq M_n(f_1, f_2, \cdots, f_n) \leq \bigvee_{i=1}^{n} f_i \quad (2)$$

Function like M_n that satisfies the above condition in (2) is called Additive Standard Multifactorial (ASM) functions [14]. Using such an ASM function M_n, we make a multifactorial evaluation as follows:

$$V' = (v_1, v_2, \cdots, v_m) = (M_n(v_{11}, v_{12}, \cdots, v_{n1}), M_n(v_{12}, v_{22}, \cdots, v_{n2}), \cdots, M_n(v_{1m}, v_{2m}, \cdots, v_{nm})) \quad (3)$$

Note that all v_{ij}'s are in [0, 1] and since M_n is ASM-func, all v_i's (i = 1, 2... m) in V' are also in [0, 1]. Let v_k be the maximum over all the v_i's, i.e. $v_k = max\{v_1, v_2, ..., v_n\}$ then evaluation corresponding to v_k (i.e. the k-th column in matrix v) indicates the optimal solution. However, we try to find several v_k's to achieve the best possible segmentation for the touching characters, details of which presented in the next.

2.2 Segmentation of Touching Characters

The approach for selecting cut-points for separating the touching characters is based on five fuzzy factors. Design of these factors is based on two criteria: (i) factors should be effective enough to find the right cut-points and (ii) factor-evaluation must be computationally very fast. The factors are as follows:

- f_{ic} : inverse crossing-count = c^{-1} where c is the vertical crossing-count for a pixel column.

In touching characters, character parts overlap each other to form a small black blob is formed at the touching zone. Hence, the column scan at the touching point, in general, encounters a single black run. This is reflected by f_{ic}. On the other hand, the vertical thickness of the black blob at the touching point is always small compared to the thickness of the other character parts; the second factor (f_{mt}) is defined as follows:

- f_{mt} : measure of blob thickness = 1 - t/T, where t (see fig. 2) is the number of black pixels (measured in the middle zone of a character) encountered in one column scan and T (see fig. 2) is the height of the character's middle zone.

Unlike in Roman script, the characters in Devnagari and Bangla scripts have a tendency to touch each other at the middle region of character's middle zone. Factor f_{dm} takes care of this tendency. Mathematically f_{dm} is defined as:

- f_{dm} : degree of middleness $= \dfrac{\min(l_1, l_2)}{\max(l_1, l_2)}$, where l_1 and l_2 are diagrammatically shown in fig. 2.

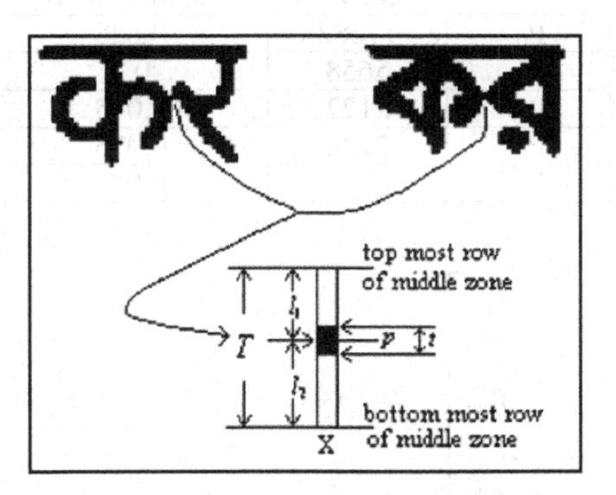

Fig. 2. Features extracted during column scan. X: a pixel column, *t*: thickness of the black run, *p*: mid-point of the black run, l_1: distance of point p from the topmost row of the middle zone, l_2: distance of point *p* from the bottom most row of the middle zone, *T*: height of the middle zone.

Pattern	Shape	Pattern	Shape
P_{u1}		P_{l1}	
P_{u2}		P_{l2}	
P_{u3}		P_{l3}	
P_{u4}		P_{l4}	
P_{u5}		P_{l5}	

Fig. 3. Different shapes detected above and below the touching points: (a) u-patterns and (b) l-patterns.

The other two factors are designed according to a fuzzy statistical analysis of character shapes. Stroke patterns above and below the touching points are examined and five different stroke patterns are detected above (*u-pattern*) and below (*l-pattern*) the touching points. These patterns (see fig. 3) are treated as fixed elements $u_0 \in U$ (universe). We compute the frequency of u_0 against the training set A_* (consists of *n* number of samples) by

$$\text{Frequency of } u_0 \text{ in } A_* = \frac{\text{number of observations of } u_0 \in A_*}{n}$$

As *n* increases, the frequency tends to stabilize or converge to a fixed number called the *degree of membership* of u_0 in A_* . Table-1 shows the *degree of membership* for the patterns listed in fig. 3.

The evaluation of our fourth and fifth factors f_{up} and f_{low} is done by the membership functions $\tilde{A}_u$ and $\tilde{A}_l$ for the u-patterns and the l-patterns, respectively.

- f_{up} : u-pattern's occurrence rate $= \tilde{A}_u$ ($u_0 = P_{ui}$; i=1, 2... 5) = frequency of P_{ui} given in Table-1, and
- f_{low}: l-pattern's occurrence rate $= \tilde{A}_l$ ($u_0 = P_{lj}$; j = 1, 2, 3) = frequency of P_{lj} given in Table-1.

The above 5-factors (f_{ic}, f_{mt}, f_{dm}, f_{up}, and f_{low}) are evaluated for each of the m columns of a touching character image and a 5 X m evaluation matrix is formed as follows:

Table 1. The degree of membership for the u-patterns and l-patterns

u-patterns	count (total: 28291)	frequency (degree of membership)	l-patterns	Count (total: 28291)	frequency (degree of membership)
P_{u1}	15305	0.541	P_{l1}	17682	0.625
P_{u2}	8402	0.297	P_{l2}	5658	0.2
P_{u3}	2292	0.081	P_{l3}	2122	0.075
P_{u4}	1528	0.054	P_{l4}	2122	0.075
P_{u5}	764	0.027	P_{l5}	707	0.025

$$V_s = \begin{bmatrix} f_{ic}^1 & f_{ic}^2 & \cdots & f_{ic}^m \\ f_{mt}^1 & f_{mt}^2 & \cdots & f_{mt}^m \\ f_{dm}^1 & f_{dm}^2 & \cdots & f_{dm}^m \\ f_{up}^1 & f_{up}^2 & \cdots & f_{up}^m \\ f_{low}^1 & f_{low}^2 & \cdots & f_{low}^m \end{bmatrix} \quad (4)$$

Next, these five aspects are mapped into a 1-D scalar by an ASM-func M_s and V_s is transformed into V_s' as follows:

$$V_s' = (M_s(f_{ic}^1, f_{mt}^1, \ldots, f_{low}^1), M_s(f_{ic}^2, f_{mt}^2, \ldots, f_{low}^2), \ldots, M_s(f_{ic}^m, f_{mt}^m, \ldots, f_{low}^m)) = (f_s^1, f_s^2, \ldots, f_s^m) \quad (5)$$

$$\text{where } f_s = M_s(f_{ic}, f_{mt}, f_{dm}, f_{up}, f_{low}) = \frac{1}{5}(f_{ic} + f_{mt} + f_{dm} + f_{up} + f_{low})$$

M_s gives a degree of membership, f_s^i to each column to reflect its possibility to be a cut-point.

2.3 Confirmation of Cut-Points

Researchers have proposed many alternatives approaches [1, 2, and 7] for the confirmation of the cut-points. But the procedures are, in general, computationally slow as the character classifier unnecessarily spends more time in attempting to recognize many patterns, which are not valid.

In our approach, we try to reduce the computational effort by introducing a concept of predicting the most favorable cut-point before character classifier is used to confirm it. The concept is borrowed from the *predictive parser* concept well known in the field of compiler-design for the programming languages [15]. The approach drastically reduces the number of alternatives tried. The f_s values given by the equ^n (5) are used as prediction and the column having the highest f_s value is predicted as the most favorable cut-point. The components formed by this cut-point are sent to the classifier and based on its output the algorithm goes forward to choose a new cut-point or backtracks to the previous pattern and tries with the column having the second best f_s value.

Fig. 4 illustrates the algorithm. Fig. 4(a) shows a triplet in Bangla script. Fig. 4b shows three cut-points (column) having top three f_s values. Column i having the highest f_s value is predicted as the most favorable cut-point that yields patterns P1 and P2. The classifier recognizes P2 while P1 is rejected (fig. 4c and 4d). P1 is then segmented by the cut-point (column j) having the highest f_s value within P1. This cut yields P3 and P4. Classifier rejects both P3 and P4 (fig. 4d) forcing algorithm to choose the next cut-point (column k) yielding patterns P5 and P6. Classifier accepts both of these patterns and the algorithm terminates.

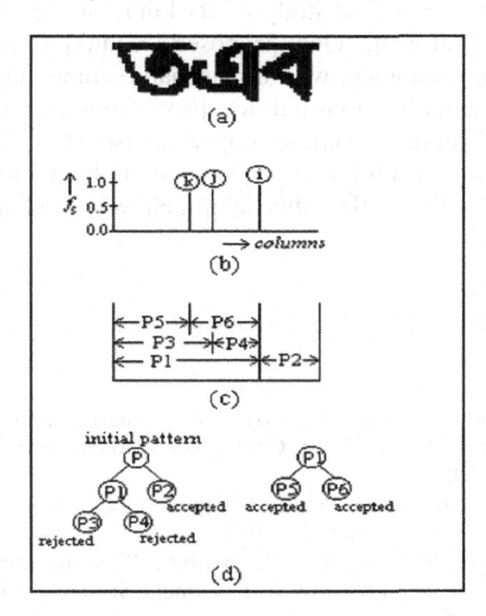

Fig. 4. Confirmation of cut-points for a triplet in printed Bangla script.

3 Test Results

The documents used in our experiment are collected from old books (some of them published in more than 20 years ago), newspapers, Xerox copies, faxed documents, etc. and scanned with a 300 dpi flat-bed scanner. All algorithms in this paper are written in C and executed on a Pentium-II 333 MHz machine.

Experimental results show an overall 98.92% (for Devnagari) and 98.47% (for Bangla) accuracy in segmenting touching characters. This empowers our original OCR system to process degraded documents with much better recognition accuracy (increased by 10% to 12%).

As per as system's *throughput* is concerned it is observed that in the modified system though the overall time required for processing a document is sometimes slightly more than the time taken by our original system, but for documents having large number of touching characters there is a considerable improvement in the system *throughput*. This achievement is tested by using various types of documents where the number of touching characters varies from 1% to 20%.

We have carried out an extensive test to check how fast our technique is to locate correct cut-points. Experimental results show that our proposed technique has efficiency of 79.64% and 78.99% for Devnagari and Bangla, respectively. This indicates that out of 100 cut-points predicted by our technique, more that 79 cut-points (78 for Bangla) are valid; a fact strongly supports our idea of using fuzzy multi-factorial analysis.

4 Conclusions

We have developed an effective strategy for processing of degraded Devnagari and Bangla documents that contains large number of touching characters. To the best of our knowledge, this is the first study of its kind for these two major scripts in the entire Indian sub-continent. Our proposed technique shows high accuracy in segmenting touching characters with a reasonable computational effort. It seems this technique can be suitably modified for processing degraded documents in other scripts like Roman (English). Another important aspect of this study is the application of fuzzy multifactorial analysis in a practical problem involving document image analysis (DIA). It is likely that this approach will find application in other DIA problems.

References:

1. R.G. Casey and G. Nagy, "Recursive segmentation and classification of composite character patterns", *Proc. 6th Int. Conf. Pattern Recognition (ICPR)*, Munich, Germany, pp. 1023-1026, 1982.
2. S. Tsujimoto and H. Asada, "major Components of a Complete Text Reading System", *Proc. IEEE,* **80**(7), pp. 1133-1149, 1992.
3. H. Fujisawa, Y. Nakano, and K. Kurino, "Segmentation Methods for Character Recognition: From Segmentation to Document Structure Analysis", *Proc. IEEE*, **80**(7), pp. 1079-1092, 1992.
4. T. Nartker, ISRI 1993 Annual Report, Univ. of Nevada, Las Vegas, 1993.
5. D.G. Elliman and I.T. Lancaster, "A Review of Segmentation and Contextual Analysis Techniques for Text Recognition", *Pattern Recognition*, **23**(3/4), pp. 337-346, 1990.
6. R.G. Casey and E. Lecolinet, "A Survey of Methods and Strategies in Character Segmentation", IEEE *Trans. on Pattern Analysis and Machine Intelligence*, **18**(7), 1996.
7. S. Liang, M. Shridhar, and M. Ahmadi, "Segmentation of Touching Characters in printed Document Recognition", *Pattern Recognition*, **27**(6), pp. 825-840, 1994.
8. Y. Lu, "On the Segmentation of Touching Characters", *Proc. Int. Conf. On Document Analysis and Recognition (ICDAR)*, Japan, pp. 440-443, 1993.
9. B.B. Chaudhuri and U. Pal, "A Complete Printed Bangla OCR System", *Pattern Recognition*, **31**, pp. 531-549, 1998.
10. B.B. Chaudhuri and U. Pal, "An OCR system to read two Indian language scripts: Bangla and Devnagari (Hindi)", in *Proc. 4th Int. Conf. on Document Analysis and Recognition (ICDAR)*, Ulm, Germany, pp. 1011-1016, 1997.
11. B.B. Chaudhuri, U. Garain, and M. Mitra, "On OCR of the Most Popular Indian Scripts: Devnagari and Bangla", *Visual Text Recognition and Document Processing*, Ed: N. Murshed, World Scientific, 2000 (in press).
12. U. Garain and B.B. Chaudhuri, "On Recognition of Touching Characters in Printed Bangla Documents", in *Proc. of Indian Conf. on Computer Vision, Graphics and Image Processing*, Eds: Santanu Chaudhury and Shree K. Nayar, Viva Books Private Limited, Delhi, India, pp. 377-380, 1998.
13. P.-Z. Wang and M. Sugeno, "The factor fields and background structure for fuzzy subsets", *Fuzzy Mathematics*, **2**(2), pp. 45-54, 1982.
14. H.X. Li and V.C. Yen, *Fuzzy sets and fuzzy decision-making*, CRC Press, 1995, USA.
15. Aho, Sethi and Ullman, *Compilers: Principles, Techniques, and Tools*, Addison-Wesley Publishing Co., 1986.

A Bootstrapped Modular Learning Approach for Scaling and Generalisation of Grey-Level Corner Detection

Rajeev Kumar[1] and Peter Rockett[2]

[1] Department of Computer Science & Engineering,
Indian Institute of Technology, Kharagpur 721 302, India
rkumar@cse.iitkgp.ernet.in
[2] Department of Electronic & Electrical Engineering
Mappin St, University of Sheffield, Sheffield S1 3JD, England
p.rockett@sheffield.ac.uk

Abstract. In this paper, we study a bootstrapped learning procedure applied to corner detection using synthetic training data generated from a grey-level model of a corner feature which permits sampling of the pattern space at arbitrary density as well as providing a self-consistent validation set to assess the classifier generalisation. Since adequate learning of the whole mapping by a single neural network is problematic we partition data across modules using bootstrapping and which we then combine by a meta-learning stage. We test the hierarchical classifier on real images and compare results with those obtained by a monolithic network.

1 Introduction

During the process of learning from examples, a neural network approximates the functional relationship of the (sub-)domain covered by training set. An important factor in the *scaling* of connectionist models for solving complex problems is that larger networks require increasing amounts of training time and data; eventually the complexity of the optimisation task can become computationally unmanageable - the learning complexity in a feedforward network is approximately $O(N^3)$ [1]. Another phenomenon contributing to the problem of slow/difficult training is cross-talk, the presence of conflicting information in the training data that retards learning. Increasing the number of hidden units is a popular solution to problematic training but this increases the number of free-parameters in the model which can again lead to poor generalisation.

In the context of addressing complex learning domains, two basic approaches are emerging as possible solutions to the scalability of neural networks: ensemble based and modular systems. See [2] for a partial survey. The family of ensemble-based approaches relies on combining predictions of multiple models, each of which is trained

N.R. Pal and M. Sugeno (Eds.): AFSS 2002, LNAI 2275, pp. 395–400, 2002.

on the same data; in general the emphasis is on improving the accuracy for better generalisation, not on simplifying the function approximators.

On the other hand, modularity can take advantage of function decomposition. A divide-and-conquer strategy splits a problem into a series of sub-problems and then assigns a set of function approximators to each sub-problem such that each module learns to specialise only in sub-domain. The advantage of partitioning is that the complexity of each stage is less than a single unpartitioned computation and the problem is better constrained and more tractable. Nonetheless decomposition has its own difficulties – partitions in the absence of *a priori* knowledge of the pattern space are not unique. Modularity requires an implicit or explicit way to decompose a complex problem and to allow the desired sub-function mappings to emerge from the learning of subtasks. Many methods for task-decomposition have been suggested *e.g.* the learning-follows-decomposition [3] strategy partitions the task using genetic algorithms as a pre-processor to neural learning.

In this work we have selected corner detection as the test problem because reliable corner labelling is a challenging problem in machine vision. We have generated a very large training dataset using a grey-level image corner model and employed a bootstrap procedure to partition this data across neural modules. We find that the bootstrap procedure results in lower validation error and therefore superior generalisation ability compared to a neural detector trained on the whole unpartitioned dataset. We present real image labelling comparison of our bootstrap approach with a monolithic network which in turn has been shown to be superior to conventional corner detection approaches [5].

2 Data Description

Various attempts have been made in the past to apply neural networks to feature labelling using exemplars obtained from real images but these have been largely disappointing, due principally to use of inadequate training sets.

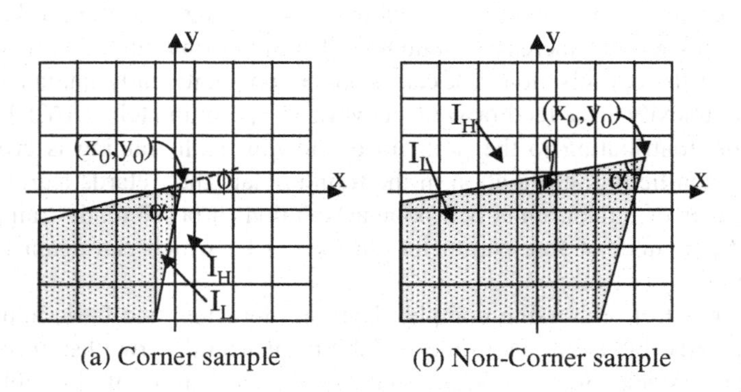

(a) Corner sample (b) Non-Corner sample

Fig. 1. Corner Training Data

In order to address the issue of the sufficiency of the training data and other issues, Chen *et al.* [4], [5] used grey-level models of edge and corner features to generate training data which adequately spanned the pattern space and were thus able to obtain good labelling results. In this work, we use the grey-level corner model of Chen. Taking a 7 × 7 lattice of pixels – Figure 1 – we define a pattern as a corner if the two intersecting lines meet in the central cell, as depicted in Figure 1(a). Conversely, if the point of intersection falls outside the central cell we label this as a non-corner – Figure 1 (b). We refer to this later situation as an *unobvious* non-corner to distinguish it from *obvious* non-corner examples such as edges and uniform patches. Thus an instance of a corner/unobvious non-corner is described by its opening angle (α), inclination to the positive x-axis (ϕ), the point of intersection (x_o,y_o) and high-low intensities(I_H,I_L). Additionally, 'dark' corners on a light ground are distinct from 'light' corners on a dark ground so both 'polarities' of corners are represented. Taking suitable ranges and increments of the model parameters and employing elementary plane analytic geometry we can compute the corresponding vector of grey-level pixel intensities. We process the 7 × 7 pixel patches to exploit the known invariance properties of corners, namely; invariance to rotation and illumination. Further, since corners are – by definition – regions of high intensity gradient we have used the Sobel gradient operator to compute a 5 × 5 vector of image gradients since this operator incorporates some degree of image smoothing at no additional computational cost – see [5] for further details.

A key factor in constructing the training set is that to train an *unbiased* classifier, the prior probability of corners has to be accurately reflected in the composition of training set [6]. Across a range of typical image we observe that the corner prior is 50-1000: 1 but for the sake of convenience we have taken a constant and representative value of 100: 1 in this work. Thus for n corner examples we require 100n non-corner examples. Such a highly unbalanced set produces learning difficulties but we surmounted this problem by sub-sampling the non-corner data to modify the prior to 24:1 and subsequently remapping the network output back to a prior of 100:1 using a technique due to Lacey *et al* [7]. Our final training set for each corner polarity and each corner tip angle (α) comprises 180 corner samples and 4320 non-corners. The advantages of the present corner data model can be summarised as:

- The training data can be arranged to span the whole pattern space with arbitrary resolution. We have observed from studies of training neural detectors that using data from real images leads to poor generalisation [4], [5].
- The data are guaranteed free from both noise and outliers.
- For the purposes of classification, the composition of the training set can be adjusted to reflect the prior probabilities of occurrence of the features in real images so the labels approximate Bayesian posterior probability [6].
- As well as generating the training data, the same model can be used to generate a dense validation set which can be used to probe the network's generalisation properties.

3 Module Training by Bootstrapping

We would like to detect corners in the range of $0^\circ < \alpha < 180^\circ$, however this requires an extremely large training set. The experience of Chen [5] is that learning the necessary mapping for all corners is difficult so we have employed a modular approach whereby we have stratified the corner data according to the opening angles, α. Starting with a single module trained on 90° corners alone, we have used essentially a bootstrap procedure to partition the training data.

From the response of this first trained module to a verification set containing corners with all opening angles – see Figure 2 – we selected the *single* opening angle for the next module such that its minimum response would rise to 0.5 at the point where the minimum response of the 90° module fell to 0.5. Thus we obtain a modular composite of overlapping responses. In fact, of course, this procedure gives two angles, one below 90° (70°) and the one above 90° (110°). This bootstrap procedure was repeated resulting in five modules each trained on single opening angles of: 50°, 70°, 90°, 110° and 130° respectively. Adding further modules is unlikely to be of value since there is genuine confusion between the patterns; as the opening angle increases towards 180° there is increasing confusion with straight lines. Similarly, trying to locate corners with very acute angles is genuinely problematic.

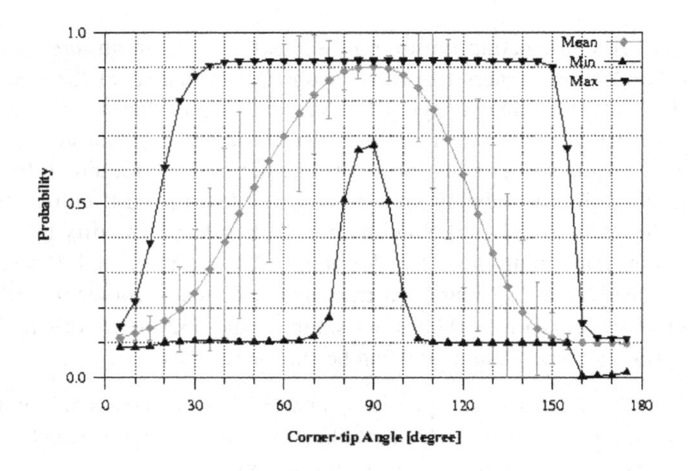

Fig. 2. Validation response of network trained on 90° corners alone

We used multilayer feedforward networks with a sigmoid activation function at each node. The network for each corner module comprised 25 input nodes, 4 hidden nodes and a single output node trained with the error backpropagation learning algorithm. The point of optimal classifier generalisation was assessed using a validation set generated with parameter values which were not used for the training data and with smaller parameter increments for a dense sampling. We experimented with networks with different numbers of hidden units but found that four hidden units were sufficient.

There are potentially many approaches for integrating multiple learned modules - see [8] for a review. Here we combined the meta-pattern data by training a fusing network which in turn approximates the final class probability. This approach has the advantage of reducing the influence of false positives in the final labelling stage which otherwise have degraded the performance of a winner-takes-all strategy. The combined response of the whole system to the validation set is shown in Figure 3; the upper curves come from corners near the centre of the lattice while the lower curves are from corners whose tips are at the edge of the central lattice cell and therefore near their respective decision boundaries. The middle curves are average responses with standard error bars. In this way we were able to obtain a much wider labelling response as shown in Figure 3.

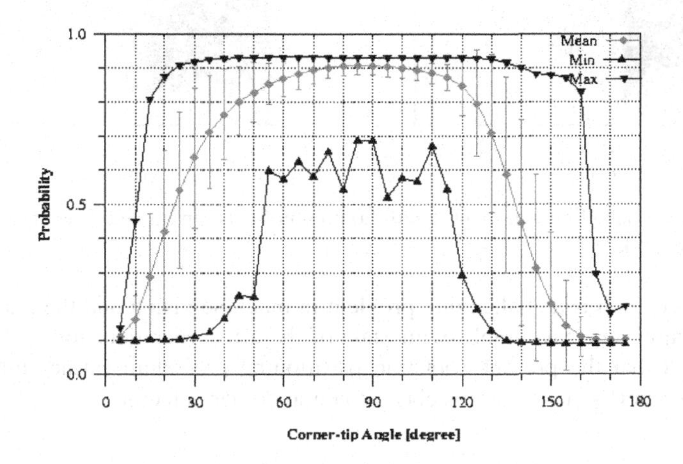

Fig. 3. Validation response – the composite response from the whole system

We observed consistently from repeated trainings that the modular system exhibited far lower validation errors than were attainable with the monolithic network demonstrating that the partitioning of the problem produced superior generalisation and more accurate learning of this challenging classification task.

4 Results and Conclusions

We tested the results on real images of 2D objects and on complex scenes. The monolithic detector was trained on the whole unpartitioned dataset. Since our network outputs approximate Bayesian probabilities on a two-class problem, we have taken pixel sites where the detector outputs > 0.5 to be corners in the sense of a maximum *a posteriori* (MAP) classifier. Typical results for real images are shown in Figure 4 where the performance of the modular network is superior to both those obtained from a single monolithic network and conventional corner detectors in that greater numbers of true positives are labelled as well as fewer false positives. Interestingly, many false positives are due to image aliasing effects which produce arguably corner-like pixel patterns.

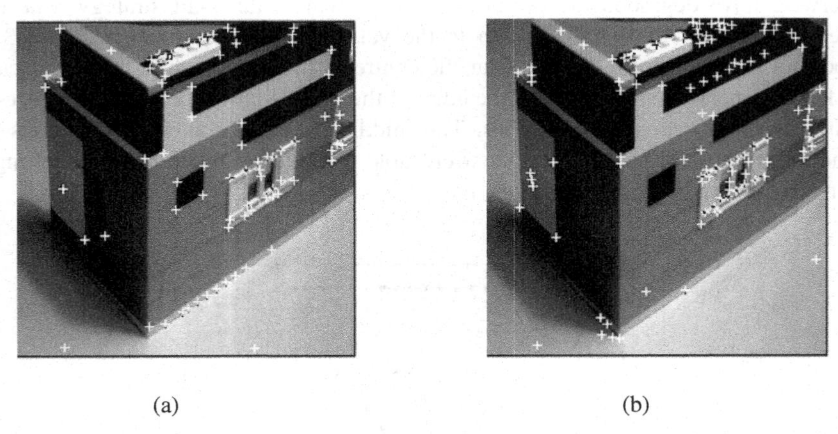

(a) (b)

Fig. 4. Results of labelling real images with (a) present bootstrap trained detector and (b) single monolithic network.

Corner detection is a challenging problem in machine vision and the results in Figure 4 compare very favourably with state-of-the-art corner detectors [5]. We have demonstrated that the present bootstrap partitioning procedure further improves the generalisation performance of this class of neural corner detector.

References

1. Hinton, G.E.: Connectionist Learning Procedures. Artificial Intelligence 40 (1989) 185-234
2. Auda, G, Kamel, M.: Modular Neural Networks – a Survey. Int. J. Neural Systems 9 (1999) 129-151.
3. Kumar, R., Rockett, P.I.: Multiobjective Genetic Algorithm Partitioning for Hierarchical Learning of High-Dimensional Pattern Spaces: A Learning-follows-Decomposition Strategy. IEEE Trans. Neural Networks 9 (1998) 822-830
4. Chen, W.C., Thacker, N.A., Rockett, P.I.: A Neural Network for Probabilistic Edge Labelling with a Step Edge Model. IEE Proc. – Vision, Image & Signal 143 (1986) 41-50
5. Chen, W.C.: Probabilistic Labelling of Edge and Corner Image Features with Neural Networks. Ph.D. Thesis. Dept. Electronic & Electrical Engineering, Sheffield Univ., 1996.
6. Richard, M.D., Lippmann, R.P.: Neural Network Classifiers Estimates Bayesian *a posteriori* probabilities. Neural Computation 3 (1991) 461-483
7. Lacey, A.J., Thacker, N.A., Seed, N.L.: Smart Feature Detector Using an Invariance Network Architecture. Proc. 6th British Machine Vision Conf. (1995) 327-336
8. Jacobs, J.A.: Methods for Combining Experts' Probability Assessments. Neural Computations 7 (1995) 867-888

Towards Fuzzy Calibration

C.V. Jawahar and P.J. Narayanan

International Institute of Information Technology
Gachibowli, Hyderabad 500019, India
{jawahar,pjn}@iiit.net

Abstract. A framework for fuzzy calibration is introduced here. Fuzzy calibration is necessary to account for the imprecision of the camera model that can be computed during calibration. It is the first step of a fuzzy vision framework in which the uncertainties are propagated forward through the different levels of processing until precise values are absolutely necessary. We present the fuzzy calibration framework for weak and strong calibration. We also present some thoughts outlining how such calibration can be used in higher levels of vision processing.

1 Introduction

Calibration is most fundamental to stereo vision. Calibration is the process of computing the parameters of a standard mathematical model that represents the imaging process of a camera. The pin-hole camera model is most popular and adequately represents the projection process of a real camera. For applications that use lenses of low-focal lengths, the lens distortion parameters can be recovered and used to correct distortions such as the radial lens distortion. Different calibration algorithms have been proposed and implemented by Computer Vision researchers in the past couple of decades.

The calibration as used in Computer Vision can be broadly classified into two categories: Strong and Weak [3]. Strong calibration fits a pin-hole model to the measured projection characteristics of the camera. That is, given a point $\bar{X}$ in the 3D space, its projection $\bar{x}$ to the image space can be computed via a matrix M, called the *Projection Matrix*, calculated by strong calibration. Weak calibration, on the other hand, provides the relative geometry between two or more views of the same scene. The *Fundamental Matrix* $\mathbf{F}$ of a pair of views encodes the relations of how the mapping of a pixel in one image is constrained in the other. In presence of more than two views, trifocal tensors and other linear relationships can provide the weak calibration.

Calibration is a hotly debated area among Computer Vision researchers. One school of thought believes strong calibration is essential and relies on it for applications such as 3D metric measurements. Another school discards a priori calibration because of its unreliability. We believe that a middle path that combines the advantages of each can yield the best results. We will use the best available calibration tools, but will not trust their output completely. The ambiguities and uncertainties in the calibration will be modelled and used in the

N.R. Pal and M. Sugeno (Eds.): AFSS 2002, LNAI 2275, pp. 401–407, 2002.

subsequent levels of processing. Fuzzy set theory can provide a theoretical framework for modelling the uncertainty associated with the ambiguity and vagueness of image measurements and is best suited to model the ambiguity of calibration.

Fuzzy notions have been employed extensively in image processing and pattern recognition. However, their applications to higher-level vision have been quite limited in the literature. [1,6,7]. The uncertainty incurred in the initial stages of camera calibration can cause errors down the line if propagated as such. We believe that a fuzzy calibration followed by processing of a fuzzy set of correspondences can preserve the ambiguities in the low-level information so that a better decision can be taken at a higher level, if more precise information is available.

Section 2 presents a brief description of the notion of fuzzy calibration and how it differs from the conventional calibration. Strong and weak camera calibration schemes are extended to fuzzy notions in Section 3. In section 4, we describe the scope and applicability of the fuzzy calibration procedures. Section 5 presents a few concluding remarks.

2 Fuzzy Calibration: The First Step

Calibration errors have a great impact on stereo vision systems that use calibration. The calibration parameters are in practice estimated from a large number of observations, typically using a non-linear optimization procedure. The parameters are used as precise values representing the camera, though the estimation process is aware of the mismatch between the model and the measurements. Thus, the uncertainty in the calibration process is ignored completely. This can result in further errors in the subsequent stages of processing. Thus, it is essential for the calibration step to preserve the uncertainties that are found. For a task at a higher level such as stereo vision, a fuzzy set of correspondences can be computed, using the fuzzy calibration data. The uncertainties in the correspondence should reflect those in the calibration as well as new uncertainties introduced by the matching step. The depth values computed from these correspondences will take into account the known uncertainties till then. This process should be carried forward at all levels till a crisp decision is unavoidable. For example, crisp depth values could be calculated when the depth map has to be converted to a triangulated model for display.

Better results can thus be expected if the uncertainty involved is modelled explicitly and used by all levels of processing. Fuzzy sets and numbers provide an excellent framework for this. It has already been applied quite successfully to many low-level vision tasks. Algorithms that take fuzzy inputs can produce suitable fuzzy outputs representing the uncertainty for the subsequent level of processing. Most vision algorithms can modified easily to handle fuzzy data by replacing each operation by its fuzzy equivalent. Well defined processes are available to convert fuzzy results into crisp ones. The process of handling the uncertainties in vision should rightly begin with *fuzzy calibration*, as it is typically the first step in a vision process.

The right place to start fuzzy vision processing is calibration. Calibration is computed from a few known points specified manually or computed using an appropriate algorithm. It is also possible to boot-strap the process starting with a few manually selected points, as an algorithm can find more known points if an approximate calibration exists. Calibration data consists of a few parameters for the camera model. Conceptually, fuzzy calibration uses fuzzy numbers for each of those parameters. The uncertainties encoded by them will result in different, and more general, constraints. The fuzzy numbers and their associated membership functions can be computed from the error measures used by the linear or non-linear optimization algorithm used for conventional calibration.

3 Fuzzy Calibration Models

In this section, we briefly introduce the models of weak and strong calibration mathematically and discuss how they can be modified to fuzzy models.

3.1 Weak Calibration

The weak calibration of two cameras is given by the fundamental matrix **F** that encodes the constraints between the images of the same world point in two views [3,4]. If $\bar{x}$ and $\bar{x}'$ are the homogeneous coordinates of the projections of the same point in the left and right views, the following relation holds about them.

$$\bar{x}'^{\mathrm{T}} \mathbf{F} \bar{x} = 0 \tag{1}$$

F is 3×3 matrices of rank 2 and can be computed from 8 or more pairs of matching points using a linear algorithm. In practice, there are linear, nonlinear and statistically optimal algorithms that use a large number of points matching points for computing the fundamental matrix [4,8].

Let us look at a simple arrangement of two cameras, called a parallel, rectified camera configuration. The fundamental matrix for such a configuration is given by [4]

$$F = \begin{bmatrix} f_{11} & f_{12} & f_{13} \\ f_{21} & f_{22} & f_{23} \\ f_{31} & f_{32} & f_{33} \end{bmatrix} = \begin{bmatrix} 0 & 0 & 0 \\ 0 & 0 & 1 \\ 0 & -1 & 0 \end{bmatrix} \tag{2}$$

The arrangement may only be approximately parallel in practice. Equation 1 can be expanded for two matching points $[x', y', 1]^{\mathrm{T}}$ and $[x, y, 1]^{\mathrm{T}}$ as

$$x'(xf_{11} + yf_{12} + f_{13}) + y'(xf_{21} + yf_{22} + f_{23}) + (xf_{31} + yf_{32} + f_{33}) = 0. \tag{3}$$

Since the overall scale is unimportant, we can safely assume $f_{23} = 1$ for the parallel rectified situation, since we know that number is non-zero for the ideal situation. Equation 3 reduces to $y' - y = 0$ ideally. However, the parallel, rectified model could be slightly incorrect in a few ways in a practical situation, in spite of the best efforts at making the cameras parallel and rectified.

1. Incorrect alignment of scan lines. That is, Equation 3 reduces to $y'-y+\delta=0$.
2. Incorrect magnification. Equation 3 reduces to $y'-y(1+\gamma)=0$ in this case.
3. Non-parallel situation. Equation 3 reduces to $\epsilon x'+y'-y=0$ in this case.

In all these cases, the values δ, ϵ, and γ are very small, ideally zero. The fundamental matrix for a practical parallel, rectified camera arrangement can account for such uncertainties in measurements and can be given by

$$F=\begin{bmatrix}0 & 0 & \epsilon\\ 0 & 0 & 1\\ 0 & -(1+\gamma) & \delta\end{bmatrix}. \tag{4}$$

At this stage, two important questions are to be answered: (a) Can we estimate $\delta, \epsilon, \text{and} \gamma$ from observations? (b) Can a precise set of crisp values model the uncertainty in the calibration?

The answer to the first question is in the affirmative. The number of image measurements is typically more than the number of unknowns and one could employ a least mean/median square solution [2,3,4]. The end results give crisp values for the entries of **F**, forcing the corresponding point to lie on an epipolar line in the second image. A better way to capture such uncertainty is to consider the image measurements as fuzzy measurements, obtained on a discrete 2D grid using projection of a 3D point. The values of δ, ϵ, and γ can be computed as fuzzy numbers using fuzzy regression techniques [5]. We can derive a fuzzy set in the second image where the corresponding feature point should lie using the fuzzy **F** matrix. It will be interesting to investigate what the correspondence implies in this case. We will discuss this problem in the next section. The notion of fuzzy feature measurements allow us to handle the points "in and around" the conventional epipolar line.

3.2 Strong Calibration

The projection equation given the strong calibration matrix M is given by $\bar{x}=M\bar{X}$. The matrix M is a 3×4 matrix as homogeneous coordinates are used to represent points in both the 2D and 3D spaces due to the mathematical ease of dealing with rotations and translations uniformly using them. When the point $\bar{X}$ is specified in a projective or affine space instead of a Euclidean space, corresponding projective or affine calibration matrices can be used. The projection matrix obeys appropriate constraints in each of these spaces and can be computed from the world and image coordinates of a number of points.

We can decompose the projection matrix into the *intrinsic* and *extrinsic* components

$$M=PV=\begin{bmatrix}f_x & 0 & 0 & 0\\ 0 & f_y & 0 & 0\\ 0 & 0 & 1 & 0\end{bmatrix}\begin{bmatrix}R\,;\,T\\ \mathbf{0}\quad 1\end{bmatrix}, \tag{5}$$

where P is the perspective projection matrix for points specified in a coordinate frame rooted at the camera center, V is the view transformation matrix that transforms the world coordinate frame to the camera coordinate frame. f_x and f_y are the focal lengths in the two directions, and R and T are the rotation matrix and the translation vector respectively that align the world coordinate system to the camera coordinate system.

A typical calibration algorithm [9] gives crisp values to the camera parameters. Achieving precise parameters from measurements has been a difficult task [3,4]. However, We can treat the camera parameters as fuzzy quantities to account for the uncertainties. A discussion on the complete mathematical formulation of the framework is beyond the scope of this paper. However, we provide a glimpse of the geometric interpretation of the fuzzy calibration parameters now.

Let us look at the region of the world (i.e., the subset of $\mathbb{R}^3$) that maps to a pixel. For that we have to "invert" the projection matrix. We have that $M^{-1} = V^{-1}P^{-1}$. The inverse of the view transformation matrix is given by

$$V^{-1} = \begin{bmatrix} R^{\mathrm{T}} \,;\, \mathbf{0} \\ \mathbf{0} \quad 1 \end{bmatrix} \begin{bmatrix} \mathbf{I} \,;\, -T \\ \mathbf{0} \quad 1 \end{bmatrix},$$

since R is an orthonormal rotation matrix. The inverse of P is not uniquely defined. We can decompose P as $[Q;0]$ where Q is a 3×3 matrix consisting of the left 3 columns of P (see Equation 5). We now can see that,

$$Q^{-1} = \begin{bmatrix} 1/f_x & 0 & 0 \\ 0 & 1/f_y & 0 \\ 0 & 0 & 1 \end{bmatrix} \tag{6}$$

Now, we can "invert" the projection matrix as

$$M^{-1}\bar{x} = V^{-1}P^{-1}\bar{x} = V^{-1} \begin{bmatrix} Q^{-1}\bar{x} \\ k \end{bmatrix} = \begin{bmatrix} R^{\mathrm{T}} \,;\, \mathbf{0} \\ \mathbf{0} \quad 1 \end{bmatrix} \begin{bmatrix} \mathbf{I} \,;\, -T \\ \mathbf{0} \quad 1 \end{bmatrix} \begin{bmatrix} Q^{-1}\bar{x} \\ k \end{bmatrix} \tag{7}$$

where k is a constant that cannot be computed. Together, the above equation gives the parametric equation of the imaging ray for the pixel $\bar{x}$, the parameter being k. Equation 7 gives the unbounded volume of space that projects to the pixel in the global coordinate frame. If P and V give precise projection and transformation, the volume that projects to the pixel is a ray called the *imaging ray.* If uncertainties in the focal lengths or the rotation and translation estimates can be encoded using fuzzy numbers in the corresponding matrices, these equations specify an *imaging volume*, which is the generalization of the imaging ray. We can compute a membership functions in the fuzzy framework for each ray in the imaging volume.

Comments We presented fuzzy models for weak and strong calibration in this section. Fuzzy fundamental matrix can be computed from crisp or fuzzy points in multiple images, relaxing the strong epipolar assumption. This can account for the possible inaccuracies of imaging models and the rectification process. Fuzzy strong calibration can account for uncertainties in estimation of parameters such as the focal length, rotations, and translations. We can compute the geometrical imaging volume with associated distribution of uncertainties within it as the possible locations of the 3D scene point that projects to each pixel.

4 Fuzzy Vision: Beyond Calibration

In this section, we outline how the fuzzy calibration can be used for other kinds of vision processing. There are multiple ways to handle the imprecision of the calibration process and to develop algorithms using it. We present simple approaches to some problems in this section. We use the notion of *fuzzy correspondence* for the examples given here. Fuzzy correspondence gives a set of points in the right image with associated membership functions for each pixel in the left image, based on fuzzy feature measurements that encode similarity [7]. We can study fuzzy correspondences and fuzzy depth maps with the help of the rich literature available on fuzzy arithmetic.

Fuzzy Correspondences and Weak Calibration: In a typical situation, we have two problems: obtaining the fuzzy correspondences and estimating the fundamental matrix from them. Here, we outline an iterative algorithm which carries out both simultaneously starting with a set of crisp pixel matches given by a standard matching algorithm.

Let $\{\mathbf{x_1}, \mathbf{x_2}, \ldots \mathbf{x_n}\}$ be a set of pixels in the first image and $\{\mathbf{y_1}, \mathbf{y_2}, \ldots, \mathbf{y_n}\}$ be the pixels in the second image. We can assign memberships to the corresponding pairs of matches based on a similarity measure for the feature points. Such correspondences need not be geometrically valid and the second point may not lie "near" the epipolar line. If μ_i is the membership of the pair $(\mathbf{x_i}, \mathbf{y_i})$ in the fuzzy correspondence, weak calibration can be obtained by minimizing $J = \sum_i \mu_i(d(x_i, Fy_i) + d(y_i, F^T x_i))$, where $d(x_i, Fy_i)$ gives the distance of the point x_i to the epipolar line given by Fy_i. The minimization of the above objective function is carried out in two stages. First, compute the fundamental matrix based on the eigenvectors of $X^{\mathrm{T}}MX$ where M is a diagonal matrix with fuzzy memberships as the diagonal elements and X is a measurement matrix. More detailed description of a non-fuzzy implementation of this computation can be seen in [4]. Second, assign memberships based on the nearness of the point to the epipolar line. $\mu_i = \frac{k}{(d(x_i, Fy_i) + d(y_i, F^T x_i))}$, for a constant k. It can be shown that the above iterative algorithm converges and provides an optimal estimate of the fuzzy correspondence and the weak calibration simultaneously.

Fuzzy 3D Point from Stereo We saw how fuzzy strong calibration can compute the imaging volume corresponding to a pixel in Section 3.2. When the correspondence between two pixels of an image pair is known, their imaging volumes can be intersected like in stereo vision. The correspondence used itself could be fuzzy as described above, adding an additional level of uncertainty. The three levels of imprecision are: (a) the pixel of the left image represents a set of imaging rays with associated membership values μ_i^{L}, (b) the correspondence pairs it up with a set of pixels in the right image with its own membership values μ_{ij}^c, and (c) each corresponding pixel represents a set of imaging rays with memberships μ_j^R using the right camera's calibration data. We can intersect the viewing volumes taking into account the three uncertainty measures. The result will be a region in space (a fuzzy subset over $\mathbb{R}^3$) with an associated confidence measure for each point in the region. The distribution of the confidence measure is a function of the three membership functions given above. An appropriate T-norm may be used for this purpose. A more detailed treatment of this is beyond the scope of this paper. This results in a disparity/depth map where the disparity/depth estimate at each point is a fuzzy set, or otherwise we have multiple depth estimates with varying amount of certainty.

5 Conclusions and Future Work

We proposed a fuzzy calibration scheme to incorporate the uncertainty associated in the imaging process into the subsequent levels of processing. Calibration is indeed the first step in many vision processes. It is important to keep track of the uncertainties in calibration and use it at higher levels. A fuzzy framework is ideally suited for this. We intend to investigate fuzzy notions for higher level vision processes such as modelling using multiple cameras, new view generation, etc. We believe the fuzzy framework will yield better results to such problems.

References

1. M. Ben-Ezra, S. Peleg, and M. Werman. Efficient Computation of Most Probable Motion from Fuzzy Correspondences. In *IEEE Workshop on Applications of Computer Vision*, Oct. 1998.
2. Z. Zhang *et. al.* A Robust Technique for Matching Two Uncalibrated Images Through the recovery of unknown epipolar geometry. *Artificial Intelligence Journal*, 78:87 – 119, 1995.
3. O. D. Faugeras. *Three-Dimensional Computer Vision: A Geometric Viewpoint.* MIT Press, 1993.
4. R. I. Hartley and A. Zisserman. *Multiple View Geometry in Computer Vision.* Cambridge University Press, ISBN: 0521623049, 2000.
5. K. Jajuga. Linear Fuzzy Regression . *Fuzzy Sets and Systems*, 20:343 – 353, 1986.
6. C. V. Jawahar. Stereo Correspondence Based on Correlation of Fuzzy Texture Measures. In *Indian Conference on Computer Vision, Graphics and Image Processing*, pages 273–278, 1998.
7. C. V. Jawahar, A. M. Namboodiri, and P. J. Naryanan. Integration of Stereo Correspondence Based on Fuzzy Notions. In *International Conference on Advances in Pattern Recognition and Digital Techniques (ICAPRDT)*, December 1999.
8. K. Kanatani. Optimal fundamental matrix computation: algorithm and reliability analysis . In *Proc. of the SSII2000*, pages 14–16, Jun. 2000.
9. R. Tsai. A versatile camera calibration technique for high-accuracy 3D machine vision metrology using off-the-shelf tv cameras and lenses. *IEEE Journal of Robotics and Automation*, 3(4):323 – 344, 1987.

Fuzzy-Similarity-Based Image Noise Cancellation

Gustav Tolt and Ivan Kalaykov

Center for Applied Autonomous Sensor Systems
Department of Technology
Örebro University
SE-701 82 Örebro, Sweden

Abstract. We introduce a new approach for image noise cancellation based on fuzzy similarity. The proposed method allows for simple tuning of fuzzy filter properties and is very convenient for high-speed real-time processing. An example structure with estimated execution time is presented. Comparisons with other image noise cancellation techniques show the advantages of the method.

1 Introduction

Removing noise is a basic operation in image processing. Two of the most widely examined image noise types are Gaussian and impulse noise. Linear low pass filters deal quite well with Gaussian noise, but fail in removing impulse noise and blur edges. Therefore, non-linear filters are mostly applied in practice.

In the last years, filters based on fuzzy models have shown to be effective tools for noise reduction. Russo [6] proposed a FIRE-fuzzy filtering structure that evaluates the intensity differences between the pixel to be processed and some of its neighbors. If the differences satisfy some of the rules in the rule base, the pixel's intensity is corrected accordingly; otherwise the pixel is left unchanged. Many fuzzy filters can be seen as extensions of conventional non-fuzzy filters. In the fuzzy weighted mean filter [8], weights are evaluated by means of fuzzy membership functions. The fuzzy cluster filter [3] can be seen as a fuzzy weighted mean filter applied iteratively, and whose weights depend on the distance to the cluster center. In other fuzzy filters, such as those presented in [1] and [7], a fuzzy weighted sum of the outputs of several non-fuzzy operators, like median, mean, identity and midpoint filters, is computed. The weights are obtained by a fuzzy system depending on some local features, which require prior knowledge or an estimate of the noise variance. A good overview of different fuzzy approaches for image filtering can be found in [5]. The last few years, fuzzy filtering in hardware have shown to provide real-time operation. In [2], Delva et *al.* proposed a FPGA-based fuzzy filter for removal of mixed noise. In this paper, we introduce a fuzzy-similarity-based (FSB) noise cancellation method. It allows a simple tuning of the filter properties and is convenient for hardware implementation for high-speed image processing. In Section 2, we present the proposed new method, and

N.R. Pal and M. Sugeno (Eds.): AFSS 2002, LNAI 2275, pp. 408–413, 2002.

in Section 3, we illustrate the properties of the filter compared to other methods. Finally, some concluding remarks are given in Section 4.

2 Fuzzy-Similarity-Based Image Noise Cancellation

Let X be a noise corrupted grayscale image of size $m \times n$ in which the pixel (i, j) has the intensity $x(i,j) = s(i,j) + n(i,j)$, where $n(i,j)$ is the noise signal. The task is that given $x(i,j)$, find a good estimate of the uncorrupted signal $s(i,j)$. We assume the following hypothesis with a fuzzy context:
H_0: The pixels within a local neighborhood in an image have similar intensities,
where local neighborhood refers to a number of equidistant pixels to any randomly taken pixel, and similarity corresponds to a fuzzy relation "A is similar to B", as presented further in Fig. 2. H_0 expresses the local properties of the image; hence H_0 appears to be more true when the neighborhood size is small and less true as the size increases. When the image is corrupted by noise, many pixels change their values and H_0 is disregarded to a degree depending on the noise intensity. Therefore, the noise cancellation task is to analyze the pixel similarity within all local neighborhoods by scanning the entire image. Each local neighborhood can be represented by a window with $w \times w$ pixels, where w is small enough to limit the window to the local image properties and odd in order to allocate a central point. As the true pixel intensities are unknown a priori, their distribution within the window is unknown too. Hence, it is straightforward to analyze how the intensity is distributed within a number of templates. A template $\chi = \{x_k, x_l, x_m, \ldots, x_n\}$ has a number of window pixels x_i and a central pixel x_0. The variety of the templates corresponds to the variety of the image local properties. Therefore, we must consider several different template configurations when analyzing the need to reduce the noise by correcting the intensity of pixel x_0. Better is to have equal size templates, such that a possible hardware implementation becomes regular. Some example templates in a 3×3 window are shown in Fig. 1.

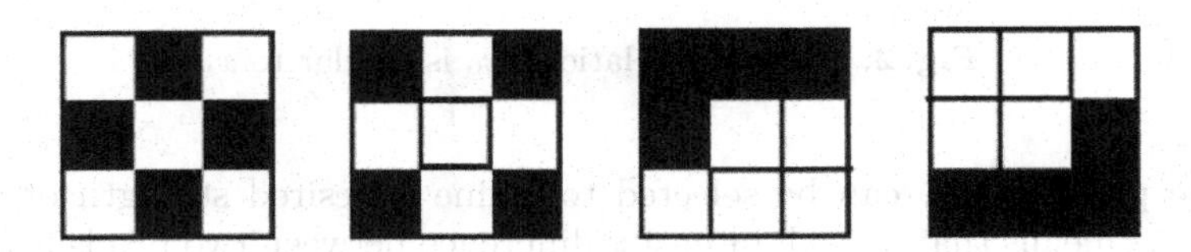

Fig. 1. (a) Four example templates (shadowed pixels). (b) The fuzzy relation "x_0 *is similar to* x_i"

We verbalize the FSB concept of noise cancellation by the following related rules:

R_1: IF the central window pixel and all templates have similar intensities
THEN the window has nearly uniform intensity distribution
AND do not change the central pixel;

R_2: IF the window has not uniform intensity distribution
THEN the max similarity templates best portrays the local image properties
AND change the central pixel according to the intensity of the max similarity templates;

R_3: IF the central window pixel and a template have no similar intensity
THEN the template does not contribute to the noise cancellation.

All these rules require evaluating the similarity between the intensity of pixel x_0 and the templates. Therefore, x_0 is compared to the templates χ_j by a fuzzy relation $x_0 \simeq \chi_j$ meaning that "*x_0 and χ_j have similar intensities*" with a degree of membership obtained by the fuzzy relation $R_{x_0 \simeq \chi_j} = R(x_0, x_k, x_l, x_m, \ldots, x_n)$. As the arguments are independent this is an intersection of two-dimensional relations

$$R_{x_0 \simeq \chi_k} = R(x_0, x_k) \cap R(x_0, x_l) \cap \ldots \cap R(x_0, x_n). \tag{1}$$

Each of them depicts the relation $x_0 \simeq x_i$, meaning that "*x_0 is similar to x_i*" (Fig. 2).

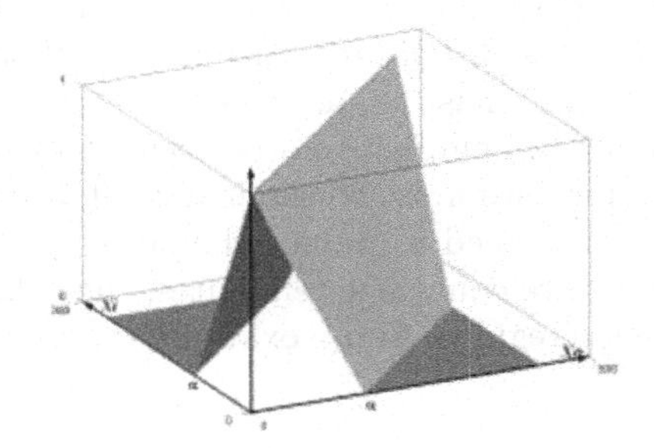

Fig. 2. The fuzzy relation "x_0 is similar to x_i"

The parameter α can be selected to define a desired strength of $x_0 \simeq x_i$. Smaller α means that a small intensity difference between two pixels is necessary for higher strength of the the relation, whereas large α gives high similarity for larger differences. The triangular profile of the membership function simplifies the computations and the hardware implementation. In this case the value of the fuzzy similarity relation $x_0 \simeq x_i$ comes directly from Fig. 2:

$$\mu_{x_0 \simeq x_i} = \begin{cases} 1 - |x_0 - x_i|/\alpha, & |x_0 - x_i| < \alpha \\ 0, & \text{elsewhere} \end{cases} \tag{2}$$

Low similarities between the central pixel and the templates indicate a long-tailed noise distribution in the window. Therefore, the median value of the intensities of the central pixel and its 4 closest neighbors (left-most templates in Fig. 1) is used in an additional rule. The strength of this rule corresponds to the degree of *dissimilarity*, and is computed as

$$\mu_{dis} = 1 - \max_j (\mu_j), \tag{3}$$

where μ_j is the degree of activation of rule R_2 associated with template j.

To check the above rules we first evaluate the similarity of x_0 and all predefined templates. When a template shows no similarity with the central pixel (R_3), it does not contribute to the filtering. In case all templates provide a high level of similarity (R_1), x_0 is unchanged. Generally, the correction is given by

$$x_0^{new} = f(\mu_1, \ldots, \mu_n, \mu_{dis}, g(\chi_1), \ldots, g(\chi_n), g_{dis}), \tag{4}$$

where n is the number of templates, $g(\chi_j)$ is the output of the consequent part of rule R_2 associated with template j and g_{dis} is the median of the intensities of the central pixel and its 4 closest neighbors. The form of the function $g(\cdot)$ can be chosen, e.g. on the apriori information about noise properties. It can be for example the mean or median value of the pixel intensities.

3 Experiments

The method was evaluated in MATLAB on a number of images, contaminated with Gaussian noise, $N(0, \sigma^2)$ and impulse noise of probability P. Tests were made with "salt & pepper" noise as well as impulse noise with random amplitude (uniformly distributed on [0,255]). The mean square error (MSE) was used for comparison with some other algorithms (Table 1). A series of images illustrating the filtering is shown in Fig. 3. The test was performed with the min-operator for the intersections and with center average defuzzification. Hence, the function f in (4) has the following form:

$$f = \frac{\sum_{i=1}^{n} \mu_i g(\chi_i) + \mu_{dis} g_{dis}}{\sum_{i=1}^{n} \mu_i + \mu_{dis}}, \tag{5}$$

where all g:s are the median values of the intensities of the corresponding pixels.

Table 1. MSE results for the 'Lena' image with different noise distributions.

Noise type	Noisy	FSB filter	Delva99	Russo00
σ^2=100, P=0.1 (salt & pepper)	2071.3	63.5	78.1	43.6
σ^2=100, P=0.1 (unif. distr.)	552.6	49.6	72.9	50.3
σ^2=100, P=0	99.6	41.0	61.8	34.8

All filters were implemented recursively in the sense that the updated pixel values were used directly for further processing, resulting in lower MSE values. The values of the tunable parameters in the Delva99 [2] and Russo00 [6] filters, respectively, were selected to minimize the MSE for the Lena image contaminated with Gaussian noise σ^2=100, and salt&pepper noise (P=0.1), in order to study the sensitivity to changes in the noise distribution.

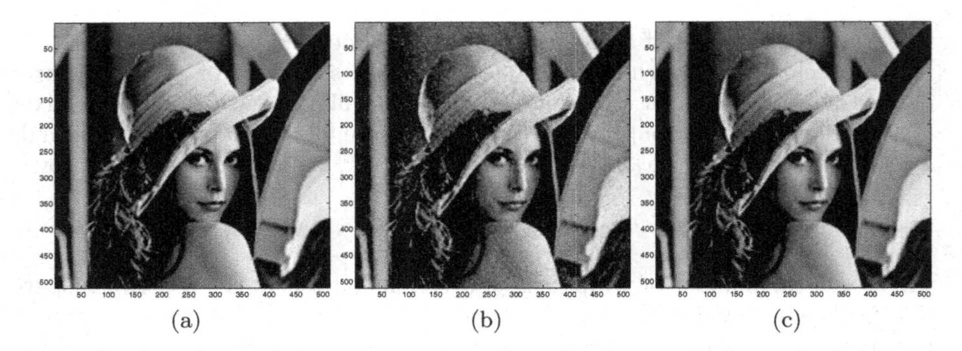

Fig. 3. (a) Original test image. (b) Noise-corrupted image. (c) Image filtered with α=32.

Russo's method applies internally triple filtering in the sense that the noise cancellation acts like three filters applied sequentially. This results in somewhat lower MSE values, but at the cost of higher complexity. The Delva99 filter was designed to be implemented on a FPGA, in order to operate at high speed. Although the Delva99 and the Russo00 filters were originally tested on a mixture of Gaussian and salt&pepper noise, they performed quite well with other types of noise as well.

Table 2. Performance of the FSB filter implemented on a Xilinx Virtex FPGA chip.

Frame size	Time per frame	Frame rate
1024 × 1024	16-19 ms	52-62 fps
768 × 560	6.5-8 ms	125-150 fps
512 × 512	4-5 ms	200-250 fps
320 × 256	1.3-1.5 ms	650-750 fps

From the tests, we see that the FSB filter provides efficient filtering for different noise distributions. Its main advantages are its simplicity and suitability for high-speed noise cancellation. The filter was implemented on a single Xilinx Virtex FPGA chip using the previously designed very high-speed fuzzy processor [4]. The fuzzy similarity evaluation takes only part of the chip as the other is needed to execute preprocessing and connections for the templates. The simulations show quite good performance (Table 2), useful for real-time operation.

4 Conclusions

The FSB method gives quite good overall results for $\alpha \in [16, 128]$, so the tuning of the filter is not a critical process. The set of templates used can be optimized if analyzing how the MSE varies with different templates. The overall properties of the FSB method indicate that it is possible to perform fuzzy filtering of noise-corrupted images at very high frame rate and at high resolution. The ongoing research of the authors concerns further development of algorithms for image processing tasks and their integration on a system with a high-speed vision sensor and FPGA-based very fast fuzzy processor.

References

1. Y. Choi and R. Krishnapuram. A robust approach to image enhancement based on fuzzy logic. *IEEE Transactions on Image Processing*, 6(6):808–825, 1997.
2. J. G. R. Delva, A. M. Reza, and R. D. Turney. FPGA implementation of a non-linear two dimensional fuzzy filter. In *Proc. IEEE Conf. on Acoust., Speech, Signal Processing, ICASSP'99*, pages 2143–2146, Piscataway, NJ, 1999.
3. M. Doroodchi and A. M. Reza. Fuzzy cluster filter. In *Proc. of IEEE Conference on Image Processing, ICIP'96*, pages 939–942, Lausanne, Switzerland, 1996.
4. I. Kalaykov. Parallelism for very fast fuzzy hardware. In *Proc. of IASTED Conf. on Artificial Intelligence and Soft Computing, ASC'2001*, Cancun, Mexico.
5. F. Russo. Recent advances in fuzzy techniques for image enhancement. *IEEE Transactions on Instrumentation and Measurement*, 47(6):1428–1424, 1998.
6. F. Russo. A technique for image restoration based on recursive processing and error correction. In *Proc. of IEEE Instrumentation and Measurement Technology Conference IMTC'00*, volume 3, pages 1232–1236, Piscataway, NJ, 2000.
7. A. Taguchi and M. Meguro. Adaptive L-filters based on fuzzy rules. In *Proc. of IEEE Symposium on Circuits and Systems., ISCAS'95*, pages 961–964, Seattle, WA, 1995.
8. A. Taguchi, H. Takashima, and F. Russo. Data-dependent filtering using the fuzzy inference. In *Proc. of IEEE Instrumentation and Measurement Technology Conference, IMTC'95*, pages 752–756, Waltham, MA, 1995.

Genetic-Fuzzy Approach in Robot Motion Planning Revisited: Rigorous Testing and towards an Implementation

Akshay Mohan and Kalyanmoy Deb

Kanpur Genetic Algorithms Laboratory (KanGAL),
Department of Mechanical Engineering, Indian Institute of Technology – Kanpur,
Kanpur – 208016, India
{mohana,deb}@iitk.ac.in

Abstract. This study endeavors to analyze the GA-Fuzzy approach and is an attempt to make it more practical, real-time and less computationally intensive. A new hybrid technique of using the GA-Fuzzy approach with a local search technique is presented and its efficacy viz a viz the GA-Fuzzy approach alone is demonstrated. It makes the existing technique more robust, efficient and computationally less expensive. The GA-Fuzzy approach converts an online problem of finding an obstacle-free, time optimal path for a mobile robot to an offline problem of optimizing a fuzzy rulebase using a genetic algorithm. Simulation results demonstrating the efficacy of the new design are presented.

1 Introduction

Fuzzy computation and genetic algorithms [1] have found its way into the repertoire of researchers and engineers alike, for handling imprecision in data along with traditional and cutting-edge optimization in different areas [1]. The study aims to analyze, criticize and modify the GA-Fuzzy approach in the domain of path planning, to make it more suitable for actual deployment. It introduces reforms to the existing techniques, seeks the rationale for doing so and presents the improved results for such an algorithm. A new hybrid technique of using the GA-Fuzzy approach with a local search is presented and its efficacy is demonstrated in making the technique more robust, efficient and computationally less expensive. Suggestions are also given for forming the test cases for such algorithms.

The following sections precisely define the problem and introduce the reader to the GA-Fuzzy approach. It is followed by a detailed analysis of the algorithm based on exhaustive simulations, highlights its weakness and suggests remedies. The improvements are tested against the existing technique by rigorous simulations. Finally topics of future research in the area are discussed followed by conclusions.

2 The Fundamental Problem

The problem of motion planning being considered can be stated as to find a time optimal obstacle-free path by an autonomous robot to reach its goal from its start

N.R. Pal and M. Sugeno (Eds.): AFSS 2002, LNAI 2275, pp. 414–420, 2002.

point in an unknown static environment. Here the robot is considered to be a point robot with equal-sized square shaped obstacles.

3 The Solution

Considerable work has been done in the area of motion planning [2] and various approaches have been developed that use fuzzy logic for the purpose [3]. Among the various techniques developed, [3] proposed a GA-Fuzzy approach where a genetic algorithm is used for developing an optimal fuzzy logic controller. This effectively converts an online path planning problem to an offline optimization problem. The scheme is depicted in fig. 1.

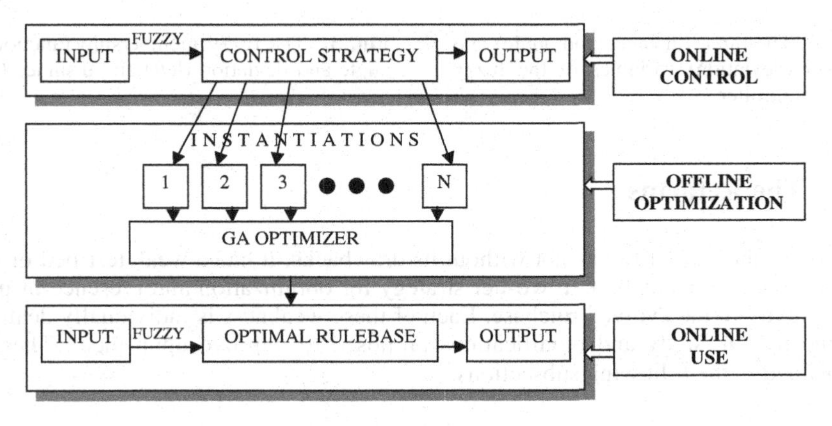

Fig. 1. The GA-Fuzzy approach for robot motion planning

Instantiations of the control strategy to be implemented online are randomly generated and act as individuals for the GA population. The GA then optimizes the control strategy offline. The optimized rulebase as obtained by the GA is then used online for path navigation and obstacle avoidance. This allows the computationally intensive optimization to be done offline and leads to an efficient online implementation that gives an optimal path, yet which is able to do it in real-time and only using the scarce memory and computational resources available on an autonomous robot.

The rules were represented as a set of condition-action pairs. The first condition is the distance of the nearest obstacle forward from the robot. Here forward means within the field of view of the robot. As our obstacles are square shaped, the distance is calculated from the line joining the current position of the robot to the point of intersection of line joining the robot to the center of the obstacle and the nearest edge of the obstacle. The second condition is the angle between the path joining the robot and the target point and the path from the robot to the nearest obstacle forward. The action is represented by the angle of deviation for the robot from the path joining the robot to the target point. This is depicted in fig. 2. Thus angle and distance form the fuzzy inputs and deviation forms the output. Here the fuzzy values for distance, angle and deviation are depicted in fig. 3. Triangular membership functions were used with the relative spacing of the membership distributions being constant. The membership functions were kept same for angle and deviation.

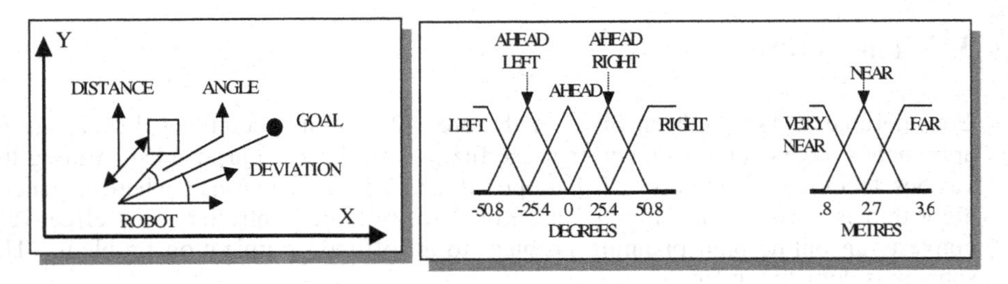

Fig. 2. The input (distance and angle) and output (deviation) variables of the fuzzy logic controller.

Fig. 3. The fuzzy membership functions for angle and deviation (left) and distance (right)

4 The Reforms

However the study [3] was not without its drawbacks. It had a weak test-bed of cases for testing and evaluation, a two-tier strategy for optimization and presence of undue complexity in the evolved rulebase. Each of these weakness is individually dealt with in the present study and modifications proposed to improve upon them. These are explained in the following subsections.

4.1 A Rigorous Test-Bed

In [3], the evolved rules were not rigorously tested and only three to eight obstacle cases were considered. Moreover, their position was decided based on human preference. The fitness evaluation was done on the basis of ten cases with each case having three to eight obstacles in the environment of the robot.

In contrast, our study uses a solid test-bed of cases having eight to more than two hundred obstacles and provides a benchmark for future studies in the field. In each of the cases, the desired number of obstacles is positioned randomly on the workspace. The fitness of the individual is determined as a function of the total time taken by the individual to reach a predetermined goal in each of the one hundred cases considered. Suitable penalties are imposed, if the robot collides with an obstacle, is not able to reach the goal in the allowed maximum time or goes outside its work area. During the optimization of the fuzzy logic controller, our aim is to minimize the total time taken. For this the time is suitably converted to fitness values. A count is also made for each individual (rulebase) of the number of cases for which the robot is successfully able to reach its goal. Thus a perfect individual would be the rulebase using which the robot is successfully able to reach its goal in all the cases in a near time optimal manner. In this context our proposed hybrid algorithm is successfully able to solve each of the cases in a time optimal manner.

On the basis of extensive tests we establish that it is very crucial for obstacles to be placed directly on the line joining the start point of the robot to the target point. Otherwise defective rulebase develop which are not robust enough to handle all the possible cases. In fact, if this is not done, the optimal rulebase is simple: just one rule!!! Go towards the goal without deviating. Thus it is ensured that a fixed minimum number of obstacles lie on path joining the start point of the robot and the goal point.

4.2 Simultaneous Optimization of Action and Rulebase

Among the various schemes proposed for the development of the optimal fuzzy logic controller [7], one of them was the two-stage design of the fuzzy rulebase using GA where in the first stage the action (deviation) is determined for each of the condition (angle and distance) followed by a reduction in the number of rules using a GA. This introduced undue complexity in the algorithm. We use another strategy where the determination of the condition-action pairs and the optimal number of rules is done simultaneously. This simplifies the procedure while at the same time reducing computational effort. Here the scaling factors of the state variables are assumed constant. The values are as depicted in fig. 3.

4.3 A New Hybrid Approach

Using a GA alone is not without its potential drawbacks. Although the GA was able to evolve the perfect individual (rulebase) that was able to solve all the 100 cases, however it was seen that the evolved models of the optimal fuzzy logic controller (even the perfect individual) were complicated and included a large number of rules that seemed redundant. The evolved optimal rulebase for cases when the number of obstacles was changed to 8, 16, 32, 40, 64, 128 and 256 came out to be different. This should not be the case as ultimately the final rulebase to be implemented on the robot would be unique irrespective of the number of obstacles in the environment. After all if one can know the number of obstacles in an environment, one can also know their position and then the environment is no longer unknown!!

Thus a new scheme was devised where local search was performed on the optimal evolved rulebase that was able to solve all the cases. Each of the possible subsets of the evolved rulebase was tested and its fitness evaluated. Typically the rulebase consisted of seven to nine rules. This meant an evaluation of 128 to 512 more different individuals (combinations of valid rules in a rulebase), which is not a significant overhead. We feel an approximate, yet faster algorithm would also perform well.

The hypothesis proved correct as the developed fuzzy logic controller had far less number of rules and was more intuitive. The shortened rulebase obtained after local search was still able to solve to all the cases. Significantly, each of the rulebases that had this shortened rulebase as a subset and the perfect individual evolved by the GA as the superset was also optimal and the robot successfully navigated through all the obstacles using those rulebase as well. The shortened rulebase also performed well in cases with different number of obstacles. This agreed well with the redundancy theory and the hypothesis that one can obtain a single optimal rulebase valid across all conditions. A significant advantage that is obtained with having less number of rules is during the real-time implementation where understandably significant benefits are obtained both in terms of memory and computation requirements needed for executing the fuzzy logic controller. This increases the usefulness of the algorithm tremendously. A specific case is considered in the following section and the obtained rulebase explained.

5 The Rulebase Obtained

The rulebase was evolved with a test-bed of 100 different cases. Each case has 40 obstacles of which 10 were placed on the path joining the start point of the robot to the end point. Here the workspace was of size 80x80 with the initial point being the origin and the goal point being (76,76). Goal was attained if the robot came within unit distance from the goal.

The GA evolved the perfect individual after 1581 fitness evaluations. The population size of GA was kept 100, mutation probability 0.02 and crossover probability 0.9. Length of the string was kept 45 with the maximum number of allowed rules in the rulebase being fifteen. The evolved rulebase is shown in table.1. Here the first rule can be interpreted as: if the distance of the nearest obstacle forward from the robot is *very near* and the angle between the line joining the current robot position to the goal point and the line joining the current robot position to the center of the obstacle is *left*, then the robot should not deviate and go ahead.

Table 1. Perfect rulebase evolved by GA

Distance	Angle	Deviation
Very near	Left	Ahead
Very near	Ahead left	Ahead right
Very near	Ahead	Right
Very near	Right	Ahead left
Near	Left	Right
Near	Ahead left	Right
Near	Ahead	Right
Near	Right	Ahead right
Far	Ahead	Right

However when we apply local search, the rule base is shortened to the one given in table.2. This was due to the redundant rules that were present in the rulebase evolved by the GA got eliminated by using local search. The number of rules in the rulebase is reduced from nine to four without any reduction in the fitness. Both of the rulebase are able to solve all the 100 cases in 11738 seconds, thus taking on an average 117.38 seconds for each case as against the allowed maximum time of 400 seconds.

Table 2. Perfect rulebase shortened by local search

Distance	Angle	Deviation
Very near	Right	Ahead left
Near	Ahead left	Right
Near	Right	Ahead right
Far	Ahead	Right

The reader would agree that the above rulebase is greatly simplified, and is more intuitive with respect to what a human would do in a similar situation. Fig. 4 depicts the path taken by the robot using the above rulebase in one of the cases. The path obtained by the robot is indeed near time optimal and the robot does not deviate in excess and only by an amount that is necessary to maintain a safe distance between itself and the obstacles.

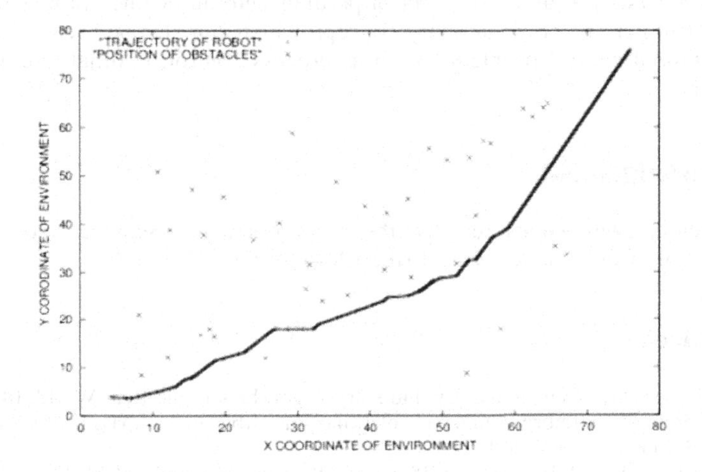

Fig. 4. The position of obstacles and the path of the robot obtained using the shortened rulebase

Thus we see that the robot is able to effectively navigate through obstacles and reach its goal. The path of the robot in the other ninety-nine cases is similar to the above curve and proves the efficacy of the technique.

6 Future Research

In the present study, obstacles were considered to be of constant size and a point robot. It would be interesting to develop an optimal rulebase for the case of varying obstacle size. For such an approach, the size of the obstacle in the form of the area of the obstacle visible to the robot or the bounding angle would also have to be considered as a fuzzy input variable apart from the existing inputs of angle and distance. Automatic scaling of the parameters and using different membership functions can also to be explored to further improve performance.

After successful computer simulations and development of an optimal rulebase, KanGAL in collaboration with Robotics Lab, IIT – Kanpur, has developed a hexapod PaNtHER (Path Navigator Having Evolutionary Resources) for experimentation.

7 Conclusions

A new hybrid algorithm of using a GA-Fuzzy approach along with a local search technique was examined. It effectively converted an online path planning problem to an offline optimization problem. It proved to be an efficient method capable of providing an optimal rulebase using which the robot is successfully able to avoid the obstacles. The rulebase developed were optimal, intuitive and simple enough to be used for real-time implementation. Finally future research topics were discussed, which on their implementation would provide insights into the GA-Fuzzy approach and move it closer to being suitable for actual implementation.

The conversion of an online control problem into an offline optimization problem as suggested here is a revolutionary concept, which can also be used in other optimal online optimal control problems, such as process control, optimal aircraft guidance, and others.

Acknowledgements

The research was sponsored by the Department of Science and Technology, Government of India under project III.5 (204)/99-ET.

References

1. Goldberg, D.E.: Genetic and Evolutionary Algorithms in the Real World. Illinois Genetic Algorithms Laboratory, University of Illinois at Urbana-Champaign, IlliGAL Report No. 99013, March 1999 (1999).
2. Latombe, J.C.: Robot Motion Planning. Kluwer Academic Publishing, Norwell, MA (1991).
3. Pratihar, D.K.: Path and Gait Generation of Legged Robots Using GA-Fuzzy Approach. PhD Disseration, Department of Mechanical Engineering, Indian Institute of Technology – Kanpur, May 1999 (1999).

Evolutionary Approaches to Rule Extraction for Fuzzy Logic Controllers

Tandra Pal*

Electronics and Communication Sciences Unit,
Indian Statistical Institute,
203 B. T. Road, Calcutta 700035, India
tandra_v@isical.ac.in

Abstract. We first discuss some limitations of Chan et al.'s [5] method and propose some modifications on their 'optimized fuzzy logic controller' (OFLC) to eliminate those limitations. Then we propose a new method to reduce the number of rules in a symmetric rulebase which reduces the search space as well as the design time. Our fitness function can reduce the number of rules maintaining the performance of the rulebase. It requires no prior knowledge about the system. Applying this procedure to the inverted pendulum problem, we get a rulebase containing less than 3% of all possible fuzzy rules and it takes about 42 steps on average to balance over the entire input space. Our results are compared with those of Lim et al.'s [4] method.

1 Introduction

Selection of an appropriate rulebase is the most important task for successful design of a fuzzy control system [1], [2]. In this paper, we first propose some modifications of OFLC of Chan et al. [5], named simplified OFLC (SOFLC) and then we present a new scheme ISOFLC (improved SOFLC) with symmetric rulebase where genetic algorithm (GA) derives n fuzzy control rules. Here, n is not fixed. The variable length rulebase will be able to grow or shrink according to the need of the problem. Our proposed evaluation function itself keeps a balance between the number of rules and the performance of the rulebase on the system. The performance of the scheme is demonstrated on the inverted pendulum problem, and to show its superiority, we compare it with a work of Lim et al. [4].

2 Previous Work (OFLC) and Some Remarks

Symmetric Rule Table and Mirror Action: The system is assumed to be symmetric and hence the rulebase (a chromosome represents a rulebase) is taken as symmetric, which reduces the search space. Due to symmetric nature of the

* On leave from R.E. College, Durgapur 713209.

N.R. Pal and M. Sugeno (Eds.): AFSS 2002, LNAI 2275, pp. 421–428, 2002.

system, the second half of a chromosome is taken as the mirror image of the first half with respect to the line of symmetry located at $(length\ of\ chromosome + 1)/2$. So, crossover and mutation operations are done on the first half of the chromosome and then it is mirrored to the second half. The mirror image of allele value NS (3) is PS (5). The index, d, of the fuzzy set representing the mirror image of a fuzzy set with the index k is governed by the relation, $d = (number\ of\ fuzzy\ subsets\ (l) + 1) - k$.

Initial Population, Reproduction, Parents Selection and Crossover: Chan et al. [5] initialized the population using a random generator and/or used an expert provided rulebase. The reproduction module consists of parents selection, crossover, and mutation. Unlike simple genetic algorithms where mutation follows crossover, this algorithm performs *only one of the two operations* on selected parents. One point crossover is used and if the crossover rate is 0.7 then the mutation rate is 0.3. Roulette wheel selection is used in the OFLC to select parents. To avoid the effect of super individuals, the maximum fitness of a generation is added to the fitness value of each chromosome for linearization of fitness values.

Mutation: Chan et al. [5] used 'one step change mutation scheme', which changes a randomly selected allele value to either its next or previous value with equal probability. Mutation is done on a site selected randomly on the first half of the chromosome and then it is mirrored to the second half. When the allele value at the site is 1, it will not be decreased, also when it is l, it will not be increased further. Let n_m be the number of mutations to be done on a chromosome. Then the mutation operation is repeated n_m times on both the parent chromosomes, selected by the parents selection module.

Generation Selection: It selects the next generation population. In the OFLC, steady state without duplicates (SSWOD) [3] is used to select the best fit individuals between the parents and children for the new generation. If the population size is N, then the reproduction module will produce N children using mutation and crossover. Next, the existing population and their N children, *i.e.*, total $2N$ individuals are passed to the generation selection module to select the best fit population of N individuals for the next generation.

If none of the new N offspring is selected for the next generation, then the number of mutations n_m is increased by one in the next generation. And the whole population is replaced by another randomly generated population, except for the best string that is kept in the new population.

3 Some Remarks on OFLC

If the controller has k inputs and one output each with d linguistic values, the possible number of rule tables is d^{d^k}. For a symmetric controller, there exists an equilibrium point and the search space is reduced. Chan et al. [5] did not consider reduction of the number of rules in a rulebase. This can further reduce the search space as well as the computation time in each generation. Above all, a rulebase with fewer rules will always have a low computational cost when the controller is operating.

Though, 'one step change mutation scheme' of Chan et al. [5] can improve local search, it can greatly influence the maintenance of diversity. For this reason, we think, the OFLC of Chan et al. [5] required to check whether a new offspring was transferred to the next generation. This is a very time consuming operation.

As mentioned earlier, when there is no offspring in a new population, the number of mutations in a chromosome is increased by one in the next generation. And the whole population is replaced by another randomly generated population, except for the best chromosome which is kept. This is nothing but a new run with added information and with more mutation probability to restore diversity. In their 'one step change mutation scheme', the exploration property of GA may get significantly affected.

4 Proposed Schemes

4.1 Simplified OFLC (SOFLC)

In SOFLC, we do some modifications on OFLC of Chan et al. [5] and show that it can avoid some problems and unnecessary complexities in the design of OFLC.

Mutation: We substitute 'one step change mutation' of OFLC [5] by random mutation. If the number of linguistic terms characterizing the output variable is l, in random mutation, an allele is randomly replaced with a random number between 1 to l. This can help maintaining the diversity of the population at any state and can thereby increase the convergence rate. It can also help the search algorithm to escape from local minima. Moreover, in the generation selection module, there is no need to check the existence of an offspring in the new population. There is no need to increase the number of mutations, n_m, to be done on a chromosome and to generate a new random population as discussed in Section 2.

Simulation Results of SOFLC: The inverted pendulum problem is used to illustrate the effectiveness of the proposed scheme (SOFLC). The objective of the control problem is to apply forces to the cart until the pole is balanced in the vertical position (*i.e.*, $\theta = 0$ and $\dot{\theta} = 0$). The control rules are of the form

$$\text{if } (\theta \text{ is } \{\text{NB}, \ldots, \text{PB}\}) \text{ and } (\dot{\theta} \text{ is } \{\text{NB}, \ldots, \text{PB}\}) \text{ then } (u \text{ is } \{\text{NB}, \ldots, \text{PB}\}).$$

Here θ is the angular displacement, $\dot{\theta}$ is the angular velocity of the pole, and u is the force applied on the cart. For simplicity, we have ignored the cart positioning part and constrained ourselves only to pole balancing.

We have used the following computational protocols: pole length 0.5 m; pole mass 0.1 Kg; cart mass 2 Kg; the maximum allowable angular deviation is 0.18 rad and angular velocity 1.8 $rad/second$. Each of the linguistic variables θ, $\dot{\theta}$, and u has seven linguistic values or fuzzy sets: NB, NM, NS, Z, PS, PM, and PB. We used isosceles triangles as membership functions with 50% overlap with the neighboring triangles. All membership functions have equal base-length. This is possibly the most natural and unbiased choice for the membership functions.

There are 49 (=7 × 7) alleles in a chromosome. The allele value is 1 for NB, 2 for NM, and so on. In our simulation, each chromosome is tested with the 20 different initial positions, each of which is simulated for T=200 time steps. Parameters of the genetic algorithms are: *population size*=50; *mutation rate*=0.3; *crossover rate* =0.7; *maximum number of generations*=30 (for proposed Schemes) and=100 (for OFLC) and *number of rules in a chromosome in the initial population (in ISOFLC) lies between* 7 *and* 10 *(generated randomly).*

The pole is considered to be balanced if $\theta <= 0.001$ rad and $\dot{\theta} <= 0.01$ $rad/sec.$ within 200 time steps. We generated 1369 (= 37 × 37) samples (initial states) uniformly distributed over the product space $\theta \times \dot{\theta}$ which are used to examine the stability and performance of the rulebase extracted. On SOFLC we use the same fitness function as in [5]:

$$fitness = 1/(1 + ITAE). \tag{1}$$

Here, ITAE of a rulebase is defined as $ITAE = \sum_{i=1}^{d} (ITAE)_i$.

For OFLC, we use the same initial states as used in SOFLC. For comparison of the two models, we illustrate a typical variation of fitness with number of generations in Figure 1(a). As in Figure 1(a), usually the fitness for SOFLC is found to increase faster than that in OFLC. Rule tables generated by OFLC and SOFLC after 50 and 25 generations respectively are shown in Tables 1(a) and 1(b) respectively. Though the number of generations required for OFLC is twice that of SOFLC, the ITAE and the average number of time steps (to balance the system) are less for SOFLC than those of OFLC.

Table 1(a). A rule-set generated by OFLC after 50 generations, (having ITAE=1.7017, average time steps=45).

$\dot{\theta}$ / θ	NB	NM	NS	Z	PS	PM	PB
NB	1	1	1	1	2	1	6
NM	1	2	2	1	1	6	6
NS	1	3	1	2	6	2	3
Z	2	1	4	4	4	7	6
PS	5	6	2	6	7	5	7
PM	2	2	7	7	6	6	7
PB	2	7	6	7	7	7	7

Table 1(b). A rule-set generated by SOFLC after 25 generations (having ITAE=1.0313, average time steps=19).

$\dot{\theta}$ / θ	NB	NM	NS	Z	PS	PM	PB
NB	2	1	2	2	1	2	4
NM	5	3	2	3	3	2	6
NS	7	2	2	3	1	6	5
Z	5	3	2	4	6	5	3
PS	3	2	7	5	6	6	1
PM	2	6	5	5	6	5	3
PB	4	6	7	6	6	7	6

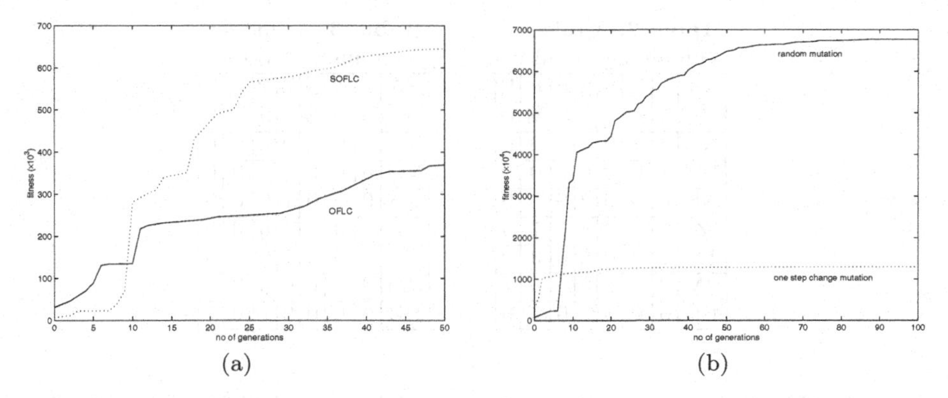

Fig. 1. Plot of fitness as a function of number of generations for OFLC and SOFLC for (a)some typical run (b)'one step change mutation' and 'random mutation'.

To show that the random mutation of SOFLC outperforms the 'one step change mutation' of OFLC in a situation where all the chromosomes refer to bad locations in the search space, we take a random population but with first allele having value farthest from the desired one (i.e., 7 (PB) instead of 1 (NB)) in the first half of the symmetric chromosome. In this situation also SOFLC is able to select a good rulebase within few iterations but OFLC could not. This is illustrated in Figure 1(b).

4.2 Improved SOFLC (ISOFLC)

SOFLC and OFLC do not allow rule deletion. The scheme proposed here, named improved SOFLC (ISOFLC) entertains rule deletion also and thereby reduces the search space exponentially. To achieve this, we need some modifications and changes described next.

Chromosome Representation: The allele value at each location in a chromosome contains either the label of an output linguistic value to be used for a given rule or *zero*. If a chromosome contains an allele value of *zero* at position i, it represents no rule with the antecedent clause corresponding to that site.

Initial Population: For each chromosome, a random number is generated between q (lower limit) and Q (upper limit), which gives the initial number of rules in the chromosome. Positions of these rules in the chromosome are also selected randomly. The allele value for each of these rules is also selected randomly from the set $\{1, 2, ..., l\}$. We use the same crossover operation of SOFLC in ISOFLC and allow the number of rules in the rulebase to vary.

Fitness Function: We should look for a rulebase with fewer rules but with good performance. So, we introduce a new fitness function which uses number of rules

Table 2. A rule-set containing 19 rules.

$\dot{\theta}$ / θ	NB	NM	NS	Z	PS	PM	PB
NB	2	1	0	1	0	0	0
NM	1	0	3	0	0	0	0
NS	0	5	0	2	0	0	0
Z	0	1	3	4	5	7	0
PS	0	0	0	6	0	3	0
PM	0	0	0	0	5	0	7
PB	0	0	0	7	0	7	6

Table 3. A rule-set containing 9 rules.

$\dot{\theta}$ / θ	NB	NM	NS	Z	PS	PM	PB
NB	1	0	1	0	0	0	0
NM	0	0	0	0	0	0	0
NS	0	0	0	1	0	0	0
Z	0	0	2	4	6	0	0
PS	0	0	0	7	0	0	0
PM	0	0	0	0	0	0	0
PB	0	0	0	0	7	0	7

as well as ITAE to measure the performance. The simplified fitness function that decides on the trade off between the number of rules and the quality of control of the rule-set, is

$$fitness = 1/(n_r + ITAE). \tag{2}$$

Here, n_r is the number of rules in the rule-set and ITAE is as defined earlier.

Mutation: We use random mutation. If the number of linguistic terms associated with the output variable is l, in this mutation scheme, an allele is randomly replaced with a random number between *zero* to l. So, it can insert a rule, delete a rule and modify a rule in the rulebase. It helps the learning process to get a rulebase having fewer rules due to its rule deletion property.

So, ISOFLC attempts to achieve the following: (i) to minimize the number of rules (considering the variable n_r in the fitness function and allowing deletion in the mutation), (ii) to minimize the average time to reach the set point, i.e., settling time (by using ITAE to evaluate the fitness) and (iii) to ensure that the controller operates over the entire input space (by proper choice of initial states).

Simulation Results: Keeping all computational protocols the same, we have implemented the proposed method (ISOFLC) on the inverted pendulum problem with several initial populations. In most cases we get very few rules in the rulebase within 25 generations without any guidance from human experts. In order to show the effectiveness of our fitness function, we consider rule deletion but with the fitness function of Chan et al. [5] described in (1). We did several simulations and a typical rule-set is shown in Table 2 having 19 rules. Whereas, using our fitness function (2), ISOFLC gets rule-sets having much less number of rules. We report only one in Table 3 having only 9 rules.

In order to ascertain the quality of the rule-sets extracted by ISOFLC, we have examined each of the 1369($= 37 \times 37$) initial conditions over the entire input space. Table 4 shows the performance comparison of the results obtained by the proposed ISOFLC and OFLC of Chan et al. [5], averaged over 10 runs. Table 5 shows the inverse relation between the number of rules and the quality

Table 4. Performance comparison of ISOFLC and OFLC, averaged over 10 runs.

	ISOFLC	OFLC
No. of rules	11.4	49 (fixed)
Avg. no. of time steps	42	28
Avg. ITAE	3.9732	1.0609
No. of balanced positions among 1369	1369	1369
Failure (%)	0	0

Table 5. Number of rules and the corresponding ITAE and fitness of different rule-sets, generated by ISOFLC.

No. of rules	ITAE	Fitness
9	6.5314	0.0644
9	6.8115	0.0632
11	4.0702	0.0664
11	4.1371	0.0661
11	4.2023	0.0658
11	4.5851	0.0642
13	2.0513	0.0664
13	2.1703	0.0659
13	2.4203	0.0648
13	2.7528	0.0635

(measured by ITAE) of a rulebase produced by ISOFLC, where all the rulebases considered in Table 5 are able to bring the system to the target state.

5 Comparison with Lim et al.'s [4] Method

Lim et al. [4] used fixed number, R, of rules in the rule-set and hence, special type of crossover and mutation operators. They used the inverted pendulum for simulation and the fitness function used is $f = \sum_{i=1}^{d}(bT_{\theta i} + (1-b)T_{ei})/dT$. Here, d =total number of different initial conditions used for testing the chromosome, $T_{\theta i}$ =number of time steps the pole remains within 1 degree from the vertical position and T_{ei} =number of time steps elapsed before the poll falls or the specified value of $T(=200)$ time steps is reached. The fitness value of all chromosomes in the population are scaled to avoid the effect of super individuals.

Table 6. Performance comparison of ISOFLC and Lim et al.'s [4] method.

	ISOFLC	Lim et al.[4]
No. of rules	11.4	20 (fixed)
Avg. no. of time steps considering only balanced positions	42	40*
Avg. ITAE	3.9732	300
No. of balanced positions among 1369	1369	1311*
Failure (%)	0	3.5*

* In [4], the balanced condition is: $\theta <= 0.005$ at least for 5 time steps.

In [4] the pole is considered balanced if $\theta <= 0.005$ for at least 5 time steps. It cannot do a better balancing probably because the fitness function does not care for the set point. We use a much more strict condition for balancing. They set $b = 0.6$ in their simulation. Table 6 shows a comparison of our result with their result, averaged over 10 runs.

6 Conclusions

We simplified OFLC of Chan et al. [5] and also proposed an improved version of OFLC. Both schemes work quite well. The ISOFLC can reduce the number of rules in the rulebase maintaining good performances. The fitness function used is able to generate a rule-set having a very small number of rules which can bring the system to the set point over the entire domain of input variables.

Acknowledgements

Author greatly acknowledges R.E.College, Durgapur, West Bengal, India for partial support to complete this work.

References

1. C. C. Lee, "Fuzzy logic in control system: fuzzy logic controller - Part I & Part II", *IEEE Trans. on Systems Man Cybernetics*, vol. SMC-20, no. 2, pp. 404-435, 1990.
2. D. Driankov and H. Hellendoorn, and M. Reinfrank, *An Introduction to fuzzy Control*, New York: Springer-Verlag, 1993.
3. L. Davis, *Handbook of Genetic Algorithms*, Reinhold:Van Norstrand, 1991.
4. M.H. Lim, S. Rahardja, and B.H. Gwee, "A GA paradigm for learning fuzzy rules", *Fuzzy Sets and Systems*, vol. 82, pp. 177-186, 1996.
5. P.T. Chan, W.F. Xie, and A.B. Rad, "Tuning of fuzzy controller for an open-loop unstable system: A genetic approach" *Fuzzy Sets and Systems*, vol. 111, pp. 137-152, 2000.

Regular Grammatical Inference: A Genetic Algorithm Approach

Pravin Pawar and G. Nagaraja

Department of Computer Science and Engineering,
Indian Institute of Technology, Bombay, Mumbai - 400 076, India
{pravinp,gn}@cse.iitb.ac.in

Abstract. Grammatical inference is the problem of inferring a grammar, given a set of positive samples which the inferred grammar should accept and a set of negative samples which the grammar should not accept. Here we apply genetic algorithm for inferring regular languages. The genetic search is started from maximal canonical automaton built from structurally complete sample. In view of limiting the increasing complexity as the sample size grows, we have edited structurally complete sample. We have tested our algorithm for 16 languages and have compared our results with previous works of regular grammatical inference using genetic algorithm. The results obtained confirm the feasibility of applying genetic algorithm for regular grammatical inference.

1 Introduction

Grammatical inference is an instance of the inductive learning problem [2], which can be formulated as the task of discovering common structures in examples, which are supposed to be generated by the same process. In this particular case, the examples are sentences defined on a specific alphabet and the common structures are represented by a grammar or an equivalent machine. Once the grammar has been inferred from the learning data, the induced language is identified. A set of negative information, i.e. a set of examples not respecting the common structures, may also help the language induction [2]. Here, we apply genetic algorithm for regular grammatical inference. Previous work done for genetic algorithm for inferring regular grammar is described in section 4. Other approaches for regular grammatical inference are uv^kw algorithm [1] based on pumping lemma, K-Tails [7] and Tail Clustering [1] based on merging of states of maximal canonical grammar and Error Correcting Grammatical Inference (ECGI) [8]. Here, we concentrate in using genetic algorithm for regular grammatical inference.

2 The Regular Grammatical Inference Problem

This section reviews the formal theory of regular inference problem and we state it as a search through a Boolean lattice.

N.R. Pal and M. Sugeno (Eds.): AFSS 2002, LNAI 2275, pp. 429–435, 2002.

2.1 Definitions and Notations

A regular language *L* may be referred to by *L(R), L(A) or L(G),* depending on whether this language is respectively, but equivalently, specified by a regular expression *R*, a finite automaton *A* or a regular grammar *G*. Let $MCA(I_+)$ denote the maximal canonical automaton with respect to I_+. It is the automaton having the largest number of states with respect to which I_+ is structurally complete. Let *UA* denote the universal automaton which is the smallest automaton with respect to which every sample of Σ^* is structurally complete where $L(UA) = \Sigma^*$. Let *mDFA(L)* denote the unique minimal deterministic automaton accepting the language *L*. The fundamental theorem of regular grammatical inference states that let I_+ be a positive sample of any regular language *L* and *A* be any automaton accepting exactly *L*. If I_+ is structurally complete with respect to *A*, then *A* may be derived from the $MCA(I_+)$ for some partition Π [1]. If I_+ consists of N letters, the number of states of $MCA(I_+)$ is of the order of N and the number of solutions is of the order of the number of partitions of their set of N elements. This number is given by recurrence relation [1]

$$E(0) = 1 \tag{1}$$

$$E(N) = \sum_{j=0}^{N} \binom{N}{j} E j \tag{2}$$

Order of magnitude is $E(10) \approx 10^5$ and $E(20) \approx 10^{15}$ [1].

2.2 Statement of the Regular Inference Problem

Given the positive and negative samples I_+ and I_-, a finite automaton *A* is a solution of the regular inference problem if the three following conditions are satisfied: I_+ is structurally complete with respect to *A*, $I_- \subseteq \Sigma^* - L(A)$ and *A* is an automaton having the fewest number of states, which fulfils the two preceding conditions [1].
We may also restrict the possible solutions to deterministic automata. In that particular case, we search for the *minimal DFA* consistent with I_+ and I_-. That is we take the simplicity of the inferred automaton, as generality criterion but there is no guarantee that the identified language includes all languages agreeing with I_+ and I_-.
The *minimal DFA* problem can be considered as an optimization problem, that is, the search through the lattice $P(I_+)$ for the partition Π, which generates *mDFA*.

3 Why Genetic Algorithms (GA)?

GA is a parallel search technique [4]. GA offers a way to numerically optimize data in symbolic form. Since GA aims at continuously improving potential solutions, it should offer the possibility to rapidly converge to a new solution after updating the learning samples [2]. It is found to be helpful to overcome problems of local minima [4]. GA is a robust search technique, which is found to be useful for very large search space. For our problem, search space is large as denoted by equation 2. The elements of genetic algorithm are initial population, crossover, mutation, selection, reproduction and number of generations [4].

4 Previous Work towards RGI Using Genetic Algorithm

The influential work in this direction is due to Dupont [2]. The work done towards RGI using GA in [2] uses semi-incremental procedure to infer regular grammar. This procedure consists of adding learning samples from the set $I+$ to build $MCA(I+)$ until all the negative strings are accepted or all the strings of $I+$ had been considered. The work done in [3] consists of exploring the search space in a top to bottom fashion with the search started from universal automaton.

5 Our Approach: Regular Grammatical Inference from $MCA(I_+)$

[2,3] have mentioned about not initiating search using $MCA(I_+)$ as the size of search space dramatically increases with the size of $MCA(I_+)$. This fact motivated us to apply genetic algorithm for regular grammatical inference initiating search from $MCA(I_+)$. We have used elitism to carry best chromosome obtained in the next generation. We use roulette wheel selection technique. We have used classical partition representation scheme [2] to represent partitions on the states of $MCA(I_+)$.

5.1 Genetic Operators

We have used four operators for structural mutation. *Structural mutation 1* [2] consists of a randomly selecting a state in some block of a given partition and putting in into another existing partition. The resulting automaton is in the same depth as the parent automaton in the lattice. *Structural mutation 2* [2] consists of a randomly selecting a state in some block of a given partition and putting in into another partition containing only that element. The resulting automaton is one level away from universal automaton and one level closer to $MCA(I_+)$ than the parent automaton in the lattice. We have developed two new mutation operators. *Structural mutation 3* consists of selecting all the states having some randomly chosen partition number and putting each of them into another existing partition randomly chosen. The resulting automaton is one level closer to universal automaton and one level away from $MCA(I_+)$ than the parent automaton in the lattice. *Structural mutation 4* consists of a randomly selecting a chromosome from the population and replacing it with some new partition generated randomly. This operator was used to overcome the saturation of population during the later generations of genetic algorithm. We have used structural crossover operator used in [2]. The structural crossover consists of the union in both parent partitions of a randomly selected block. For example, let the two following partitions be selected for crossover viz. $\{\{1,4\}, \{2, 3,5\}\}$ and $\{\{1,3\}, \{2\}, \{4\}, \{5\}\}$ and let the first block be randomly selected in both parents. The union of the selected blocks is $\{1,3\} \cup \{1,4\} = \{1,3,4\}$ and the two resulting partitions are $\{\{1,3,4\}, \{2,5\}\}$ and $\{\{1,3,4\}, \{2\}, \{5\}\}$. The derived automata obtained after crossover share a common block, i.e. a common subset of states merged together. Moreover, the depth of each new automaton is either equal to the original one or increased by 1.

5.2 Generation of Sample Set

For some parameter *M=1*, training sample was randomly generated with the restriction that the length of the string should not exceed 12 with no repetition of strings. The next string accepted by *DFA* is added to the structurally complete sample if and only if it covers new transition or new state, which are not covered by the previous strings. For the parameter *M = 3*, the training sample was created as follows. Let $| I_+ |c$ be the sample size required to randomly get structurally complete sample. The samples were *generated randomly* up to three times $| I_+ |c$. No repetition of strings was allowed. For each language we generated *250 positive* and *negative testing samples randomly* with *no repetition* of the string allowed and maximum length of the string not exceeding 12. For some languages number of positive tasting samples is less than 250 as total number of strings accepted by these languages are less than 250.

5.3 Experimental Assessment

We tested 16 languages defined below. Languages *L1 to L16* are set as benchmarks for regular grammatical inference problem [2]. The 16th language is defined in [3]. Languages *L1* to *L7* are those defined by *Tomita* [2].

(L1) a*, *(L2)* (ab)*, *(L3)* any sentence without an odd number of consecutive a's after an odd number of consecutive b's, *(L4)* any sentence over the alphabet a,b without more than two consecutive a's, *(L5)* any sentence with an even number of a's and an even number of b's, *(L6)* any sentence such that the number of a's differs from the number of b's by 0 modulo 3, *(L7)* a*b*a*b*, *(L8)* a*b, *(L9)* (a* + c*)b, *(L10)* (aa)*(bbb)*, *(L11)* any sentence with an even number of a's and an odd number of b's, *(L12)* a(aa)*b, *(L13)* any sentence over the alphabet a,b with an even number of a's, *(L14)* (aa)*ba*, *(L15)* bc*b + ac*a, *(L16)* ca*b + c(aa)*b.

5.4 Results Obtained and Analysis

The *table 1* shows results for *M = 1*. Columns from left to right indicate language code, number of states of target *mDFA* and number of transitions for that *DFA,* number of states of inferred DFA and number of transitions for inferred *DFA* in [3], number of states of inferred DFA and number of transitions for inferred *DFA* by us, percentage of positive test samples accepted by inferred automaton in [2], percentage of negative test samples rejected by inferred automaton in [2], percentage of positive test samples accepted by inferred automaton in [3], percentage of negative test samples rejected by inferred automaton in [3], percentage of positive test samples accepted by the inferred automaton in our experiment, percentage of negative test samples rejected by the inferred automaton in our experiment and number of generation during which inferred automaton was obtained. The last row gives the percentage of positive test sample accepted and negative test samples rejected under respective columns for [2], [3] and our experiment. Table 2 shows results for *M=3* for the same parameters. Columns from left to right in table 3 indicate language code, number of training samples and maximum length of the strings of training samples for *M = 1*, number of training samples and maximum length of the strings of training samples for *M = 3*, number of positive testing examples respectively. Maximum number of negative testing examples is 250 for *M = 1* and *M = 3* except language *L1*.

Table 1 : Results for *M* = 1

L	*s/tr*	Ris/tr	*is/tr*	*+Pd*	*-Pd*	+Rcl	-Rcl	*+cl*	*-cl*	*gen*
L1	**1/1**	1/1	**1/1**	*100.0*	*100.0*	100.0	100.0	**100.0**	-	**1**
L2	**2/2**	2/2	**2/2**	*99.7*	*100.0*	100.0	100.0	**100.0**	**100.0**	**1**
L3	**4/7**	2/3	**10/13**	*91.9*	*98.8*	100.0	70.8	**42.4**	**73.2**	**1**
L4	**3/5**	6/17	**5/8**	*68.0*	*79.7*	100.0	50.0	**100.0**	**100.0**	**12**
L5	**4/8**	14/48	**4/8**	*92.0*	*86.3*	76.2	40.0	**100.0**	**75.2**	**11**
L6	**3/6**	3/8	**15/30**	*92.0*	*86.3*	100.0	10.0	**50.0**	**61.6**	**26**
L7	**4/7**	5/10	**9/15**	*96.3*	*93.3*	100.0	100.0	**88.8**	**73.6**	**3**
L8	**2/2**	2/2	**2/2**	*100.0*	*100.0*	100.0	100.0	**100.0**	**100.0**	**1**
L9	**4/7**	3/6	**6/10**	*99.4*	*98.5*	100.0	31.0	**100.0**	**99.2**	**1**
L10	**5/6**	15/25	**6/7**	*93.9*	*99.1*	90.9	100.0	**100.0**	**97.2**	**4**
L11	**4/8**	6/11	**6/12**	*78.6*	*42.7*	100.0	47.8	**100.0**	**70.0**	**14**
L12	**3/3**	2/3	**3/3**	*100.0*	*99.6*	100.0	100.0	**100.0**	**100.0**	**1**
L13	**2/4**	5/22	**2/4**	*98.8*	*97.7*	100.0	9.1	**100.0**	**100.0**	**5**
L14	**3/4**	3/8	**8/12**	*89.2*	*96.4*	100.0	60.0	**100.0**	**97.2**	**6**
L15	**4/6**	2/5	**4/6**	*94.2*	*98.7*	100.0	41.6	**100.0**	**100.0**	**2**
L16	**5/7**	2/4	**4/6**	-	-	100.0	86.6	**100.0**	**100.0**	**2**
%				*92.93*	*91.80*	97.94	65.43	**92.58**	**90.45**	

We obtained comparable results with [2,3] for *M = 1* in terms of accepting positive testing samples and rejecting negative samples. For languages *L4, L5, L10* and *L13* our results are better than [3] because of *crossover* and *structural mutation 3* operator, which help in minimizing the number of partitions. But our method proved less efficient in obtaining target *DFA* size as less as possible for L6, L9, L14 as compared to [3]. This is because since we have extended the maximum length of strings in positive and negative testing set from 9 to 12, search space size increases exponentially. For *M=3*, the search space has become dramatically large and set samples for training is three times larger than that for *M=1*. Still, we were able to find good acceptance rate and rejection rate compared to [3] for some languages. For languages *L7* and *L13* we were not able to obtain good classification results as since languages *L7* and *L13* have a large number of widely distributed positive sample space and proper classification of positive samples from negative samples is difficult since the search space is very huge. For some languages, we have obtained very good results in first generation of genetic algorithm compared to [3]. As we start our search in between the lattice of automata unlike in [2,3], so it is possible that the best DFA is found in the first generation.

Table 2 : Results for M = 3

L	st/tr	Ris/tr	is/tr	+Pd	-Pd	+Rcl	-Rcl	+cl	-cl	gen
L1	**1/1**	1/1	**1/1**	100.0	100.0	100.0	100.0	**100.0**	-	**1**
L2	**2/2**	2/2	**20/30**	99.7	100.0	100.0	100.0	**100.0**	**100.0**	**1**
L3	**4/7**	2/3	**22/38**	91.9	98.8	100.0	70.8	**56.5**	**78.8**	**5**
L4	**3/5**	6/17	**29/50**	68.0	79.7	100.0	50.0	**100.0**	**100.0**	**2**
L5	**4/8**	14/48	**10/29**	92.0	86.3	76.2	40.0	**51.2**	**72.8**	**1**
L6	**3/6**	3/8	**5/9**	92.0	86.3	100.0	10.0	**50.0**	**61.6**	**26**
L7	**4/7**	5/10	-	96.3	93.3	100.0	100.0	**100.0**	**0.0**	-
L8	**2/2**	2/2	**8/8**	100.0	100.0	100.0	100.0	**100.0**	**100.0**	**4**
L9	**4/7**	3/6	**10/27**	99.4	98.5	100.0	31.0	**100**	**31.2**	**8**
L10	**5/6**	15/25	**6/8**	93.9	99.1	90.9	100.0	**100**	**96.0**	**2**
L11	**4/8**	6/11	**22/43**	78.6	42.7	100.0	47.8	**74.4**	**24.4**	**1**
L12	**3/3**	2/3	**10/15**	100.0	99.6	100.0	100.0	**100**	**99.6**	**3**
L13	**2/4**	5/22	**5/10**	98.8	97.7	100.0	9.1	**100**	**9.2**	**55**
L14	**3/4**	3/8	**10/18**	89.2	96.4	100.0	60.0	**100**	**87.6**	**1**
L15	**4/6**	2/5	**13/22**	94.2	98.7	100.0	41.6	**100**	**98.8**	**5**
L16	**5/7**	2/4	**3/5**	-	-	100.0	86.6	**100**	**94.4**	**3**
%								**89.50**	**65.90**	

5.5 How Is Our Approach Distinct?

We started search from *MCA(I+)*. Our approach doesn't involve incremental procedure as in [2,3]. Maximum string size in our experiment is 12 while it was 9 in [2,3]. Our experiment calculates results for both $M = 1$, and $M = 3$ and we have given separate table of results for them. We have mentioned the number of samples and maximum length of the samples in training set. Since this number and length is a crucial parameter (as can be proven by good results obtained for small sample size), comparison without these values will be simply blindness towards calculating performance of the algorithm. Maximum size of positive and negative testing sample set is not given anywhere in [3].

6 Conclusions

If training sample size is less, then we were able to find quickly the *DFA* accepting positive samples and rejecting negative samples for all the languages. Since our approach is not incremental, it takes less time to converge. For eight languages, the *inferred DFA* has size equal or less than the target *minimal DFA*. For $M=3$, we found good acceptance rate for positive testing samples and good rejection rate for negative testing samples in the case of simple languages for which the positive sample space is not widely distributed. In certain respects, it is not possible to compare the performance with other implementations since some of the parameters such as sample size, number of samples etc. are not known.

Table 3: length and number of samples

L	N1	L1	N3	L3	+Nt
L1	2	1	6	10	12
L2	3	4	9	12	6
L3	4	12	12	12	250
L4	4	8	12	8	250
L5	1	10	3	10	250
L6	4	11	12	12	250
L7	7	6	21	12	250
L8	2	2	6	7	12

L	N1	L1	N3	L3	+Nt
L9	5	4	15	7	24
L10	5	11	15	11	18
L11	3	11	9	11	250
L12	2	4	6	8	6
L13	4	10	12	12	250
L14	4	8	12	8	42
L15	2	3	6	5	22
L16	4	6	8	11	11

References

1. L. Miclet: Structural methods in pattern recognition. North Oxford Academic publication (1984)
2. Pierre Dupont, Regular Grammatical Inference from positive and negative samples by genetic search. Grammatical Inference and Applications, Second International Colloquium, ICGI-94, Proceedings, Berlin. Springer (1994)
3. S. Ramakrishnan, Application of genetic algorithm for regular grammatical inference. Mtech Dissertation, IIT-Bombay (1996)
4. D.E. Goldberg, Genetic Algorithms in search, optimization and machine learning. Addisson Wesley publication (1989)
5. Melanie Mitchell.: Introduction to genetic algorithms. MIT press (1996)
6. K.E. Man, K.S. Tang, S. Kwong.: Genetic algorithms: Concept and Design. Springer (1999)
7. A.W. Biermann and J.A. Feldmann: On the synthesis of Finite-State Machines from Samples of their behavior. IEEE Trans. Compt. C-21 592-597, (1972)
8. Rulot H and Vidal E.: Modeling (sub)string length based constraints through a Grammatical Inference method. Eu Pattern Recognition: Theory and Applications (451-459), Springer Verlag (1987)

Parallelized Crowding Scheme Using a New Interconnection Model

P.K. Nanda[1,*], D.P. Muni[2], and P. Kanungo[1]

[1] Department of Electrical Engineering,
Regional Engineering College, Rourkela,
Orissa, India-769008
pknanda@rec.ren.nic.in

[2] Electronics and Communication Science Unit,
Indian Statistical Institute, 203 B. T. Road,
Calcutta, India-700 035

Abstract. In this article, a new interconnection model is proposed for Parallel Genetic Algorithm based crowding scheme. The crowding scheme is employed to maintain stable subpopulations at niches of a multi modal nonlinear function. The computational burden is greatly reduced by parallelizing the scheme based on the notion of coarse grained parallelization. The proposed interconnection model with a new crossover operator known as Generalized Crossover (GC) was found to maintain stable subpopulation for different classes and its performance was superior to that of the with two point crossover operators. Convergence properties of the algorithm is established and simulation results are presented to demonstrate the efficacy of the scheme.

1 Introduction

Genetic Algorithms (GAs) and Evolutionary Computation have been extensively used in different fields for solving complex optimization problems[1,?,3]. GA based class models have been developed to maintain stable sub-populations at the niches of a multi modal function[4] . Usually these class models are developed based on the notion of crowding and sharing. Although satisfactory results have been obtained by using GA, the major bottleneck is the high computational burden. Hence, the objective of designing parallel GA is two fold: (i) reducing the computational burden and (ii) improving the quality of the solutions. The design of parallel GAs (PGAs)involves choices of multiple populations where the size of the population must be decided judiciously. These populations may remain isolated or they may communicate exchanging individuals. This process of dividing the entire population into sub-populations and then providing the mechanism of interaction between them is known as coarse grained parallelism. The process of communications between individual demes is known as migration.

* This work is supported by the MHRD project on Biomedical Image Processing Using Genetic Algorithm

N.R. Pal and M. Sugeno (Eds.): AFSS 2002, LNAI 2275, pp. 436–443, 2002.

The coarse grained PGA is broadly based on the island model and stepping stone model. In an island model the population is partitioned into small subpopulations by geographic isolation and individuals can migrate to *any* other subpopulation [5]. The takeover times in case of coarse grained Parallel genetic Algorithms have been investigated in [7].

In this article, a new interconnection model for the demes is proposed while attempting to parallelize the GA based crowding scheme. Our proposed PGA is based on the notion of coarse grained parallelization. This topology of the proposed model allows both intra demes and inter demes migration. Besides, a new crossover operator known as Generalized Crossover (GC) is proposed. The convergence analysis is carried out for the proposed scheme. Although, effect of migration policies, rate of migration, number of demes and size of the demes on the quality of the solution has been investigated, for the sake of illustration simulation results are presented only for the migration policy where the Good migrants of a demes replaces the bad migrants of other demes.

2 GA Class Models

Usually GA are used for function optimization and hence determining the global optimal solutions. In case of nonlinear multi modal function optimization, the problem of determining the global optimal solution as well as the local optimal solution reduces to determining the niches in the multi modal function. Thus the problem boils down to clustering the population elements around the given niches. Some effort has been directed in this direction for last couple of years where new strategies and algorithms are proposed[4,6,7].

2.1 Crowding Method

In the deterministic crowding, sampling occurs without replacement [4]. We will assume that an element in a given class is closer to an element of its own class than to elements of other classes. A crossover operation between two elements of same class yield two elements of that class, and the crossover operation between two elements of different class will yield either: (i) one element from both the classes, (ii) one element from two hybrid classes. For example, for a four class problem, the crossover operation between two elements of class AA and BB may result in elements either belonging to the set of classes AA, BB or AB, BA. Hence, the class AB offspring will compete against the class AB parents, the class BA offspring will compete with class BA parents. Analogously for a two class problem, if two elements of class A get randomly paired, the offspring will also be of class A, and the resulting tournament will advance two class A elements to the next generation. The random pairing of two class B elements will similarly result in no net change to the distribution in the next generation. If an element of class A gets paired with an element of class B, one offspring will be from class A, and the other from class B. The class A offspring will compete against the class A parent, the class B offspring against the class B parent. The

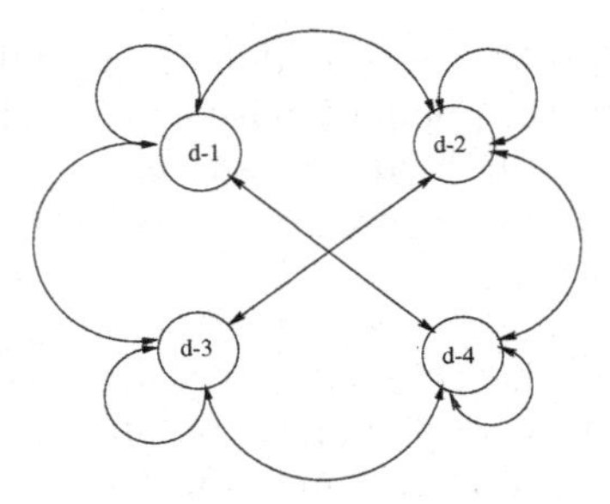

Fig. 1. Proposed Interconnection Model for 4 demes

end result will be that one element of the both classes advances to the next generation no net change.

3 Interconnection Model

Besides migration policy, migration rate also affects the rate of convergence. A good migration policy with optimum migration rate may not always yield optimum solutions because the rate of convergence and the quality of solution also depends upon type of interconnection structure of the island model.

Figure 1 shows the proposed interconnection model for a four deme model where the self loop allows intra deme migration and the other connections among demes allows inter deme migration. The new model is fully interconnected in the sense that intra deme and inter deme exchanges are allowed. The intra deme migration accelerates the convergence because it allows the proportion of the good individuals to grow rapidly. In a model consisting of more than four demes, each deme is connected to every other deme in the interconnection topology. Thus the proposed model is a fully connected hybrid model based on the notion of Island model with the exception that the neighboring demes take part in migration.

3.1 Generalized Crossover Operator

We propose a new crossover operator known as *Generalized Crossover Operator (GC)*, which when applied to two parents produces one offspring instead of two offsprings in the Basic Genetic Algorithm. The operator can be described as follows. The two parents p_1 and p_2 are selected at random and the two crossover points are also selected at random. In between the two crossover points, two bits of the respective positions of the two selected parents are now passed through a switching function to produce one output. For two variables case, a switching function is selected at random from the 16 possible functions and the two bits are impressed as the input and the corresponding output is stored in the same bit position of one of the parents. Analogously, all other bits are generated by selecting the other respective bits from the two parents and passing through the randomly selected switching function. Hence, a stream of bits between the two crossover points is generated that replaces one of the parents to generate one offspring. The motivation is two fold: (i) it helps to examine the diversity of solutions in the solution space, (ii) this model is more plausible from the evolutionistic sense that two parents produce one offspring at a time. Same GC operator is applied to the same two parents with the two new randomly chosen crossover points and the necessary switching function to produce one more offspring. As a result of this operation two offsprings are produced from the two parents by applying the GC operator twice. This process may be repeated to produce M offsprings form N parents. In order to maintain the total population of elements constant over generations M is equal to N.

4 Algorithm

The steps are the parallelized Crowding scheme are the following.

1. Initialize randomly population elements of size N.
2. Divide the population space into fixed number of sub-populations and determine the class of individual in each sub-population.
3. i. In the given sub-population, choose two elements at random for Generalized Crossover (GC) and mutation operation.
   ii. Evaluate fitness of each parents and offspring.
   iii. The tournament selection mechanism is a *binary tournament* selection. Among the two parents and offsprings, the set which contains the individual having highest fitness among the four cements is selected to be the set of parents for the next generation.
   iv. Repeat steps (i), (ii) and (iii) for all the elements in the sub-population.
   v. Repeat step (i), (ii), (iii) and (iv) for a fixed number of generation.
4. Step (3) is repeated for each sub-population.
5. Migration is allowed from each deme to every other deme. The individuals are migrated based on the selected migration policy. Numbers of elements to migrate are determined from the selected rate of migration. The elements migrate with migration probability P_{mig}. At last some percentage of individuals of one deme replace the same percentage of individuals of the same deme, this self migration is valid for all demes with a probability of migration P_{smig}. The individuals migrated in self-loop are based on the selected migration policy.
6. Repeat steps 3, 4 and 5 till convergence is achieved. The algorithm stops when the average fitness of the total population is above preselected threshold.

Theorem 1 *Assume P_{k-1} to be the proportion of good individuals after $(k-1)^{th}$ migration. Then for any arbitrary initial condition with P_0, the algorithm converge for*

$$P_{k-1} = (1 - \delta_k)^{\frac{1}{N}}$$

where, $N = s^n$, $s =$ *Tournament size of tournament selection method,* $n =$ *Number of generations between two consecutive migrations and,* $\delta_k =$ *Proportion of good individuals taking part in* k^{th} *migration.*

Proof: In the whole population of mixed fitness, we assume an element to be a *good* individual if its fitness is above a threshold and *bad* if the threshold is below a threshold. Thus in the whole population each individual may be either good or bad. Let the individuals be selected to the next generation using tournament selection. In tournament selection a random sample of s individuals is selected and out of these s participants one best individual is selected. If all the s participants are bad and since one individual is to be selected so the selected individual is a bad individual. Thus a bad individual will survive only if all the s individuals are bad.

If the initial proportion of good and bad individuals are P_0 and Q_0 respectively, then the proportion of bad individuals in the next generation is:

$$Q_1 = {Q_0}^s \tag{1}$$

(1) implies that $Q_2 = (Q_1)^s = ({Q_0}^s)^s = {Q_0}^{s^2}$. Therefore, at the n_{th} generation, $Q_n = {Q_0}^{s^n}$. Let the first migration be allowed after n generations. Then the Proportion of bad individuals after first migration or in other words after n generations can be expressed as $Q_{1n} = {Q_0}^{s^n} - \delta_1$. Where $\delta_1 =$ Proportion of bad individuals replaced by good migrated individuals after first migration. It can be shown that the Proportion of bad individuals after k^{th} migration or kn generations.

$$Q_{kn} = Q_{k-1}^{s^n} - \delta_k \tag{2}$$

Where $\delta_k =$ Proportion of bad individuals replaced by good migrated individuals by k^{th} migration.

Since there are only two types of individuals i.e. good and bad, so the sum of proportion of good and bad individuals is always unity.

The algorithm will converge to the desired solution when all individuals are good individuals or the proportion of good individuals P_{kn} is unity . This implies that the proportion of bad individuals is zero. Thus for convergence $Q_{kn} = 0$ Since, δ_k is the proportion of good individuals taking part in k_{th} generation, so

$$\delta_k \geq 0 \tag{3}$$

Substituting (3) in (2),we have

$$Q_{kn} \leq Q_{k-1}^{s^n} \tag{4}$$

Since Q_{k-1} is a proportion ,from (4) it is evident that the population of bad individuals has a monotonically decreasing trend. This implies that the population of good individuals will have an increasing trend. From(2), we have $Q_{k-1}^{s^n} = \delta_k$. This implies that $P_{k-1}^{s^n} = 1 - \delta_k$ or $P_{k-1} = (1 - \delta_k)^{\frac{1}{N}}$. Hence, proved. The theorem provided a bound on the proportion of good individuals taking part in migration among the demes.

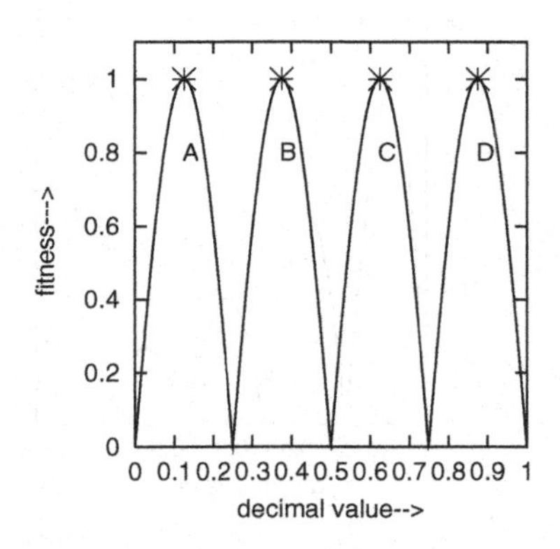

Fig. 2. Good-Bad migration policy, A = 93, B = 93, C = 106, D = 108, generation = 20

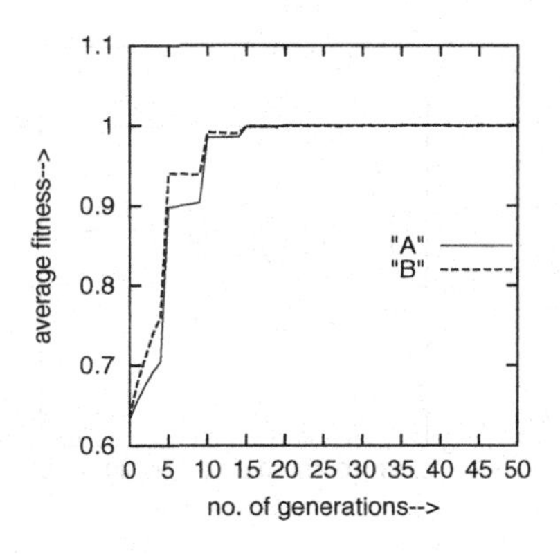

Fig. 3. Good-Bad Migration policy A for GC and B for Twopoint Crossover, migration rate: 8%interdeme 8%intrademe

5 Simulation

For the sake of illustration, We have considered the four class problem given by the following functions; $f(x) = \mid Sin4\pi x \mid, \quad 0 \leq x \leq \quad 1$ and $f(x) = \mid e^{2.0(x-0.125)} Sin4\pi x \mid, \quad 0 \leq x \leq \ 1$. The parameters used are: Total Number of population elements N = 400, Number of demes = 4, Probability of Crossover = 0.8, Probability of mutation = 0.001, Probability of migration $p_{mig} = 0.9$, probability of self migration $P_{smg} = 0.9$, rate of migration=20%, and the threshold of fitness for the stopping criterion = 0.98. In our simulation we have employed only Good-Bad migration policy. Simulation was carried out 40 times with different initial sampling and the average of the 40 experiments is presented. The population of element converged to their respective peaks as shown in Figure 2. The performance of the algorithm with the new model and with the proposed GC operator was compared with model employing two point crossover operator as shown in Figure 3. It is clear form Figure 3 that the algorithm converges faster than that of the model employing two point crossover operator. The performance of the algorithm depends upon the proper choice of rate of migration as shown in Figure 4. From Figure 4 it is clear that as the migration rate increases from 8% to 40% the convergence time decreases and again increases as the migration rate is increased to 56%. The population distribution for the decaying sinusoidal function is shown in Figure 5. It is clear from Figure 5 that the algorithm maintained stable sub-populations in the respective peaks even if the niches are of different heights. In this case also the model with GC operator outperforms to that of the two point crossover operator. This effect for each class is exhibited in Figure 6. The effect of rate of migration is also presented in Figure 7 where it is observed that as the rate increases form 8% to 20% the time of convergence decreases and it shows again a decreasing trend if the rate is further increased to 32%.

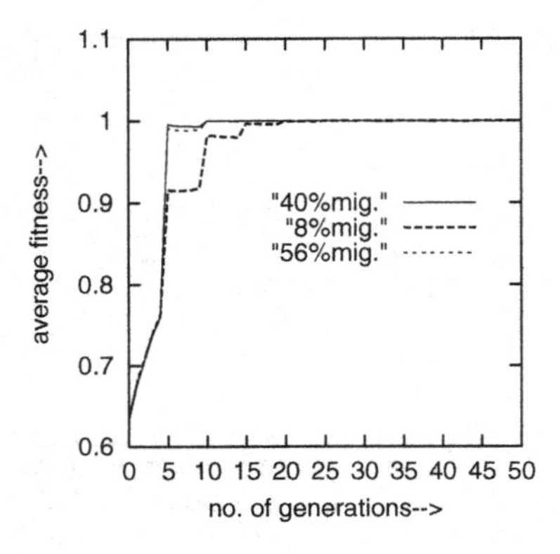

Fig. 4. Effect of rate of migration for Good-Bad Migration policy

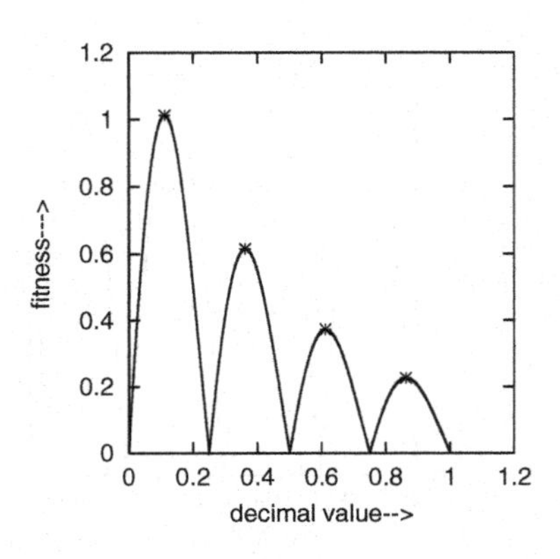

Fig. 5. Good-Bad Migration policy, A = 90, B = 113, C = 106, D = 91, GC operator.

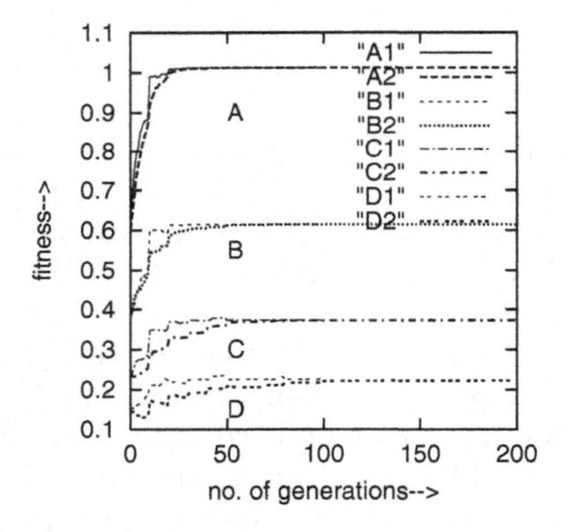

Fig. 6. Good-Bad migration policy, A1, B1, C1, D1 for GC operator, A2, B2, C2, D2 for Two point Crossover.

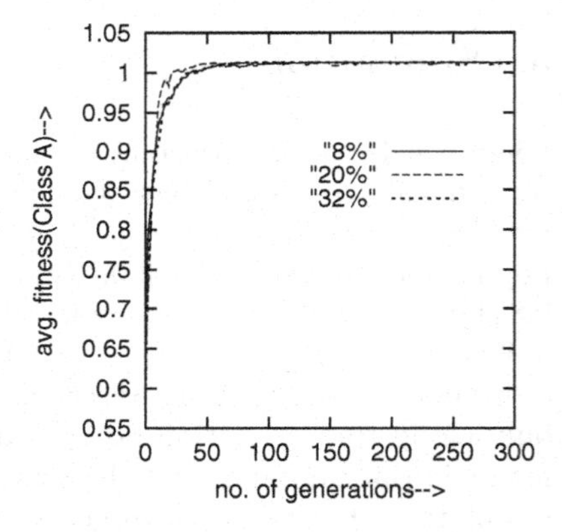

Fig. 7. Effect of rate of migration, Good-Bad Migration policy, 8% intrademe migration.

6 Conclusions

A new interconnection model with a new crossover operator is proposed for parallelizing the crowding scheme for maintaining the sub-populations in respective classes. Thus a new parallel Genetic Algorithm Based Class model is proposed for classification. The efficacy of proposed algorithm is better than that of the algorithm with other models. Convergence analysis is carried out and is shown that the algorithm converges for an optimum rate of migration. The results presented are the serial implementation of the parallel algorithms. Attempts are made to obtain results with parallel implementation.

References

1. Goldberg,D.E. *Genetic Algorithm in Search, Optimization and machine learning*, Addison-Wesley, (1989).
2. T. Back, D. B. Fogel and T. Michalewicz, (Ed.): Evolutionary Computation 1; Basic Algorithms and operators, Institute of Physics publishing, Bristol and Philadelphia (2000)
3. Chellapilla Kumar: Combining Mutation Operators in Evolutionary Programming, IEEE Transaction on Evolutionary Computation, Vol.2, No.3, (1998), pp.91-96.
4. Mahfoud,Samir.W.: Simple Analytical Models of Genetic Algorithms for Multi modal Function Optimization,Technical Report .Illinois Genetic Algorithm Laboratory, IlliGAL Report No.93001, Feb.(1993).
5. Cantu-Paz,E.: A summary of Research on Parallel Genetic Algorithms, IlliGAL Report No.95007, July (1995).
6. Cant-Paz,E.: Designing Efficient and Accurate Parallel Genetic Algorithms, parallel Genetic Algorithms, Ph.D. dissertation, Illinois Genetic Algorithm Laboratory, UIUC, USA, (1999).
7. Cant-Paz,E.: Migration policies and takeover times in parallel Genetic Algorithms, Proceedings of the International Conference on Genetic and Evolutionary Computation, San Francisco, CA, (1999), pp.775-779.

A GA-FUZZY Approach to Evolve Hopfield Type Optimum Networks for Object Extraction

Susmita Ghosh (nee De)[1] and Ashish Ghosh[2]

[1] Department of Computer Science & Engineering,
Jadavpur University, Kolkata 700 032, India
susmita_de@hotmail.com

[2] Machine Intelligence Unit,
Indian Statistical Institute, Kolkata 700 035, India
ash@isical.ac.in
http://www.isical.ac.in/~ash

Abstract. Earlier attempts are made to design Hopfield type neural network architecture for object extraction using Genetic Algorithms (GAs). Energy value of the neural network was taken as the index of fitness of the GA. In the present article fuzzy logic reasoning is incorporated into this Neuro-GA hybrid framework to remove some of the drawbacks of earlier attempts. Here, GAs have been used to evolve Hopfield type optimum neural network architecture for object background classification. Each chromosome of the GA represents an architecture. The output status of the neurons at the converged state of the network is viewed as a *fuzzy set* and measure of fuzziness of this set is taken as a measure of fitness of the chromosome. The best chromosome of the final generation represents the optimum network configuration. When the input images are less noisy, the evolved networks are found to have less (compared to the corresponding energy based objective evaluation) connectivity for providing comparable outputs.

1 Introduction

Neural networks (NNs) are designated by the network topology, connection strengths between pairs of neurons/nodes, node characteristics and rules for updating status. In spite of the wide range of applicability of NNs, there is no formal procedure to design an optimum neural network for a given problem. Recently, some attempts are made in this regard using Genetic Algorithms (GAs) [1]- [4]. Each chromosome of a GA represents a network architecture. In one of our earlier attempts, optimum Hopfield type networks [5] were evolved using GAs for extracting object regions from gray images [6,7]. In that work, the energy value (which involves the connectivity also) of the converged network was taken as a measure of fitness of the corresponding chromosome. Thus, two different output images may have the same energy value and also the same set of output images may correspond to networks having different energy values (due

N.R. Pal and M. Sugeno (Eds.): AFSS 2002, LNAI 2275, pp. 444–449, 2002.

to different connectivity configuration). This may cause problems if energy value is considered as a fitness measure. For example, a network with more number of connections will provide more fitness value than the network with less number of connections even if they produce the same output image. This drawback can be overcome if the output status of the neurons in the converged state of each network is considered as a *fuzzy set* and a *fuzziness measure* of this set is taken as a measure of error/instability of the network which in turn reflects the fitness of the corresponding chromosome. In the present work, different types of fuzziness measures, namely, index of fuzziness, entropy of a fuzzy set and fuzzy correlation are used [8]. The working principle of object extraction using this network is similar to that used in [7]. The proposed technique is compared with its energy based counterpart [7]. It is found that in case of less corrupted images, to produce comparable segmented output, fuzziness measure based objective function evolves networks having less connectivity than the corresponding networks evoluted by energy based model.

2 Measures of Fuzziness of a Fuzzy Set

A fuzziness measure [8,9] of a fuzzy set expresses the average amount of ambiguity in making a decision whether an element belongs to the set or not. Several authors have made attempts to define such measures. A few measures relevant to the present work are described here.

2.1 Index of Fuzziness

The index of fuzziness (*f1*) of a fuzzy set A having n supporting elements is defined as

$$\begin{aligned}\gamma_p(A) &= \frac{2}{n^{\frac{1}{p}}}\, l^p(A,\overline{A}) \\ &= \frac{2}{n^{\frac{1}{p}}} \left[\sum_{i=1}^{n} \{\min(\mu_A(x_i),\ 1-\mu_A(x_i))\}^p\right]^{\frac{1}{p}},\end{aligned} \tag{1}$$

when $l^P(A,\overline{A})$ denotes the distance between fuzzy set A and its nearest ordinary set $\overline{A}$. An ordinary set $\overline{A}$ nearest to the fuzzy set A is defined as:

$$\mu_{\overline{A}}(x) = \begin{cases} 0 \text{ if } & \mu_A(x) \leq 0.5 \\ 1 \text{ if } & \mu_A(x) > 0.5. \end{cases} \tag{2}$$

We have considered quadratic index of fuzziness *i.e.*, $p = 2$.

2.2 Entropy

Entropy of a fuzzy set as defined by De Luca and Termini (*f2*) is given by

$$H(A) = \frac{1}{n \ln 2} \sum_{i=1}^{n} \{S_n(\mu_A(x_i))\}, \tag{3}$$

with

$$S_n(\mu_A(x_i)) = -\ \mu_A(x_i)\ \ln\{\mu_A(x_i)\} \\ -\ \{1-\mu_A(x_i)\}\ \ln\ \{1-\mu_A(x_i)\}; \tag{4}$$

and that of Pal and Pal (*f3*) is given by

$$H(A) = \frac{1}{n(\sqrt{e}-1)} \sum_{i=1}^{n} \{S_n(\mu_A(x_i)) - 1\} \tag{5}$$

with

$$S_n(\mu_A(x_i)) = \mu_A(x_i)\ e^{1-\mu_a(x_i)} + \{1-\mu_A(x_i)\}\ e^{\mu_A(x_i)}. \tag{6}$$

Another definition of entropy which involves the distance of the fuzzy sets from its furthest ordinary set is given by Bart Kosko (*f4*). It says

$$R_p(A) = \frac{l^p(A,\overline{A})}{l^p(A,\underline{A})} = \frac{\left[\sum_{i=1}^{n} \{\min(\mu_A(x_i), 1-\mu_A(x_i))\}^p\right]^{\frac{1}{p}}}{\left[\sum_{i=1}^{n} \{\max(\mu_A(x_i), 1-\mu_A(x_i))\}^p\right]^{\frac{1}{p}}} \tag{7}$$

where $\underline{A}$ is an ordinary set furthest to the fuzzy set A, defined by

$$\mu_{\underline{A}}(x) = \begin{cases} 0 \ \text{if} & \mu_A(x) \geq 0.5 \\ 1 \ \text{if} & \mu_A(x) \leq 0.5\ . \end{cases} \tag{8}$$

2.3 Fuzzy Correlation

A concept of correlation (*f5*), giving a measure of relationship between two representatives of an image was given by Pal and Ghosh. Correlation between a fuzzy property and its nearest two tone version represents the degree of closeness of the fuzzy set to the nearest ordinary set. In other words, correlation also provides a measure of information about the distance of a fuzzy set (represented by μ_1) from its nearest ordinary set (represented by μ_2) and is expressed as

$$C(\mu_1, \mu_2) = 1 - \frac{4}{X_1 + X_2} \sum_{i=1}^{n} \{\mu_1(i) - \mu_2(i)\}^2 \tag{9}$$

with

$$X_1 = \sum_{i=1}^{n} \{2\mu_1(i) - 1\}^2 \tag{10}$$

$$X_2 = \sum_{i=1}^{n} \{2\mu_2(i) - 1\}^2, \tag{11}$$

$0 \leq C(\mu_1, \mu_2) \leq 1.$

3 Optimum Architecture Evolution Using GAs

To use a Hopfield type neural network for object background classification, a neuron is assigned corresponding to every pixel. Each neuron can be connected to its neighbors (over a window) only. The connection can be full or can be partial. In a third order connectivity, the maximum number of connections of each neuron with its neighbors is 8. In practice, all these connections may not exist. The status updating rules are similar to those of Hopfield's model. The energy function of this model consists of two parts. In terms of images, the first part can be viewed as the impact of gray levels of the neighboring pixels and the second part can be attributed to the gray value of the pixel under consideration. The expression of energy for this problem takes the form : $E = -\sum_i \sum_j W_{ij} V_i V_j - \sum_i I_i V_i$, where V_i, V_j are the status of ith and jth neurons, respectively, W_{ij} represents the connection strength between these two neurons and I_i is the initial input bias (which is taken to be proportional to the actual gray level for the corresponding pixel). From a given initial state, the status of a neuron is modified iteratively to attain a stable state. The stable states of the network are made to correspond to the partitioning of a scene into compact regions.

In the present work, each chromosome of the GA represents a network architecture. For an $m \times n$ image, each pixel (neuron) being connected to at most k of its neighbors, the length of the chromosome is $m \times n \times k$ bits. If a neuron is connected to any of its neighbors, the corresponding bit of the chromosome is set to 1, else 0. The initial population is generated randomly. Each network is then allowed to run for object extraction as described in [4]. Each pixel of the extracted output image is considered as an element of a fuzzy set. The fuzziness measure of this set is taken as an index of fitness of the corresponding chromosome (minimum value corresponds to maximum fitness, except the case of using fuzzy correlation) for its selection for the next generation. Crossover and mutation operations are performed on these selected chromosomes to get new offspring (architectures). The whole process is continued for a number of generations until the GA converges. The best chromosome of the final population is considered as the optimum architecture for object extraction.

4 Implementation and Results

The effectiveness of the proposed technique has been demonstrated using some synthetic images of different types which are generated by adding N(0, σ_i^2) noise (σ =10, 20, 32) to each pixel of the binary (two-tone) image. The size of each image is 40×40.

A chromosome is represented as a binary string of length $40 \times 40 \times 8$ (we have taken 8 neighbors for a pixel). The population size is kept fixed at 30. Generational replacement technique and linear normalization selection procedure [10] are adopted. Number of copies produced by the ith chromosome with linear normalized fitness value f_i in a population of size n is taken as round(c_i); where

Table 1. *pcc* for the input images

Image with	*pcc* with fitness evaluation based on											
	Energy		fuzziness measure									
			$f1$		$f2$		$f3$		$f4$		$f5$	
	NE	E	NE	E	NE	E	NE	E	NE	E	NE	E
$\sigma = 10$	99.13	99.25	98.81	98.94	99.44	99.44	99.06	99.44	98.81	98.94	98.81	98.94
$\sigma = 20$	97.88	96.56	96.94	96.88	96.94	96.19	96.81	96.63	96.81	96.62	96.38	96.63
$\sigma = 32$	93.44	92.75	90.06	90.69	89.94	92.25	90.19	88.31	89.44	88.19	90.25	87.69

Table 2. Total number of connections (*noc*) of the evolved networks

Image with	*noc* with fitness evaluation based on											
	Energy		fuzziness measure									
			$f1$		$f2$		$f3$		$f4$		$f5$	
	NE	E	NE	E	NE	E	NE	E	NE	E	NE	E
$\sigma = 10$	9155	8910	7351	7848	7433	7441	7412	7480	7360	7477	7436	7795
$\sigma = 20$	8871	8592	7384	7804	7495	7738	7344	7754	7392	7722	7372	7837
$\sigma = 32$	8387	8488	7425	7716	7487	7903	7452	7834	7424	7943	7460	8049

$c_i = \frac{n \times f_i}{\sum_{i=1}^{n} f_i}$. Both the elitist model (E) and the standard GA (non-elitist model, NE) are implemented. The crossover and mutation probabilities are taken as 0.8 and 0.002, respectively. The algorithm has been run for 200 iterations. It is seen that the average value of the percentage of correct classification of pixels (*pcc*) over the population does not change much after this. The experiment is performed on a SPARC Classic Workstation (frequency 59 MHz). To execute each generation, CPU time requires 2 minutes (approximately).

The percentage of correct classification of pixels of the best chromosome in the last generation using energy and fuzziness measures based fitness evaluations for different noisy versions of the synthetic image is depicted in Table 1. The number of connections (*noc*) of these evolved architectures is put in Table 2. It is seen from Table 1 that the energy based fitness measure performs better than the corresponding fuzziness based ones when the inputs are highly corrupted by noise, and in other cases the results are comparable. On the other hand, it is found from Table 2 that for all the cases, fuzziness measure based evaluations evolve networks with less connectivity (which implies less cost) than those evolved by energy based one.

5 Conclusion

An investigation is carried out to design Hopfield type networks for object extraction using fuzziness measure as fitness criterion. It is found that when the inputs are less corrupted this technique has been able to evolve networks whose connectivity is less compared to the corresponding energy based one in order to produce comparable segmented output. Hence the proposed technique evolves better networks than those evolved using energy based evaluation.

References

1. Mitchell, M.: An Introduction to Genetic Algorithms. The MIT Press, Cambridge, Massachusetts (1996)
2. Whitley, D., Starkweather, T., Bogart, C.: Genetic algorithms and neural networks: Optimizing connections and connectivity. Parallel Computing. **14** (1990) 347–361
3. Pal, S. K., Bhandari, D.:Selection of optimal set of weights in a layered network using genetic algorithms. Information Sciences. **80** (1994) 213–234
4. De, S., Ghosh, A., Pal, S. K.:An application of genetic algorithms to evolve Hopfield type optimum network architectures for object extraction. Proceedings 1995 IEEE International Conference on Evolutionary Computation, ICEC'95, Perth, Western Australia. **1** (1995) 504–508
5. Hopfield, J. J.: Neural network and physical systems with emergent collective computational abilities. Proceedings National Academy of Science, USA. **79** (1982) 2554–2558
6. Ghosh, A., Pal, N. R., Pal, S. K.: Object background classification using Hopfield type neural network. International Journal of Pattern recognition and Artificial Intelligence. **6** (1992) 989–1008
7. Pal, S. K., De, S., Ghosh, A.: Designing Hopfield type networks using genetic algorithms and its comparison with simulated annealing. International Journal of Pattern Recognition and Artificial Intelligence. **11** (1997) 447–461
8. Ghosh, A.: Use of fuzziness measures in layered networks for object extraction : a generalization. Fuzzy Sets and Systems. **72** (1995) 331–348
9. Klir, G. J., Folger, T.: Fuzzy Sets, Uncertainty and Information. Prentice-Hall, Englewood, Cliffs (1988)
10. Davis, L. (ed.): Handbook of Genetic Algorithms. Van Nostrand Reinhold, New York (1991)

Soft Computing in E-Commerce

Raghu Krishnapuram, Manoj Kumar, Jayanta Basak, and Vivek Jain

IBM India Research Lab, Block 1, Indian Institute of Technology
Hauz Khas, New Delhi 110016, India
{kraghura,manoj,bjayanta,jvivek}@in.ibm.com

Abstract. Electronic commerce (or e-commerce for short) is a new way of conducting, managing, and executing business using computer and telecommunication networks. There are two main paradigms in e-commerce, namely, business-to-business (B2B) e-commerce and business-to-consumer (B2C) e-commerce. In this paper, we outline the various issues involved in these two types of e-commerce and suggest some ways in which soft computing concepts can play a role.

1 Introduction

Electronic commerce (or e-commerce for short) is a new way of conducting, managing, and executing business using computer and telecommunication networks [1]. E-commerce research involves many diverse fields such as software engineering, databases, communication networks, security and cryptography, operations research, business policies, psychology, social behavior, law and politics. On the buying side, the Internet provides us the ability to obtain detailed information about the goods that we want to procure, as well as the trust-worthiness and financial standings of the selling parties. On the selling side, the Internet allows us to reach potential buyers quickly in remote geographical areas at a low cost. Processing and storage capabilities at improved price/performance ratios allow us to organize, analyze, and store this information about goods and past transactions. This can result in better decision support for the trading parties to negotiate favorable deals that maximize the sum of profitability for both parties [1,2,3,4,5,6]. Many of the computational tasks in e-commerce use concepts that cannot be described precisely, e.g., trust-worthiness of trading partners, brand-loyalty of customers, and product preferences of buyers in terms of various product attributes. Moreover, on-line learning plays an important role. In this article, we discuss how soft computing tools can provide the necessary computational framework for dealing with such technical challenges. More details may be found in two recent articles [7,8].

2 E-Commerce Environments: B2B and B2C

The nature of interactions between different players or intermediaries give rise to different models of business-to-business (B2B) e-commerce. We may have buy-side and sell-side market places which are operated by buyer and seller consortia

N.R. Pal and M. Sugeno (Eds.): AFSS 2002, LNAI 2275, pp. 450–458, 2002.

respectively, as well as neutral marketplaces. There can be different models of markets such as "single buyer - multiple sellers," "single buyer - single seller," "single seller - multiple buyers," and "multiple buyers with multiple sellers". B2B deals with catalog management, market mechanisms (such as auctions, exchanges, payment support and fulfilment) and decision support. Since the number of buyers and sellers is relatively small in B2B, game theoretic analysis, and consequently access of market information, are critical for decision support.

B2C or business-to-consumer e-commerce, generally refers to the transactions between a single merchant (seller) and multiple consumers (buyers). An example is a storefront. The transactions at a storefront, however, are not necessarily neutral as in the case of a marketplace, and may depend on the nature of the consumers and the merchant. In order to provide better service and increased satisfaction among customers, personalization or customization is highly desirable in B2C [9]. Based on the individual profile of a customer, the enterprise can design the product offering, price the product, offer it for use and provide service to the customer [9].

3 Soft Computing in B2B E-Commerce

Here we discuss a few research issues in B2B e-commerce that can be benefit by using soft computing techniques. There are many other problems involving fraud and anomaly detection [10,11] as well as prediction [12,13,14] which we do not discuss here.

Forecasting: Forecasting plays an important role in the B2B market place. Since suppliers to businesses do not have an interface with the end customer, the feedback from the final consumer takes time to percolate up the value chain. Small forecasting errors can, therefore, cause wild swings in demand. Forecasting area includes prediction of market conditions, trend detection and analysis, forecasting of demand/supply, catastrophic events, impending shortages, and other factors that play a role in the market place. Since mathematical models such as TSCR (Trend, Seasonal, Cyclical and Random Effect) are very precise and susceptible to errors, soft-computing tools that learn models from data can be useful. Examples are multilayer perceptrons, reinforcement learning models, fuzzy rule-based systems, and evolutionary algorithms.

Bidding and Auctions: In electronic markets and Internet auctions, prices of various items (or bids and offers placed on them) change with time [15,16]. These changes occur as various market participants factor in the market information as it is created and revise their valuation or business strategy. Therefore, given the past history and market conditions (which are defined in uncertain terms, e.g., a product is in *high* demand, a product has a *low* price elasticity and so on), a prediction model can be learned using soft computing techniques. Based on the embedded learned models that take into account the imprecision in market conditions, pricing and bidding strategy from the auctioneer's and bidder's points of view can be derived.

Negotiations: Two party negotiation is an important problem where the price of a product is negotiated over the Internet between a buyer and a seller [17,18,19]. In this problem, the seller usually has a reserved price for a product below which he will not sell the product, and initially he asks for a high price for the product. The buyer also has an estimate about the actual price of the product and has a price above which he does not like to buy. Thus every time the buyer or seller makes a bid, the other may or may not change his price. The optimum price is usually the one where both the buyer and the seller fulfill their objectives to the maximum extent. In this particular paradigm, it is required for both the buyer and seller to model or estimate the other person or agent with reasonable accuracy. Bad estimation will lead to failure in closing a deal. Learning techniques can be applied in this paradigm in order to build certain models for both the buyer and the seller.

4 Soft Computing in B2C

Personalization lies at the heart of B2C e-commerce. In an on-line environment, customers can be individually identified, observed, analyzed and addressed easily. Personalization includes customization of all interaction between the customer and the Internet intermediary. It can be at the individual level or segment level. For example, on a Web page containing sports news (say cricket), a banner advertisement or coupon for a related sports item (e.g. sports shoes) can be shown to individuals that are interested in sports, with a link to an on-line store.

Figure 1 illustrates the iterative refinement of interactions between a merchant and his customers in a typical B2C storefront. Here the processes driven by soft-computing kernels are shown as ellipses. They interact with the merchant, customer, knowledge database, and product space. The knowledge base contains information about customer models/profiles, business intelligence, policies and expertise in terms of rules, and customer segment hierarchies. In the remainder of this section, we briefly discuss some aspects of the B2C e-commerce paradigm shown in Figure 1.

4.1 Customer Model

The customer model or profile [31-33] is a repository of all demographic as well as psycho graphic information about the customer. In addition, it also stores goals (long term), preferences about various product categories and attributes, interests, context (short-term) and action (previous purchase history, click stream/browsing pattern) of the customer. The user profile is the interpretation of the explicitly stated and implicitly derived information.

Explicitly inquiring the users about their attributes, preferences and interests is invasive. Supervised learning methods for learning attributes [20,21,22] imposes an undue burden on the consumer or the merchant, and should be avoided whenever possible. Reinforcement learning and unsupervised learning

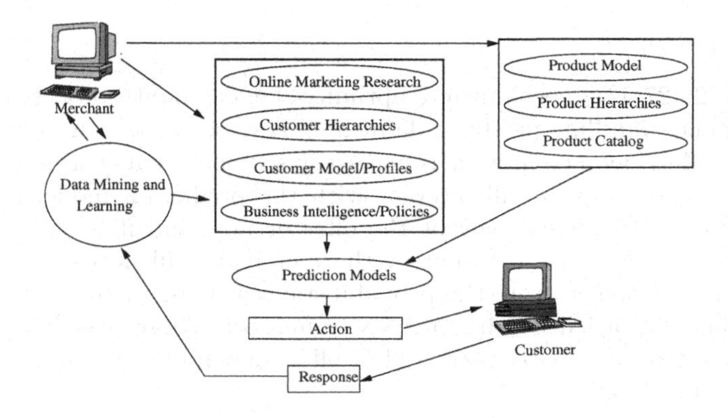

Fig. 1. Data Flow in Personalization

techniques are preferable, and some have been used in the literature [23]. Another aspect of customer profiles is the aggregation (clustering) of similar profiles to create customer "segments" [24].

Most customer attributes such as "quality-conscious" and "calorie-conscious" are inherently fuzzy. Also, customer segments are not crisp. Thus, fuzzy sets can play a major role in customer profile representation [25,26,27]. Bautista [28] uses a genetic algorithm to build an adaptive consumer profile based on documents retrieved by the user.

4.2 Online Marketing Research

Surveys provide an insight into customer perceptions and are useful in targeting customers based on segmentation and sub-segmentation[29]. In the brick-and-mortar world, customer opinions expressed in surveys and the actual purchases cannot be correlated. However, on the Internet, individual record of customer activity enables identification of even micro-segments and customer specific responses. Fuzzy approaches can be used to summarize the survey results in the form of linguistic knowledge that can be understood by merchants easily [30].

4.3 Product Model

The product model is a repository of all product-related information. A different set of attributes influence the consumer behavior for each product category. Therefore, the product in question determines which attributes of the customer profile shall be used for targeting. The attributes and customer's perception about the product determines the categorization, but the imposition of crisp boundaries between categories is arbitrary. Viswanathan and Childers [31] consider product categories as fuzzy sets. Products are said to have degrees of membership in specific attributes. The memberships at the attribute level are then combined to obtain an overall degree of membership of a product in a category.

4.4 Targeting

Targeting [26,27] is a mechanism to optimize the merchant's objective (e.g., gross sales, revenue, bundling products, launching new products). In order to target a particular offer, say a coupon, it is necessary to generate a customer model. The customer model in turn will interact with the prediction model that specifies whether the customer will accept the offer or not, and if he accepts the offer then how much benefit the seller or the merchant will derive. The prediction model can be designed from the probabilistic reasoning or fuzzy modeling from the past history, or learned in an ANN framework. Alternatively, each segment can be targeted for a particular kind of offer, depending on the business policy set by the merchant.

4.5 Personalized Catalogue and Online Shopping Assistance

In B2C storefronts, it is desirable to reorganize product catalogues in a dynamic fashion to reflect the interests and preferences of the customer. To achieve this, we need to capture a customer's interests and preferences, and relate them to the products. Then online classification or hierarchical clustering may be performed over product space to generate a personalized catalogue. A customer's preferences may be captured by relevance feedback or the navigational pattern in the product space. In the case of online shopping assistance, imprecision creeps into many aspects of the problem. First, it is difficult to characterize the interests of a customer in crisp terms. Second, the product attributes are often fuzzy (e.g., a *red* car, with *high* horsepower). Third, the matching process is not exact. Once the product attributes and customer preferences are modeled, we still need to define a similarity measure in order to recommend products to the customer for cross-sell or up-selling.

4.6 Data Analysis Tools

Data mining, clustering, classification, and prediction algorithms are central to addressing the problems in personalization. Association analysis [32,33,34,35,36,37] between consumer attributes and products is useful for targeting a given a product to the right consumer segments. Association analysis has been extended to discovery of fuzzy association rules [38,39,40,41,42]and sequential patterns [43,44]. Collaborative filtering is based on clustering customers into groups based on the products they purchase [45,46,47]. A new customer is mapped to one of the clusters, and the products that have been bought by other customers in the clusters are recommended to the new customer. This form of collaborative filtering does not take into account the contents or attributes of the products. It simply determines which products tend to be bought together. This fact limits its generalization capability. Moreover, since customer groups are inherently fuzzy, ideally fuzzy clustering should be used for collaborative filtering. Prediction models [48,49] have been widely used in the marketing literature for understanding consumer characteristics that influence purchase. Consumers aggregate the degrees to which a product satisfies various criteria in different ways to evaluate and rank the product. Fuzzy set theory provides a host of parametrized operators that can be used to model various aggregation strategies [50,51,52]. Once the consumer purchase behavior is modeled (i.e, the parameters of the operator are learned), it can be used to create a ranked list of products for recommendation. Mela and Lehmann [53] establish a parametric link between fuzzy set theoretic techniques and commonly used preference formation rules in psychology and marketing. Yager [54] presents an approach for the construction of decision functions to represent the probabilistic information as well as the decision style of the decision-maker.

Fuzzy clustering can be used to extract rules for target selection. Setnes and Kaymak [55,56] describe an application of a fuzzy clustering algorithm to extract fuzzy rules from consumer response data collected by a sampling procedure. The rules are used to rank customers and the top n customers are considered targets. Fuzzy Adaptive Resonance (ART) can also be used for clustering customers in groups for targeting [57]. Fuzzy rule-based systems can also be used to decide which ad or coupon to show to a viewer, based on the viewer's characteristics [58,59]. This approach is attractive since the merchant can easily incorporate his/her business knowledge about customers in the form of fuzzy rules.

5 Summary

In e-commerce, the number of attributes and parameters required to address the problems effectively tends to be extremely large. At the same time, the amount of data available for learning/building the models is fairly small. Privacy concerns may further restrict the collection of data by marketers. Therefore, approaches that can combine information from sparse data with business-specific knowledge that merchants can provide are perhaps the only viable option. Since fuzzy sets can model uncertainty associated with attributes/customer segments, they are ideally suited for this application. Similarly, neural nets and genetic approaches can be useful in the learning aspects of e-commerce. However, the algorithms need to be highly scalable and they need to perform in real time. These requirements are not always satisfied by soft computing approaches.

References

1. R. Kalakota and A. B. Whinston, *Frontiers of Electronic Commerce*, Addison Wesley, Reading, 1999.
2. M. Klush, "Information agent technology for the internet : A survey," *Nature*, vol. 36, no. 3, pp. 337–372, 2001.
3. K. Decker, K. Sycara, and M. Williamson, "Intelligent adaptive information agents," *J. of Intelligent Information Systems*, vol. 9, pp. 239–260, 1997.
4. V. N. Gudivada, "Information retrieval on the world wide web," *IEEE Internet Computing*, vol. 1, no. 5, 1997.

5. R. Guttman, A. Moukas, and P. Maes, "Agents as mediators in electronic commerce," in *Intelligent Information Agents, M. Klush (Ed)*, Springer, Berlin, 1999, p. Chapter 6.
6. P. Noriega and C. Sierra (Eds), "Proc. int. conf. agent mediated electronic trading (amet-98)," in *Lecture Notes in Artificial Intelligence*, Springer, Berlin, 1998, vol. 1571.
7. V. Jain and R. Krishnapuram, "Applications of fuzzy sets in personalization for e-commerce," in *Proc. IFSA-NAFIPS Conference*, Vancouver, Canada, 2001.
8. J. Basak and M. Kumar, "E-commerce and soft computing : Scopes of research," *IETE Technical Review*, vol. 18, July-August 2001.
9. D. Riecken-Guest-Editor, "Special issue on personalization," *Comm. of the ACM*, vol. 43, no. 9, Sept. 2000.
10. M. J. A. Berry and G. S. L. Linoff, *Mastering Data Mining : the Art and Science of Customer Relationship Management*, John Wiley, New York, 2000.
11. G. Adomavicious and A. Tuzhilim, "Using data mining methods to build customer profiles," *IEEE Computer*, vol. 34, no. 2, pp. 74–82, 2001.
12. D. Hackerman J.S. Breese and C. Kadie, "Empirical analysis of predictive algorithms for collaborative filtering," in *Proc. 14th Conf. Uncertainty in Artificial Intelligence*, Madison, WI, 1998.
13. M. Pazzani, "A framework for collaborative, content-based and demographic filtering," *Artificial Intelligence Review*, pp. 393–408, December 1999.
14. K. S. Natarajan et al., "A framework for collaborative, content-based and demographic filtering," *IBM Technical Disclosure Bulletin*, vol. 29, no. 10, pp. 4468–4471, 1987.
15. R. P. McAfee and J. McMillan, "Auctions and bidding," *Journal of Economic Literature*, vol. 25, 1987.
16. M. Kumar and S. I. Feldman, "Internet auctions," in *Proc. Third USENIX workshop on Electronic Commerce*, Boston, 1998, pp. 49–60.
17. M. Kumar and S. I. Feldman, "Business negotiations on the internet," in *Proc. Inet-98*, Geneva, Switzerland, 1998.
18. N. Matos, C. Sierra, and N. Jennings, "Determining successful negotiation strategies : An evolutionary approach," in *Proc. ICMAS-98*, Paris, 1998.
19. K. Sycara and D. Zheng, "Benefits of learning in negotiation," in *Proc. AAAI*, Providence, 1997.
20. M. Pazzani, L. Nguyen, and S. Mantik, "Learning from hotlists and coldlists: Towards a WWW information filtering and seeking agent," in *Proc. of Tools with Artificial Intelligence*, Washington, DC, 1995, pp. 492–495.
21. A. Pretschner and S.Gauch, "Personalization on the web," *Technical Report, Department of Elecrtical Engineering and Computer Science, The University of Kansas*, vol. ITTC-FY2000, no. TR-13591-01, pp. 1–31, Dec. 1999.
22. P.H. Chan, "A non-invasive learning approach in building web user profiles," in *5th KDD-99 Workshop on Web Usage Analysis and User Profiling*, San Diego, USA, Aug. 1999, ACM.
23. D.H. Widyantoro, T.R. Ioerger, and J. Yen, "An adaptive algorithm for learning changes in user interests," in *Proc. CIKM*, Kansas City, Nov. 1999, pp. 405–412.
24. B. Mobasher, H. Dai, T. Luo, Y. Sung, M. Nakagawa, and J. Wiltshire, "Discovery of aggregate usage profiles for web personalization," in *Proc. of the Web Mining for E-Commerce Workshop, WebKDD*, Boston, USA, Aug. 2000.
25. R. Krishnapuram, O. Nasraoui, A. Joshi, and L.Yi, "Low-complexity fuzzy relational clustering algorithms for web mining," *IEEE Trans. on Fuzzy Systems*, vol. 9, no. 4, pp. 595–607, 2001.

26. O. Nasraoui and R. Krishnapuram, "Mining web access logs using a relational clustering algorithm based on a robust estimator," in *Proc. of the Eighth Intl. WWW Conf.*, Toronto, 1999, pp. 40–41.
27. O. Nasraoui, R. Krishnapuram, and A. Joshi, "Relational clustering based on a new robust estimator with application to web mining," in *Proc. of the NAFIPS Workshop Intl. WWW Conf.*, New York City, 1999, pp. 705–709.
28. M.J. Martin-Bautista, M.A.Vila, and H.L. Larsen, "Building adaptive user profiles by a genetic fuzzy classifier with feature selection," in *The Ninth IEEE Intl. Conf. on Fuzzy Systems*, 2000, vol. 1, pp. 308–312.
29. P.E. Green, D.S. Tull, and G. Albaum, *Research for Marketing Decisions*, Prenctice Hall, N.J., USA, 1988.
30. T.H. Hsu, K.M.Chu, and H.C.Chan, "The fuzzy clustering on market segment," in *The Ninth IEEE Intl. Conf. on Fuzzy Systems*, May 2000, vol. 2, pp. 621–626.
31. M. Viswanathan and T. L. Childers, "Understanding how product attributes infuence product categorization: development and validation of fuzzy set-based measures of gradednesss in product categories," *Journal of Marketing Research*, vol. XXXVI, pp. 75–94, Feb. 1999.
32. R. Agrawal, T. Imielinski, and A. Swami, "Mining association rules between sets of items in large databases," in *Proc. of ACM SIGMOD Int'l Conf. Management of Data*, Washington, D.C, May 1993, pp. 207–216.
33. R. Agrawal and R. Srikant, "Fast algorithms for mining association rules," in *Proc. of Int'l Conf. Very Large Data Bases*, Santiago, Chile, Sept. 1994, pp. 487–499.
34. R. Agrawal and J.C. Shafer, "Parallel mining of association rules: Design, implementation, and experience," *IEEE. Trans. Pattern Analysis and Machine Intelligence*, vol. 8, pp. 962–969, 1996.
35. R.J. Bayardo-Jr. and R. Agrawal, "Mining the most interesting rules," in *KDD*, San Diego, USA, 1999.
36. C.C. Aggarwal, J.L. Wolf, and P.S. Yu, "A method for similarity indexing of market basket data," in *ACM SIGMOD Conf. on Management of Data*, Philadelphia, PA USA, May 1999, pp. 407–418.
37. B. Kitts, D. Freed, and M. Vrieze, "Cross-sell: A fast promotion-tunable customer-item recommendation method based on conditionally independent probabilities," in *KDD*, Boston, USA, Aug. 2000, pp. 437–446.
38. W.H. Au and K.C.C. Chan, "An effective algorithm for discovering fuzzy rules in relational databases," in *Proc. of The IEEE Intl. Conf. on Fuzzy Systems*, 1998, vol. 2, pp. 1314–1319.
39. W.H. Au and K.C.C. Chan, "Farm: a data mining system for discovering fuzzy association rules," in *Proc. of the IEEE Intl. Fuzzy Systems Conf.*, 1999, vol. 3, pp. 1217 –1222.
40. J.J. Mazlack, "Approximate clustering in association rules," in *19th Intl. Conf. of the North American Fuzzy Information Processing Society*, 2000, pp. 256 –260.
41. C.M. Kuok, A. Fu, and M. H.Wong, "Mining fuzzy association rules in databases," *SIGMOD Record*, vol. 27, no. 1, pp. 41–46, March 1998.
42. A.W.C. Fu, M.H. Wong, S.C. Sze, W.C. Wong, W.L. Wong, and W.K. Yu, "Finding fuzzy sets for the mining of fuzzy association rules for numerical attributes," in *1 st Intl. Symposium on Intelligent Data Engineering and Learning (IDEAL'98)*, Oct. 1998, vol. 2, pp. 263–268.
43. R. Agrawal and R. Srikant, "Mining sequential patterns," in *Proc. of Int'l Conf. Data Eng*, Taipei, Taiwan, March 1995, pp. 3–14.

44. T.P. Hong, C.S. Kuo, and S.C. Chi, "Mining fuzzy sequential patterns from quantitative data," in *IEEE Intl. Conf. on Systems, Man, and Cybernetics*, 1999, vol. 3, pp. 962 –966.
45. P. Resnick, N. Iacovou, M. Suchak, P. Bergstrom, and J. Riedl, "Grouplens: an open architecture for collaborative filtering of netnews," in *Proc. of ACM Conf. on Computer Supported Cooperative Work*, Chapel Hill, USA, Oct. 1994, pp. 175–186.
46. H. Kautz, B. Selman, and M. Shah, "Referal Web: Combining social networks and collaborative filtering," *Comm. of the ACM*, vol. 40, no. 3, pp. 63–65, 1997.
47. U. Shardanand and P. Maes, "Social information filetering: Algorithms for automating 'word of mouth'," in *Proc. of CHI'95 Conf. on Human Factors in Computing Systems*, New York, 1995, ACM Press.
48. P.M. Guadagni and J.D.C. Little, "A logit model of brand choice calibrated on scanner data," *Marketing Science*, vol. 2, no. 3, pp. 203–238, Summer 1983.
49. P. E. Rossi, R. E. McCulloch, and G. M. Allenby, "The value of purchase history data in target marketing," *Marketing Science*, vol. 15, no. 4, pp. 321–340, Winter 1996.
50. R. Krishnapuram and J. Lee, "Fuzzy-connective-based hierarchical aggregation networks for decision making," *Fuzzy Sets and Systems*, vol. 46, no. 1, pp. 11–27, Feb. 1992.
51. M. Grabisch, "On equivalence classes of fuzzy connectives: The case of fuzzy integrals," *IEEE Trans. on Fuzzy Systems*, vol. 8, no. 1, pp. 96–109, 1995.
52. Y. Choi, D. Kim, and R. Krishnapuram, "Relevance feedback for content-based image retrieval using the choquet integral," in *IEEE Conf. on Multimedia and Expo*, New York City, July-Aug 2000.
53. C.F. Mela and D.R.Lehmann, "Using fuzzy set theoretic techniques to identify preference rules from interactions in the linear model: an empirical study," *Fuzzy Sets and Systems*, vol. 71, no. 2, pp. 165–181, 1995.
54. R.R. Yager, "Fuzzy modeling for intelligent decision making under uncertainty," *IEEE Trans. on Systems, Man and Cybernetics, Part B Cybernetics*, vol. 30, no. 1, pp. 60–70, Feb. 2000.
55. M. Setnes, U. Kaymak, and H.R.V.N. Lemke, "Fuzzy target selection in direct marketing," in *Proc. of the IEEE/IAFE/INFORMS Conf. on Computational Intelligence for Financial Engineering (CIFEr)*, 1998, pp. 92–97.
56. M. Setnes and U. Kaymak, "Fuzzy modeling of client preference from large data sets: an application to target selection in direct marketing," *IEEE Trans. on Fuzzy Systems, to appear in*, 2001.
57. S. Park, "Neural networks and customer grouping in e-commerce: a framework using fuzzy art," in *IEEE Academia/Industry Working Conf. Proc. on Research Challenges*, 2000, pp. 331–336.
58. R.R. Yager, "Fuzzy methods in e-commerce," in *18th Intl. Conf. of the North American Fuzzy Information Processing Society*, 1999, pp. 5–11.
59. R.R. Yager, "Targeted e-commerce marketing using fuzzy intelligent agents," *IEEE Intelligent Systems*, vol. 15, no. 6, pp. 42–45, Nov. 2000.

Info-Miner: Bridging Agent Technology with Approximate Information Retrieval

Vincenzo Loia, Paolo Luongo, Sabrina Senatore, and Maria I. Sessa

Dipartimento Matematica e Informatica – Universita' di Salerno,
via S. Allende, 84081 Baronissi (SA), Italy
{loia,pluongo,ssenatore,misessa}@unisa.it

Abstract. In this paper we discuss our research in developing an application for data gathering. The key ideas are to realize an agent-based architecture which merges different technologies and paradigms in order to discover and extract consistent patterns of knowledge from Web pages, that fully describe the user requests. This goal is achieved by embedding into deductive agents the ability to process approximate knowledge.

1 Introduction

Historically, most inferential approaches to Information Retrieval (IR) [1] have been pursued by the academic and industrial communities within the probabilistic framework. Although the obtained results make possible to manage the uncertainty in IR, the nature itself of probability theory limits its applicability to a pure statistical knowledge. Human-defined knowledge may be better handled with the fuzzy linguistic approach, especially for Internet IR, where the vagueness, incompleteness, and complexity of the data sources make difficult (in some case impracticable) the applicability of probabilistic approaches.

This paper presents the Info-Miner architecture, where the advantages of agent and mobile computation, are joined with a more flexible management of logic-based deductive models. The "reasoning capability" has been implemented with an extended version of SLD Resolution which overcomes failure situations in the unification process by exploiting the Similarity Relation. It allows us to compute approximate solutions when failures of the exact inference process occur. The rest of the paper is organized as follows: Section 2 outlines theoretical aspects of the similarity-based deductive model. Section 3 highlights our architecture proposal, and then, in Section 4, we present a simple application example of Web searching. Related works and conclusions close the paper.

2 Theoretical Aspects

The mathematical notion of *Similarity relation* is a many valued extension of the equality, and it is widely exploited in any context where a weakening of the equality constraint is useful [2] [3]. We summarize some results concerning this notion. At first, let us recall that a $T-norm$ $\wedge$ in $[0,1]$ is a binary operation

N.R. Pal and M. Sugeno (Eds.): AFSS 2002, LNAI 2275, pp. 459–465, 2002.

$\wedge : [0,1] \times [0,1] \rightarrow [0,1]$ associative, commutative, non-decreasing in both the variables, and such that $x \wedge 1 = 1 \wedge x = x$ for any $x \in [0,1]$. In the sequel, we assume that $x \wedge y$ is the *minimum* between the two elements $x, y \in [0,1]$.

Definition 1. *A* Similarity *on a domain* $\mathcal{U}$ *is a fuzzy subset* $\mathcal{R} : \mathcal{U} \times \mathcal{U} \rightarrow [0,1]$ *of* $\mathcal{U} \times \mathcal{U}$ *such that the following properties hold*
i) $\mathcal{R}(x,x) = 1$ *for any* $x \in \mathcal{U}$ *(reflexivity)*
ii) $\mathcal{R}(x,y) = \mathcal{R}(y,x)$ *for any* $x, y \in \mathcal{U}$ *(symmetry)*
iii) $\mathcal{R}(x,z) \geq \mathcal{R}(x,y) \wedge \mathcal{R}(y,z)$ *for any* $x, y, z \in \mathcal{U}$ *(transitivity).*

Similarity relations are strictly related with equivalence relations and, then, to closure operators, as stated by the following property.

Proposition 1. *Let* $\mathcal{U}$ *be a domain and* $\mathcal{R} : \mathcal{U} \times \mathcal{U} \rightarrow [0,1]$ *a Similarity on* $\mathcal{U}$*. Then, for any* $\lambda \in [0,1]$*, the relation* $\cong_{\mathcal{R},\lambda}$ *in* $\mathcal{U}$*, named* cut of level λ *(in short* λ-cut*) of* $\mathcal{R}$*, defined as*

$$x \cong_{\mathcal{R},\lambda} y \iff \mathcal{R}(x,y) \geq \lambda,$$

is an equivalence relation. Then, a closure operator can be defined by setting $H_\lambda : \mathcal{P}(\mathcal{U}) \rightarrow \mathcal{P}(\mathcal{U})$ *such that* $\forall X \in \mathcal{P}(\mathcal{U})$

$$H_\lambda(X) = \{z \in \mathcal{U} \mid \exists x \in X : \mathcal{R}(z,x) \geq \lambda\}.$$

The equivalence $\cong_{\mathcal{R},\lambda}$ can be considered as a generalization of the identity relation. In [4] the exact matching between different entities is relaxed by introducing a Similarity relation in the set of constant and predicate symbols in the language of a function free logic program P. We briefly recall that a logic program P is a set of universally quantified Horn clauses on a first order language $\mathcal{L}$, denoted with $H \longleftarrow B_1, ..., B_k$, and a goal is a negative clause, denoted with $\longleftarrow A_1, ..., A_n$ [5]. We denote with $B_\mathcal{L}$ the set of ground atomic formulae in $\mathcal{L}$, i.e. the Herbrand base of $\mathcal{L}$, and with T_P the immediate consequence operator $T_P : \mathcal{P}(B_\mathcal{L}) \rightarrow \mathcal{P}(B_\mathcal{L})$ defined by

$$T_P(X) = \{a \mid a \longleftarrow a_1, ..., a_n \in ground(P) \text{ and } a_i \in X, \quad n \geq i \geq 1\}$$

where *ground*(P) denotes the set of all ground instances of clauses in P. The application of Tarski's fixpoint theorem yields a characterization of the semantics of P, which is the least Herbrand model M_P of P given by

$$M_P = lfp(T_P) = \bigcup_{n \geq 0} T_P^n(\emptyset).$$

In order to deal with the approximation introduced by a similarity relation $\mathcal{R}$ defined between predicate and constant symbols, the program P is extended by adding new clauses which are "similar" at least with a fixed degree $\lambda \in [0,1]$ to the given ones.

2.1 Similarity-Based SLD Resolution

In [6] a new modified version of SLD Resolution, named *Similarity-based SLD*, enables to perform these kinds of extended computations exploiting the original program P. The failure of the unification between different function or predicate symbols is avoided by relaxing the equality constraint with the Similarity relation. It leads to the notion of *unification-degree* associated to a substitution, and a *weak most general unifier* (in short *weak m.g.u.*) is a more general substitution which provides the best unification-degree. By using this notion of weak m.g.u., when exact solutions do not exist, it is possible to obtain *approximate computed answer substitutions* with an associated *approximation-degree.* In general, a computed answer substitution can be obtained with different SLD refutations and different approximation-degrees, then the maximum of this values characterizes the best refutations of the goal. In particular, a refutation with approximation-degree 1 provides an exact solution. Formally, we introduce the following generalization of the SLD derivation.

Definition 2. *Given a Similarity $\mathcal{R}$, a program P in a first order language and a goal G_0, a* Similarity-based *SLD* derivation *of* $P \cup \{G_0\}$*, denoted by* $G_0 \Rightarrow_{C_1,\theta_1,U_1} G_1 \Rightarrow \ldots \Rightarrow_{C_m,\theta_m,U_m} G_m \Rightarrow \ldots.$*, consists of a sequence* G_0, G_1, ... *of negative clauses, together with a sequence* C_1, C_2, ... *of variants of clauses from* P*, a sequence of substitution* θ_1, θ_2, ... *and a sequence* U_0, U_1, ... *of values in (0,1], such that for all* $i \geq 1$:
i) G_i *is a resolvent of* G_{i-1} *and* C_i *by using the idempotent m.g.u.* θ_i *with unification-degree* U_i up to $\mathcal{R}$,
ii) C_i *has no variables in common with* G_0, C_0, ... C_{i-1}.
When one of the resolvents G_i *is the empty clause* $\square$*, the derivation is called a Similarity-based SLD refutation up to R. A Similarity-based SLD derivation is called failed if it is not a refutation.*

It is easy to see that when the Similarity is the identity, the previous definition provides the classical notion of SLD refutation. In the classical case, the restriction of $\gamma = \theta_1 \ldots \theta_m$ to the variables of the initial goal G_0 provides a computed answer substitution for $P \cup \{G_0\}$. When a Similarity-based SLD refutation is considered, at any derivation step, the unification algorithm provides the m.g.u. θ_i with an associated unification degree U_i, $i = 1, \ldots, m$. We generalize the definition of computed answer substitution in the frame of Similarity-based SLD Resolution as follows.

Definition 3. *Given a Similarity* $\mathcal{R}$*, a program* P *in a first order language, a goal* G_0*, and a Similarity-based SLD refutation* D *for* $P \cup \{G_0\}$*, denoted with* $\theta_1 \ldots \theta_m = \{x_1/u_1, \ldots, x_k/u_k\}$ *the restriction of the composition of the m.g.u.'s in* D *to the variables of* G_0 *with associated approximation-degree* λ*, we consider the family* $\Psi = \{\sigma_j\}_{j \in I}$ *of substitutions such that for any* $j \in I$

$$\sigma_j = \{x_1/t_1, \ldots, x_k/t_k\} \qquad \text{with} \quad t_h \in H_\lambda(u_h), \quad k \geq h \geq 1$$

Any element in the family Ψ *is named* computed answer substitution for $P \cup \{G_0\}$ with approximation-degree λ up to $\mathcal{R}$, *and is denoted with* $\langle \sigma_j, \lambda \rangle$.

In [6] several properties of the Similarity-based SLD Resolution are proved.

3 Info-Miner Architecture

Info-Miner is composed of the following agent classes.

- *Similarity Agent.* It collects the several similarity relations, defined upon the input dictionaries and, when the user does not specify the current similarity relation, proactively loads the dictionary and relative similarity that better match the input clause (user's query).

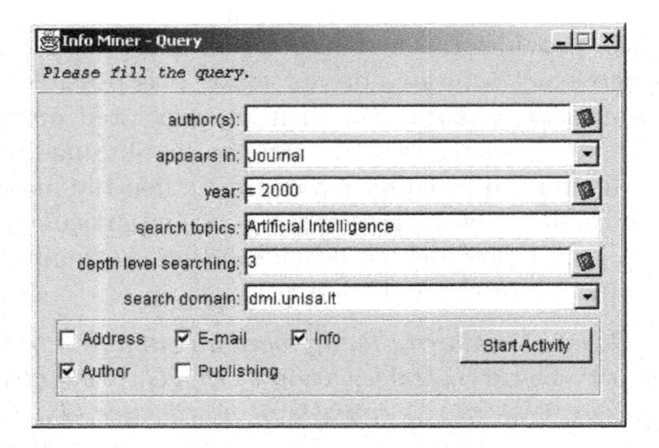

Fig. 1. InfoMiner - User Query Panel

- *Assistant Agent.* It provides the user interface service, in order to gather the input data and transform them in a logic goal, suitable to be processed by the system.
- *Discovery Agent.* The agent owns a knowledge extractor and a logic-based reasoning model. It can move on the net: its goal is to reach prefixed sites and to prove the logic query corresponding to the user's request. Through the similarity-based reasoning and thanks to the acquired knowledge, the agent triggers a resolution procedure returning, as result, answers to the query.
- *Collector Agent.* This agent gathers all returned information from the Discovery Agents. It filters the messages by assuring the data consistency and returns the refined information to the Assistant Agent which, at its turn, sends the results in a digest form to the user.

Info-Miner provides an appropriate module designed to interact with the user by considering the Similarity information. The Similarity Agent handles the similarity among the terms in different sets of terms (or dictionaries). The proof-based query and the current Similarity are hence injected into the Discovery Agents, becoming part of its internal knowledge. The Discovery Agent, with this baggage, moves towards the destination site, where it parses the web documents by extracting the useful knowledge. While it examines the local page, it clones itself, every time that an external link is founded to allow the Discovery Agents to cover a portion of the Web input domain. The collection of the different results is done by the Collector Agent, active on the user's machine, designed to perform filtering and fusion techniques.

4 Web Searching Example

Figure 1 gives a user request panel to formulate the query. The user is interested in searching scientific documents: some information may be specified (authors, proceedings or journal contribution, publication date, topic(s), ...). Other input values serve to specify search mode features (the web domain to be analyzed, depth level searching, ...). As shown in the example of Figure 1, the user looks

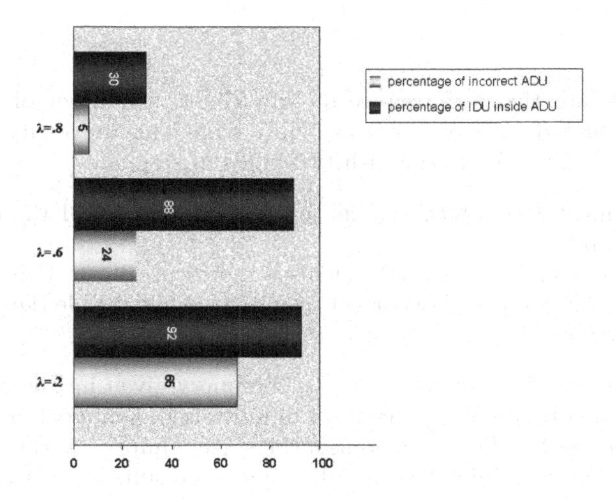

Fig. 2. Experimental results graph

for papers appeared in a *Journal* in the year *2000*, dealing with the topic *Artificial Intelligence*. The Assistant Agent translates the input query in the following logic-based goal:

$$? - \text{doc(IdPage, IdDoc, Author, Title), topic(IdDoc, ['Artificial Intelligence']),}$$
$$\text{doc\_reference(IdDoc, 'Journal'), year(IdDoc, Date, 2000),} \ldots,$$
$$\text{photo(IdPage, Author, File).} \quad (1)$$

The Discovery Agent is sent on the specified Web area (limited to our university area), that is composed of different hosts. On these Web sites the Discovery Agents tries to prove the previous goal, by using its predefined rule-base and its new knowledge, produced after the analysis of the Web files. As an example, let us consider the following rule, inside the agent rule-base:

$$\text{topic(IdDoc, Arg) :- doc(IdPage, IdDoc, Author, Title), is\_in(Arg, Title).} \quad (2)$$

The subgoal *is_in(Arg, Title)* in this rule succeeds when in the list of terms, *Title* appears a term which is similar (in the best case equal) to the term *Arg*. As an example, if we consider *['Merging', 'Fuzzy Logic', 'Neural Network', ...]* (values of *Title*) and *'Artificial Intelligence'* (value of *Arg*), the subgoal does not fail, as in the case of "standard" SLD Resolution, thanks to an existing non zero similarity values (subjectively-defined) between the different terms *'Artificial Intelligence'* and *'Fuzzy Logic'*, *'Neural Network'*,... Thus the subgoal *topic(IdDoc, 'Artificial Intelligence'])* succeeds with a approximation degree which provides a measure of the "weakening" of the equality exploited to avoid unification failure. The results are finally returned to the Collector Agent, charged to glue the different feedbacks in a single, efficient visualization.

5 Testbed

In order to validate the performance of Info-Miner, a number of experiments have been conducted on web sites available from our university web portal (http://www.unisa.it). We distinguish two kinds of results:

- Ideal Document Urls (**IDU**). This set is composed of all the correct Urls computed a-priori.
- Actual Document Urls (**ADU**). This set represent all the Urls returned by the system. It is composed of correct results (elements inside IDU) and wrong results (elements not in IDU).

Our experiments consider as user request, the query given in Figure 1. Considering Figure 2, we observe the percentage of unwanted documents in case of low value of the threshold λ. The behaviour of the system improves when we increase the values of λ. The remaining loss of right expected results is due to two reasons: (1) formats of HTML documents not supported (2) intrinsic expressiveness of natural language which gives ambiguous meanings to same words. In particular, in the case of $\lambda = 0.8$ we note the reduction of errors (unwanted documents). The intermediate level ($\lambda = 0.6$) characterizes a good compromise by balancing correct documents with unwanted results.

6 Concluding Remarks

In these recent years different works have witnessed the importance of the soft computing role as key technique to incorporate human-defined knowledge in the inference process. From a wider perspective of agent modeling with soft computing methodology (see [7], [8] for recent literature), there is an emerging interest in combining soft computing and agents to solve IR-based problems. For sake of short discussion, we cite few works.

- TalkMine [9] is a recommendation systems designed over principles of Fuzzy Set and Evidence Theories (an extension of fuzzy sets) with basic concepts of Distributed Artificial Intelligence. TalkMine is able to characterize information resources and model linguistic categories in the identification of user interests during IR activity.
- FQUERY [10] allows Internet users to use imprecise information in their querying activities over WWW-level services. The fuzzy query engine exploits fuzzy values, fuzzy relations and linguistic quantifiers, embedded in a distributed architecture compliant with popular Web browsers.
- In the [11] the Linguistic Weighted Averaging (LWA) operator is used by computational agents to aggregate linguistic weighted information for Internet information gathering. The LWA operator improves the flexibility of the information gathering process through the management of the degrees of importance that the user assigns to the terms of the query and by reducing the effect of low satisfactions of some terms within the overall performance of the information agent system.

- In [12] a fuzzy system modeling is introduced in order to help the agents in the determination of the appropriateness of displaying or not a given advertisement to a Web user.

This work presents a fundamental part of a larger research project framed in the area of advanced tools for Web searching. In order to provide a robust approach to treat the difference between the information available and the information necessary to the agent to make the best decision, we have embedded into the agents the Similarity-based Reasoning as key issue for a flexible interpretation of information. As the first prototype we report the positive evaluation of a system which is able to detect the right subset of potential unknown readers to whom an email message can be targeted [5].

References

1. Lancaster, F.W.: Information Retrieval Systems: Characteristics, Testing and Evaluation. Wiley, New York (1968)
2. Zadeh: L.A., Similarity Relations and Fuzzy Orderings. Information Sciences **3** (1971) 177–200
3. Ruspini E.H.: On the semantics of fuzzy logic. J . of Approximate Reasoning **5** (1991) 45–88
4. Gerla G., Sessa M.I.: Similarity in logic programming. In: G. Chen, M. Ying, K.-Y. Cai (Ed.s). Fuzzy Logic and Soft Computing, Kluwer Acc. Pub., Norwell. (1999) 19–31
5. Apt R.K. In: Logic Programming, in: J. van Leeuwen (Ed.). Volume B. ((Elsevier, Amsterdam, 1990)) 492–574
6. Sessa M.I.: Translations and similarity-based logic programming. Soft Computing Journal **5 issue 2** (2001) 160–170
7. Loia, V., Sessa, S.: A soft computing framework for adaptive agents. In: Soft Computing Agents : New Trends for Designing Autonomous Systems. Volume 75., Studies in Fuzziness and Soft Computing, Springer-Verlag (June 2001)
8. Martín-Bautista M.J., Vila M.A., Sánchez D, Larsen H.L.: Building adaptive user profiles by a genetic algorithm approach to an adaptive information retrieval agent. In: Proceedings of the IEEE conference on Fuzzy System, S. Antonio, Texas. Volume 1. (2000) 308–312
9. Rocha L.: Talkmine and the adaptive recommendation project. In: the Proceedings of the Association for Computing Machinery (ACM) - Digital Libraries 99. U.C. Berkely (August 1999) 242–243
10. Zadrozny S., Kacprzyk J: Implementing fuzzy querying via the internet/www: Java applets, activex controls and cookies. In: Proc. Third Int. Conference on Flexible Query Answering Systems 98 Lecture Notes in Computer Science. Volume 1495. (1999) 382–393
11. Delgado M., Herrera F., Herrera-Viedma E., Martin-Bautista M.J., Vila M.A.: Combining linguistic information in a distributed intelligent agent model for information gathering on the internet. Computing with Words, P.P. Wang, Ed., John Wiley & Son (2001)
12. Yager R.R.: Intelligent agents for world wide web advertising decisions. International J. of Intelligent Systems **12** (1997) 379–390

Fuzzy Points and Fuzzy Prototypes

Joseph M. Barone

Datatek, Inc., 721 Route 202-206,
Bridgewater, NJ 08807 USA
jbarone@datatekcorp.com

Abstract. There has been considerable interest in the nature of fuzzy points and in their relationship to crisp points. This paper describes a very general approach to the characterization of points which treats them as generalized maps from one object to another in a 2-categorical framework. It will be shown that fuzzy points are also amenable to this treatment, and that, if the category (of locales) is assumed to be equipped with a KZ monad, the locales of fuzzy sets exhibit behavior which allows for such useful results as the justification of the Zadeh Extension Principle and the unification of change of base and change of set in fuzzy set theory.

1 Introduction

This paper explores the conditions under which various kinds of algebraic structures can be said to have "points" and when such points can most reasonably be construed as "fuzzy points."

There are two "standard" definitions of points with which we shall begin our exposition: D1. For a poset (a set P with a reflexive, antisymmetric, and transitive binary relation for all elements x, y in P), a point is any element whose height is 1, that is, any element which covers the universal lower bound 0 (see, e.g., [1], p. 472); D2. For any object X in a category C, a point is a morphism from the terminal object into X (see, e.g., [2], p. 475).

For ordinary sets, (the category **Set** of (small) sets and functions between them), this corresponds exactly to the conventional notion of point, and the same is true for a topological space X in **Top** (replace "morphism" by "continuous map" and "object" by "space" in the definition). For a locale L (in **Loc**), then, the categorical approach tells us that a point is a morphism from the terminal locale to L, which is equivalent to a frame morphism to the initial frame {0, 1} ([2], p. 476).

There is also the "standard" definition of a fuzzy point, due to Warner [3]: D3. Given a set X, a frame L, and the locale L(X) of all fuzzy subsets of X (functions from X to L), a fuzzy point of L(X) is a frame homomorphism $p : L(X) \rightarrow \{0, 1\}$.

Obviously, D3 is just a specific instance of D2. It is important to remember, as Warner notes ([3], p. 337), that the principal ideal generated by a prime element p of a topological space consists of the open sets to which p does *not* belong. Now the prime

N.R. Pal and M. Sugeno (Eds.): AFSS 2002, LNAI 2275, pp. 466–470, 2002.

elements of the (fuzzy) locale L(X) are based, as it were, on the prime elements of the underlying locale L; each prime element is in fact a fuzzy subset $P : X \to L$ where for some x in the underlying set X (Warner calls it x_0) $P(x_0) = \lambda$ (λ prime in L) and for all other x in X, $P(x) = 1$. Each fuzzy point, therefore, can be characterized unambiguously by the pair (x, λ), where $x \in X$, $\lambda \in L$ and prime in L.

2 A Categorical Approach to Points

There are reasons, in certain instances, why the "global" points of a locale (i.e., the homomorphisms just described above) may be insufficient. It is necessary, therefore, to look at more "generalized" points, i.e., maps from an arbitrary locale (rather than from {0, 1}) to the locale whose points are being enumerated (see esp. [4], and see [2], p. 164 for a succinct description of generalized elements). This paper attempts to draw out the connection between fuzzy points, as defined in [3], and these generalized points. In the Kripke-Joyal semantics, as outlined in [4], a point of type D (D an object of a category C) is simply a morphism (*any* morphism) targeted at D. Note that generalized points are local in the sense that they are "meaningful" only relative to some other particular object of the containing category; in this sense, these points may be thought of as prototypes.

3 Locales and Monads

Following Vickers [4], we also assume that **Loc** is order-enriched, that is, that the set of morphisms between any two locales is a member of **Pos**; we shall use the symbol $\Leftarrow$ to denote the "specialization" order in the object (set of morphisms, poset, category) $\mathcal{M}$. Furthermore, we assume, again with Vickers, that **Loc** is equipped with a powerobject KZ monad Given these assumptions, Vickers argues for (as he puts it) a "reasoning style" using points. We shall see that this same approach allows us to construct a somewhat different view of fuzzy powerobjects of the kind carefully elucidated by Rodabaugh in [5]; in effect, we take the argument one step farther, reasoning not from $f^{\to} : P(X) \to P(Y)$ to $f_L^{\to} : L^X \to L^Y$ but from $f_L^{\to} : L^X \to L^Y$ to $\mathcal{L}$ $f_L^{\to}: \mathcal{L}L^X \to \mathcal{L}L^Y$, lifting the fuzzy locales to their corresponding fuzzy powerlocales. The key observation is that the powerobject of an object in **Loc** may have distinct points which cannot be distinguished by *global* points of the object ([4], comments after Axiom 3.4).

KZ monads arose out of the work of A. Kock on the free completion of categories [6]. Four structures and their interrelationships lead to a very general notion of a monadic endo-2-functor on a category **C**. These include in particular the canonical singleton functor y_C, which maps each object C of **C** to the singleton functor $C : 1 \to$ **C**, and which serves as the unit of the monad, the functor $T = U \bullet F$, which takes (via the forgetful functor U) the free (co-)completion of **C** back to **C**, and which serves as the monad's functor, and $m : T \bullet T \rightarrow T$, the monad's multiplication. Finally, there is the 2-cell λ_C, which takes each 1-cell $T(y_C)$ to the 1-cell $y_{T(C)}$.

4 Powerlocales

The lower (Hoare) powerlocale is defined by ([4], Def. 4.1,and see also [7-8]):

$$\Omega P_L D = \mathrm{Fr} < \Diamond a \ (a \in \Omega D) \mid \Diamond \vee_i a_i = \vee_i \Diamond a_i >$$

where D is a locale, ΩD is its corresponding frame, and ◊ is the symbol for the free frame generator qua suplattice (see esp. [8], Sec. 6, and for the general idea of free frame generation [9], pp. 57-59). If one thinks of the ◊ operator as merely denoting "set of," then the powerframe of ΩD is simply the frame whose elements (opens) are sets of elements (opens) of ΩD where the set operation itself preserves joins.

Given this brief exposition, we can ask how the lower powerlocale of a locale whose elements are fuzzy sets (hereafter referred to as a fuzzy locale) can be constructed (we use the notation here of [3]). Given a complete lattice L and a set X, the elements of the fuzzy locale L(X) are the maps $A : X \rightarrow L$. This collection is a locale with respect to the partial ordering $A \leq B$ iff $A(x) \leq B(x)$ for all x in X. For any two fuzzy sets A and B in L(X), meet and join are defined by:

$$(A \wedge B)(x) = A(x) \wedge B(x)$$
$$(A \vee B)(x) = A(x) \vee B(x)$$

As usual, morphisms are functions preserving arbitrary joins and finite meets. Since we are thinking of **Loc** as a poset-enriched category, we know that the morphisms themselves (for a given object locale) are ordered. In 2-categorical terms, of course, each morphism is a 1-cell. The reader should be reminded that the morphisms we are referring to include both endomorphisms (elements of **Loc**(L(X) → L(X))) and "unrestricted" localic 2-categorical 1-cells, i.e., morphisms in **Loc**(L(X) → M(Y)) which allow both change of basis and change of set (as described in [5], esp. pp. 185 ff.). Morphisms between fuzzy locales are (partially) ordered just like morphisms between any locales, i.e., by finding inserters. Inserters, for fuzzy locales, are fuzzy locales constructed by application of the formula (from [4]):

$$\Omega I = \mathrm{Fr} < \Omega D \ (\text{qua Fr}) \mid \Omega f(b) \leq \Omega g(b) \ (b \in \Omega E) >$$

where f and g are morphisms from D to E (in other words, the inserter I orders $f, g : D \rightarrow E$ by $f \bullet i \Leftarrow g \bullet i$, $i : I \rightarrow D$). Notice that in our very general 2-categorical context, the inserter morphism i is just a point of D, whether f and g involve change of set, change of basis, both, or neither (are endomorphic).

What we want to do is describe the connection between the morphisms into (the locale of fuzzy sets) D (which, of course, are also the generalized points of D) and the (global) points of $\Omega P_L D$, and, by doing so, provide a justification for the compositional rule of inference different from the one in [5]. We also want to understand whether the assumption of order-enrichment mentioned above plays any important role in fuzzy locales. Finally, we want to understand whether the "prototypes" mentioned earlier may have a role in fuzzy logic.

A bit more explanation of the lower or Hoare powerlocale and the nature of its (global) points may be helpful here (see [10] pp. 169 ff. for details). If we think of the elements of the frame ΩD as being the "opens" of the locale (topological system) D (i.e., we think "topologically" for the time being), then an element (call it a^c) of the Hoare "powerframe" may be identified with the closed set which is the complement of the corresponding open a (again keeping in mind that an "open" here may be a set of elements of the locale D, a sequence of elements of the locale D, or whatever reflects its role as an open in the topological system under consideration). The Plotkin powerlocale of a locale D ($P_P D$ - also known as the Vietoris locale) is defined by (important details are omitted here, since this material is introduced purely for its explanatory value - see [10], pp. 171 ff. for the omissions):

$$\Omega P_P D = \mathrm{Fr} < \Diamond a \ (a \in \Omega D) \mid$$
$$(1)\ \Box(\wedge S) = \vee\{\Box a : a \in S\},$$
$$(2)\ \Diamond(\vee S) = \vee\{\Diamond a : a \in S\},$$
$$(3)\ \Box(a \wedge b) \leq \Box a \vee \Diamond b,$$
$$(4)\ \Box a \vee \Diamond b \leq \Diamond(a \wedge b)$$

5 The Zadeh Extension Principle

For notational simplicity, we denote "fuzzy" morphisms (mappings) simply by f, g, ..., keeping in mind of course that what we really mean is $f_L^{\rightarrow}$, $g_L^{\rightarrow}$, ..., and that these mappings are morphisms in **LOC** (or **LOCML** - see [5], esp. pp. 99 ff. and [11], esp. pp. 284-285) which lift corresponding morphisms in **SET**. We now state the following theorem:

In **Loc** equipped with a KZ monad, The Zadeh Extension Principle is correct.

Summary of proof. (We work in the 2-categorical context outlined above.)
The crucial property of (lower) powerlocales for our purposes is ([4], Prop. 4.4):

$$\forall X : P_L D \,.\, \forall Y : P_L E \,.\, (P_L f(X) \subseteq Y \leftrightarrow \forall x : D \,.\, (x \in X \rightarrow f(x) \in Y))$$

This proposition says that if $f : D \rightarrow E$ is a map of locales, and if for all points X in the powerlocale of D and all points Y in the powerlocale of E, the lifting of f applied to X precedes Y (in the specialization order), then for all points x of D in X, f(x) must be in Y, and vice versa. We need in particular the $\leftarrow$ direction of the proposition. Essentially, the proof follows the lines of the proof of Proposition 4.4 in [4]. For it to be true that if $\forall x : D \,.\, (x \in X \rightarrow f(x) \in Y)$ implies $P_L f(X) \subseteq Y$ it must be the case that f applied to the principal ideal generated by the morphism (point) x must retain its "shape" when it is lifted to the powerlocale. But this can only be the case if f is consistent with the ZEP.

6 Conclusions and Extensions

Though this study must be considered preliminary, it seems to be clear that a 2-categorical approach to fuzzy sets as localic objects may be of considerable value in generalizing the notion of fuzzy function, fuzzy point, and such (possibly artificial) distinctions as change of basis and change of set in fuzzy mappings. In this more general context, fuzzy points, fuzzy prototypes, and fuzzy mappings (morphisms) may all be seen as specific (perhaps sublocalic) realizations of 2-categorical constructs and operations.

There may be another way to justify the ZEP using bicategories or 2-categories, which is to consider fuzzy sets not as objects in **Loc**, as we have done here, but as 1-cells in the bicategory **Rel** (sets, relations, and inclusions - see [12]) enriched in **Cslat** (complete lattices with sup-preserving functions). Morphisms between 1-cells (fuzzy sets) would then be Rodabaugh's $f_L^{\leftarrow}$ functions ([5], Th. 1.1.4) and the argument proceeds at a different level (see [12], p. 198, example (3)). It may even be possible to treat fuzzy sets, in this connection, as a subcategory of a self-enriched **Cslat** (**Cslat** enriched in **Cslat** - see [12], p. 210, example (2)). These matters will, hopefully, be considered in future work.

References

1. Mac Lane, S. and Birkhoff, G.: Algebra (Third Edition). AMS Chelsea Publishing, Providence, Rhode Island (1999)
2. Mac Lane,S. and Moerdijk, I.: Sheaves in Geometry and Logic: A First Introduction to Topos Theory. Springer-Verlag, New York (1992)
3. Warner, M.W.: Frame-fuzzy points and membership. Fuzzy Sets and Systems **42** (1991) 335-344
4. Vickers, S.: Locales are not pointless. In: Hankin, C. et. al. (eds): Theory and Formal Methods 1994. Imperial College Press, London (1995) 199-216
5. Rodabaugh, S.: Powerset Operator Foundations For Poslat Fuzzy Set Theories And Topologies. In: Hohle, U. and Rodabaugh, S. (eds.): Mathematics of Fuzzy Sets: Logic, Topology, and Measure Theory. Kluwer Academic Publishers, Boston (1999) 91-116
6. Kock, A.: Monads for which structures are adjoint to units. Journal of Pure and Applied Algebra **104** (1995) 41-59
7. Vickers, S.: Information Systems for Continuous Posets. Theoretical Computer Science **114** (1993) 201-229
8. Vickers, S.: Localic Completion of Quasimetric Spaces. Research Report DoC97/2, Dept. of Computing, Imperial College, London
9. Johnstone, P.: Stone Spaces. Cambridge University Press, Cambridge (1982)
10. Vickers, S.: Topology via Logic. Cambridge University Press, Cambridge (1989)
11. Rodabaugh, S.: Categorical Foundations Of Variable-Basis Fuzzy Topology. In: Hohle, U. nd Rodabaugh, S. (eds.): Mathematics of Fuzzy Sets: Logic, Topology, and Measure Theory. Kluwer Academic Publishers, Boston (1999) 273-388
12. Koslowski, J.: Monads and Interpolads in Bicategories. Theory and Application of Categories **8** (1997) 182-212

Some Injective Cogenerators in Fuzzy Topology

Arun K. Srivastava

Department of Mathematics, Faculty of Science,
Banaras Hindu University, Varanasi-221005, India
aks@banaras.ernet.in

Abstract. We study here essentially two injective cogenerators, respectively in the categories of T_o-fuzzy topological spaces and T_o-fuzzy bitopological spaces, and discuss their relationships with T_o-ness, injectivity, and sobriety in these categories.

1 Introduction

Injective cogenerators in a category can be interesting objects (see e.g., [1]). In particular, in the category $\mathbf{TOP_o}$ of T_o- topological spaces, the well-known two-point Sierpinski space, denoted here as 2_S, is a very *prolific* injective cogenerator. Among other things, it 'gives' all $\mathbf{TOP_o}$-objects, all sober topological spaces, and all injective $\mathbf{TOP_o}$-objects. An analogous injective cogenerator in the category $\mathbf{BTOP_o}$ of T_o- bitopological spaces, found by Giuli and Salbany [5] and called *quad*, plays an almost identical role there (cf. Salbany [13] also). The purpose of this contribution is to demonstrate that almost perfectly analogous injective cogenerators exist in appropriate categories in fuzzy topology also.

2 Preliminaries

We first recall that a fuzzy topological space (referred to as a *fuzzy space* also) (X, Δ) is called T_o if for all distinct pairs of elements $x, y \in X, \exists u \in \Delta$ with $u(x) \neq u(y)$ (cf. [10]). A fuzzy bitopological space (X, Δ_1, Δ_2) (referred to as a *fuzzy bispace*) is called T_o, if $\forall x, y \in X, x \neq y, \exists u \in \Delta_1 \cup \Delta_2$ such that $u(x) \neq u(y)$.

Let **FTS** (resp. $\mathbf{FTS_o}$) and **BFTS** (resp. $\mathbf{BFTS_o}$) respectively denote the categories of fuzzy (resp. T_o-fuzzy) topological spaces and fuzzy (resp. T_o-fuzzy) bitopological spaces.

We put $I = [0,1]$. $1_A : X \to [0,1]$ shall denote the characteristic function of $A \subseteq X$. For $\alpha \in I$, the α-valued constant fuzzy set will also be denoted as α.

Let $\mathbf{C}$ be a category (e.g., $\mathbf{FTS}, \mathbf{FTS_o}, \mathbf{BFTS}$, or $\mathbf{BFTS_o}$). An object $X \in \text{ob}\mathbf{C}$ is called *injective* if $\forall$ $\mathbf{C}$-morphism $f : Y \to X$ and $\forall$ $\mathbf{C}$-embedding $e : Y \to Z, \exists$ a $\mathbf{C}$-morphism $g : Z \to X$ such that $f = g \circ e$.

Given any category $\mathbf{C}$, recall that $X \in ob\mathbf{C}$ is called a *cogenerator* (or a *coseparator*) if for each pair of distinct $f, g \in \mathbf{C}(Y, Z), \exists h \in \mathbf{C}(Z, X)$ with $h \circ f \neq h \circ g$.

For all other categorical concepts, used here, one can refer [2].

N.R. Pal and M. Sugeno (Eds.): AFSS 2002, LNAI 2275, pp. 471–477, 2002.

3 An Injective Cogenerator in $\mathbf{FTS_o}$ and Its Relation to T_o, Injective T_o, and Sober Fuzzy Spaces

Consider $(I, S) \in ob\mathbf{FTS}$, where S is the fuzzy topology on I generated by $\{id : I \to I\}$ (cf. [14]). Thus,

- $S = \{0, id, 1\}$ (for fuzzy topology in the sense of Chang [4]), or
- $S = \{(\alpha \wedge id) \vee \beta \mid \alpha, \beta \in I\}$ (for fuzzy topology in the sense of Lowen [9]).

(I, S) has been called *fuzzy Sierpinski space* in [14] for reasons outlined therein. We shall frequently denote (I, S) as just I_S.

Theorem 1. *$I_S = (I, S)$ is an injective cogenerator in* $\mathbf{FTS_o}$.

Proof: As $id \in S, I_S$ is clearly T_o. Next note that for any $(X, \Delta) \in ob\mathbf{FTS}, u \in \Delta$ iff $u : (X, \Delta) \to (I, S)$ is fuzzy continuous. Now, given any **FTS**-morphism, $f : (Y, \Delta) \to (I, S)$ and an **FTS**-embedding $e : (Y, \Delta) \to (Z, \Omega)$, we see that $f \in \Delta$, whereby $\exists g \in \Omega$ with $g \wedge 1_Y = f$. Clearly, $g : (Z, \Omega) \to (I, S)$ is fuzzy continuous with $g \circ e = f$, showing the injectivity of (I, S). Finally, given distinct $f, g \in \mathbf{FTS_o}((Y, \Delta), (Z, \Omega)), \exists y \in Y$ with $f(y) \neq g(y)$, whereby $u(f(y)) \neq u(g(y))$, for some $u \in \Omega$. But then $u \in \mathbf{FTS_o}((Z, \Omega), (I, S))$ and $u \circ f \neq u \circ g$, showing that I_S is a cogenerator in $\mathbf{FTS_o}$.

We now begin to give some interesting properties of I_S. We first recall some concepts and notations due to Herrlich, Salicrup, and Strecker [6].

In a category $\mathbf{C}$, define a relation $\sigma \subseteq mor\mathbf{C} \times$ ob$\mathbf{C}$ by :

$$(e, Y) \in \sigma \text{ iff } \forall \text{ pairs } f, g : \bullet \to Y \text{ of } \mathbf{C}\text{-morphisms }, f \circ e = g \circ e \Rightarrow f = g.$$

Given classes $\mathcal{E} \subseteq$ mor$\mathbf{C}$ and $\mathcal{A} \subseteq$ ob$\mathbf{C}$, we put

$$\mathcal{E}\text{-sep} = \{Y \mid e\sigma Y, \forall e \in \mathcal{E}\} \text{ and } \mathcal{A}\text{-epi} = \{e \mid e\sigma Y, \forall Y \in \mathcal{A}\}.$$

Elements of $\mathcal{E}$-sep and $\mathcal{A}$-epi are respectively called *$\mathcal{E}$-separated* objects and *$\mathcal{A}$-epimorphisms* of $\mathbf{C}$. (cf. [6])

Theorem 2 (Srivastava and Mishra [16]).

1. $\mathbf{FTS_o}$*-epis* $= \{I_S\}$*-epi* $= \{e : X \to Y \in ob\mathbf{FTS_o} \mid u \circ e = v \circ e \Rightarrow u = v,$ *for all open fuzzy sets* u, v *in* $Y\}$.
2. $\mathcal{E}$*-sep* $= ob\mathbf{FTS_o}$, *where* $\mathcal{E} = \{I_S\}$*-epi.*

Remark 1. A much more useful characterization of epimorphisms in $\mathbf{FTS_o}$ (involving I_S again!) has been obtained by Alderton and Castellini[1] in terms of a 'b-closure operator' for $\mathbf{FTS_o}$.

We next proceed to point out that I_S is capable of 'giving' all $\mathbf{FTS_o}$-objects and all *injective* $\mathbf{FTS_o}$-objects.

Recall that the smallest epireflective subcategory of a category $\mathbf{C}$, containing $A \in \mathbf{C}$, is called the *epireflective hull of* A in $\mathbf{C}$. We denote it as $\mathbf{C}(A)$.

Theorem 3 (Lowen and Srivastava [10]). $\mathbf{FTS_o} = \mathbf{FTS}(I_S)$.

The proof requires one to verify that each $(X, \Delta) \in \text{ob}\mathbf{FTS_o}$ is identifiable as a subspace of a product of copies of $I_S = (I, S)$. This is done by checking that the evaluation map $e : (X, \Delta) \to (I_S)^S = (I^S, S^S)$ is a continuous embedding, whereby $X \cong e(X)$.

Remark 2. An internal description of this epireflection is as follows:
Given $(X, \Delta) \in \text{ob}\mathbf{FTS}$, declare $x \sim y$ iff $u(x) = u(y), \forall u \in \Delta$.
Then $\sim$ is an equivalence relation on X and the resulting quotient map $q : (X, \Delta) \to (X^*, \Delta^*) = (X, \Delta)/\sim$ is the desired epireflection (and is 'initial' too).

The proof of Theorem 3 also shows that:

Theorem 4 (Srivastava and Mishra [16]). *$X \in ob\mathbf{FTS_o}$ is injective iff it is a retract of a product of copies of I_S.*

In topology, it is known that

$$\mathbf{TOP}(2_D) = \mathbf{ZTOP_o},$$

where 2_D denotes the two-point discrete space and $\mathbf{ZTOP_o}$ denotes the category of zero-dimensional T_o-topological spaces. Moreover, 2_D is an *injective cogenerator* in $\mathbf{ZTOP_o}$. We show below the existence of an analogous situation in fuzzy topology.

Consider $I_D = (I, D) \in \text{ob}\mathbf{FTS}$ where D is the fuzzy topology on I generated by $\{id, 1 - id\}$. Call $(X, \Delta) \in \text{ob}\mathbf{FTS}$ *zero-dimensional* if it has a basis of Δ-clopen fuzzy sets (e.g., I_D is *both* zero-dimensional and T_o). Let $\mathbf{ZFTS}$ denote the resulting category. We then have:

Theorem 5. *1. I_D is an injective cogenerator in $\mathbf{ZFTS_o}$.*
2. $\mathbf{ZFTS_o} = \mathbf{ZFTS}(I_D)$.
3. $X \in ob\mathbf{ZFTS_o}$ is injective iff it is retract of a product of copies of I_D.

We next show that I_S is can also give all *sober* fuzzy spaces.

Definition 1. *$X \in ob\mathbf{TOP_o}$ is called* sober *if $\forall$ irreducible nonempty closed $F \subseteq X, \exists x \in X$, with $F = clx$ (F is* irreducible *if $F \subseteq A \cup B$, A, B closed $\Rightarrow F \subseteq A$ or $F \subseteq B$).*

It is known that:

- $\mathbf{SOB} = \mathbf{TOP_o}(2_S)$ where $\mathbf{SOB}$ denotes the category of all sober topological spaces (cf. Nel and Wilson, [11]).

Recall that a *frame* is a complete lattice (0 and 1 denoting its lower and upper bounds) satisfying the infinite distributive law $a \wedge (\vee b_i) = \vee (a \wedge b_i)$. A *frame-map* is a map between frames preserving arbitrary $\vee$ and finite $\wedge$.

Following Johnstone [7] (where sobriety in $\mathbf{TOP}$ has been described in terms of frame-maps also), Rodabaugh [12] introduced sobriety in $\mathbf{FTS}$ as follows:

For $(X, \Delta) \in ob\mathbf{FTS}$, let $X^S = \{\text{all frame-maps } f : \Delta \to I\}$ and let

$$\Psi = \Psi_X : X \to X^S$$

be defined as $\Psi(x)(u) = u(x), x \in X, u \in \Delta$.

Definition 2. $(X, \Delta) \in ob\mathbf{FTS}$ *is called* **sober** *if* Ψ *is bijective.*

(Example: I_S is sober.)
FSOB shall denote the category of sober fuzzy spaces.
Given $(X, \Delta) \in ob\mathbf{FTS_o}$ and $u \in \Delta$, define $u^S : X^S \to I$ as $u^S(p) = p(u)$ and let $\Delta^S = \{u^S : u \in \Delta\}$. Then

- $(X^S, \Delta^S) \in ob\mathbf{FSOB}$ and $\Psi : (X, \Delta) \to (X^S, \Delta^S)$ is continuous and open ,(cf. Rodabaugh [12]).

(X^S, Δ^S), or rather the map $\Psi : (X, \Delta) \to (X^S, \Delta^S)$, shall be referred to as the *soberification* of $(X, \Delta) \in$ ob$\mathbf{FTS_o}$.

Theorem 6 (Srivastava and Khastgir [15]). FSOB *is epireflective in* $\mathbf{FTS_o}$ *with the soberification of* $(X, \Delta) \in ob\mathbf{FTS_o}$ *acting as its epireflection.*

Another description of the soberification, *in terms of* I_S, can be given as follows, using the '*b*-closure operator' of Alderton and Castellini [1] (see [1] for the definition of the '*b*-closure operator').

Given $(X, \Delta) \in ob\mathbf{FTS_o}$, consider the evaluation map $e : X \to {I_S}^{\Delta}$, given by $e(x)u = u(x)$. Then $e : X \to b(e(X))$ and $\Psi : X \to X^S$, as $\mathbf{FTS_o}$-morphisms, turn out to be the *same* (up to isomorphism), giving thereby an *external* description of the soberification.

Theorem 7 (Srivastava and Khastgir [15]).

(a) **FSOB** *is epireflective* [1] *in* $\mathbf{FTS_o}$.
(b) **FSOB** $= \mathbf{FTS_o}(I_S)$.

Remark 3. It is interesting to point out here that the soberification $\Psi : (X, \Delta) \to (X^S, \Delta^S)$ of $(X, \Delta) \in$ ob$\mathbf{FTS_o}$ turns out to be an 'epi-injective hull' of (X, Δ). This follows from a rather purely category theoretic argument; cf. [3] (Prop. 2.10).

[1] In fact, **FSOB** is a 'firm' epireflective subcategory of $\mathbf{FTS_o}$ in the sense that each epireflection $r : X \to X^S$ is an embedding and for every epi-embedding $f : X \to Y \in$ ob**FSOB**, the unique **FSOB**-morphism $f^* : X^S \to Y$ (with $f^* \circ r = f$) is a homeomorphism.

4 An Injective Cogenerator in $\mathbf{BFTS_o}$ and Its Relation to T_o, Injective T_o, and Sober Fuzzy Bispaces

Denote $(I \times \{0\}) \cup (\{0\} \times I)$ as $2I$ and consider on it two fuzzy topologies

$P_i = \{0, p_i, 1\}, i = 1, 2$, where $p_1, p_2 : 2I \to I$ are defined as

$$p_1(x) = \begin{cases} a \text{ if } x = (a, 0) \in I \times \{0\} \\ 0 \text{ otherwise} \end{cases}$$

$$p_2(x) = \begin{cases} a \text{ if } x = (0, a) \in \{0\} \times I \\ 0 \text{ otherwise} \end{cases}$$

We next define another object similar to $2I$.

Denote $I \times I$ as I^2. Consider on I^2 the two fuzzy topologies $\Pi_i = \{0, p_i, 1\}, i = 1, 2$, where $p_1, p_2 : I^2 \to I$ are the two *projection* maps. Then $I^2 = (I^2, \Pi_1, \Pi_2) \in ob\mathbf{BFTS_o}$.

Theorem 8 (Khastgir and Srivastava [8]).

1. *Both $2I$ and I^2 are cogenerators for* $\mathbf{BFTS_o}$.
2. $\mathbf{BFTS_o} = \mathbf{BFTS}(2I) = \mathbf{BFTS}(I^2)$.
3. *I^2 is injective in $\mathbf{BFTS_o}$ whereas $2I$ is not (hence I^2 is an injective cogenerator in $\mathbf{BFTS_o}$).*

We next point out the relationship of the injective cogenerator I^2 with injective $\mathbf{BFTS_0}$-objects.

Theorem 9. *$X = (X, \Delta_1, \Delta_2) \in ob\mathbf{BFTS_0}$ is injective iff it is a retract of a product of copies of (I^2, Π_1, Π_2).*

Finally, we point out the relationship of I^2 with, what we shall call, *sober* and *absolutely T_o-closed* fuzzy bitopological spaces.

Let $(X, \Delta_1, \Delta_2) \in$ ob$\mathbf{BFTS}$. Put X^*={all frame-maps from $\Delta_1 \vee \Delta_2$ to I} and define

$$\Phi = \Phi_X : X \to X^* \text{ by } \Phi(x)(u) = u(x),\ x \in X, u \in \Delta_1 \vee \Delta_2.$$

For, $u \in \Delta_1 \vee \Delta_2$, define $u^* : X^* \to I$ by $u^*(p) = p(u)$, $p \in X^*$. Let

$$\Delta_i^* = \{u^* \mid u \in \Delta_i\}, i = 1, 2.$$

Then Δ_1^* and Δ_2^* are fuzzy topologies on X^* under which $\Phi : (X, \Delta_1, \Delta_2) \to (X^*, \Delta_1^*, \Delta_2^*)$ becomes fuzzy bicontinuous.

Definition 3. *$(X, \Delta_1, \Delta_2) \in ob\mathbf{BFTS}$ is called* **sober** *if $\Phi : (X, \Delta_1, \Delta_2) \to (X^*, \Delta_1^*, \Delta_2^*)$ is a* $\mathbf{BFTS}$*-isomorphism.*

BFSOB shall denote the resulting category of sober fuzzy bispaces.

It turns out that:

Theorem 10 (Khastgir and Srivastava [8]).

1. *The cogenerators* $2I$ *and* I^2 *are* **BFSOB**-*objects.*
2. **BFSOB** *is epireflective in* $\mathbf{BFTS_o}$.

Actually, $\Phi : (X, \Delta_1, \Delta_2) \to (X^*, \Delta_1^*, \Delta_2^*)$ turns out to be the desired epireflection of $(X, \Delta_1, \Delta_2) \in \text{ob}\mathbf{BFTS_o}$, as we now indicate. Given any **BFTS**-morphism $f : (X, \Delta_1, \Delta_2) \to (Y, \Omega_1, \Omega_2) \in ob\mathbf{BFSOB}$, define $f^* : (X^*, \Delta_1^*, \Delta_2^*) \to (Y, \Omega_1, \Omega_2)$ as follows:
Given any $p \in X^*$, the map $p' : \Omega_1 \vee \Omega_2 \to I$, given by $p'(v) = p(v \circ f), v \in \Omega_1 \vee \Omega_2$, turns out to be a frame-map, so that $p' \in Y^S$. The sobriety of $S(Y) = (Y, \Omega_1 \vee \Omega_2)$ then produces a unique element $y \in Y$ with $\Psi_Y(y) = p'$.
Put $f^*(p) = y$. The f^*, so defined, turns out to be the unique **BFTS**-morphism with $f^* \circ \Phi = f$.

Remark 4. **BFSOB**, however, does not turn out to be the epireflective hull of I^2 (or of $2I$) in $\mathbf{BFTS_o}$, as we shall see next.

For each $M \subseteq (X, \Delta_1, \Delta_2) \in \text{ob}\mathbf{BFTS}$, put

$$[M] = \cap\{Eq(f,g) \mid f,g : (X, \Delta_1, \Delta_2) \to (Y, \Omega_1, \Omega_2) \in \text{ob}\mathbf{BFTS_o} \text{ are } \mathbf{BFTS}\text{-morphisms with } f/M = g/M\},$$

Definition 4. $X = (X, \Delta_1, \Delta_2) \in ob\mathbf{BFTS_o}$, *is called* **absolutely $2T_o$-closed** *if* $\forall\mathbf{BFTS_o}$-*embedding* $e : X \to Y$, $[e(X)] = e(X)$.

$AC - \mathbf{BFTS_o}$ denotes the category of absolutely T_o-closed $\mathbf{BFTS_o}$-objects.

Theorem 11 (Khastgir and Srivastava [8]). AC-$\mathbf{BFTS_o} = \mathbf{BFTS_o}(I^2)$.

Proof-outline: We check that $\mathbf{BFTS_o}$ is a well-powered, and complete (Epi, Extremal mono)-category. Next, as I^2 cogenerates $\mathbf{BFTS_o}$ and is injective in $\mathbf{BFTS_o}$, the result follows from known categorical results.

An *explicit description* of the absolutely $2T_o$-closed epireflection, *in terms of* I^2, is as follows. Given $X = (X, \Delta_1, \Delta_2) \in \mathbf{BFTS_o}$, put $\mathcal{C} = \mathbf{BFTS_o}(X, I^2)$. As $I^2 = (I^2, \Pi_1, \Pi_2)$ cogenerates $\mathbf{BFTS_o}$, X can be embedded into the product $(I^2)^{\mathcal{C}}$ via the evaluation map $e : X \to (I^2)^{\mathcal{C}}$. Let RX denote $[e(X)]$, with the relative fuzzy topologies on $[e(X)]$ induced by the product fuzzy topologies $\Pi_1^{\mathcal{C}}$ and $\Pi_2^{\mathcal{C}}$. Noting that I^2 is absolutely $2T_o$-closed and that RX is a [ ]-closed subspace of a product of copies of I^2, it turns out that RX (in fact, $rX : X \to RX$, where $rX = e$), so obtained, turns out to be the absolutely $2T_o$-closed epireflection of X.

Concluding Remarks: Out of the several directions in which the study presented here can be pursued, we point out two. For the injective cogenerator 2_S, it is known that the functor $\mathbf{TOP}(-,2_S)$: $\mathbf{TOP}^{op} \to \mathbf{SET}$ is *monadic* and the category of associated *T-algebras* is the category of frames and frame-maps. It would be interesting to find out what happens when 2_S and **TOP** are respectively replaced by I_S and **FTS**. Also, as injective T_o topological spaces are known to correspond to *continuous lattices*, an interesting question to answer would be: To what do injective T_o *fuzzy* topological spaces correspond?

References

1. I.W. Alderton and G. Castellini, Epimorphisms in categories of separated fuzzy topological spaces, *Fuzzy Sets and Systems* **56** (1993), 323-330.
2. F. Borceux, *Handbook of Categorical Algebra 1*, Cambridge University Press, 1994.
3. G.C.L. Brümmer, E. Giuli and H. Herrlich, Epireflections which are completions, *Cahiers Topologie Geom. Diff. Categoriques* **33** (1992), 71-93.
4. C.L. Chang, Fuzzy topological spaces, *Jour. Math. Anal. Appl.* **24** (1968), 182-189.
5. E. Giuli and S. Salbany, $2T_o$ spaces and closure operators, *Seminarberichte aus dem Fachbereich Mathematik und Informatik, Fernuniverität, Hagen* **29** (1988), 11-40.
6. H.Herrlich, G. Salicrup and G.E. Strecker, Factorizations, denseness, separation, and relatively compact objects, *Topology Appl.* **27** (1987), 157-169.
7. P.T. Johnstone, *Stone Spaces*, Cambridge University Press, 1982.
8. A.S. Khastgir and A.K. Srivastava, On T_o-objects in **FTS** and **BFTS**, *Fuzzy Sets and Systems* **109** (2000), 301-304.
9. R. Lowen, Fuzzy topological spaces and fuzzy compactness, *Jour. Math. Anal. Appl.* **56** (1976), 621-633.
10. R. Lowen and A.K. Srivastava, $\mathbf{FTS_o}$: The epireflective hull of the Sierpinski object in **FTS**, *Fuzzy Sets and Systems* **29** (1989), 171-176.
11. L.D. Nel and R.G. Wilson, Epireflections in the category of T_o-spaces, *Fund. Math.* **75** (1972), 69-74.
12. S.E. Rodabaugh, Point-set lattice theoretic topology, *Fuzzy Sets and Systems* **40** (1991), 297-345.
13. S. Salbany, On Injective topological spaces and bispaces, *Research Report*, no. 268/98(**14**), Sept. 1998 (Dept. of Math., University of South Africa).
14. A.K. Srivastava, Fuzzy Sierpinski space, *Jour. Math. Anal. Appl.* **103** (1984), 103-105.
15. A.K. Srivastava and A.S. Khastgir, On fuzzy sobriety, *Information Sciences* **110** (1998), 195-205
16. A.K. Srivastava and S.P. Mishra, Epis and injectives in $\mathbf{FTS_o}$, *Jour. of Fuzzy Math.* **1** (1993), 389-393.

Derivative and Differential of Convex Fuzzy Valued Functions and Application*

Wu Congxin[1,**], Wang Guixiang[1], and Wu Cong[2]

[1] Department of Mathematics, Harbin Institute of Technology, Harbin, 150001, People's Republic of China

[2] School of Manage, Harbin Institute of Technology, Harbin, 150001, People's Republic of China

Abstract. In this paper we put forward the concepts of directional derivative, differential and subdifferential for fuzzy valued functions, discuss the characterizations of them and the relations among them. We also define the gradients of a fuzzy valued function and study the relation between gradients and partial derivatioves, the relation between the directional a.e. cut-derivatives and the subgradients of the convex fuzzy mappings. At the end, we give two results of application in convex fuzzy programming.

1 Introduction

In 1972, S. S. L. Chang, L. A. Zadeh [1] introduced the concept of fuzzy numbers with the consideration of the properties of probability functions. With the development of theories of fuzzy numbers and its application the concept of fuzzy numbers becomes more and more important.

On the other hand, it is well known that convex analysis is an important branch of mathematics, and has also wide application in convex programming. If the values of the target function for which an optimal solution is sought are crisp real numbers, the programming is a general crisp programming. But in reality, sometimes the values of the target function are estimated values, so it is more suitable to express the values by fuzzy numbers. In 1997, Butnariu [12] studied the problem of the fuzzy convex programming for convex continuous fuzzy mappings. In [6], with the criteria for convex fuzzy mappings, Nanda and Kar also discussed fuzzy convex programming. The purpose of this paper is to study some fundamental concepts for convex fuzzy valued functions so that the problem of convex fuzzy programming can be considered.

2 Preliminaries

Let us denote by $E = \{u | u : R \to [0,1]$, u is normal and fuzzy convex, $u(x)$ is upper semi-continuous, and $[u]^0$ is a compact set$\}$, where $[u]^0 = cl\{x \in R :$

* This paper is supported by HIT. MD. 2000. 21.

** Corresponding author.

N.R. Pal and M. Sugeno (Eds.): AFSS 2002, LNAI 2275, pp. 478–484, 2002.

$u(x) > 0\}$. For any $u \in E$, u is called a fuzzy number and E is called fuzzy number space. Obviously, $[u]^r$ are nonempty bounded closed intervals (denoted $[\underline{u}(r), \bar{u}(r)]$) for any $u \in E$ and $r \in [0,1]$, where $[u]^r = \{u \in R : u(x) \geq r\}$ when $r \in (0,1]$.

For any $a \in R$, define a fuzzy number $\hat{a}$ by $\hat{a}(t) = 1$ whenever $t = a$, $\hat{a}(t) = 0$ whenever $t \neq a$.

For $u, v \in E$ and $\lambda \in R$, we define

$$(u+v)(x) = \sup_{y+z=x} \min[u(y), v(z)],$$

$$(\lambda u)(x) = \begin{cases} u(\lambda^{-1}x) & if\ \lambda \neq 0 \\ \hat{0} & if\ \lambda = 0 \end{cases}$$

It is well known that for any $u, v \in E$, $\lambda \in R$, $u+v$, $\lambda u \in E$ and $[u+v]^r = [u]^r + [v]^r$, $[\lambda u]^r = \lambda [u]^r$.

For $x = (x_1, x_2, \cdots, x_n)$, $y = (y_1, y_2, \cdots, y_n) \in R^n$ we define $x \leq y$ iff $x_i \leq y_i$ $(i = 1, 2, \cdots, n)$.

For $u, v \in E$, we define $u \leq v$ iff $[u]^r = [\underline{u}(r), \bar{u}(r)] \leq [v]^r = [\underline{v}(r), \bar{v}(r)]$ for any $r \in [0,1]$. And $[u]^r \leq [v]^r$ iff $\underline{u}(r) \leq \underline{v}(r)$ and $\bar{u}(r) \leq \bar{v}(r)$.

If for $u, v \in E$, there exists $w \in E$ such that $u = v + w$, then we say the Hukuhara difference (in short, H- difference) $u - v$ to exist, and denote $u - v = w$.

For $u, v \in E$, define $D(u,v) = \sup_{r \in [0,1]} \max(|\underline{u}(r) - \underline{v}(r)|, |\bar{u}(r) - \bar{v}(r)|)$.

A mapping $F : M(\subset R^n) \to E$ is said to be a fuzzy valued function. For any $r \in [0,1]$, denote $[F(x)]^r = [\underline{F(x)}(r), \overline{F(x)}(r)]$. We call F to be a convex fuzzy valued function if for any $x, y \in K$, a convex subset of R^n, $t \in (0,1)$, $F(tx + (1-t)y) \leq tF(x) + (1-t)F(y)$.

3 Main Results

In [5], Puri and Ralescu defined the H-derivative of fuzzy mappings from an open subset of a normed space into n-dimension fuzzy number space E^n by using H-difference. Using H-difference, we also can define the directional H-derivative of fuzzy mappings from R^n into E.

In order to avoid the appearance of the difficulty that H-differences bring to us, in what follows, we introduce Definition 3.1.

Definition 3.1. Let $F : M(\subset R^n) \to (E, D)$ be a fuzzy valued function, $x \in M$. If for $y \in R^n$, there exists $\delta > 0$ such that $x + hy \in M$ (resp. $x - hy \in M$) for any $h \in (0, \delta)$, and there exists $u^+ \in E$ (resp. $u^- \in E$) such that $\lim_{h \to 0^+} \frac{D(F(x+hy), F(x)+hu^+)}{h} = 0$ (resp. $\lim_{h \to 0^+} \frac{D(F(x), F(x-hy)+hu^-)}{h} = 0$), then we say F to be right (resp. left) differentiable at x in the direction y, and call the unique (the uniqueness can be proved) u^+ (resp. u^-) (denote $F'_+(x,y) = u^+$ (resp. $F'_-(x,y) = u^-$)) the right (resp. left) derivative of F at x in the direction y. And if $F'_+(x,y) = F'_-(x,y)$, then we say F to be differentiable at x

in the direction y, denote $F'(x,y) = F'_+(x,y) = F'_-(x,y)$, and call $F'(x,y)$ the derivative of F at x in the direction y.

In the following, we give a characterization of directional differentiability of fuzzy mapping $F : M(\subset R^n) \to (E,D)$.

Theorem 3.2. Let $F : M(\subset R^n) \to (E,D)$ be a fuzzy valued function, $x \in M$, $y \in R^n$. Then

(1) $F(x)$ is right (resp. left) differentiable at x in the direction y $\Leftrightarrow$ for any fixed $r \in [0,1]$, $\underline{F(x)}(r)$ and $\overline{F(x)}(r)$ are right (resp. left) differentiable at x in the direction y, $\underline{F'_+}(x,y)(r)$ and $\bar{F}'_+(x,y)(r)$ (resp. $\underline{F'_-}(x,y)(r)$ and $\bar{F}'_-(x,y)(r)$) satisfy the conditions (1)-(4) of Theorem 2.1 in part 1 of [10], and $\underline{G_+^{x,y}(h)(r)} = \frac{\underline{F(x+hy)(r)}-\underline{F(x)(r)}}{h}$, $\overline{G_+^{x,y}(h)}(r) = \frac{\overline{F(x+hy)}(r)-\overline{F(x)}(r)}{h}$ (resp. $\underline{G_-^{x,y}(h)(r)} = \frac{\underline{F(x)(r)}-\underline{F(x-hy)(r)}}{h}$, $\overline{G_-^{x,y}(h)}(r) = \frac{\overline{F(x)}(r)-\overline{F(x-hy)}(r)}{h}$) uniformly (for $r \in [0,1]$) converge to $\underline{F'_+}(x,y)(r)$ and $\overline{F'_+}(x,y)(r)$ (resp. $\underline{F'_-}(x,y)(r)$ and $\overline{F'_-}(x,y)(r)$) as $h \to 0^+$, respectively. And if the condition is satisfied, then $[F'_+(x,y)]^r = [\underline{F'_+}(x,y)(r), \overline{F'_+}(x,y)(r)]$ (resp. $[F'_-(x,y)]^r = [\underline{F'_-}(x,y)(r),$ $\overline{F'_-}(x,y)(r))]$, where $\underline{F'_+}(x,y)(r)$ and $\overline{F'_+}(x,y)(r)$, (resp. $\underline{F'_-}(x,y)(r)$ and $\overline{F'_-}(x,y)(r)$) are respectively the right (resp. left) derivatives of $\underline{F(x)}(r)$ and $\overline{F(x)}(r)$ at x in the direction y.

(2) $F(x)$ is differentiable at x in the direction y $\Leftrightarrow$ for any fixed $r \in [0,1]$, $\underline{F(x)}(r)$ and $\overline{F(x)}(r)$ are differentiable at x in the direction y, $\underline{F'}(x,y)(r)$ and $\overline{F}'(x,y)(r)$ satisfy the conditions (1)-(4) of Theorem 2.1 in part 1 of [10], and $\underline{G_+^{x,y}(h)(r)} = \frac{\underline{F(x+hy)(r)}-\underline{F(x)(r)}}{h}$, $\overline{G_+^{x,y}(h)}(r) = \frac{\overline{F(x+hy)}(r)-\overline{F(x)}(r)}{h}$ uniformly (for $r \in [0,1]$) converge to $\underline{F'}(x,y)(r)$ and $\overline{F'}(x,y)(r)$ as $h \to 0$, respectively. And if the condition is satisfied, then $[F'(x,y)]^r = [\underline{F'}(x,y)(r), \overline{F'}(x,y)(r)]$, where $\underline{F'}(x,y)(r)$ and $\overline{F'}(x,y)(r)$ are respectively the derivatives of $\underline{F(x)}(r)$ and $\overline{F(x)}(r)$ at x in the direction y.

In the following, we give the definition of partial derivative of fuzzy mapping $F : M(\subset R^n) \to (E,D)$, which is equivalent to the concept of the partial derivatives touched upon by Buckley and Feuring in [13].

Definition 3.3. If $F(x)$ is differentiable at x in the direction e_i ($e_i = (a_1, a_2, \cdots, a_i, \cdots, a_n)$ with $a_i = 1$ and $a_j = 0$ for $j \neq i$), then we say that $F(x)$ is partially differentiable at x with respect to x_i, called $F'(x, e_i)$. The partial derivative of $F(x)$ at x with respect to x_i, and denote $F'_{x_i}(x) = F'(x, e_i)$. The partially H-differentiability and $(H)F'_{x_i}(x)$ can be defined similarly.

In what follows, we give the definition of directional a.e cut-differentiability for fuzzy number valued functions so that we can more conveniently discuss convex fuzzy mappings and convex fuzzy programming. According the definition of directional a.e cut-differentiability, the convex fuzzy mappings satisfying the condition of possessing H-difference are directional right and left a.e cut-differentiable (see theorem 3.5).

Definition 3.4. Let $F : M(\subset R^n) \to (E, D)$ be a fuzzy valued function, $x \in M$. If for given $y \in R^n$, there exists $\delta > 0$ such that $x + hy \in M$ (resp. $x - hy \in M$), and the H-difference $F(x + hy) - F(x)$ (resp. $F(x) - F(x - hy)$) exists for any $h \in (0, \delta)$, and there exists $u^+ \in E$ (resp. $u^- \in E$) such that $\lim_{h\to 0^+} \left[\frac{F(x+hy)-F(x)}{h}\right]^r = [u^+]^r$ (resp. $\lim_{h\to 0^+} \left[\frac{F(x)-F(x-hy)}{h}\right]^r = [u^-]^r$) (in the Hausdorff metric d) almost everywhere holds for r on $[0, 1]$, then we say F to be right (resp. left) a.e. cut-differentiable at x in the direction y, call u^+ (resp. u^-) is unique, denote $F^*_+(x, y) = u^+$ (resp. $F^*_-(x, y) = u^-$)) the right (resp. left) a.e. cut-derivative of F at x in the direction y. And if $F^*_+(x, y) = F^*_-(x, y)$, then we say F to be a.e. cut-differentiable at x in the direction y, denote $F^*(x, y) = F^*_+(x, y) = F^*_-(x, y)$, and call $F^*(x, y)$ the a.e. cut-derivative of F at x in the direction y.

Theorem 3.5. Let K be a convex subset of R^n, $F : K \to E$ be a convex fuzzy valued function, and $x \in K$. If for given $y \in R^n$, there exists $\delta > 0$ such that $x + hy, x - hy \in K$, and the H-differences $F(x + hy) - F(x)$, $F(x) - F(x - hy)$ exist for any $h \in (0, \delta)$, then $F^*_+(x, y)$ and $F^*_-(x, y)$ exist, and $F^*_+(x, y) = \inf_{h\in(0,\delta)} G^{x,y}_+(h) \geq \sup_{h\in(0,\delta)} G^{x,y}_-(h) = F^*_-(x, y)$. (about the definitions of supremum and infimum, we can see [11].

In what follows, we define a kind of differential, it is different from the de Blasi differential that Diamond and Kloeden defined in [3], by our definition we can give the concept of gradient.

Definition 3.6. Let $F : M(\subset R^n) \to E$ be a fuzzy valued function, $x^0 = (x^0_1, x^0_2, \cdots, x^0_n) \in \text{ int } M$. If there exist $u_1, u_2, \cdots, u_n \in E$ such that $\lim_{x\to x^0} \frac{D(F(x)+\sum_{i=1}^n \Delta(x^0_i,x_i)|x_i-x^0_i|u_i,F(x^0)+\sum_{i=1}^n \Delta(x_i,x^0_i)|x_i-x^0_i|u_i)}{d(x,x^0)} = 0$, then we say F to be differentiable at x^0, and call $(u_1, u_2, \cdots, u_n)$ (denote $\bigtriangledown F(x^0) = (u_1, u_2, \cdots, u_n)$) the gradient of F at x^0, where $x = (x_1, x_2, \cdots, x_n)$, and for any $t, s \in R$, $\Delta(t, s) = 1$ as $t \geq s$, $\Delta(t, s) = 0$ as $t < s$.

The following Theorem 3.7 reveals the relation between gradients and partial derivatives.

Theorem 3.7. Let $F : M(\subset R^n) \to E$ be a fuzzy valued function, $x = (x_1, x_2, \cdots, x_n) \in \text{ int } M$.

(1) If F is differentiable at x, then $F'_{x_i}(x)$ exists, and $F'_{x_i}(x) = u_i$ ($i = 1, 2, \cdots, n$), where $(u_1, u_2, \cdots, u_n) = \bigtriangledown F(x)$.

(2) If F is differentiable at x, then for any $y = (y_1, y_2, \cdots, y_n)(\neq 0) \in R^n$ with the H-difference $\sum_{\substack{y_i\geq 0\\ i=1,2,\cdots,n}} y_i F'_{x_i}(x) - \sum_{\substack{y_i<0\\ i=1,2,\cdots,n}} |y_i| F'_{x_i}(x)$ exists, then $F'(x, y)$ exists, and $F'(x, y) = \sum_{\substack{y_i\geq 0\\ i=1,2,\cdots,n}} y_i F'_{x_i}(x) - \sum_{\substack{y_i<0\\ i=1,2,\cdots,n}} |y_i| F'_{x_i}(x)$.

The following Theorem 3.8 gives a characterization of differentiability for fuzzy mapping $F : M(\subset R^n) \to E$.

Theorem 3.8. Let $F : M(\subset R^n) \to E$ be a fuzzy valued function, $x^0 = (x^0_1, x^0_2, \cdots, x^0_n) \in \text{ int } M$. Then F is differentiable at x^0 if and only if $\underline{F(x)}(r)$

and $\overline{F(x)}(r)$ are differentiable at x^0, $\underline{F'_{x_i}}(x^0)(r), \overline{F'_{x_i}}(x^0)(r)$ $(i = 1, 2, \cdots, n)$ satisfy conditions (1)-(4) of theorem 2.1 in part 1 of [10], and

$\underline{T_{x^0}}(x, r) = \frac{\underline{F(x)}(r) - \underline{F(x^0)}(r) - \sum_{i=1}^{n}(x_i - x_i^0)\underline{F'_{x_i}}(x^0)(r)}{d(x, x^0)}$,

$\overline{T_{x^0}}(x, r) = \frac{\overline{F(x)}(r) - \overline{F(x^0)}(r) - \sum_{i=1}^{n}(x_i - x_i^0)\overline{F'_{x_i}}(x^0)(r)}{d(x, x^0)}$, uniformly (for $r \in [0, 1]$) converge to 0 as $x \to x^0$.

In the following, we give the definition of subdifferential of fuzzy mapping $F : M(\subset R^n) \to E$.

Definition 3.9. Let $F : M(\subset R^n) \to E$ be a fuzzy valued function, $x^0 = (x_1^0, x_2^0, \cdots, x_n^0) \in M$. If there exist $u_1, u_2, \cdots, u_n \in E$ and $\delta > 0$ such that $F(x) + \sum_{\substack{x_i < x_i^0 \\ i=1,2,\cdots,n}} |x_i - x_i^0| u_i \geq F(x^0) + \sum_{\substack{x_i \geq x_i^0 \\ i=1,2,\cdots,n}} (x_i - x_i^0) u_i$ holds for all $x \in U(x^0, \delta) \cap M$ (where $U(x^0, \delta) = \{x \in R^n : d(x, x^0) < \delta\}$), then we call $(u_1, u_2, \cdots, u_n)$ a subgradient of F at x^0, and say the set of all subgradients of F at x^0 to be subdifferential of F at x^0 (denoted $\partial F(x^0)$), i.e. $\partial F(x^0) = \{(u_1, u_2, \cdots, u_n) : u_i \in E\ (i = 1, 2, \cdots, n)$ and there exists $\delta > 0$ such that $F(x) + \sum_{\substack{x_i < x_i^0 \\ i=1,2,\cdots,n}} |x_i - x_i^0| u_i \geq F(x^0) + \sum_{\substack{x_i \geq x_i^0 \\ i=1,2,\cdots,n}} (x_i - x_i^0) u_i$ for any $x \in U(x^0, \delta) \cap M\}$.

The following Theorem 3.10 explains the relation between the directional a.e cut-derivatives and the subgradients of the convex fuzzy mappings.

Theorem 3.10. Let K be a convex subset of R^n, $F : K \to E$ be a convex fuzzy valued function, $x = (x_1, x_2, \cdots, x_n) \in K$.

(1) If for $y = (y_1, y_2, \cdots y_n) \in R^n$, there exists $\delta > 0$ such that $x + hy \in K$, and the H-difference $F(x + hy) - F(x)$ exists, then $(u_1, u_2, \cdots, u_n) \in \partial F(x)$ implies $F_+^*(x, y) + \sum_{\substack{y_i < 0 \\ i=1,2,\cdots,n}} |y_i| u_i \geq \sum_{\substack{y_i \geq 0 \\ i=1,2,\cdots,n}} y_i u_i$.

(2) Let $u_1, u_2, \cdots, u_n \in E^1$. If there exists $\delta > 0$ such that for any $y \in U(0, \delta) \cap (K - x)$ (where $K - x = \{x - x : x \in K\}$), $F_+^*(x, y) + \sum_{\substack{y_i < 0 \\ i=1,2,\cdots,n}} |y_i| u_i \geq \sum_{\substack{y_i \geq 0 \\ i=1,2,\cdots,n}} y_i u_i$, then $(u_1, u_2, \cdots, u_n) \in \partial F(x)$.

(3) If F is differentiable at $x (x \in \text{ int } K)$, and $F'_{x_i}(x)$ $(i = 1, 2, \cdots, n)$ is a crisp fuzzy number (i.e. there exists a $A \in R^1$ such that $F'_{x_i}(x)_{(t)} = \begin{cases} 1\ if\ t = A \\ 0\ if\ t \neq A \end{cases}$), and there exists $\delta_0, \delta_1 > 0$ such that for any $h \in (0, \delta_1)$, $y \in U(0, \delta_0) \cap (K - x)$, the H-difference $F(x + hy) - F(x)$ exists, then $\partial F(x) = \{\nabla F(x)\} = \{(F'_{x_1}(x), F'_{x_2}(x), \cdots, F'_{x_n}(x))\}$.

In what follows, we give the problem of convex fuzzy programming.

Let K be a convex subset of R^n, $F : K \to E$ be a convex fuzzy valued function. The following problem

$$\begin{cases} \min F(x) \\ x \in K \end{cases} \qquad \text{(FCP)}$$

is called a convex fuzzy programming. If there exists $x^0 \in K$ such that $F(x) \geq F(x^0)$ for any $x \in K$, then x^0 is called a global minimum solution of (FCP) in K. If there exists $x^0 \in K$ and $\delta > 0$ such that $F(x) \geq F(x^0)$ for any $x \in U(x^0, \delta) \cap K$, then x^0 is called a local minimum solution of (FCP).

Remark 3.11. We can show that local minimum solution and global minimum solution are equivalent, so we will not distinguish between local minimum solution and global minimum solution, and call it minimum solution.

The following Theorem 3.12 characterizes the existence of minimum solutions of (FCP) with the conditions of subgradient.

Theorem 3.12. Let K be a convex subset of R^n, $F : K \to E$ be a convex fuzzy valued function. Then $x^0 = (x_1^0, x_2^0, \cdots, x_n^0)$ is a minimum solution of (FCP) if and only if there exists a subgradient $(u_1, u_2, \cdots, u_n)$ of F at x^0 such that $\sum\limits_{\substack{x_i \geq x_i^0 \\ i=1,2,\cdots,n}} (x_i - x_i^0)u_i \geq \sum\limits_{\substack{x_i < x_i^0 \\ i=1,2,\cdots,n}} |x_i - x_i^0|u_i$ for any $x \in K$.

The following Theorem 3.13 indicates the relation between minimum solutions of (FCP) and gradients.

Theorem 3.13. Let K be a convex subset of R, $F : K \to E$ be a convex fuzzy valued function, $x^0 \in$ int K.

(1) If F is differentiable at x^0, and $\bigtriangledown F(x^0) = (\overbrace{\hat{0}, \hat{0}, \cdots, \hat{0}}^{n})$, and there exists $\delta_0, \delta_1 > 0$ such that the H-difference $F(x^0 + hy) - F(x^0)$ exists for any $h \in (0, \delta_1)$, $y \in U(0, \delta_0) \cap (K - x^0)$, then x^0 is a minimum solution of (FCP).

(2) If F is differentiable at x^0, and $F'_{x_i}(x^0)$ $(i = 1, 2, \cdots, n)$ is a crisp fuzzy number, and there exists $\delta_0, \delta_1 > 0$ such that for any $h \in (0, \delta_1)$, $y \in U(0, \delta_0) \cap (K - x^0)$, the H-difference $F(x^0 + hy) - F(x^0)$ exists, then x^0 being a minimum solution of (FCP) implies $\bigtriangledown F(x^0) = (\overbrace{\hat{0}, \hat{0}, \cdots, \hat{0}}^{n})$.

Example 3.14. If $F : R^n \to E^1$ is defined by the following expression

$$F(x)(t) = \begin{cases} 1 + (t - \|x\|^2) \; if & t \in [\|x\|^2 - 1, \|x\|^2] \\ 1 - (t - \|x\|^2) \; if & t \in [\|x\|^2, \|x\|^2 + 1] \\ \quad 0 & if \; t \notin [\|x\|^2 - 1, \|x\|^2 + 1] \end{cases}$$
for any $x = (x_1, x_2, \cdots, x_n) \in R^n$,

i.e. it is determined by $[F(x)]^r = [\|x\|^2 - 1 + r, \|x\|^2 + 1 - r]$ for any $r \in [0, 1]$, then F is a convex fuzzy mapping from R^n into E^1, and is differentiable at $x = (x_1, x_2, \cdots, x_n) \in R^n$, $\bigtriangledown F(x) = (2\hat{x}_1, 2\hat{x}_1, \cdots, 2\hat{x}_1)$. From Theorem 3.12, we know that $x = 0 = (\overbrace{0, 0, \cdots, 0}^{n})$ is a minimum solution of (FCP).

References

1. S. S. L. Chang, L. A. Zadeh: On fuzzy mapping and control, IEEE Trans. Systems Man Cyberet., **2** (1), 30-34 (1972).
2. D. Dubois and H. Prade: Towards fuzzy differential calculus - Part 1, 2, 3, Fuzzy Sets and Systems **8**, 1-17, 105-116, 225-234(1982).
3. P. Diamond, P. Kloeden: Metric Space of Fuzzy Sets, World Scientific, Singapore, 1994.
4. O. Kaleva: Fuzzy differential equations, Fuzzy Sets and Systems **24**, 301-317(1987).
5. M. L. Puri and D. A. Ralescu: Differentials for fuzzy functions, J. Math. Anal. Appl. **91**, 552-558(1983).
6. S. Nanda, K. Kar: Convex fuzzy mappings, Fuzzy Sets and Systems **48**, 129-132(1992).
7. R. T. Rockafellar: Convex analysis, Princeton University Press, Princeton, New Jersey, 1970.
8. D. Sarkar:Concavoconvex fuzzy set: Fuzzy Sets and Systems **79**, 267-269(1996).
9. Yu-Ru Syau: On convex and concave fuzzy mappings, Fuzzy Sets and Systems **103**, 163-168(1999).
10. Wu Congxin, Ma Ming: On embedding problem of fuzzy number space: Part 1, 2, Fuzzy Sets and Systems **44**, 33-38(1991), **45**, 189-202(1992).
11. Wu Congxin, Wu Cong: The supremum and infimum of the set of fuzzy numbers and its application, J. Math. Anal. Appl. **210**, 499-511(1997).
12. D. Butnariu: Methods of solving optimization problems and linear equations in the space of fuzzy vectors, Libertas Mathematica, **17**, 1-7(1997).
13. J.J. Buckley, T. Feuring, Introduction to fuzzy partial differential equations, Fuzzy Sets and Systems **105**, 241-248(1999).

A Topology for Fuzzy Automata*

Arun K. Srivastava and S.P. Tiwari

Dept. of Mathematics
Banaras Hindu University
Varanasi, India
aks@banaras.ernet.in

Abstract. This work is an introductory investigation, on the lines of [9] and [10], into some topological aspects of fuzzy machines (studied in [4–8]), wherein we introduce a topology on the state-set of a fuzzy automaton and use it, together with some standard topological results, to deduce some fuzzy automata theoretic results given in [4–8].

1 Introduction and Preliminaries

In [9] and [10], it was demonstrated that several standard topological concepts and ideas can often be used in automata theory to obtain certain results therein, pertaining particularly to their connectivity and separation properties. We initiate here a similar programme for fuzzy automata (machines) studied by Malik, Mordeson, Sen and Nair [4–8]. For lack of space, not all results could be accommodated here and some proofs are sketchy, or even absent altogether.

The following concept of fuzzy automata resembles the concept of fuzzy machines, as given, e.g., in [4].

Definition 1. *A* **fuzzy automaton** *is a triple $M = (Q, X, \delta)$, where Q is a set (of* **states** *of M), X is a monoid (the* **input monoid** *of M), whose identity shall be denoted as e, and δ is a fuzzy subset of $Q \times X \times Q$, i.e., a map $\delta : Q \times X \times Q \to [0,1]$, such that $\forall q, p \in Q, \forall x, y \in X$,*

1. *$\delta(q, e, p) = 1$ or 0, according as $q = p$ or $q \neq p$,*
2. *$\delta(q, xy, p) = \vee\{\delta(q, x, r) \wedge \delta(r, y, p) : r \in Q\}$.*

A **map** from a fuzzy automaton $M = (Q, X, \delta)$ to a fuzzy automaton $N = (R, Y, \lambda)$ is a pair (f, g) where $f : Q \to R$ is a function and $g : X \to Y$ is a homomorphism such that

$$\forall (q, x, p) \in Q \times X \times Q, \quad \lambda(f(q), g(x), f(p)) \geq \delta(q, x, p).$$

The class of all fuzzy automata and their maps obviously form a category, say **FuA** (under obvious composition of maps). We shall denote the class of all fuzzy automata also by **FuA**.

* The authors acknowledge with thanks the support received through a research grant, provided by the Department of Science and Technology, New Delhi, under which this work has been carried out.

N.R. Pal and M. Sugeno (Eds.): AFSS 2002, LNAI 2275, pp. 485–491, 2002.

Definition 2. *([7]) Given $(Q, X, \delta) \in$ **FuA** and $A \subseteq Q$, the* **source** *and the* **successor** *of A, are respectively the sets*

$$\sigma_{Q}(A) = \{q \in Q : \delta(q, x, p) > 0, \text{ for some } (x, p) \in X \times A\}, \text{ and}$$
$$s_{Q}(A) = \{p \in Q : \delta(q, x, p) > 0, \text{ for some } (x, q) \in X \times A\}.$$

We shall frequently write $\sigma_{Q}(A)$ and $s_{Q}(A)$ as just $\sigma(A)$ and $s(A)$, and $\sigma(\{q\})$ and $s(\{q\})$ as just $\sigma(q)$ and $s(q)$.

Definition 3. *$(R, X, \lambda) \in$ **FuA** is called a* **subautomaton** *of $(Q, X, \delta) \in$ **FuA** if $R \subseteq Q$, $s_{Q}(R) = R$, and $\delta_{/R\times X\times R} = \lambda$. Further, this subautomaton is called* **separated** *if $s_{Q}(Q - R) \cap R = \phi$.*

Definition 4. *$M = (Q, X, \delta) \in$ **FuA** is called*

1. **strongly connected** *if, $\forall q, p \in Q, q \in s(p)$,*
2. **connected** *if M has no separated proper subautomaton,* and
3. **retrievable** *if, $\forall q \in Q, \forall x \in X, \delta(q, x, p) > 0$, for some $p \in Q \Rightarrow \delta(p, y, q) > 0$, for some $y \in X$.*

2 Topologies on State-Sets

Proposition 1. *Let $(Q, X, \delta) \in$ **FuA**. Then*

(a) The source and the successor, viewed as functions $\sigma, s : 2^{Q} \to 2^{Q}$, are Kuratowski closure operators on Q, inducing two topologies, say $T(Q)$ and $T^{*}(Q)$ respectively, on Q.
(b) Both $T(Q)$ and $T^{*}(Q)$ are 'saturated', i.e., closed under arbitrary intersections also.
(c) Each $A \subseteq Q$ is $T(Q)$-open iff A is $T^{*}(Q)$-closed.

Proof: The proof can be given on the same lines as for the corresponding results for 'crisp' automata in [9] and so we leave it to the reader.

Remark 1. By sending $M = (Q, X, \delta) \in$ **FuA** to $(Q, T(Q))$ and map (f, g) from M to $N = (R, Y, \lambda) \in$ **FuA** to $f : (Q, T(Q)) \to (R, T(R))$, we get a faithful functor $s :$ **FuA** $\to$ **TOP**, from **FuA** to the category **TOP** of topological spaces and continuous functions.

The next proposition is easy to establish.

Proposition 2. *Let $M = (Q, X, \delta)$ and $N = (R, Y, \lambda)$ be fuzzy automata. Then*

(a) N is subautomaton of M iff R is $T(Q)$-open.
(b) N is a separated subautomaton of M iff R is $T(Q)$-clopen (i.e., $T(Q)$-open as well as $T(Q)$-closed).

Before stating the next proposition, we first recall that a topological space X is called R_0 if $\forall x, y \in X, x \in cl(y) \Rightarrow y \in cl(x)$; cf., e.g., [3].

Proposition 3. *Let $M = (Q, X, \delta)$ and $N = (R, X, \lambda)$ be fuzzy automata. Then*

(a) M is strongly connected iff $T(Q)$ is an indiscrete topology.
(b) M is connected iff $T(Q)$ is connected topology.
(c) M is retrievable iff $T(Q)$ is an R_0-topology.

Proof: (a) and (b) are proved easily. For (c), just note that $\forall p, q \in Q, [q \in \sigma(p) \Rightarrow p \in \sigma(q)] \Rightarrow [\delta(q, y, p) > 0$, for some $y \in X \Rightarrow \delta(p, x, q) > 0$, for some $x \in X]$.

A number of results proved earlier, e.g., in [6] and [8], turn out to be easy translations of some well known results in topology concerning the topological concepts involved in Propositions 2 and 3 above. The next proposition illustrates this (other similar illustrations pertain to, e.g., Theorems 35, 41, and 43 in [6] and Theorems 24 and 25 in [8].

Proposition 4. *Let $M = (Q, X, \delta) \in$ **FuA**. Then*

(a) M is connected iff $\forall$ proper subautomaton $N = (R, X, \delta)$ of $M, \exists r \in R$ and $q \in Q - R$, with $s_Q(r) \cap s_Q(q) \neq \phi$.
(b) M is retrievable iff $\forall$ subautomaton $N = (R, X, \delta)$ of $M, \sigma_Q(R) = R$.
(c) M is strongly connected iff M has no proper subautomaton.

An examination of the 'categorical product' in the category **FuA** leads to a concept of 'product' of fuzzy automata, which we illustrate here for two fuzzy automata $M = (Q, X, \delta)$ and $N = (R, Y, \lambda)$ as follows.

Define $\quad \nu : (Q \times R) \times (X \times Y) \times (Q \times R) \rightarrow [0, 1]$
as $\quad \nu((q, r), (x, y), (q', r')) = \delta(q, x, q') \wedge \lambda(r, y, r')$.

Then it can be verified that $(Q \times R, X \times Y, \nu)$ is a fuzzy automaton. We shall denote it as $M \times N$ and refer to it as the **product** of the fuzzy automata M and N. It can be further verified that $M \times N$ does turn out to be the categorical product of M, $N \in$ **FuA**. (This product may be interpreted as the 'parallel composition' of M and N)

Proposition 5. *Let (Q, X, δ) and (R, Y, λ) be two fuzzy automata. Then $T(Q \times R)$ is the product topology $T(Q) \times T(R)$ of the topologies $T(Q)$ and $T(R)$.*

Proof: We leave the proof to the reader.

An immediate consequence of above proposition is

Proposition 6. *The functor s :* **FuA** $\rightarrow$ **TOP** *preserves categorical products.*

As the topological properties of connectedness, indiscreteness, and R_0-ness are known to be closed under topological products, we easily get the following

Proposition 7. *The product of connected (resp. strongly connected, resp. retrievable) fuzzy automata are connected (resp. strongly connected, resp. retrievable).*

3 Core Operator

Throughout this section, Q denotes the state-set of a fuzzy automaton $M = (Q, X, \delta)$. Following Bavel [2], we now introduce another operator on Q, referred to as the **core operator**.

Definition 5. *The* **core** *of any subset R of the state-set Q of a fuzzy automaton is the set*

$$\mu(R) = \{q \in R : \sigma(q) \subseteq R\}.$$

We shall frequently write $\mu\{q\}$ as just $\mu(q)$.

Proposition 8. *The map $\mu : 2^{Q} \to 2^{Q}$, which sends each $R \in 2^{Q}$ to $\mu(R)$, is an interior operator on Q, inducing the topology $T^*(Q)$ on Q.*

Proof: The proof can be given on the same lines as for the corresponding results for 'crisp' automata in [9].

4 Primaries

The concept of a primary of a fuzzy automaton $M = (Q, X, \delta)$ has been introduced in [5]. We show that it has a nice topological interpretation through the concept of a *regular closed set* in topology (a subset of a topological space is called **regular closed** if it equals the closure of its interior, cf., [11]). For the remaining part of the paper, we shall assume that Q is equipped with the topology $T^*(Q)$.

Definition 6. *A subset $R \subseteq Q$ is called*

1. **genetically closed** *if $\exists P \subseteq R$ such that $\sigma(P) \subseteq s(P)$ and $s(P) = R$* and
2. *a* **primary** *of Q if R is a minimal genetically closed subset of Q.*

Definition 7. *A* **primary** *of a fuzzy automaton $M = (Q, X, \delta)$ is a subautomaton of M whose state-set is a primary of Q.*

Proposition 9. *$R \subseteq Q$ is genetically closed iff R is a regular closed subset of Q.*

Proof: Let R be genetically closed. Then $\exists P \subseteq R$ such that $\sigma(P) \subseteq s(P) = R$. We note that $q \in P \Rightarrow \sigma(q) \subseteq R$ (as $\sigma(P) \subseteq R$), whereby $q \in \mu(R)$). Thus $P \subseteq \mu(R)$, whereby $s(P) \subseteq s(\mu(R))$, or that $R \subseteq s(\mu(R))$. Hence R is a regular closed subset of Q.

Conversely, let R be regular closed subset of Q, so that $s(\mu(R)) = R$. Put $P = \mu(R)$, then $P \subseteq R$. Also, $q \in \sigma(P) = \sigma(\mu(R) \Rightarrow \delta(q, x, p) > 0$, for some $(r, p) \in X \times \mu(R) \Rightarrow q \in \sigma(p)$, for some $p \in \mu(R)$. But $p \in \mu(R) \Rightarrow \sigma(p) \subseteq R$. Thus, $q \in R = s(\mu(R)) = s(P)$. This shows that $\sigma(P) \subseteq s(P) = R$ and so R is genetically closed.

Proposition 10. *A state-set of a primary of a fuzzy automaton (Q, X, δ) is a minimal regular closed subset of Q.*

Proposition 11. *For every $R \subseteq Q, s(\sigma(R))$ is a genetically closed subset of Q.*

Proof: The proof follows from Proposition 9 and the definition of regular closed sets.

Proposition 12. *For each $q \in Q$,*

(a) $s(\mu(q))$ is a primary of Q and
(b) $s(q)$ is a primary of Q iff $q \in \mu(s(q))$.

Proof: We omit the proof once again and refer to [2] and [10] for guidelines.

5 Compact Automata and Their Primary Decomposition

A well-known result about primaries of finite automata is the so called *"Primary Decomposition Theorem"*. Its counterpart for fuzzy automata, having *finite* state-sets, has already been proved in [5]. It turns out that this theorem need not hold for arbitrary fuzzy automata having *infinite* state-sets. However, we show that a counterpart of this result holds for fuzzy automata, i.e., having even *infinite* state-sets, provided their state-set topologies are *compact.* This leads us to the following definition.

Definition 8. *A fuzzy automaton (Q, X, δ) is called* **compact** *if the topology $T(Q)$ is compact.*

Given a saturated topological space (X, T), sending each $R \subseteq X$ to $\hat{R}$, where $\hat{R} = \cap\{U \in T : R \subseteq U\}$, defines a closure operator on X, giving rise to another topology on X, say T^* (cf. Lorrain [3]). It turns out that

Theorem 1. *(Lorrain [3]) A subset R of saturated topological space (X, T) is compact iff there exists a finite subset of X, which is dense in X with respect to the topology T^*.*

The following is now immediate.

Proposition 13. *A fuzzy automaton (Q, X, δ) is compact iff there exists a finite subset Q' of Q such that $s(Q') = Q$ or, equivalently, iff $\sigma(Q)$ is finite.*

Lemma 1. *Let $N = (R, X, \delta)$ be a primary of $M = (Q, X, \delta) \in$ **FuA**. Then $p \in \mu(R) \Rightarrow s(\sigma(p)) = R$.*

Lemma 2. *Let (R, X, δ) be a nonempty primary of a fuzzy automaton (Q, X, δ). Then for every finite subset T of R, $\exists r \in R$ such that $T \subseteq s(r)$.*

Proof: This can be proved, e.g., by induction on the number of elements in T.

Proposition 14. *A nonempty primary of a compact fuzzy automaton is a maximal 'singly generated' subautomaton (i.e., having state-set of the form $s(q)$ for some state q).*

Proof: Let (R, X, δ) be a nonempty primary of a compact fuzzy automaton (Q, X, δ). Let $p \in R$. Then $p \in s(\mu(R))$. So $\exists q \in \mu(R)$ such that $\delta(q, x, p) > 0$, for some $x \in X$. Clearly, $\sigma(q) \subseteq R$. Now $\sigma(q)$ is finite, owing to the compactness of (Q, X, δ). So $\exists q' \in R$ such that $\sigma(q) \subseteq s(q')$ (by Lemma 2). Hence $s(\sigma(q)) \subseteq s(q')$. Also, $q \in s(q') \Rightarrow q' \in \sigma(q) \Rightarrow s(q') \subseteq s(\sigma(q'))$. Thus, $s(\sigma(q')) = s(q')$, whereby $s(q')$ is a genetically closed subset of R (as $q' \in R$). So, by the minimality of R, $s(q') = R$. Hence the primary (R, X, δ) is singly generated. So let $R = s(m)$. Also, let $S = s(t)$ be a singly generated subset of Q such that $R = s(m) \subseteq S$. To prove that R is maximal, it is enough to show that $t \in s(m)$. Now $s(m) \subseteq S = s(t) \Rightarrow m \in s(t)$, so that $t \in \sigma(m) \subseteq s(m)$ (as $s(m)$ is a primary (Proposition 12(b))).

Proposition 15. *A compact fuzzy automaton $M = (Q, X, \delta)$ has only a finite number of distinct primaries.*

Proof: As (Q, X, δ) is compact, $\exists$ a finite subset $Q' = \{q_1, q_2, ..., q_k\}$ of Q such that $s(Q') = Q$. It can now be shown that each primary is of the form $s(q_i)$ for some $q_i \in Q'$.

We close with the following theorem which is not dificult to prove.

Theorem 2. *(Primary Decomposition Theorem) Let $P_1, P_2, ..., P_n$ be all the disinct primaries of a compact fuzzy automaton $M = (Q, X, \delta)$. Then*

(1) $M = \cup_{i=1}^{n} P_i$ and
(2) for any $j, 1 \leq j \leq n, M \neq \cup_{i=1, i\neq j}^{n} P_i$.

References

1. M.A. Arbib and E.G. Manes (1975) *Arrows, Structures, and Functors: The Categorical Imperative*, Academic Press, New York.
2. Z. Bavel (1967) On the decomposibility of monadic algebras and automata, *Proceedings of the 8th Annual Symp. on Switching and Automata Theory*, 322–335.
3. F. Lorrain (1969) Notes on topological spaces with minimum neighbourhoods, *Amer. Math. Monthly*, **76**: 616–627.
4. D.S. Malik, J.N. Mordeson, and M.K. Sen (1994) On subsystems of fuzzy finite state machines, *Fuzzy Sets and Systems*, **68**: 83–92.
5. D.S. Malik, J.N. Mordeson, and M.K. Sen (1994) Submachines of fuzzy finite state machines, *Jour. of Fuzzy Maths.*, **2**: 781–792.
6. D.S. Malik, J.N. Mordeson, and M.K. Sen (1995) The cartesian composition of fuzzy finite state machines, *Kybernetics*, **24**: 98–110.
7. J.N. Mordeson and P. Nair (1996) Successor and source of (fuzzy) finite state machines and (fuzzy) directed graphs, *Inform. Sc.*, **95**: 113–124.

Fuzzy Number Linear Programming: A Probabilistic Approach

Hamid Reza Maleki[1,*], Hassan Mishmast N.[2], and Mashaallah Mashinchi[1]

[1] Faculty of Math. and Computer Sciences, Kerman University, Kerman, Iran
{maleki,Mash}@arg3.uk.ac.ir
[2] Department of Mathematics,
Sistan and Baluchestan university, Zahedan, Iran
hmnehi@hamoon.usb.ac.ir

Abstract. In real world there are many problems which have linear programming models where all decision parameters are fuzzy numbers. There are some approaches which are using different ranking functions for solving these problems. Unfortunately all these methods when there exist alternative optimal solutions, usually with different fuzzy value of the objective function for these solutions, can not specify a clear approach for choosing a solution. In this paper using the concept of expectation and variance as ranking functions, we propose a method to remove the above shortcomings in solving fuzzy number linear programming problems.

1 Introduction

Since the pioneering work on the concept of decision making in fuzzy environments by Bellman and Zadeh [2], to solve fuzzy linear programming problems, several approaches proposed by different authors [9]. Some of them used the concept of comparison of fuzzy numbers for solving fuzzy number linear programming problem (FNLP). One of the most convenient of these methods is based on the concept of comparison of fuzzy numbers by using ranking functions [5,7]. Usually, in such methods authors define a crisp model which is equivalent with FNLP and then use optimal solution of this model as the optimal solution of FNLP. When crisp model has alternative optimal solutions authors concluded that fuzzy problem also has alternative optimal solutions. But unfortunately, fuzzy value of the objective function for these solutions are not necessarily the same as functions and it is not evident which optimal solution should be choosen. In this paper, at first we associate a probability density function to each fuzzy number. Then using the expectation of these density functions as a ranking function, we introduce a method for solving FNLP. Finally if FNLP has alternative optimal solutions, in order to choose a solution we use the variance of the density functions associated to the fuzzy values of the objective function corresponding to these solutions to find solution which has minimum variance.

* Corresponding author: Fax:++98-341-263244

N.R. Pal and M. Sugeno (Eds.): AFSS 2002, LNAI 2275, pp. 491–496, 2002.

2 Ranking Function

Many ranking methods can be found in fuzzy literature. Bass and Kwakernaak [1] are among the pioneers in this area. Bortolan and Degani [3] have reviewed different ranking methods. According to Chen and Hwang [4] the methods are categorized into four different groups one of which is a probabilistic approach. To examine the methods of each group see [4]. In spite of the existence of a variety of methods, no one can rank fuzzy numbers satisfactorily in all cases and situations [6]. This is our motivation to use the probabilistic approach to be able to rank the fuzzy numbers for identifying the proper optimal solutions of a fuzzy number linear programming problem discussed in the next section.

Now let $\tilde{a}$ be a fuzzy number, i.e. a convex normalized fuzzy subset of the real line whose membership function is piecewise continuous. In this paper we denote the set of all fuzzy numbers by $F(\mathbb{R})$. There are two important topics in the real world applications of the fuzzy set theory: arithmetic operations on fuzzy numbers and comparison of fuzzy numbers which usually follow arithmetic operations. Thanks to extension principle we have no problem with the first topic. However there is no common approach for the comparison of fuzzy numbers. Indeed there are different approaches for ranking fuzzy numbers. A simple but efficient approach for ordering of the elements of $F(\mathbb{R})$ is to define of a ranking function $R : F(\mathbb{R}) \to \mathbb{R}$ which maps each fuzzy number into the real line, where a natural order exists, and defining order on $F(\mathbb{R})$ by

- $\tilde{a} \underset{R}{\geq} \tilde{b}$ iff $R(\tilde{a}) \geq R(\tilde{b})$,
- $\tilde{a} \underset{R}{>} \tilde{b}$ iff $R(\tilde{a}) > R(\tilde{b})$,
- $\tilde{a} \underset{R}{=} \tilde{b}$ iff $R(\tilde{a}) = R(\tilde{b})$,

where $\tilde{a}$ and $\tilde{b}$ belong to $F(\mathbb{R})$. Also we write $\tilde{a} \underset{R}{\leq} \tilde{b}$ if and only if $\tilde{b} \underset{R}{\geq} \tilde{a}$.

Several ranking functions have been proposed by researchers to suit their requirements of the problems under consideration. Some examples are in [3]. In the sequel, by the aid of membership function of a fuzzy number $\tilde{a}$, we associate a probability density function to $\tilde{a}$. Then we use the expectation and variance of this density function as the value of ranking function.

One way for converting a membership function into a density function is by the help of a linear transformation $f_{\tilde{a}}(x) = c\,\tilde{a}(x)$, where c is a proportional constant satisfying $\int f_{\tilde{a}}(x)dx = 1$. For example, if $\tilde{a} = (a^L, a^U, \alpha, \beta)$ is a trapezoidal fuzzy number, then $c = \frac{2}{2(a^U - a^L) + \alpha + \beta}$. See [7].

Since we have a density function associated with each fuzzy number, we can use methods of probability theory to determine the expectation and variance of this density function. From now on for simplicity, we use the terms expected value and variance of $\tilde{a}$ instead of expected value and variance of the density function associated with $\tilde{a}$. In following we use Mellin transform to find the expectation and variance [8].

Definition 2.1. The Mellin transform $M_X(s)$ of a probability density function (pdf) $f(x)$, where x is positive, is defined as

$$M_X(s) = E(X^{s-1}) = \int_0^{+\infty} x^{s-1} f(x) dx,$$

whenever the integral exists.

Hence $E(X^s) = M_X(s+1)$. Thus, the expectation and variance of a random variable X are $E(X) = M_X(2)$ and $Var(X) = M_X(3) - (M_X(2))^2$.

Remark 2.1. The table of Mellin transforms related to some of fuzzy numbers are given in table (2) of [8].

For example if $\tilde{a} = (a^L, a^U, \alpha, \beta)$ is a trapizoidal fuzzy number such that $a^L - \alpha > 0$, then

$$E(\tilde{a}) = \frac{1}{3}\left[2(a^L + a^U) + (\beta - \alpha) + \frac{a^L(a^L - \alpha) - a^U(a^U + \beta)}{2(a^U - a^L) + (\beta + \alpha)}\right],$$

$$Var(\tilde{a}) = \frac{1}{6(2(a^U - A^L) + \beta + \alpha)}\left[\frac{(a^U + \beta)^4 - (a^U)^4}{\beta} - \frac{(a^L)^4 - (a^L - \alpha)^4}{\alpha}\right] - [E(\tilde{a})]^2 .$$

Note that expectation ranking function is a linear ranking function, if we take $R(\tilde{a}) = E(\tilde{a})$. See [7].

3 Fuzzy Number Linear Programming Problem

In this section we define fuzzy number linear programming problems and propose a method for solving them.

Definition 3.1. The model

$$\max : \tilde{z} = \tilde{\mathbf{c}}\mathbf{x} , \qquad \text{s.t.} \qquad \tilde{\mathbf{A}}\mathbf{x} \underset{R}{\leq} \tilde{\mathbf{b}} , \qquad \mathbf{x} \geq \mathbf{0} , \tag{1}$$

where $\tilde{\mathbf{A}} = (\tilde{a}_{ij})_{m\times n}$, $\tilde{\mathbf{c}} = (\tilde{c}_1, \cdots, \tilde{c}_n)$, $\tilde{\mathbf{b}} = (\tilde{b}_1, \cdots, \tilde{b}_m)'$ and $\tilde{a}_{ij}$, $\tilde{b}_i$ and $\tilde{c}_j \in F(\mathbb{R})$ for $i = 1, 2, \cdots, m$, $j = 1, 2, \cdots, n$, is called a *fuzzy number linear programming problem* (FNLP) [5,7].

Definition 3.2. Any $\mathbf{x}$ which satisfies the set of constraints of FNLP is called a feasible solution for FNLP. Let Q be the set of all feasible solutions of FNLP. We shall say that $\mathbf{x}^o \in Q$ is an optimal feasible solution for FNLP if $\tilde{\mathbf{c}}\mathbf{x} \underset{R}{\leq} \tilde{\mathbf{c}}\mathbf{x}^o$ for all $\mathbf{x} \in Q$.

The following theorem which is an extension of Lemma 3.1 in [5] shows that we can reduce any FNLP to a linear programming problem in the classical form.

Theorem 3.1. The following problem and FNLP are equivalent:

$$\max : z = \mathbf{cx} \ , \qquad \text{s.t.} \qquad \mathbf{Ax} \leq \mathbf{b} \ , \qquad \mathbf{x} \geq \mathbf{0} \ , \tag{2}$$

where $\mathbf{A} = (a_{ij})_{m \times n}$, $\mathbf{c} = (c_1, \cdots, c_n)$, $\mathbf{b} = (b_1, \cdots, b_m)'$ and a_{ij}, b_i, c_j are real numbers corresponding to the fuzzy numbers $\tilde{a}_{ij}$, $\tilde{b}_i$, $\tilde{c}_j$, with respect to a given linear ranking function R ,respectively, i.e. $a_{ij} = R(\tilde{a}_{ij})$, $b_i = R(\tilde{b}_i)$ and $c_j = R(\tilde{c}_j)$.

Remark 3.1. It follows that if the model (2) does not have a solution, then FNLP also does not have a solution. If there are alternative optimal solutions for the model (2), then FNLP will also have alternative optimal solutions.

Example 3.1. Consider the following FNLP

$$\begin{aligned} \max : \tilde{z} = & \quad (1, 3, 1, 1)x_1 + (1.5, 2.5, 1, 1)x_2 \ , \\ \text{s.t.} & \\ & x_1 + x_2 \leq 8 \ , \quad 2x_1 + x_2 \leq 10 \ , \quad x_1, x_2 \geq 0, \end{aligned}$$

where $(a^L, a^U, \alpha, \beta)$ is a trapezoidal fuzzy number. We may apply the expectation ranking function for solving the above FNLP. Then the problem reduces to the following:

$$\begin{aligned} \max : z = & \quad 2x_1 + 2x_2 \ , \\ \text{s.t.} & \\ & x_1 + x_2 \leq 8 \ , \quad 2x_1 + x_2 \leq 10 \ , \quad x_1, x_2 \geq 0. \end{aligned}$$

Now solving the above problem, we see that $x_1 = 2$, $x_2 = 6$ is an optimal basic feasible solution with $\tilde{z}_1 = (11, 21, 8, 8)$, for FNLP. Moreover, $x_1 = 0$, $x_2 = 8$ is another optimal basic feasible solution with $\tilde{z}_2 = (12, 20, 8, 8)$. Although $R(\tilde{z}_1) = R(\tilde{z}_2) = 16$, But we can not say that $\tilde{z}_1(x) = \tilde{z}_2(x)$ for all $x \in \mathbb{R}$.

Remark 3.2. It is important to note that when the model (1) has alternative optimal solutions, then the value of ranking function for the fuzzy value of the objective function corresponding to all optimal solutions is the same. However the fuzzy value of the objective function from one optimal solution to another is not necessarily the same. Hence it may be asked if problem (1) has two optimal solutions with different fuzzy values for the objective function, which of these solutions must be preferred to the other?

In this regard consider the problem in the standard form:

$$\max : \tilde{z} = \tilde{\mathbf{c}}\mathbf{x} \ , \qquad \text{s.t.} \qquad \mathbf{Ax} = \mathbf{b} \ , \qquad \mathbf{x} \geq \mathbf{0} \ , \tag{3}$$

where $\mathbf{x} \in R^n$, $\mathbf{A}$ is a $m \times n$ matrix, $\tilde{\mathbf{c}}' \in (F(\mathbb{R}))^n$ and $\mathbf{b} \in R^m$. Then we have a fuzzy version of a classical theorem of the past literature:

Theorem 3.2. If $\mathbf{x}_1, \cdots, \mathbf{x}_k$ are k different extreme points of the set of feasible solutions of the model (3) which are optimal, then any convex combination of these solutions is an optimal solution for the model (3).

Proof. Let $\mathbf{x} = \sum_{i=1}^{k} \mu_i \mathbf{x}_i$ where $\mu_i \geq 0, \quad i = 1, \cdots, \quad k$ and $\sum_{i=1}^{k} \mu_i = 1$. It is clear that $\mathbf{x}$ is a feasible solution of the model (3). If $\tilde{\mathbf{c}}\mathbf{x}_i \underset{R}{=} \tilde{\mathbf{c}}\mathbf{x}_1$, $i = 2, \cdots, k$, then the fuzzy value of the objective function for $\mathbf{x}$ is:

$$\tilde{\mathbf{c}}\mathbf{x} \underset{R}{=} \sum_{i=1}^{k} \mu_i \tilde{\mathbf{c}}\mathbf{x}_i \underset{R}{=} \sum_{i=1}^{k} \mu_i \tilde{\mathbf{c}}\mathbf{x}_1 \underset{R}{=} \tilde{\mathbf{c}}\mathbf{x}_1 \sum_{i=1}^{k} \mu_i \underset{R}{=} \tilde{\mathbf{c}}\mathbf{x}_1.$$

Hence $\mathbf{x}$ is an optimal solution for the model (3).

4 Identifying Optimal Solution

Now we are ready to study the problem of alternative optimal solutions for model (3) raised in Remark 3.2. When the model (3) has alternative optimal solutions, then as mentioned in Remark 3.2, the value of ranking function for the fuzzy value of the objective function corresponding to all optimal solutions is equal. However from one optimal solution to another the fuzzy value of the objective function may not be equal. Subject to this condition, if we apply the expectation ranking function, it is natural to identify the optimal solution with least variance of the fuzzy value of the objective function. Especially, when we have $\mathbf{x}_1, \cdots, \mathbf{x}_k$ as k different optimal extreme points, then using Theorem 3.2 and the above idea we may solve the following model to determine optimal solution of (3) with least variance of the fuzzy value of the objective function:

$$\begin{aligned} \min : Var \left[\textstyle\sum_{j=1}^{k} \lambda_j \tilde{\mathbf{c}}\mathbf{x}_j\right] \ , \\ s.t. \qquad\qquad \\ \textstyle\sum_{j=1}^{k} \lambda_j = 1 \ , \\ \lambda_j \geq \mathbf{0} \ . \end{aligned} \tag{4}$$

Note that $\sum_{j=1}^{k} \lambda_j \tilde{\mathbf{c}}\mathbf{x}_j$ in (4) is a fuzzy number.

Example 4.1. In order to choose an optimal solution from those found in Example 3.1, we may solve the following problem:

$$\min : Var((11, 21, 8, 8)\lambda_1 + (12, 20, 8, 8)\lambda_2) \ ,$$

s.t.

$$\lambda_1 + \lambda_2 = 1 \quad , \quad \lambda_1, \lambda_2 \geq 0,$$

now solving the above problem, we find $\lambda_1 = 0$ and $\lambda_2 = 1$. Thus we can choose $\mathbf{x} = (0, 8)'$ as optimal solution of our problem.

References

1. Bass S. M. and Kwakernaak H.: Rating and Ranking of Multiple Aspent Alternative Using Fuzzy Sets. Automatica, **13** (1977), 47-58.
2. Bellman R. E. and Zadeh L. A.:Decision Making in a Fuzzy Environment. Management Sci., **17** (1970), 141-164.
3. Bortolan G. and Degani R.: A Review Of Some Methods For Ranking Fuzzy Numbers. Fuzzy Sets and Systems, **15** (1985), 1-19.
4. Chen S. J. and Hwang C. L.: Fuzzy Multiple Attribute Decision Making, Methods and Applications. Springer, Berline, (1992).
5. Maleki H. R., Tata M. and Mashinchi M.: Linear Programming With Fuzzy Variables. Fuzzy Sets and Systems, **109** (2000), 21-33.
6. Modarres M. and Sadi-Nezhad S.: Ranking Fuzzy Numbers by Preference Ratio. Fuzzy Sets and Systems, **118** (2001), 429-436.
7. Yazdani Peraei E., Maleki H.R. and Mashinchi M.: A Method For Solving A Fuzzy Linear Programming. Korean J. Comput. & Appl. Math., **8** (2001), No. 2, 347-356.
8. Yoon K.P.: A Probabilistic Approach To Rank Complex Fuzzy Numbers. Fuzzy Sets and Systems, **80** (1996) 167-176.
9. Zimmermann H. J.: Applications of Fuzzy Sets Theory to Mathematical Programming. Inform. Sci. **36** (1985), 29-58.

Composition of General Fuzzy Approximation Spaces

Jusheng Mi and Wenxiu Zhang

[1] Institute for Information and System Science, Faculty of Science,
Xi'an Jiaotong University, Xi'an, Shaan'xi, 710049, P.R. China
mijsh@263.net
[2] College of Mathematics and Information Science,
Hebei Normal University, Hebei, Shijiazhuang, 050016, P.R. China

Abstract. In this paper, the properties of general fuzzy approximation operators are studied, and the concept of composition of two general fuzzy approximation spaces in the rough set theory is given. The relationships between the approximation operators in the composite space and in the two fuzzy approximation spaces are discussed. It is proved that the approximation operators in the composite space are just the composition of the approximation operators in the two fuzzy approximation spaces.

1 Introduction

The theory of rough set [1,2] is motivated by practical needs in classification, data mining and concept formation with insufficient and incomplete information. The main idea of rough approximations is of finding a lower and upper bounds for a subset of the universe [3-7]. In real life, the decision and condition concepts are often fuzzy and can be described by fuzzy sets. There have been several attempts to integrate fuzzy sets with rough sets [5,8].

Many applications of the rough set theory only consider information obtained from unique source [1,4]. But when one deals with complex situations with several systems, especially in the case that the attribute values of one system are just the objects of another system, one should study the composition of these systems. How to compose two approximation spaces? What are the relationships between the approximation operators in the composite space and in the two fuzzy approximation spaces? In this paper, we generalize the concept of approximation operators, and give the properties of general fuzzy approximation operators. Moreover, we study the composition of fuzzy approximation spaces, and prove that the approximation operators in the composite space are just the composition of the approximation operators in the two fuzzy approximation spaces.

2 General Fuzzy Rough Set Models

Let U, V, W denote nonempty finite sets called universes. The set of all fuzzy subsets of U is denoted by $\mathcal{F}(U)$. If $R \in \mathcal{F}(U \times V)$, we then say that R is a

N.R. Pal and M. Sugeno (Eds.): AFSS 2002, LNAI 2275, pp. 497–501, 2002.

fuzzy relation from U to V. In particular, if $U = V$, we say that R is a fuzzy relation on U.

A binary operation T on the unit interval $I = [0, 1]$ is said to be a triangular norm [9], if $\forall a, b, c, d \in I$, we have

(1) $T(a, b) = T(b, a)$;
(2) $T(a, 1) = a$;
(3) $a \leq c, b \leq d \Rightarrow T(a, b) \leq T(c, d)$;
(4) $T(T(a, b), c) = T(a, T(b, c))$.

We say that a binary relation S on the unit interval I is the dual of a triangular norm T, if $\forall a, b \in I$, we have $S(a, b) = 1 - T(1 - a, 1 - b)$.

For every $R \in \mathcal{F}(U \times V)$, we can define a pair of fuzzy set-valued operators L_R and H_R from $\mathcal{F}(U)$ to $\mathcal{F}(U)$: $\forall X \in \mathcal{F}(V)$, $x \in U$,

$$L_R X(x) = \min_{y \in V} S(1 - R(x, y), X(y)),$$

$$H_R X(x) = \max_{y \in V} T(R(x, y), X(y)).$$

The pair $(L_R X, H_R X)$ is referred to as a general fuzzy rough set of X induced by R. And at the same time, L_R and H_R are referred to as general fuzzy lower and upper approximation operators respectively. The triple (U, V, R) is called a general fuzzy approximation space. When R is an ordinary binary relation, (U, V, R) is just the general approximation space [6]. It is easy to prove the following results:

Property 2.1. L_R and H_R are a pair of dual general fuzzy approximation operators, that is, $\forall X, Y \in \mathcal{F}(V)$,
(1) $L_R X = -H_R(-X); H_R(X) = -L_R(-X)$;
(2) $L_R V = U; H_R \emptyset = \emptyset$;
(3) $L_R(X \cap Y) = L_R X \cap L_R Y; H_R(X \cup Y) = H_R X \cup H_R Y$;
(4) If $X \subseteq Y$, then $L_R X \subseteq L_R Y; H_R X \subseteq H_R Y$.

When R is a fuzzy relation on U, we have the following results:

Property 2.2. If R is a reflexive fuzzy relation on U, i.e., $R(x, x) = 1$ for all $x \in U$, then $\forall X \in \mathcal{F}(V)$, we have that

(1) $L_R X \subseteq X$ and $H_R X \supseteq X$.
(2) If R is a T transitive fuzzy relation on U, that is, for $\forall x, y, z \in U$,$R(x, z) \geq T(R(x, y), R(y, z))$, then $\forall X \in \mathcal{F}(U)$, we have that

$$L_R X \subseteq L_R L_R X, \quad H_R H_R X \subseteq H_R X.$$

3 Composition of General Fuzzy Approximation Spaces

In this section, we study the composition of two general fuzzy approximation spaces.

Definition 3.1. Let $A = (U, V, R_1)$ and $B = (V, W, R_2)$ be two general fuzzy approximation spaces, the composition of A and B is defined by $A \otimes B = (U, W, R)$, where R is the composition of R_1 and R_2, that is, $\forall (x, z) \in U \times W$,

$$R(x, z) = (R_1 \circ R_2)(x, z).$$

The following theorem illustrates that the approximation operators in the composite space are just the composition of the approximation operators in the two approximation spaces.

Theorem 3.1. Let $A = (U, V, R_1)$ and $B = (V, W, R_2)$ be two general fuzzy approximation spaces, $A \otimes B = (U, W.R)$ the composition of A and B, then

$$(1) H_R = H_{R_1} \circ H_{R_2}; \quad (2) L_R = L_{R_1} \circ L_{R_2}.$$

Proof. (1) By virtue of the property of triangular norm T and the definition of upper approximation operator, we have that $\forall X \in \mathcal{F}(W)$, $x \in U$,

$$\begin{aligned}(H_{R_1} \circ H_{R_2})X(x) &= H_{R_1}(H_{R_2}X)(x) = \max_{y \in V} T(R_1(x, y), H_{R_2}X(y)) \\ &= \max_{y \in V} T(R_1(x, y), \max_{z \in W} T(R_2(y, z), X(z))) \\ &= \max_{y \in V} \max_{z \in W} T(R_1(x, y), T(R_2(y, z), X(z))) \\ &= \max_{z \in W} \max_{y \in V} T(T(R_1(x, y), R_2(y, z)), X(z)) \\ &= \max_{z \in W} T(\max_{y \in V} T(R_1(x, y), R_2(y, z)), X(z)) \\ &= \max_{z \in W} T((R_1 \circ R_2)(x, z), X(z)) \\ &= \max_{z \in W} T(R(x, z), X(z)) = H_R X(x)\end{aligned}$$

Consequently, we have $H_R = H_{R_1} \circ H_{R_2}$.

(2) For every $X \in \mathcal{F}(W)$, by Property 2.1 and the above result (1), we have

$$\begin{aligned}(L_{R_1} \circ L_{R_2})X &= L_{R_1}(L_{R_2}X) = -H_{R_1}(-L_{R_2}X) = -H_{R_1}(H_{R_2}(-X)) \\ &= -H_R(-X) = L_R(X).\end{aligned}$$

Hence, we have $L_R = L_{R_1} \circ L_{R_2}$.

Theorem 3.2. Let R_1 and R_2 be two reflexive fuzzy relations on U, $A = (U, R_1)$ and $B = (U, R_2)$, $A \otimes B = (U, R)$, where R is the composition of R_1 and R_2, then R is reflexive and for every $X \in \mathcal{F}(U)$, we have

$$L_R X \subseteq L_{R_i} X \subseteq X \subseteq H_{R_i} X \subseteq H_R X, \quad i = 1, 2.$$

Proof. For every $x \in U$, we have

$$\begin{aligned}R(x, x) &= (R_1 \circ R_2)(x, x) = \max_{y \in U} T(R_1(x, y), R_2(y, x)) \\ &\geq T(R_1(x, x), R_2(x, x)) = T(1, 1) = 1.\end{aligned}$$

Then $R(x, x) = 1$, and R is reflexive.

By Property 2.2 we get that

$$L_{R_i}X \subseteq X \subseteq H_{R_i}X, i = 1, 2.$$

By Theorem 3.1 and Property 2.2 we obtain that

$$L_R X \subseteq L_{R_1}(L_{R_2}X) \subseteq L_{R_i}X, i = 1, 2,$$
$$H_R X = H_{R_1}(H_{R_2}X) \supseteq H_{R_i}X, i = 1, 2.$$

Therefore, the proof of the theorem is complete.

The following example is presented to illustrate the basic ideas developed in Theorem 3.1.

Example 3.1. Let $U = \{1, 2, 3, 4\}$, $V = \{a, b, c\}$ and $W = \{Y, N\}$. If $R_1 \in \mathcal{F}(U \times V)$ and $R_2 \in \mathcal{F}(V \times W)$ are two fuzzy relations defined respectively by

$$R_1 = \begin{pmatrix} 0.3\ 0.7\ 0.2 \\ 1.0\ 0.0\ 0.4 \\ 0.0\ 0.5\ 1.0 \\ 0.6\ 0.7\ 0.8 \end{pmatrix}, R_2 = \begin{pmatrix} 0.1\ 0.9 \\ 0.9\ 0.1 \\ 0.6\ 0.4 \end{pmatrix}$$

then

$$R = R_1 \circ R_2 = \begin{pmatrix} 0.7\ 0.3 \\ 0.4\ 0.9 \\ 0.6\ 0.4 \\ 0.7\ 0.6 \end{pmatrix} \in \mathcal{F}(U \times W).$$

Suppose $A = 0.2/Y + 0.6/N \in \mathcal{F}(W)$, then by the definitions of general fuzzy lower and upper approximation operators defined in Section 2, we can easily compute

$$L_R A = 0.3/1 + 0.6/2 + 0.4/3 + 0.3/4;$$
$$L_{R_2} A = 0.6/a + 0.2/b + 0.4/c;$$
$$L_{R_1}(L_{R_2}A) = 0.3/1 + 0.6/2 + 0.4/3 + 0.3/4.$$

We can see that $L_R A = L_{R_1}(L_{R_2}A)$.

Similarly, we have that

$$H_R A = 0.3/1 + 0.6/2 + 0.4/3 + 0.6/4,$$
$$H_{R_1}(H_{R_2}A) = 0.3/1 + 0.6/2 + 0.4/3 + 0.6/4.$$

Which implies that $H_R A = H_{R_1}(H_{R_2}A)$.

References

1. Pawlak Z.: Rough sets. International Journal of Computer and Information Sciences **11** (1982) 205-218
2. Zhang Wenxiu, Wu Weizhi et al.: Rough Set Theory and Approac. Science Press, Beijing (2001)
3. Yao Y. Y. and Lin T. Y.: Generalization of rough sets using modal logic. Intelligent Automation and Soft Computing, an International Journal **2** (1996) 103-120
4. Zhang Wenxiu and Wu Weizhi: Rough set models based on random sets (I). Journal of Xi'an Jiaotong University **34** (2000) 75-79
5. Morsi N. N. and Yakout M. M.: Axioms for fuzzy rough sets. Fuzzy Sets and Systems, **100** (1998) 327-342
6. Yao Y. Y.: Constructive and algebraic methods of the theory of rough sets. Information Sciences **109** (1998) 21-47
7. Thiele H.: On axiomatic characterizations of crisp approximation operators. Information ciences **129** (2000) 221-226
8. Dubois D. and Prade H.: Rough fuzzy sets and fuzzy rough sets. International Journal of General Systems **17** (1990) 191-208
9. Zhang Wenxiu, Wang Guojun et al.: Introduction to Fuzzy Set Theory. Xi'an Jiaotong University Press, Xi'an (1991)

The Lower and Upper Approximations of Fuzzy Sets in a Fuzzy Group

Degang Chen[1,2] and Wenxiu Zhang[2]

[1] Department of Mathematics,Jinzhou Teacher's College,
Jinzhou, 121003, P.R. China
chengdegang@263.net.cn

[2] Institute for Information and System Science, Faculty of Science,
Xi'an Jiaotong University, Xi'an, Shaan'xi, 710049, P.R. China

Abstract. In this paper we define the lower and upper approximations of fuzzy sets in a group with respect to a fuzzy normal subgroup and study their product properties. We introduce the notion of a rough fuzzy subgroup of a group with respect to a fuzzy normal subgroup and study the relation between the fuzzy subgroup and the rough fuzzy subgroup. We also study the homomorphic properties of the rough fuzzy subgroups.

1 Introduction

The notion of rough sets is introduced by Pawlak in his paper [1] and is motivated by practical needs in classification, data mining and concept formation with in sufficient and incomplete information. Some authors have studied the fuzzy rough sets [2,3,4], and some authors have studied the algebra structure of rough sets with respect to fuzzy sets [6,7]. In this paper we study the lower and upper approximations of fuzzy sets in a group. We also define the rough fuzzy subgroup of a group with respect to a fuzzy normal subgroup and study its properties.

2 Preliminaries

Let T be a triangular norm. The following binary operation on $I = [0,1]$

$$\vartheta_T(\alpha,\gamma) = \sup\{\theta \in I : \alpha T \theta \leq \gamma\},\ \alpha,\gamma \in I$$

is called the residnation implication of T, or the T-residnated implication. If $T = \text{Min}$, then $\vartheta_{\text{Min}}(\alpha,\gamma) = \begin{cases} 1, \alpha \leq \gamma \\ \gamma, \alpha > \gamma \end{cases}$. The properties of ϑ_T are referred to [4].

Definition 2.1. Let X be a set. A fuzzy binary relation R on X is called a fuzzy equivalent relation if it satisfies for all $x,y,z \in X$ the following conditions:

1) $R(x,x) = 1$; 2) $R(x,y) = R(y,x)$; 3) $R(x,z) \wedge R(x,y) \leq R(z,y)$.

(X,R) is called a fuzzy approximation space.

N.R. Pal and M. Sugeno (Eds.): AFSS 2002, LNAI 2275, pp. 502–508, 2002.

Definition 2.2. [4] Let (X, R) be a fuzzy approximation space. Define an operator $\bar{A}_R$ on I^X by $(\bar{A}_R\mu)(x) = \sup_{u \in X} \min\{R(u,x), \mu(u)\}, \forall x \in X, \mu \in I^X$, then $\bar{A}_R$ is called the upper approximation operator of (X, R). Define an operator $\underline{A}_R$ on I^X by $(\underline{A}_R\mu)(x) = \wedge_{u \in X} \vartheta_{\text{Min}}(R(u,x), \mu(u)), \forall x \in X, \mu \in I^X$, then $\underline{A}_R$ is called the lower approximation operator of (X, R).

The properties of $\bar{A}_R$ and $\underline{A}_R$ are referred to [4].

Definition 2.3. [5] Let G be a group, $A \in I^G$. If A satisfies the following conditions:

1) $A(xy) \geq A(x) \wedge A(y)$; 2) $A(x^{-1}) \geq A(x)$; 3) $A(e) = 1$,

then A is called a fuzzy subgroup of G. If $A(xy) = A(yx), x, y \in G$, then A is called a fuzzy normal subgroup of G.

Definition 2.4. [5] Let G be a group, $A, B \in I^G$ then $A \circ B$ is defined as

$$A \circ B(x) = \sup_{y \in G} A(xy^{-1}) \wedge B(y), \ x \in G.$$

3 The Lower and Upper Approximations of Fuzzy Sets with Respect to a Fuzzy Normal Group

Let G be a group, B a fuzzy normal subgroup of G, then it is clear the binary relation R_B defined as $R_B(x,y) = B(xy^{-1})$ is a fuzzy equivalent relation on G, so we can introduce the upper approximation operator $\bar{A}_{R_B}$ and the lower approximation operator $\underline{A}_{R_B}$ with respect to B on G. For the upper approximation operator $\bar{A}_{R_B}$, we have the following lemma.

Lemma 3.1. $\bar{A}_{R_B}\mu = B \circ \mu, \mu \in I^G$.

Proof. For $\forall x \in G$,

$$\begin{aligned}\bar{A}_{R_B}\mu(x) &= \sup_{u \in G} R_B(u,x) \wedge \mu(x) = \sup_{u \in G} B(ux^{-1}) \wedge \mu(u) \\ &= \sup_{u \in G} B(xu^{-1}) \wedge \mu(u) = B \circ \mu(x),\end{aligned}$$

so $\bar{A}_{R_B}\mu = B \circ \mu, \mu \in I^G$. □

Proposition 3.2. Let B be a fuzzy normal subgroup of a group G, $\mu, \gamma \in I^G$, then

$$\bar{A}_{R_B}\mu \circ \bar{A}_{R_B}\gamma = \bar{A}_{R_B}(\mu \circ \gamma).$$

Proof. Since B is normal, we have $B \circ \mu = \mu \circ B$ and $B \circ B = B$. Then

$$\begin{aligned}\bar{A}_{R_b}\mu \circ \bar{A}_{R_B}\gamma &= (B \circ \mu) \circ (B \circ \gamma) = B \circ (\mu \circ B) \circ \gamma \\ &= (B \circ B) \circ (\mu \circ \gamma) = B \circ (\mu \circ \gamma) = \bar{A}_{R_B}(\mu \circ \gamma).\end{aligned}$$

□

Proposition 3.3. Let B be a fuzzy normal subgroup of G, $\mu, \gamma \in I^G$, then

$$(\underline{A}_{R_B}\mu) \circ (\underline{A}_{R_B}\gamma) \subseteq \underline{A}_{R_B}(\mu \circ \gamma).$$

Proof. For $x \in G$, we have

$$\begin{aligned}((\underline{A}_{R_B}\mu) \circ (\underline{A}_{R_B}\gamma))(x) &= \sup_{y\in G}(\underline{A}_{R_B}\mu)(xy^{-1}) \wedge (\underline{A}_{R_B}\gamma)(y) \\ &\leq \sup_{y\in G}(\underline{A}_{R_B}\mu)(xy^{-1}) \wedge \gamma(y) \\ &= \sup_{y\in G}(\wedge_{u\in G}\vartheta_{\text{Min}}(B(uyx^{-1}), \mu(u))) \wedge \gamma(y) \\ &= \sup_{y\in G}(\wedge_{v\in G}(\vartheta_{\text{Min}}(B(vx^{-1}), \mu(vy^{-1})) \wedge \gamma(y))) \\ &= \sup_{y\in G} \wedge_{v\in G}\vartheta_{\text{Min}}(B(vx^{-1}), \mu(vy^{-1}) \wedge \gamma(y)) \\ &\leq \wedge_{v\in G}\vartheta_{\text{Min}}(B(vx^{-1}), \sup_{y\in G} \mu(vy^{-1}) \wedge \gamma(y)) \\ &= \wedge_{v\in G}\vartheta_{\text{Min}}(B(vx^{-1}), \mu \circ \gamma(v)) = \underline{A}_{R_B}(\mu \circ \gamma)(x).\end{aligned}$$

Hence $(\underline{A}_{R_B}\mu) \circ (\underline{A}_{R_B}\gamma) \subseteq \underline{A}_{R_B}(\mu \circ \gamma)$. □

Proposition 3.4. Let H, N be fuzzy normal subgroups of G, $\mu \in I^G$. Then we have:
i) $\bar{A}_{R_{H\cap N}} \subseteq \bar{A}_{R_H}\mu \cap \bar{A}_{R_N}\mu$;
ii) $\underline{A}_{R_{H\cap N}}\mu \supseteq \underline{A}_{R_H}\mu \cap \underline{A}_{R_N}\mu$.

Proof. i) For any $x \in G$, we have

$$\begin{aligned}(\bar{A}_{R_H}\mu \cap \bar{A}_{R_N}\mu)(x) &= ((H \circ \mu) \cap (N \circ \mu))(x) \\ &= (H \circ \mu)(x) \wedge (N \circ \mu)(x) \\ &= (\sup_{y\in G} H(y) \wedge \mu(y^{-1}x)) \wedge (\sup_{z\in G} N(z) \wedge \mu(y^{-1}x)) \\ &\geq \sup_{y\in G}(H(y) \wedge \mu(y^{-1}x)) \wedge (N(y) \wedge \mu(z^{-1}x)) \\ &= \sup_{y\in G}(H(y) \wedge N(y)) \wedge \mu(y^{-1}x)) \\ &= (H \cap N) \circ \mu(x) = \bar{A}_{R_{H\cap N}}\mu(x).\end{aligned}$$

ii) For any $x \in G$, we have

$$\begin{aligned}(\underline{A}_{R_H}\mu \cap \underline{A}_{R_B}\mu)(x) &= (\wedge_{u\in G}\vartheta_{\text{Min}}(H(ux^{-1}), \mu(u))) \\ &\quad \wedge(\wedge_{v\in G}\vartheta_{\text{Min}}(N(vx^{-1}), \mu(v))) \\ &\leq \wedge_{u\in G}(\vartheta_{\text{Min}}(H(ux^{-1}), \mu(u)) \wedge \vartheta_{\text{Min}}(N(ux^{-1}), \mu(u))) \\ &\leq \wedge_{u\in G}\vartheta_{\text{Min}}(H(ux^{-1}), \mu(u) \wedge \vartheta_{\text{Min}}(N(ux^{-1}), \mu(u))) \\ &\leq \wedge_{u\in G}\vartheta_{\text{Min}}(H(ux^{-1}), \vartheta_{\text{Min}}(N(ux^{-1}), \mu(u))) \\ &\leq \wedge_{u\in G}\vartheta_{\text{Min}}(H(ux^{-1}) \wedge N(ux^{-1}), \mu(u)) \\ &\leq \wedge_{u\in G}\vartheta_{\text{Min}}(H \cap N(ux^{-1}), \mu(u)) \\ &= \underline{A}_{R_{H\cap N}}\mu(x).\end{aligned}$$

Hence $\underline{A}_{R_{H\cap N}}\mu \supseteq \underline{A}_{R_H}\mu \cap \underline{A}_{R_N}\mu$. □

Proposition 3.5. Let H, N be fuzzy normal subgroups of a group G, $\mu \in I^G$ be a fuzzy subgroup of G, then we have

$$\bar{A}_{R_H}\mu \circ \bar{A}_{R_N}\mu = \bar{A}_{R_{H\circ N}}\mu.$$

Proof. $\bar{A}_{R_H}\mu \circ \bar{A}_{R_N}\mu = \bar{A}_{R_{H\circ N}}\mu = (H\circ\mu)\circ(N\circ\mu) = (H\circ N)\circ(\mu\circ\mu).\bar{A}_{R_{H\circ N}}\mu.$ □

Let $A, B \in I^G$, then $A\times B$ is defined as $A\times B(x,y) = A(x)\wedge B(y)$, if A, B are fuzzy normal subgroups of G, then $A\times B$ is the fuzzy normal subgroup of $G\times G$ [5]. We have the following proposition.

Proposition 3.6. Let A, B be fuzzy normal subgroups of, $\mu\gamma \in I^G$, then we have

$$\bar{A}_{R_{A\times B}}(\mu\times\gamma) = (\bar{A}_{R_A}\mu)\times(\bar{A}_{R_B}\gamma),\ \underline{A}_{R_{A\times B}}(\mu\times\gamma) \supseteq (\underline{A}_{R_A}\mu)\times(\underline{A}_{R_B}\gamma).$$

Proof. For any $x, y \in G\times G$, we have

$$\begin{aligned}
\bar{A}_{R_{A\times B}}(\mu\times\gamma)(x,y) &= (A\times B)\circ(\mu\times\gamma)(x,y)\\
&= \sup\nolimits_{x_1x_2=x,y_1y_2=y} A\times B(x_1,y_1)\wedge\mu\times\gamma(x_2,y_2)\\
&= \sup\nolimits_{x_1x_2=x,y_1y_2=y}(A(x_1)\wedge B(y_1))\wedge(\mu(x_2)\wedge\gamma(y_2))\\
&= \sup\nolimits_{x_1x_2=x,y_1y_2=y}(A(x_1)\wedge\mu(x_2))\wedge(B(y_1)\wedge\gamma(y_2))\\
&= A\circ\mu(x)\wedge B\circ\gamma(y) = (A\circ\mu\times B\circ\gamma)(x,y)\\
&= (\bar{A}_{R_A}\mu)\times(\bar{A}_{R_B}\gamma)(x,y).
\end{aligned}$$

$$\begin{aligned}
(\underline{A}_{R_A}\mu\times\underline{A}_{R_B}\gamma(x,y) &= \underline{A}_{R_A}\mu(x)\wedge\underline{A}_{R_B}\gamma(y)\\
&= (\wedge_{u\in G}\vartheta_{\text{Min}}(A(ux^{-1}),\mu(u)))\\
&\quad \wedge(\wedge_{v\in G}\vartheta_{\text{Min}}(B(vy^{-1}),\gamma(v)))\\
&= \wedge_{(u,v)\in G\times G}(\vartheta_{\text{Min}}(A(ux^{-1}),\mu(u))\\
&\quad \wedge\vartheta_{\text{Min}}(B(vy^{-1}),v(v)))\\
&\leq \wedge_{(u,v)\in G\times G}\vartheta_{\text{Min}}(A(ux^{-1}),\mu(u)\\
&\quad \wedge\vartheta_{\text{Min}}(B(vy^{-1}),\gamma(v)))\\
&\leq \wedge_{(u,v)\in G\times G}\vartheta_{\text{Min}}(A(ux^{-1}),\\
&\quad \vartheta_{\text{Min}}(B(vy^{-1}),\gamma(v)\wedge\mu(u)))\\
&= \wedge_{(u,v)\in G\times G}\vartheta_{\text{Min}}(A(ux^{-1})\\
&\quad \wedge B(vy^{-1}),\mu(u)\wedge\gamma(v))\\
&= \wedge_{(u,v)\in G\times G}\vartheta_{\text{Min}}(A\times B((u,v)(x^{-1},y^{-1})),\\
&\quad \mu\times v(u,v))\\
&= \underline{A}_{R_{A\times B}}(\mu\times v)(x,y).
\end{aligned}$$

□

4 The Lower and Upper Rough Fuzzy Subgroup of a Group

If B is a fuzzy normal subgroup of G, a fuzzy subset μ of G is called a upper rough fuzzy(normal) subgroup of G if $\bar{A}_{R_B}$ is a fuzzy(normal) subgroup of G. Similarly, a fuzzy subset μ of G is called a lower rough fuzzy(normal) subgroup of G if $\underline{A}_{R_B}\mu$ is a fuzzy(normal) subgroup of G.

Proposition 4.1. Let B be a fuzzy normal subgroup of G, μ a fuzzy(normal) subgroup of G, then μ is a upper rough fuzzy(normal) of G.

Proof. Since $\bar{A}_{R_B}\mu = B \circ \mu$, we have

1) $B \circ \mu = 1$;

2) $B \circ \mu(x) = \sup_{y \in G} B(xy^{-1}) \wedge \mu(y) = \sup_{y \in G} B(yx^{-1}) \wedge \mu(y^{-1}) = B \circ \mu(x^{-1})$.

3) $B \circ \mu(xy) = (B \circ B \circ \mu \circ \mu)(xy) = (B \circ \mu) \circ (B \circ \mu)(xy) \geq B \circ \mu(x) \wedge B \circ \mu(y)$.

If μ is normal, then for $\forall x, y \in G$, we have

$$\begin{aligned} B \circ \mu(xy) &= \sup_{z \in G} B(z) \wedge (z^{-1}xy) \\ &= \sup_{w \in G} B(xw^{-1}) \wedge \mu(yw) \\ &= \sup_{t \in G} B(t^{-1}) \wedge \mu(yxt) \\ &= \sup_{t \in G} B(t^{-1}) \wedge \mu(tyx) = B \circ \mu(yx). \end{aligned}$$

□

Proposition 4.2. Let B be a fuzzy normal subgroup of G, μ a fuzzy(normal) subgroup of G. If $B \subseteq \mu$, then μ is a lower rough fuzzy(normal) subgroup of G.

Proof. Since $B \subseteq \mu$, we have $\underline{A}_{R_B}\mu(e) = \wedge_{u \in G} \vartheta_{\mathrm{Min}}(B(u), \mu(u)) = 1$.

For any $x \in G$, we have

$$\begin{aligned} \underline{A}_{R_B}\mu(x^{-1}) &= \wedge_{u \in G} \vartheta_{\mathrm{Min}}(B(ux), \mu(u)) \\ &= \wedge_{u \in G} \vartheta_{\mathrm{Min}}(B(x^{-1}u^{-1}), \mu(u^{-1})) \\ &= \wedge_{u \in G} \vartheta_{\mathrm{Min}}(B(u^{-1}x^{-1}), \mu(u^{-1})) \\ &= \wedge_{v \in G} \vartheta_{\mathrm{Min}}(B(vx^{-1}), \mu(v)) \\ &= \underline{A}_{R_B}\mu(x). \end{aligned}$$

For $\forall x, y \in G$, we have

$$\begin{aligned} \underline{A}_{R_B}\mu(x) \wedge \underline{A}_{R_B}\mu(y) &= (\wedge_{u \in G} \vartheta_{\mathrm{Min}}(B(ux^{-1}), \mu(u))) \wedge \\ &\quad (\wedge_{v \in G} \vartheta_{\mathrm{Min}}(B(vy^{-1}), \mu(v))) \\ &\leq \wedge_{u \in G} (\vartheta_{\mathrm{Min}}(B(ux^{-1}), \mu(u)) \wedge \mu(y)) \\ &\leq \wedge_{u \in G} \vartheta_{\mathrm{Min}}(B(ux^{-1}), \mu(uy)) \\ &\leq \wedge_{v \in G} \vartheta_{\mathrm{Min}}(B(vy^{-1}x^{-1}), \mu(v)) \\ &= \underline{A}_{R_B}\mu(xy). \end{aligned}$$

If μ is normal, then for $\forall x, y \in G$, we have

$$\begin{aligned} \underline{A}_{R_B}\mu(xy) &= \wedge_{u \in G} \vartheta_{\mathrm{Min}}(B(uy^{-1}x^{-1}), \mu(u)) \\ &= \wedge_{v \in G} \vartheta_{\mathrm{Min}}(B(vx^{-1}), \mu(vy)) \\ &= \wedge_{w \in G} \vartheta_{\mathrm{Min}}(B(wy^{-1}, \mu(xw)) \\ &= \wedge_{w \in G} \vartheta_{\mathrm{Min}}(B(wy^{-1}), \mu(wx)) \\ &= \underline{A}_{R_B}\mu(yx). \end{aligned}$$

□

If $f : G \to G'$ is a homomorphism, A, A' are fuzzy subgroups of G, G' respectively, we know $f(A)$, $f^{-1}(A')$ are fuzzy subgroups of G, G' respectively. If A, A' are normal, then $f(A)$, $f^{-1}(A')$ are also normal [5].

Proposition 4.3. Let $f : G \to G'$ be a homomorphism, μ, B be fuzzy subgroups of G and B be normal, then $\bar{A}_{R_{f(B)}} = f(\bar{A}_{R_B}\mu)$.

Proof. For any $y \in G'$, we have

$$\begin{aligned}
\bar{A}_{R_{f(B)}} f(\mu)(y) &= f(B) \circ f(\mu)(y) \\
&= \sup_{y_1 y_2 = y} f(B)(y_1) \wedge f(\mu)(y_2) \\
&= \sup_{y_1 y_2 = y} (\sup_{f(x_1)=y_1} B(x_1) \wedge \sup_{f(x_2)=y_2} \mu(x_2)) \\
&= \sup_{f(x_1)f(x_2)=y} B(x_1) \wedge \mu(x_2) \\
&= \sup_{f(x)=y} \sup_{x_1 x_2 = x} B(x_1) \wedge \mu(x_2) \\
&= \sup_{f(x)=y} B \circ \mu(x) = f(\bar{A}_{R_B} \mu)(y).
\end{aligned}$$

□

Proposition 4.4. Let $f : G \to G'$ be a homomorphism, μ', B' be fuzzy subgroups of G' and B' be normal, then $f^{-1}(\bar{A}_{R'_B} \mu') = \bar{A}_{R_{f^{-1}(B')}} f^{-1}(\mu')$.

Proof. For any $x \in G$, we have

$$\begin{aligned}
\bar{A}_{R_{f^{-1}(B')}} f^{-1}(\mu')(x) &= f^{-1}(B') \circ f^{-1}(\mu')(x) \\
&= \sup_{y \in G} f^{-1}(B')(xy^{-1}) \wedge f^{-1}(\mu')(y) \\
&= \sup_{y \in G} B'(f(xy^{-1})) \wedge \mu'(f(y)) \\
&= \sup_{f(y) \in G'} B'(f(x)f(y)^{-1}) \wedge \mu'(f(y)) = B' \circ \mu'(f(x)) \\
&= f^{-1}(B' \circ \nu')(x) = f^{-1}(\bar{A}_{R'_B} \mu)(x).
\end{aligned}$$

□

5 Conclusion

In this paper we study the upper and lower approximations of a fuzzy set in a group. Our work is different from the one in [6] since they use fuzzy group to define a classical equivalent relation in [6]. By our study we know the properties of upper approximation of a fuzzy set is better than its lower approximation in a group.

Acknowledgement

This paper is supported by a grant of Liaoning province committee (20161049)

References

1. Z.Pawlak: Rough sets, Internet. J.Comput.Inform.Sci.**11**, 341-356 (1982).
2. Z.Pawlak: Rough sets and fuzzy sets, Fuzzy Sets and Systems **17**, 99-102 (1985).
3. D.Dubois and H.Prade: Rough fuzzy sets and fuzzy rough sets, International J. of General Systems **17**, 191-206 (1990).
4. N.N.Morsi, M.M.Yakout: Axiomatics for fuzzy fuzzy rough sets, Fuzzy Sets and Systems **100**, 327-342 (1998).

5. Chen Degang: Properties of T-fuzzy factor groups, Fuzzy Sets and Systems **99**, 187- 192 (1998).
6. N.Kuroki, P.P.Wang: The lower and upper approximations in a fuzzy group, Inform.Sci.**90**,203-220(1996).
7. N.Kuroki: Rough ideals in Semigroups, Inform.Sci.**100**, 139-163 (1997).

T-Fuzzy Hyperalgebraic Systems

Reza Ameri[1] and Mohammad Mehdi Zahedi[2]

[1] Departement of Mathematics
University of Mazandaran
Babolsar, Iran
ameri@umcc.ac.ir
[2] Department of mathematics
Shahid Bahonar University of Kerman
Kerman, Iran
zahedi@arg3.uk.ac.ir

Abstract. In this paper first by considering the notions of hyperalgebraic structures such as hypergroups, polygroups, hyperrings, ..., we define the notion of (sub) hyperalgebraic systems in general, also we define the notions of T-fuzzy (weak) hyperalgebraic systems, which are the generalization of fuzzy hyperalgebraic structures. Then we give some related basic results .

1 Introduction

The notion of a hypergroup was introduced by F. Marty in 1934 [10].Since then many researchers have worked on hyperalgebraic structures and developed this theory. Zadeh in 1965 [14] introduced the notion of a fuzzy subset μ of a nonempty subset X, as a function from X to $[0,1]$.In [2] Ameri and Zahedi introduced some concepts of fuzzy hyperalgebraic structures, such as fuzzy hyperideals, fuzzy hypermodules. In [3] they defined and studied the notion of hyperalgebraic system in general, not just particular systems such as hypergroups , hyperrings, ...L.Tu and W.X.Guin [12] defined and studied the fuzzy algebraic systems. Now in this paper we follow [1,12] and introduce the notions of T-fuzzy (resp. weak) hyperalgebraic systems, and then obtain some related results.

2 Preliminaries

Definition 2.1 [3,7]. (i) An n-ary hyperoperation β on the set H is a function $\beta : H^n = H \times H \times \ldots \times H \longrightarrow P_*(H)$,where $P_*(H)$ is the set of all nonempty subsets of H.

In this case we say that the size of β is n and(H,β) is called a hyperstructure.

(ii) A subset S of H is said to be closed under the n-ary hyperoperation β if $(x_1,\ldots,x_n) \in S^n$ implies that $\beta(x_1,\ldots,x_n) \subseteq S$.

Henceforth sometimes we use hyperoperation instead of the n-ary hyperoperation.

N.R. Pal and M. Sugeno (Eds.): AFSS 2002, LNAI 2275, pp. 509–514, 2002.

Definition 2.2. [3,7] A hyperstructure $(H,\cdot)$, where $\cdot$ is a 2-ary hyperoperation is called a hypergroup if the following conditions hold:

(*i*) $x\cdot(y\cdot z)=x\cdot(y\cdot z)$ for all $x,y,z\in H$,

(*ii*) $a\cdot H=H=Ha$ for all $a\in H$, where $A\cdot B=\cup_{a\in A,b\in B}a\cdot b,\quad \forall A,B\subseteq H$.

Generally, the singelton $\{a\}$ is identified with its member a.

Definition 2.3. [6,7] A hypergroup $(H,\cdot)$ is called a polygroup or a uasi-canonical hypergroup if the following axioms hold:

(*i*) There exists an element $e\in H$ such that $e\cdot x=x\cdot e=\{x\},\qquad \forall x\in H,$, here e is called a neutral element of H.

(*ii*) For every $x\in H$, there exists a unique element $x'\in H$ such that $e\in x'\cdot x\cap x'\cdot x$, here x' is called the opposite of x and it is denoted by x^{-1}.

(*iii*) $z\in x\cdot y\Longrightarrow x\in z\cdot y^{-1}\Longrightarrow y\in x^{-1}\cdot z,\quad \forall x,y,z\in H$.

Definition 2.4.[3] A hyperalgebraic system $S=\langle H,F,A\rangle$ is a set H with a set F of hyperoperations and a set A of laws (axioms) of hyperoperations.

Example 2.5 [3]. Hypergroups, hyperrings and hypermodules are hyperalgebraic systems.

Definition 2.6. Let $S_1=(H_1,F_1,A_1)$ and $S_2=(H_2,F_2,A_2)$ be two hyperalgebraic systems. Then $S_1=S_2$ if and only if $H_1=H_2$, $F_1=F_2$ and $A_1=A_2$.

Definition 2.7. Let $S=(H,F,A)$ be a hyperalgebraic system and H' be a nonempty subset of H. Then $S'=(H',F,A)$ is called a subhyperalgebraic system of S, that is H' is closed under all of hyperoperations of H.

Definition 2.8. Let T be a t-norm and $n\in N$. Then the t_n-norm extracted from T is defined as follows:

$$T_n:[0,1]^n\longrightarrow[0,1]$$

where

$$T_n(a_1,\ldots,a_n)=\begin{cases}a_1 & \text{if } n=1\\ T(a_1,a_2) & \text{if } n=2\\ T(a_i,T_{n-1}(a_1,\ldots,a_{i-1},a_{i+1},\ldots,a_n)), & \text{if } n>2\end{cases}$$

Remark 2.9. Similarly we can define the t_n-conorm extracted from a T-conorm T if we replace a t-norm with a t-conorm.

Definition 2.10 Let μ be a fuzzy subset of X.Then

(i) the level subset $\mu_t^{\geq}$ is defined by

$$\mu_t^{\geq}=\{x\in X|\mu(x)\geq t\},\qquad t\in[0,1],$$

and the strong level subset $\mu_t^{>}$ is defined by

$$\mu_t^{>}=\{x\in X|\mu(x)>t\},\qquad t\in[0,1).$$

(ii) the lower level subset $\mu_t^{\leq}$ is defined by

$$\mu_t^{\leq}=\{x\in X|\mu(x)\leq t\},\qquad t\in[0,1],$$

and the strong lowr level subset $\mu_t^<$ is defined by

$$\mu_t^< = \{x \in X | \mu(x) < t\}, \qquad t \in (0,1].$$

Lemma 2.11 Let μ be a fuzzy subset of X and $\alpha = sup_{x \in X}\mu(x)$. Then

$$\mu_t^{\geq} = \bigcap_{0 \leq r < t} \mu_r^{>}, \qquad \forall t \in [0, \alpha].$$

Proof. If $\alpha = 0$, then $\mu_t^{\geq} = \mu_0^{\geq} = X$ and $\bigcap_{0 \leq r < t} \mu_r^{>} = X$, since $\{r | 0 \leq r < t\} = \emptyset$. Hence we assume that $t \neq 0$. Let $x \in \mu_t^{\geq}$. Then $\mu(x) \geq t > r$, for all $0 \leq r < t$ and hence $x \in \mu_r^{>}$, for all $0 \leq r < t$. So $x \in \bigcap_{0 \leq r < t} \mu_r^{>}$. Therefore $\mu_t^{\geq} \subset \bigcap_{0 \leq r < t} \mu_r^{>}$. Now let $x \in \bigcap_{0 \leq r < t} \mu_r^{>}$. Then $x \in \mu_r^{>}$, for all $0 \leq r < t$ and so $\bigcap_{0 \leq r < t} \mu_r^{>} \subseteq \mu_t^{\geq}$. This completes the proof.

Notation. By FS(H) we mean the set of all fuzzy subset of H.

3 Fuzzy Hyperalgebraic Systems

Definition 3.1. Let $S = (H, F, A)$ be a hyperalgebraic system, $\mu \in FS(H)$ and T be a t-norm. If for any n-ary $f \in F$ and for any $(x_1, \ldots, x_n) \in H^n$ we have

(*i*) $\inf\limits_{z \in f(x_1, \ldots, x_n)} \mu(z) \geq T_n(\mu(x_1), \ldots, \mu(x_n))$ for all $z \in f(x_1, \ldots, x_n)$, then μ is called a T-fuzzy hyperalgebraic system of S.

(ii) $\mu(z) \geq T_n(\mu(x_1), \ldots, \mu(x_n))$ for some $z \in f(x_1, \ldots, x_n)$, then μ is called a weak T-fuzzy hyperalgebraic system of S.

Remark 3.2. (i) In Definition 3.1, if $T = \min$, then μ is called a (weak) fuzzy hyperalgebraic system.

(*ii*) If μ is a T-fuzzy hyperalgebraic system, then μ is a T'-fuzzy hyperalgebraic system for any t-norm T', where $T' \leq T$.

Definition 3.3. Let $S = (H, F, A)$ be a hyperalgebraic system, $\mu \in FS(H)$ and T be a t-conorm. Let $f \in F$ be an arbitrary n-ary and $(x_1, \ldots, x_n) \in H^n$. Then

(*i*) μ is an anti T-fuzzy hyperalgebraic system of S, if

$$\sup_{z \in f(x_1, \ldots, x_n)} \mu(z) \leq T_n(\mu(x_1), \ldots, \mu(x_n)),$$

(*ii*) μ is a weak anti T-fuzzy hyperalgebraic system of S if

$$\mu(z) \leq T_n(\mu(x_1), \ldots, \mu(x_n)), \quad \text{for some} \quad z \in f(x_1, \ldots, x_n).$$

Remark 3.4. (i) In Definition 3.3, if $T = \max$ (maximum t-conorm) then μ is called an anti fuzzy hyperalgebraic system.

(*ii*) If μ is an anti T-fuzzy hyperalgebraic system, then μ is an anti T'-fuzzy hyperalgebraic system for any t-norm T', where $T \leq T'$. Clearly every anti T-fuzzy hyperalgebraic system is weak.

Definition 3.5. An unary operation $C : [0,1] \longrightarrow [0,1]$ is called complement if (i) $x \leq y \Longrightarrow C(y) \leq C(x)$, (ii) $C(C(x)) = x$, (iii) $C(0) = 1$, for all $x, y \in [0,1]$.

Lemma 3.6. Let T be a t-norm (resp.t-conorm). Define $T^C : [0,1] \times [0,1] \longrightarrow [0,1]$, by $T^C(x,y) = C(T(C(x), C(y))), \forall x, y \in [0,1]$ · Then T^C is a t-conorm (resp. t-norm), associated to T.

Theorem 3.7. Let $S = (H, F, A)$ be a hyperalgebraic system, μ a fuzzy subset of H and T be a t-norm (resp. t-conorm). Then μ is a T-fuzzy (resp. anti T-fuzzy) hyperalgebraic system if and only if μ^C is anti T^C-fuzzy (resp. T^C-fuzz)hyperalgebraic system, where $\mu^C(x) = C(\mu(x))$, for all $x \in H$ and C is a complement.

Proof. The proof is an immediate consequence of Lemma 3.6 and Definition 3.3.

Theorem 3.8. Let μ be a fuzzy subset of a hyperalgebraic system H. Then the following statements are equivalent:

(1) μ is a fuzzy hyper-system of H,

(2) each non-empty strong level subset of μ is a subhypersystems of H.

(3) each non-empty level subset of μ is a subhypersystem of H,

Proof. (1)⇒(2). Let $\mu_t^>$ be a non-empty strong level subset of μ. Let $x_1, \ldots, x_n \in \mu_t^>$ and f be an n-ary hyperoperation. Then $\mu(x_i) > t$. $\forall i$, $1 \leq i \leq n$. Since μ is a fuzzy hypersystem of H,then

$$\inf_{z \in f(x_1,\ldots,x_n)} \mu(z) \geq \min\{\mu(x_1), \ldots, \mu(x_n)\} > t$$

Thus $z \in \mu_t^>$, $\forall z \in f(x_1, \ldots, x_n)$, and hence $f(x_1, \ldots, x_n) \subseteq \mu_t^>$.

(2)⇒(3)By lemma 2.11 we have $\mu_t = \bigcap_{t>r} \mu_r$ and $\mu_t \subseteq \mu_r^>$ for all $t > r$, then by hypothesis $\mu_r^>$ is a subhypersystem of $H, \forall t > r$. Thus μ_t as an intersection of a family of subhypersystems of H is itself a subhypersystem.

(3)⇒(1). Let $x_1, \ldots, x_n \in H$ and f be an arbitrary n-ary hyperoperation. Suppose $\mu(x_i) = t_i$ and $t_k = \inf_{1 \leq i \leq n} \{t_i\}$ and $z \in f(x_1, \ldots, x_n)$. Since μ_{t_k} is a subhypersystem of H, $z \in \mu_{t_k}$. Thus

$$inf_{z \in f(x_1,\ldots,x_n)} \mu(z) \geq t_k = min(\mu(x_1), \ldots, \mu(x_n)).$$

Theorem 3.9. (Representation Theorem) Let μ be a fuzzy subset of hyperalgebraic system H and $t_0 = \sup_{x \in H} \mu(x)$. Then the following statements are equivalent:

(i) μ is a fuzzy hyperalgebraic system of H,

(ii) Each level subset μ_t, for $t \in [0, t_0]$ is a subhyperalgebraic system of H,

(iii) Each strong level subset $\mu_t^>$, for $t \in [0, t_0]$, is a subhypersystem of H,

(iv) Each level subset μ_t, for $t \in Im(\mu)$,is a subhyperalgebraic system of H, where $Im(\mu)$ denotes the image of μ.

(v) Each strong level subset $\mu_t^>$, for $t \in Im(\mu) - \{t_0\}$ is a subhyperalgebraic system of H,

(vi) Each non-empty level subset of μ is a subhyperalgebraic system of H,

(vii) Each non-empty strong level subset of μ is subhyperalgebraic system of H,

($viii$) $1-\mu$ is an anti fuzzy hyperalgebraic system of H.

Proof. By Theorem 3.8, (i) and (viii) are equivalent. Now by considering the following diagram we will complete the processes of proof.

$$\begin{array}{ccccc} (v) & \longrightarrow & (iv) & \longleftarrow & (ii) \\ & & \downarrow & & \\ \uparrow & & (i) & & \uparrow \\ & & \downarrow & & \\ (iii) & \longrightarrow & (viii) & \longrightarrow & (vi) \end{array}$$

(i)⇒(vii). Suppose $\mu_t^>$ is a non-empty strong level subset of μ. Let $x_1,\ldots,x_n \in \mu_t^>$ and f be an arbitrary n-ary hyperoperation. Then $\mu(x_i) > t$, $\forall i$, $1 \le i \le n$. Since μ is a fuzzy hypersystem of H, we have

$$\inf_{z\in f(x_1,\ldots,x_n)} \mu(z) \ge \min\{\mu(x_1),\ldots,\mu(x_n)\} > t.$$

Thus $z \in \mu_t^>$, $\forall z \in f(x_1,\ldots,x_n)$, and hence $f(x_1,\ldots,x_n) \subseteq \mu_t^>$. Thus $\mu_t^>$ is a sub-hypersystem of H.

(vii)⇒(vi). Let $\mu_t^{\geq}$ be a non-empty level subset of μ.Now by Lemma 2.11 we have

$$\mu_t^{\geq} = \bigcap_{0\le r<t} \mu_r^>.$$

Since $\emptyset \neq \mu_t^{\geq} \subseteq \mu_r^>$ for all r, $0 \le r < t$, we get that $\mu_r^>$ is non-empty for all r, $0 \le r < t$. Hence by (vii) $\mu_r^>$ is a subhyperalgebraic system of H for all r, $0 \le r < t$.Since the intersection of a family of fuzzy hyperalgebraic system is again a fuzzy hyperalgebraic system ,then $\mu_t^{\geq}$ is a subhyperalgebraic system of H.

(vi)⇒(ii). Let $t \in [0, t_0]$. Consider the level subset $\mu_{t_0}^{\geq}$. Since $x_0 \in \mu_t^{\geq}$ and $\mu_{t_0}^{\geq} \subseteq \mu_t^{\geq}$, then $\mu_t^{\geq}$ is non-empty. Hence by (vi) $\mu_t^{\geq}$ is a subhyperalgebraic system of H.

(ii)⇒(iv). Since $Im(\mu) \subseteq [0, t_0]$, the proof is clear. The proofs of (vii)⇒(iii)⇒(v) are similar to the proofs of (vi)⇒(ii)⇒(iv).

(iv)⇒(i). Let $x_1,\ldots,x_n \in H$ and f be an n-ary hyperoperation. Suppose that $\min\{\mu(x_1),\ldots,\mu(x_r)\} = \mu(x_k) = t$ for some k, $1 \le k \le n$. Then $x_1,\ldots,x_n \in \mu_t$ and $\mu(x_i) \in Im\mu$. Hence by (iv) $\mu_t^{\geq}$ is a subhyperalgebraic system of H. Now let $z \in f(x_1,\ldots,x_n)$. Since $\mu_t^{\geq}$ is a subhyperalgebraic system of H, we have $f(x_1,\ldots,x_n) \subseteq \mu_t^{\geq}$. Thus $z \in \mu_t^{\geq}$ and hence $\inf_{z\in f(x_1,\ldots,x_n)} \mu(z) \ge t = \min\{\mu(x_1),\ldots,\mu(x_n)\}$. Therefore, μ is a fuzzy hypersystem of H.

References

1. Ameri, R., Zahedi, M.M. : Hypergroup and join spaces induced by a fuzzy subset, PU.M.A,Vol.8 (19997) 155-168.
2. Ameri,R., Zahedi, M.M. : Fuzzy subhypermodules over fuzzy hyperrings, Sixth International Congress on AHA, Democritus Univ., (1996), 1-14.
3. Ameri R., Zahedi,M.M : Hyperalgebraic Systems, Italian Journal of Pure and Applied Mathematics, No. 6 (1999), 21-32.
4. Burris, S., Sankapponavar,H.P: A Course in Universal Algebra, Springer-Verlag (1981).
5. Bhattacharya,P., Mukherjee,N.P, Fuzzy relations and fuzzy groups, Inform. Sci., 36 (1985), 267-282.
6. Comer, D.S.: Polygroups derived from cogorups,Journal of Algebra,Vol.89, (1984) 397-405.
7. Corsini,P.: Prolegomena of Hypergroup Theory, Aviani Editor, (1993).
8. Das,P.S: Fuzzy groups and level subgroups, J. Math. Anal. Appl., 84 (1981), 264-269.
9. Klement,P.E,Mesier R.: Triangular Norms,Tatra Mountains Math.Publ. 13(1997)169-193.
10. Marty,F.: Sur une généralization de la notion de groupe, Huitieme congress de Mathematiciens, Scandinaves, Stockholm, (1934), 45-49.
11. Rosenfeld, A.: Fuzzy groups, J. Math. Anal. Appl. 35 (1971), 512-517.
12. Tu,Lu.,Wen-Xiang Gu: Fuzzy algebraic systems (I): Direct products, Fuzzy Sets and Systems 61(1994), 313-327.
13. Wall,H.S.: Hypergroups, Amer. J. Math., (1937), 77-98.
14. Zadeh,L.A.: Fuzzy Sets, Inform. and Control, 8 (1965), 338-353.

On Some Weaker Forms of Fuzzy Compactness

R.N. Bhaumik

Department of Mathematics
Tripura University, Suryamaninagar,
Tripura, India, 799 130
rbhaumik@dte, vsnl.net.in

Abstract. In this paper the concepts of fuzzy isocompactness and some of its generalizations are studied. A fuzzy topological space X is said to be fuzzy isocompact if every fuzzy closed and fuzzy countably compact subspace is fuzzy compact. Every fuzzy compact space is fuzzy isocompact. Fuzzy weakly isocompactness and fuzzy nearly isocompactness are also studied as generalizations of fuzzy isocompactness. Some of the basic properties of these weaker forms of the fuzzy compactness are examined.

1 Introduction

Since the compactness is one of the most significant concepts in topology, much work of recent years in fuzzy topology has centred on the research for the theory of fuzzy compactness. This led many topologists to generalize this concept. As a result, different weaker forms of the concept of fuzzy compactness are appearing in the literature. Countably fuzzy compactness is one of the weaker forms of fuzzy compactness. Now we have the following question:

Is a space, every closed subspace of which is fuzzy countably compact, necessarily fuzzy compact?

Bhaumik and Bhattacharya [2] made an attempt to answer the above question. They introduced the concept of fuzzy isocompactness. These are the spaces in which every fuzzy closed and fuzzy countably compact subspaces are always fuzzy compact. In this paper, fuzzy isocompactness and some of its generalizations are discussed.

All the spaces considered in this paper are fuzzy Hausdorff spaces. Guojun [7] proved that in a fuzzy Hausdorff space, fuzzy compactness (Lowen [8]), strong fuzzy compactness and ultra-compactness (Lowen [9]), N-compactness (Guojun [7]) are all equivalent. In fuzzy Hausdorff space, fuzzy compactness is closed hereditary.

2 Fuzzy Isocompact Spaces

In 1968, C.L.Chang [4] introduced the concept of fuzzy compactness. It occupies a very important place in fuzzy topology and so do some of its weaker forms. In this section the concept of fuzzy isocompactness is studied.

N.R. Pal and M. Sugeno (Eds.): AFSS 2002, LNAI 2275, pp. 515–519, 2002.

Definition 2.1 *A fuzzy topological space X is said to be fuzzy isocompact if every fuzzy closed, fuzzy countably compact subspace of X is fuzzy compact.*

Obviously, every fuzzy compact space is fuzzy isocompact. In this section some properties of fuzzy isocompact space are studied.

It is clear that fuzzy isocompactness is closed hereditary. A subset is said to be fuzzy F_σ- subset if it is the union of countable collection of fuzzy closed subsets.

Theorem 2.2 *Every fuzzy F_σ- subset of a fuzzy isocompact space is fuzzy isocompact.*

Proof: Let $\mu = \cup \lambda_i$, where each λ_i is fuzzy closed, be a fuzzy F_σ - subset of a fuzzy isocompact space $(X, \mathfrak{I})$. Since fuzzy isocompactness is closed hereditary, each λ_i is fuzzy isocompact .Now we have to show that μ is fuzzy isocompact. Let β be a fuzzy closed and fuzzy countably compact subset of μ. Consider a subfamily B *of* $\mathfrak{I}_\mu$ such that $sup_{\delta \in B}\, \delta \geq \alpha$, $\alpha \in (0,1]$, i.e., B is a α-cover of μ. Then for each i, $\beta \cap \lambda_i$ *is* a fuzzy closed, fuzzy countably compact subset of λ_i. So it is fuzzy compact subset, since each λ_i is fuzzy isocompact. By fuzzy compactness of $\beta \cap \lambda_i$, there exists a finite subfamily B_i' of B such that $\sup_{\delta \in B'} \delta \geq \alpha - \varepsilon_i$. Taking the supremum on i, we have $\vee\, \varepsilon_i = \varepsilon$ and $\cup B_i' = B$. Then B has a countable subfamily B' such that $\sup_{\delta \in B_i'} \delta \geq \alpha - \varepsilon$. Since β is fuzzy countably compact, we have finite subcover B' of B. Hence β is a fuzzy compact subset which implies that μ is fuzzy isocompact.

Theorem 2.3 *If f is a fuzzy countably compact, fuzzy continuous map from a fuzzy isocompact space $(X, \mathfrak{I})$ onto a fuzzy topological space $(Y, \mathfrak{R})$, then $(Y, \mathfrak{R})$ is fuzzy isocompact.*

Proof: Let β be a fuzzy countably compact and fuzzy closed subset of $(Y, \mathfrak{R})$. Since f is fuzzy continuous and fuzzy countably compact map, $f^{-1}(\beta)$ is fuzzy closed and fuzzy countably compact subset of X. By fuzzy isocompactness of X, $f^{-1}(\beta)$ is fuzzy compact. Since f is onto fuzzy continuous and continuous image of a fuzzy compact set is fuzzy compact, $f f^{-1}(\beta) = \beta$ is fuzzy compact. Hence $(Y, \mathfrak{R})$ is fuzzy isocompact.

Theorem 2.4 *Let $f: (X, \mathfrak{I}) \rightarrow (Y, \mathfrak{R})$ be a fuzzy perfect map from a fuzzy topological space $(X, \mathfrak{I})$ onto a fuzzy isocompact space $(Y, \mathfrak{R})$. Then $(X, \mathfrak{I})$ is fuzzy isocompact.*

Proof: Let β be a fuzzy countably compact and fuzzy closed subset of $(X, \mathfrak{I})$. Since f is fuzzy closed, $f(\beta)$ is fuzzy closed. $f(\beta)$ is also fuzzy countably compact subset of $(Y, \mathfrak{R})$ as f is fuzzy continuous. Since $(Y,\mathfrak{R})$ fuzzy isocompact, $f(\beta)$ is fuzzy compact. By fuzzy perfectness of f, the inverse image $f^{-1}f(\beta)$ of fuzzy compact $f(\beta)$ is fuzzy compact. But $\beta \subset f^{-1}f(\beta)$ and is closed. So β is fuzzy compact subset of $(X, \mathfrak{I})$. Hence $(X, \mathfrak{I})$ is fuzzy isocompact.

Theorem 2.5 *The product of a fuzzy compact space and a fuzzy isocompact space is fuzzy isocompact.*

Proof: Let (X, $\Im$) is fuzzy compact space and (Y,$\Re$) be a fuzzy isocompact space and consider the projection map $\Pi_{Y:}$ $X \times Y \to Y$. Let β be a fuzzy countably compact, fuzzy closed subset of X × Y. Π_Y (β) is fuzzy countably compact subset, Π_Y being fuzzy continuous. It is also fuzzy closed, since (Y,$\Re$) is fuzzy Hausdorff space. By fuzzy isocompactness of (Y, $\Re$), Π_Y (β) is fuzzy compact. Thus $X \times \Pi_Y$ (β) is fuzzy compact. Since β is closed subset of $X \times \Pi_Y (\beta) \subset X \times Y$, β is fuzzy compact. Hence X × Y is fuzzy isocompact.

2.1 Fuzzy Closed-Complete Space and Fuzzy Isocompact Space

The concept of fuzzy closed-complete space was introduced by Bhaumik [1] as a generalization of fuzzy compact space and its different properties like closed hereditary, suprema, infima, productivity, invariant properties are studied.

Definition 2.6 *A fuzzy topological space X is said to be fuzzy closed-complete iff every fuzzy closed ultrafilter of X with countable intersection property is fixed.*

Theorem 2.7 [1] *Every fuzzy closed – complete space is fuzzy isocompact.*

3 Fuzzy Weakly Isocompact Space

Fuzzy countably almost compactness and fuzzy almost compactness are two generalizations of fuzzy compactness which were studied by various authors [5,6,10]. Bhaumik and Bhattacharya [3] introduced the concept of fuzzy weakly isocompact space as an another generalization of fuzzy compactness. In this section some properties of these spaces are studied.

Definition 3.1 *A fuzzy topological space is said to be fuzzy weakly isocompact if every fuzzy regular closed, fuzzy countably almost compact sub space is fuzzy almost compacts.*

Every fuzzy isocompact regular topological space is fuzzy weakly isocompact.

Definition 3.2 *A fuzzy regular subset β in a fuzzy topological space (X, $\Im$) is said to be strongly fuzzy regular closed if for every proper fuzzy regular closed subset α of β, $Int_\beta \alpha \subset IntX \beta$.*

It is seen that every fuzzy clopen subset of (X ,$\Im$) is strongly fuzzy regular closed and every strongly fuzzy regular closed subset is fuzzy regular closed .

Theorem 3.3 *A strongly fuzzy regular closed subspace of fuzzy weakly isocompact space (X, $\Im$) is fuzzy weakly isocompact.*

Proof: Let (η,$\Im_\eta$) be a strongly fuzzy regular closed subspace of (X, $\Im$). Let β be a fuzzy regular closed, fuzzy countably almost compact subset of η. If $\beta \neq \eta$, then β is a proper fuzzy regular closed subset of η and $Int_\eta\ \beta \subset Int_{\,X} \beta$. Hence $Int_\eta \beta$ is an open fuzzy subset of Int Xη and so is open fuzzy subset of X. Therefore $\beta = Cl\ (Int_\eta \beta)$ is a fuzzy regular closed subset of (X,$\Im$), i.e., β is a fuzzy regular closed, fuzzy countably almost compact subset of $(X,\mathfrak{I})$. As $(X,\mathfrak{I})$ is fuzzy weakly isocompact, β is fuzzy almost compact. Hence $(\eta,\mathfrak{I}_\eta)$ is fuzzy weakly isocompact.

Let $(X, \mathfrak{I})$ and $(Y, \mathfrak{R})$ be two fuzzy topological spaces. Then a map f: $(X, \mathfrak{I}) \to (Y, \mathfrak{R})$ be a fuzzy almost quasi-perfect map if it is fuzzy almost continuous, fuzzy closed and fuzzy countably compact map.

Theorem 3.4 *Let $(X, \mathfrak{I})$ and $(Y, \mathfrak{R})$ be two fuzzy topological spaces and let f: $(X, \mathfrak{I}) \to (Y, \mathfrak{R})$ be a fuzzy almost open, fuzzy almost quasi-perfect map from a fuzzy weakly isocompact space $(X, \mathfrak{I})$ onto a fuzzy topological space $(Y, \mathfrak{R})$. Then $(Y, \mathfrak{R})$ is fuzzy weakly isocompact space.*

Proof: Let β be a fuzzy countably almost compact and fuzzy regular closed subset of $(Y,\mathfrak{R})$. Since f is fuzzy almost continuous and fuzzy almost open, inverse image $f^{-1}(\beta)$ of β is a fuzzy regular closed subset of $(X,\mathfrak{I})$. By fuzzy weakly isocompactness of $(X, \mathfrak{I})$, $f^{-1}(\beta)$ is fuzzy almost compact. Since f is onto fuzzy continuous, $f f^{-1}(\beta) = \beta$ is fuzzy almost compact subset of $(Y, \mathfrak{R})$[5]. Hence $(Y,\mathfrak{R})$ is fuzzy weakly isocompact space.

Theorem 3.5 *Let $(X, \mathfrak{I})$ and $(Y, \mathfrak{R})$ be two fuzzy topological spaces and let f: $(X, \mathfrak{I}) \to (Y, \mathfrak{R})$ be a fuzzy open, fuzzy quasi-perfect map from a fuzzy topological space $(X, \mathfrak{I})$ onto a fuzzy weakly isocompact space $(Y, \mathfrak{R})$. Then $(X, \mathfrak{I})$ is fuzzy weakly isocompact space.*

Proof: Let β be a fuzzy countably almost compact and fuzzy regular closed subset of $(X,\mathfrak{I})$ Then $\beta = \text{Cl.Int}(\beta)$. So $f(\beta) = f(\text{Cl.Int}(\beta)) = \text{Cl}(f(\text{Int}\,\beta)) = \text{Cl}(\text{Int} f(\beta)$, since f is fuzzy closed, open continuous. Therefore $f(\beta)$ is fuzzy regular closed, as f / β, restriction of f to β, is fuzzy continuous. So $f(\beta)$ is also fuzzy countably almost compact subset of $(Y,\mathfrak{R})$. Since $(Y,\mathfrak{R})$ is fuzzy weakly isocompact, $f(\beta)$ is fuzzy almost compact in $(Y,\mathfrak{R})$. By fuzzy quasi –perfectness of f, $f^{-1}(f(\beta))$ is fuzzy almost compact subset of $(X,\mathfrak{I})$. Now $\beta = \text{Cl.Int}(\beta) \subset f^{-1} f(\beta)$. Thus $\beta = \text{Cl.Int}(\beta) \cap f^{-1} f(\beta) = \text{Cl.Int}(\beta)$ in $f^{-1} f(\beta)$. As Int β is fuzzy open subset of $f^{-1} f(\beta)$, β is fuzzy regular closed subset of fuzzy almost compact. Hence $(X,\mathfrak{I})$ is a fuzzy weakly isocompact space.

Theorem 3.6 The *product of a fuzzy almost compact space $(Y,\mathfrak{R})$* (*and a fuzzy weakly isocompact topological space $(X,\mathfrak{I})$* is *fuzzy weakly isocompact.*

Proof: Let Π_X: $X \times Y \to X$ be a projection map. Let β be a fuzzy countably almost compact, fuzzy regular closed subset of $X \times Y$. $\Pi_X(\beta)$ is a fuzzy countably almost compact subset of X, Π_X being fuzzy continuous. It is also fuzzy closed. So by fuzzy weakly isocompactness of $(X,\mathfrak{I})$, $\Pi_X(\beta)$ is a fuzzy compact subset of $(X,\mathfrak{I})$. Thus $X \times \Pi_Y(\beta)$ is fuzzy compact. Since β is a fuzzy regular closed subset of $X \times \Pi_Y(\beta) \subset X \times Y$, β is fuzzy almost compact fuzzy subset of $X \times Y$. Hence $X \times Y$ is fuzzy weakly isocompact space.

4 Fuzzy Nearly Isocompact Space

A.H.Es [6] introduced the concept of fuzzy nearly compactness as an another generalization of fuzzy compactness. A fuzzy topological space X is called fuzzy nearly compact if every open cover of X has a finite sub collection such that the interiors of closures of fuzzy sets in this sub collection covers X or equivalently every regular open cover of X has a finite sub cover.

In this section another generalized form of fuzzy isocompact space , called nearly fuzzy isocompact space , is introduced and its properties will be studied elsewhere .

Definition 4.1 *A fuzzy topological space (X, $\mathfrak{J}$) is said to be fuzzy nearly isocompact space if every fuzzy regular closed and fuzzy countably nearly compact subspace is fuzzy nearly compact.*

References

1. Bhaumik R.N.: Fuzzy Closed - complete Spaces, Far East J. Math. Sci. **3** (4) (2001)
2. Bhaumik R.N.,Bhattacharya J.: Fuzzy isocompact spaces, submitted
3. Bhaumik R.N., Bhattacharya J.: Weakly fuzzy isocompact spaces, J. Tri. Math. Soc. **3** (2001) 29-34
4. Chang C.L.: Fuzzy topological spaces, J. Math. Anal. Appl. **24** (1968) 182-190
5. Concilio A.D., Gerla G.: Almost compactness in fuzzy topological spaces, Fuzzy Sets & Systems **13** (1984) 187-192
6. Es A.H.: Almost compactness and near compactness in fuzzy topological spaces, Fuzzy Sets & Systems **22** (1987) 289-295
7. Guojun W.: A new fuzzy compactness defined by fuzzy nets, J. Math. Anal. Appl. **94** (1983) 1-23
8. Lowen R.: Fuzzy topological space and fuzzy compactness, J. Math. Anal. Appl. **56** (1976) 621-633
9. Lowen R.: A comparison of different compactness notions in fuzzy topological spaces, J. Math. Anal. Appl. **64** (1978) 446-457
10. Mukherjee M.N., Sinha S.P.: Almost compact fuzzy sets in fuzzy topological spaces, Fuzzy Sets & Systems **38** (1990) 388-396.

Some Results on Fuzzy Commutative and Fuzzy Cyclic Subgroups

Sandeep Kumar Bhakat

Siksha-Satra, Visva-Bharati, P.O-Sriniketan-731236, Dist.-Birbhum (W.B), India

Abstract. In this paper an attempt has been made to study different quasi-fuzzy commutativity for different fuzzy subgroups. Inter-relationship and comparative studies on these structures have been made. Interesting results have been found and discussed briefly.

1 Introduction

For the concept of $(\in, \in \vee q)$-fuzzy subgroup, the notion of q-fuzzy commutativity and two types of fuzzy cyclic subgroups have been introduced by the author in [2], [3], [4]. In this paper the concept of $(\in, \in \vee q)$-fuzzy commutativity for an $(\in, \in \vee q)$-fuzzy subgroup has been introduced. Inter-relation and comparative studies with these concepts have been studied. It is found that for a particular type of Rosenfeld fuzzy subgroup, the notion of q-fuzzy commutativity coincides with the concept of fuzzy normality of Liu [6]. The concept of $(\in \vee q)$-level subset was introduced by the author in [1]. For both types of fuzzy commutative subgroups, $(\in \vee q)$-level subgroups may not be commutative. For both types of fuzzy cyclic subgroups, if all the $(\in \vee q)$-level subgroups are cyclic, then the corresponding fuzzy subgroup may not be fuzzy cyclic. However, all the $(\in \vee q)$-level subsets of a fuzzy cyclic subgroup (both type) are cyclic.

2 Preliminaries

Let G be a group with e as the identity element. I denote the closed unit interval $[0, 1]$. M stands for min.

Definition 1. *(Ming and Ming [7]) A fuzzy point x_t is said to* **belong to** *(resp. be* **quasi-coincident with***) a fuzzy subset λ, written as $x_t \in \lambda$ (resp. $x_t\ q\ \lambda$) if $\lambda(x) \geq t$ (resp. $\lambda(x) + t \geq 1$). "$x_t \in \lambda$ or $x_t\ q\ \lambda$" will be denoted by $x_t \in \vee q\ \lambda$. "$x_t \overline{\in} \lambda, x_t \overline{\in \vee q} \lambda$" will respectively mean $x_t \in \lambda$ and $x_t \in \vee q \lambda$ do not hold.*

Definition 2. *(Bhakat [1]) For any non-empty set X and $t \in (0, 1]$, the subset $\lambda_t = \{x \in X;\ \lambda(x) \geq t$ or $\lambda(x) + t > 1\} = \{x \in X;\ x_t \in \vee q \lambda\}$ is called $(\in \vee q)$-level subset of X determined by λ and t.*

Definition 3. *(Rosenfeld [9]) A fuzzy subset λ of G is said to be a fuzzy subgroup (according to Rosenfeld) if $\forall x, y \in G$ and $t, r \in (0, 1]$,*
(i) $x_t, y_r \in \lambda \Rightarrow (xy)_{M(t,r)} \in \lambda$
(ii) $x_t \in \lambda \Rightarrow x_t^{-1} \in \lambda$.

N.R. Pal and M. Sugeno (Eds.): AFSS 2002, LNAI 2275, pp. 520–526, 2002.

Definition 4. *(Ray [8]) (i) A fuzzy subgroup η of G is said to be fuzzy abelian if $\eta(x) > 0 < \eta(y) \Rightarrow xy = yx$*
(ii) Let η be a fuzzy subgroup of G and $\eta(a) > 0$. If $\delta \in I^G$ is defined by

$$\delta(x) = \begin{cases} \eta(x) & \textit{when } x \in (a) \\ 0 & \textit{when } x \in G - (a) \end{cases}$$

then the fuzzy subgroup δ of η is said to be the fuzzy cyclic subgroup of η generated by $a \in G$.

Definition 5. *(Bhakat and Das [5]) A fuzzy subset λ of G is said to be an* $(\in, \in \vee q)$**-fuzzy subgroup** *of G if for any $x, y \in G$ and $t, r \in (0, 1]$,*
(i) $x_t, y_r \in \lambda \Rightarrow (xy)_{M(t,r)} \in \vee q\ \lambda$ and
(ii) $x_t \in \lambda \Rightarrow (x^{-1})_t \in \vee q\ \lambda$.

Henceforth, unless otherwise stated, by a fuzzy subgroup, it will mean an $(\in, \in \vee q)$-fuzzy subgroup.

Definition 6. *(Bhakat and Das [5]) A fuzzy subgroup λ of G is said to be* $(\in, \in \vee q)$**-fuzzy normal** *(or simply fuzzy normal) if for any $x, y \in G$ and $t \in (0, 1]$*

$$x_t \in \lambda \Rightarrow (y^{-1}xy)_t \in \vee q\ \lambda.$$

Definition 7. *(Bhakat [2]) A fuzzy subgroup λ of G is said to be* **q-fuzzy commutative** *if $\forall x, y \in G$,*
(i) $\lambda(x) + \lambda(y) \geq 1 \Rightarrow xy = yx$
(ii) $\lambda(x) + \lambda(y) < 1 \Rightarrow \lambda(xy) = \lambda(yx)$.

Theorem 1. *Let λ be a fuzzy subgroup of G such that $\lambda(x) < 0.5\ \forall x (\neq e) \in G$. Then λ is q-fuzzy commutative if and only if λ is $(\in, \in \vee q)$-fuzzy normal.*

Proof. Let λ be q-fuzzy commutative. Now $\forall x (\neq e), y (\neq e) \in G, \lambda(x) + \lambda(y) < 1$ and thus by q-fuzzy commutativity of λ, $\lambda(xy) = \lambda(yx)$. So λ is $\in, \in \vee q)$-fuzzy normal. Conversely, let λ be an $(\in, \in \vee q)$-fuzzy normal subgroup of G. If possible, let λ be not q-fuzzy commutative. Then there exists $x (\neq e), y (\neq e) \in G$ such that $\lambda(x) + \lambda(y) < 1$ and $\lambda(xy) \neq \lambda(yx)$ which is a contradiction, since $\lambda(xy), \lambda(yx) < 0.5$ and λ is $(\in, \in \vee q)$-fuzzy normal. So λ is q-fuzzy commutative.

Corollary 1. *If λ is a fuzzy subgroup of Rosenfeld type with $\lambda_{0.5} = \{e\}$, then λ is q-fuzzy commutative if and only if λ is $(\in, \in)$-fuzzy normal.*

3 $(\in, \in \vee q)$-Fuzzy Commutative Subgroups

Theorem 2. G is a commutative group if and only if $\forall x, y \in G$, $[x, y] = xyx^{-1}y^{-1} = e$. If λ is an $(\in, \in \vee q)$-fuzzy subgroup of G, then $\lambda([x, y]) = \lambda(e) \geq 0.5$.

Definition 8. *An $(\in, \in \vee q)$-fuzzy subgroup λ of G is said to be $(\in, \in \vee q)$-fuzzy commutative (or simply, fuzzy commutative) if $\lambda([x, y]) \geq 0.5$ $\forall x, y \in G$.* Henceforth, by a fuzzy commutative subgroup, we will mean an $(\in, \in \vee q)$-fuzzy commutative subgroup.

Remark 1. If G is commutative, then any fuzzy subgroup of G is fuzzy commutative. However, the converse may not be true.

Example 1. Consider the group $G = \{e, a, b, c, d, f\}$ where the multiplication is given below.

.	e	a	b	c	d	f
e	e	a	b	c	d	f
a	a	b	e	f	c	d
b	b	e	a	d	f	c
c	c	d	f	e	a	b
d	d	f	c	b	e	a
f	f	c	d	a	b	e

Let $\lambda : G \rightarrow I$ be defined by $\lambda = (0.8, 0.6, 0.8, 0.4, 0.4, 0.4)$. Then λ is a fuzzy commutative subgroup of G since $\lambda([x, y]) = \lambda(a)$ or $\lambda(b)$ or $\lambda(e)$ all greater than 0.5. But G is not commutative as $fd \neq df$.

Theorem 3. *If λ is fuzzy commutative , then λ is $(\in, \in \vee q)$-fuzzy normal.*

Proof. The result follows from the fact that λ is $(\in, \in \vee q)$-fuzzy normal if and only if $\lambda([x, y]) \geq M(\lambda(x), 0.5)$ $\forall x, y \in G$.

Remark 2. The converse of Theorem 3 is not necessarily true.

Example 2. Consider the group G defined in Example 1. Let $\lambda : G \rightarrow I$ be defined by $\lambda = (0.7, 0.4, 0.4, 0.4, 0.4, 0.4)$. Then λ is a $(\in, \in \vee q)$-fuzzy normal subgroup of G but since $\lambda([b, c]) = \lambda(a) = 0.4 < 0.5$, λ is not fuzzy commutative. However, λ is q-fuzzy commutative.

Remark 3. (i) Example 2. shows that q-fuzzy commutativity does not imply $(\in, \in \vee q)$-fuzzy commutativity.
(ii) Example 1. shows that λ is an $(\in, \in \vee q)$-fuzzy commutative subgroup of G and not a q-fuzzy commutative as $\lambda(f) + \lambda(d) = 0.8 < 1$ but $\lambda(fd) \neq \lambda(df)$.
(iii) λ as defined either in Example 1. or Example 2. is not a fuzzy abelian subgroup according to Definition 4 (Ray [8]).

Theorem 4. *If H is a commutative subgroup of G, then χ_H is fuzzy commutative.*

Remark 4. $(\in \vee q)$-level subgroup determined by a fuzzy commutative subgroup may not be commutative as follows from Example 1. For $0 < t \leq 0.4$, $\lambda_t = G$ and G is not commutative.

Theorem 5. *Let $f : G \rightarrow H$(group) be a group homomorphism. If λ and μ are fuzzy commutative subgroups of G and H, respectively, then $f(\lambda)$ and $f^{-1}(\mu)$ are also fuzzy commutative subgroups of $f(G)$ and G, respectively.*

Proof. $f(\lambda)$ and $f^{-1}(\mu)$ are fuzzy subgroups of G and H, respectively. Let $a, b \in f(G)$, then there exist $\forall x, y \in G$ such that $f(x) = a$ and $f(y) = b$ and $[a,b] = aba^{-1}b^{-1} = f(xyx^{-1}y^{-1}) = f([x,y])$. Now $f(\lambda)([a,b]) = sup\{\lambda(t);\ t \in f^{-1}([a,b])\} \geq \lambda([x,y]) \geq 0.5$ [since λ is fuzzy commutative]. So $f(\lambda)$ is fuzzy commutative.
Next, $\forall x, y \in G$, $f^{-1}(\mu)([x,y]) = \mu(f([x,y]) = \mu([f(x), f(y)]) \geq 0.5$ [since μ is fuzzy commutative]. So $f^{-1}(\mu)$ is fuzzy commutative.

Theorem 6. *Let λ be a fuzzy commutative subgroup of G and $\mathcal{F}$, the set of all $(\in, \in \vee q)$- fuzzy cosets of G determined by λ which is a group under multiplication defined by $\lambda_x.\lambda_y = \lambda_{xy} \forall x, y \in G$ where $\lambda_x \in I^G$ is defined by $\lambda_x(g) = M(\lambda(gx^{-1}), 0.5) = M(\lambda(x^{-1}g), 0.5)$. Let $\overline{\lambda} : \mathcal{F} \to I$ be defined by $\overline{\lambda}(\lambda_x) = M(\lambda(x^{-1}), 0.5) \quad \forall x \in G$. Then $\overline{\lambda}$ is a fuzzy commutative subgroup of $\mathcal{F}$.*

Proof. It is well established in ealier papers that λ is $(\in, \in \vee q)$-fuzzy normal and $\mathcal{F}$ is a group with respect to the given multiplication. $\overline{\lambda}$ is a fuzzy subgroup of $\mathcal{F}$. To show that $\overline{\lambda}$ is fuzzy commutative , let $\lambda_x, \lambda_y \in \mathcal{F}$. Now $\overline{\lambda}([\lambda_x, \lambda_y]) = \overline{\lambda}(\lambda_{xyx^{-1}y^{-1}}) = M(\lambda(yxy^{-1}x^{-1}), 0.5) = M(\lambda([y,x]), 0.5) = 0.5$ [since λ is fuzzy commutative]. Therefore $\overline{\lambda}$ is fuzzy commutative.

Definition 9. *A fuzzy subgroup λ of G is said to be* **q-fuzzy cyclic** *generated by $a \in G$ if*

$$\lambda(x) + \lambda(y) \geq 1 \quad \forall x, y \in (a)$$

and

$$\lambda(x) = 0 \quad \forall x \in G - (a).$$

Theorem 7. *Let $a \in G$ and $\lambda : G \to [0,1]$ be defined as follows: $\lambda(x) \geq 0.5$ if $x \in (a)$ and $\lambda(x) = 0$ if $x \in G - (a)$. Then λ is a q-fuzzy cyclic subgroup of G generated by a and is denoted by $\lambda = (a)$.* By a q-fuzzy cyclic subgroup, we mean a q-fuzzy cyclic subgroup generated by some element of G.

Theorem 8. *If λ is a q-fuzzy cyclic subgroup of G, then the $(\in \vee q)$-level subset λ_t is also a cyclic subgroup of G for all $t \in (0,1]$.*

Example 3. Let $G = \{\ a,\ b,\ c\}$ be the cyclic group and $\lambda = (0.6,\ 0.4,\ 0.4)$. Then λ_t is a cyclic subgroup of G $\quad \forall t \in (0,1]$. But λ is not a q-fuzzy cyclic subgroup of G.

Theorem 9. *If λ is q-fuzzy cyclic such that $\lambda(x) < 1 \quad \forall x \in G$ and there is atmost one element $y \in G$ with $\lambda(y) = 0$, then λ is q-fuzzy commutative.*

Remark 5. (i) The condition for λ given in Theorem 9 is not a necessary condition in case of $(\in, \in \vee q)$-fuzzy commutativity.
(ii) If λ is an $(\in, \in \vee q)$-fuzzy commutative subgroup of G, then λ is not necessarily a q-fuzzy cyclic subgroup.

Example 4. Consider the group G as defined in Example 1. Here λ is an $(\in, \in \vee q)$-fuzzy commutative subgroup but λ is not a q-fuzzy cyclic subgroup.

Definition 10. *Let $\lambda : G \to I$ be a fuzzy subgroup of G. Let $a \in G$ be such that $\lambda(x) \geq 0.5$ forall $x \in (a)$ and $\mu \leq \lambda$ defined as follows.*

$$\mu(x) \geq 0.5 \text{ if } x \in (a) \text{ and } \lambda(x) \geq 0.5 \text{ and}$$

$$\mu(x) = \begin{cases} \lambda(x) & \text{if } x \in (a) \text{ and } \lambda(x) < 0.5 \\ 0 & \text{if } x \in G - (a) \end{cases} .$$

Theorem 10. *μ is a fuzzy subgroup of λ.*

Definition 11. *μ is said to be an $(\in, \in\vee q)$-**fuzzy cyclic subgroup** of λ generated by $a \in G$.*

Theorem 11. *Let λ be a fuzzy subgroup of G and $\mu \leq \lambda$. μ is $(\in, \in\vee q)$-fuzzy cyclic subgroup of λ generated by $a \in G$ if and only if $\forall t \in (0, 1]$*

$$x_t \in \lambda \Rightarrow x_t \in \vee q\ \mu \quad \forall x \in (a)$$

$$\text{and} \qquad \mu(x) = 0 \quad \forall x \in G - (a).$$

Definition 12. *Let $\mu \leq \lambda$. Then μ is said to be an $(\in, \in)$-**fuzzy cyclic subgroup** of λ generated by $a \in G$ if*

$$\mu(x) = \begin{cases} \lambda(x) & \text{if } x \in (a) \\ 0 & \text{if } x \in G - (a). \end{cases}$$

Remark 6. (i) If μ is an $(\in, \in)$-fuzzy cyclic subgroup of λ generated by a, then μ is also an $(\in, \in\vee q)$-fuzzy cyclic subgroup of λ generated by a.
(ii) It follows from the definitions that if λ is q-fuzzy cyclic, then λ is an $(\in, \in)$-fuzzy cyclic subgroup and hence an $(\in, \in\vee q)$-fuzzy cyclic subgroup of χ_G. But not conversely.
(iii) An $(\in, \in)$-fuzzy cyclic subgroup of an $(\in, \in)$-fuzzy subgroup is a fuzzy cyclic subgroup according to Definition 4 (Ray [8]). Thus we have
(iv) A fuzzy subgroup according to Definition 4 (Ray [8]) is an $(\in, \in\vee q)$-fuzzy cyclic subgroup. However, the converse may not be true.

Theorem 12. *If μ is an $(\in, \in\vee q)$-fuzzy cyclic subgroup of λ, then μ_t is a cyclic subgroup of λ_t $\forall t \in (0, 1]$, where μ_t is the $(\in\vee q)$-level subset of G determined by μ and t.*

Remark 7. If μ_t is a cyclic subgroup of λ_t $\forall t \in (0, 1]$, then μ is not necessarily an $(\in, \in\vee q)$-fuzzy cyclic subgroup of λ.

Remark 8. If μ is an $(\in, \in\vee q)$-fuzzy cyclic subgroup of λ, then G may not be cyclic as shown by next example.

Theorem 13. *Let μ be an $(\in, \in\vee q)$-fuzzy cyclic subgroup of λ generated by a, such that $\mu(x) < 1$ $\forall x \in G$ and there is atmost one element $y \in G$ such that $\mu(y) = 0$. Then μ is*
(i) q-fuzzy commutative
(ii) an $(\in, \in\vee q)$-fuzzy normal subgroup of λ.

Remark 9. (i) The condition for μ as stated in previous Theorem 13 is not a necessary condition in case of $(\in, \in \vee q)$-fuzzy commutativity.
(ii) If λ is q-fuzzy commutative, then λ is not necessarily $(\in, \in \vee q)$-fuzzy cyclic.
(iii) If λ is an $(\in, \in \vee q)$-fuzzy commutative, then λ may not be an $(\in, \in \vee q)$-fuzzy cyclic subgroup.
(iv) If μ is an $(\in, \in \vee q)$-fuzzy cyclic subgroup of λ, then μ is not necessarily an $(\in, \in \vee q)$-fuzzy commutative.

Theorem 14. *Let λ is an $(\in, \in \vee q)$-fuzzy normal subgroup of G and $\mathcal{F}_\lambda$, the group of all fuzzy cosets of G determined by λ. Let $\mu \leq \lambda$ be an $(\in, \in \vee q)$-fuzzy normal subgroup of G. Let $\mathcal{F}_\mu$ be the group of all fuzzy cosets of G determined by μ. Let $\overline{\mu} : \mathcal{F}_\mu \rightarrow [0, 1]$ be defined by $\overline{\mu}(\mu_x) = M(\mu(x^{-1}), 0.5) \quad \forall x \in G$. If μ is $(\in, \in \vee q)$-fuzzy cyclic subgroup of λ, then $\overline{\mu}$ is also a $(\in, \in \vee q)$-fuzzy cyclic subgroup of $\overline{\lambda}$.*

Proof. By a Theorem in [4], $\overline{\lambda}$ is fuzzy normal and also $\overline{\mu}$ is a fuzzy subgroup of $\mathcal{F}_\mu$. Let μ be an $(\in, \in \vee q)$-fuzzy cyclic subgroup of λ generated by $a \in G$. Then $\mu(x) \geq 0.5$ if $x \in (a)$ and $\lambda(x) \geq 0.5$ and

$$\mu(x) = \begin{cases} \lambda(x) & \text{if } x \in (a) \text{ and } \lambda(x) < 0.5 \\ 0 & \text{if } x \in G - (a) \end{cases}$$

Now $\overline{\mu}(\mu_x) = M(\mu(x^{-1}), 0.5) \geq M(\mu(x), 0.5) \geq 0.5$, if $\mu(x) \geq 0.5$, i.e, if $x \in (a)$ and $\lambda(x) \geq 0.5$, i.e, if $\mu_x \in (\mu_a)$ and $\overline{\lambda}(\lambda_x) \geq 0.5$.
Let $x \in (a)$ and $\lambda(x) < 0.5$. Then $\mu(x) = \lambda(x) < 0.5$ and thus $\overline{\lambda}(\lambda_x) < 0.5$.
So $\overline{\mu}(\mu_x) = M(\mu(x^{-1}), 0.5$ [since $\mu(x^{-1}) < o.5] = \mu(x^{-1}) = \mu(x) = \lambda(x) = M(\lambda(x), 0.5) = M(\lambda(x^{-1}), 0.5) = \overline{\lambda}(\lambda_x) = \overline{\lambda}(\mu_x)$ [since $\lambda(x) = \mu(x)$. Let $x \in G - (a)$. Then $\mu_x \in \mathcal{F}_\mu - (\mu_a)$ and $\mu(x) = 0$ and thus $M(\mu(x^{-1}), 0.5) = 0$, i.e, $\overline{\mu}(\mu_x) = 0$. So $\overline{\mu}$ is an $(\in, \in \vee q)$-fuzzy cyclic subgroup of $\overline{\lambda}$ generated by μ_a. This completes the proof.

4 Conclusion

A notable departure can be observed from the conventional role of level subset with the fuzzy substructure of an algebraic structure. In most of the cases, the level subset determined by a fuzzy substructure ensures the classical algebraic structure of that particular fuzzy substructure and therefore it is possible to define a fuzzy substructure in terms of level subset. However, for fuzzy commutative or fuzzy cyclic subgroups defined for an $(\in, \in \vee q)$-fuzzy subgroup, the equivalence by the $(\in \vee q)$-level subset can not be obtained.

References

1. S.K. Bhakat. $(\in \vee q)$-level subset.Fuzzy Sets and Systems 103 (1999) 529-533.
2. S.K.Bhakat. On fuzzy commutativity. The Journal of Fuzzy Mathematics Vol 6, no. 4 (1998) 915-921.

3. S.K. Bhakat. q-fuzzy cyclic subgroup.. The Journal of Fuzzy Mathematics vol 7, no 2 (1999) 521-529.
4. S.K. Bhakat. $(\in, \in \vee q)$-fuzzy cyclic subgroup. The Journal of Fuzzy Mathematics vol 8, no 2 (2000).
5. S.K. Bhakat and P. Das. $(\in, \in \vee q)$-fuzzy subgroup. Fuzzy Sets and Systems 80 (1996) 359-368.
6. W.J. Liu. Fuzzy invariant subgroups and fuzzy ideals. Fuzzy Sets and Systems 8 (1982) 133-139.
7. P.P. Ming and L.Y. Ming. Fuzzy topology I: Neighbourhood structure of a fuzzy point and Moore-Smith convergence. J. Math. Anal. Appl. 76 (1980) 571-599.
8. S. Ray. Generated and cyclic fuzzy subgroup. Information Sci. 69 (1993) 185-200.
9. A. Rosenfeld. Fuzzy subgroups. J. Math. Anal. Appl. 35 (1971) 512-517.
10. L. A. Zadeh. Fuzzy sets. Inform. Control 8 (1965) 338-353.

Fuzzy Hypotheses Testing with Fuzzy Data: A Bayesian Approach

Seyed Mahmoud Taheri[1] and Javad Behboodian[2]

[1] School of Mathematics, Isfahan University of Technology, Isfahan 84154, Iran
Taheri@cc.iut.ac.ir
[2] Dept. of Stat., Shiraz University, Shiraz, Iran
Behboodian@stat.susc.ac.ir

Abstract. In hypotheses testing, such as other statistical problems, we may confront with imprecise concepts. One case is a situation in which both hypotheses and observations are imprecise. In this paper, using fuzzy set theory for formulation of imprecise hypotheses and observations, we analyze mentioned problem on the basis of a Bayesian method.

1 Introduction and Preliminaries

In ordinary hypotheses testing, we come across with two limitations:
1. We have to formulate hypotheses in exact form.
2. The classical methods are suitable only when the observations are exactly precise.

These limitations, sometimes, force us to take statistical procedures in unrealistic manner. This arises because in realistic problems there are vague concepts, but the ordinary statistics uses only exact concepts. However, if we can use the vague concepts in our analysis, then there will be proximity to problems as they really are. Fuzzy set theory provides the necessary tools, (for more details see, for example, [4], in which some methods in descriptive statistics with vague data, and some aspects of statistical inference based on the imprecise data are presented).

In this article we use some simple concepts of fuzzy sets to give a Bayesian approach to fuzzy hypotheses testing with fuzzy data. It should be mentioned that Casals [3] has worked on the problem of fuzzy hypotheses testing when the available information is fuzzy, by a Bayesian approach. The approach in present article is slightly similar to Casals's one, but in a different framework and with more advantages.

First, we present two definitions from [2], but slightly different. Let $(X, \mathcal{B}_X, P_\theta)$ be a probability space.

Definition 1 A fuzzy sample space $\mathcal{X}$ is a fuzzy partition of X by means of fuzzy events, i.e. a set of fuzzy sets $\underset{\sim}{X}$ of X whose membership functions are Borel measurable and satisfy the orthogonality constraint: $\sum_{\underset{\sim}{X} \in \mathcal{X}} \mu_{\underset{\sim}{X}}(x) = 1$, for each $x \in X$.

N.R. Pal and M. Sugeno (Eds.): AFSS 2002, LNAI 2275, pp. 527–533, 2002.

Definition 2 A fuzzy random sample (of size n) is a random sample (of size n) from $\mathcal{X}$, i.e. a n-tuple $\underset{\sim}{X}^{(n)} = (\underset{\sim}{X}_1, \ldots, \underset{\sim}{X}_n)$, $\underset{\sim}{X}_i \in \mathcal{X}$, $i = 1, \ldots, n$; whose probability is given by [7],

$$P(\underset{\sim}{X}_1, \ldots, \underset{\sim}{X}_n) = \int_{X^n} \mu_{\underset{\sim}{X}_1}(x_1) \ldots \mu_{\underset{\sim}{X}_n}(x_n) dP(x_1, \ldots, x_n)$$

where $P(x_1, \ldots, x_n)$ is the probability distribution on $(X^n, \mathcal{B}_{X^n})$ determined by P on $(X, \mathcal{B}_X)$.

Definition 3 Any hypothesis of the form "$H : \theta$ is $H(\theta)$" is called to be a fuzzy hypotheses, where "$H : \theta$ is $H(\theta)$" implies that θ is in a fuzzy set of Θ with membership function $H(\theta)$ i.e. a function from Θ to $[0, 1]$.

Note that the ordinary hypothesis "$H : \theta \in \Theta_0$" is a fuzzy hypothesis with the membership function "$H(\theta) = 1$" at $\theta \in \Theta_0$ and zero elsewhere.

Example 1 Let θ be the parameter of a Binomial distribution, and

$$H(\theta) = \begin{cases} 2\theta & 0 \le \theta < \frac{1}{2} \\ 2 - 2\theta & \frac{1}{2} \le \theta < 1 \end{cases}$$

The statement "θ is $H(\theta)$" is a fuzzy hypothesis and it means that "θ is approximately $\frac{1}{2}$".

The Main Problem Suppose that $\underset{\sim}{X}^{(n)} = (\underset{\sim}{X}_1, \ldots, \underset{\sim}{X}_n)$, $\underset{\sim}{X}_i \in \mathcal{X}$, $i = 1, \ldots, n$ be a fuzzy random sample from $f(x; \theta)$, and we want to test

$$\begin{cases} H_0 : \theta \; is \; H_0(\theta) \\ H_1 : \theta \; is \; H_1(\theta) \end{cases}$$

with given two membership functions $H_0(\theta)$ and $H_1(\theta)$. We call this problem as "fuzzy hypotheses testing with fuzzy data (observations)".

In the following we study a Bayesian approach to this problem, first without a loss function and then regarding a loss function.

2 Bayes Test without a Loss Function

Definition 4 [2] Let $\underset{\sim}{X}^{(n)} = (\underset{\sim}{X}_1, \ldots, \underset{\sim}{X}_n)$, $\underset{\sim}{X}_i \in \mathcal{X}$, $i = 1, \ldots, n$ be a fuzzy random sample from $f(x; \theta)$, and $\pi(\theta)$ be the prior density of θ. The posterior density of θ given fuzzy random sample $\underset{\sim}{X}^{(n)}$ is defined by

$$\pi(\theta | \underset{\sim}{X}^{(n)}) = \pi(\theta | \underset{\sim}{X}_1, \ldots, \underset{\sim}{X}_n) = \frac{\pi(\theta) \mathcal{P}(\underset{\sim}{X}^{(n)} | \theta)}{\int_\theta \pi(\theta) \mathcal{P}(\underset{\sim}{X}^{(n)} | \theta) d\theta} \quad (1)$$

where

$$\mathcal{P}(\underset{\sim}{X}^{(n)} | \theta) = \mathcal{P}(\underset{\sim}{X}_1, \ldots, \underset{\sim}{X}_n | \theta) = \int_{X^n} \mu_{\underset{\sim}{X}_1}(x_1) \ldots \mu_{\underset{\sim}{X}_n}(x_n) dP_\theta(x_1, \ldots, x_n)$$

Definition 5 For a problem of fuzzy hypotheses testing with fuzzy data, a Bayes test rejects H_0 iff the posterior density which is weighted by $H_0(\theta)$, is less than the posterior density which is weighted by $H_1(\theta)$; i.e. iff

$$\pi_0(\theta|\underset{\sim}{X}^{(n)}) = \int_\theta \pi(\theta|\underset{\sim}{X}^{(n)})H_0(\theta)d\theta < \pi_1(\theta|\underset{\sim}{X}^{(n)}) = \int_\theta \pi(\theta|\underset{\sim}{X}^{(n)})H_1(\theta)d\theta$$

In some cases the two values of $\pi_0(\theta|\underset{\sim}{X}^{(n)})$ and $\pi_1(\theta|\underset{\sim}{X}^{(n)})$ may be close to each other. So, from a practical point of view, specially when we operate in a fuzzy environment, it is useful to introduce a criterion regarding degree of certainty about our decision. Therefore we provide the following definitions.

Definition 6 Suppose that in above definition, a Bayes test accepts H_0. Then the value

$$\frac{\pi_0(\theta|\underset{\sim}{X}^{(n)})}{\pi_0(\theta|\underset{\sim}{X}^{(n)}) + \pi_1(\theta|\underset{\sim}{X}^{(n)})}$$

is said to be the *degree of acceptance* H_0 *versus* H_1 . A similar value can be defined, when H_1 is accepted.

Definition 7 The ratio $\frac{\pi_0(\theta|\underset{\sim}{X}^{(n)})}{\pi_1(\theta|\underset{\sim}{X}^{(n)})}$ is called the *posterior odds ratio* of H_0 to H_1, and $\frac{\int \pi(\theta)H_0(\theta)d\theta}{\int \pi(\theta)H_1(\theta)d\theta}$ is called the *prior odds ratio.*

The posterior odds ratio conveys the conclusion that H_0 is $\frac{\pi_0(\theta|\underset{\sim}{X}^{(n)})}{\pi_1(\theta|\underset{\sim}{X}^{(n)})}$ times as likely to be true as H_1. In a frequency point of view, H_0 is supported $\frac{\pi_0(\theta|\underset{\sim}{X}^{(n)})}{\pi_1(\theta|\underset{\sim}{X}^{(n)})}$ times as much by the data as is H_1.

3 Bayes Test with a Loss Function

First we provide some definitions of decision theory regarding fuzzy random sample instead of an ordinary random sample. See also [2], for slightly different approach. We will limit ourselves on the hypotheses testing problem.

Let $\mathcal{A} = \{a_0, a_1\}$ be the space of actions, and $L(\theta, a) : \Theta \times \mathcal{A} \rightarrow \mathrm{R}$ be a loss function, showing the loss if we take action a, i.e. $d(\underset{\sim}{X}) = a$, when θ is the true parameter value.

Definition 8 For a hypotheses testing problem with fuzzy data, the posterior risk w.r. to the prior density $\pi(\theta)$, is defined by

$$\underset{\sim}{R}(\pi, d) = E_{\theta|\underset{\sim}{X}}L(\theta, d) = \begin{cases} \int_\theta L(\theta, a_0)\pi(\theta|\underset{\sim}{X})d\theta \; if \;\; d(\underset{\sim}{X}) = a_0 \\ \int_\theta L(\theta, a_1)\pi(\theta|\underset{\sim}{X})d\theta \; if \;\; d(\underset{\sim}{X}) = a_1 \end{cases}$$

where $\pi(\theta|\underset{\sim}{X})$ is the posterior density given by Definition 4.

Definition 9 For a hypotheses testing with fuzzy data, given the prior π, a decision a^* is said to be a Bayes rule (a Bayes test) if

$$\underset{\sim}{R}(\pi, a^*) \;=\; Min_{a\in\{a_0,a_1\}} \; \underset{\sim}{R}(\pi, a)$$

Definition 10 For testing the fuzzy hypotheses

$$\begin{cases} H_0 : \theta \text{ is } H_0(\theta) \\ H_1 : \theta \text{ is } H_1(\theta) \end{cases}$$

we define the following loss function

$$L(\theta, a_0) = a(\theta)[1 - H_0(\theta)],\ L(\theta, a_1) = b(\theta)[1 - H_1(\theta)],$$

where $a(\theta)$ and $b(\theta)$ are arbitrary nonnegative functions choosing our sensitivity on rejection.

Theorem 1 For a fuzzy hypotheses testing problem with fuzzy data, the Bayes test accepts H_0 iff
$\int_\theta a(\theta)[1 - H_0(\theta)]\pi(\theta|\underset{\sim}{X})d\theta \leq \int_\theta b(\theta)[1 - H_1(\theta)]\pi(\theta|\underset{\sim}{X})d\theta$

Corollary 1 If $a(\theta) = C_{II}$ and $b(\theta) = C_I$, then the Bayes test accepts H_0 iff

$$\frac{1 - \int_\theta H_1(\theta)\pi(\theta|\underset{\sim}{X})d\theta}{1 - \int_\theta H_0(\theta)\pi(\theta|\underset{\sim}{X})d\theta} > \frac{C_{II}}{C_I}$$

Corollary 2 If we want to use directly the prior $\pi(\theta)$, then the criterion for accepting H_0 will be

$$\int_\theta a(\theta)[1 - H_0(\theta)]\pi(\theta|\underset{\sim}{X})d\theta \leq \int_\theta b(\theta)[1 - H_1(\theta)]\pi(\theta|\underset{\sim}{X})d\theta$$

In this case if $a(\theta) = C_{II}$ and $b(\theta) = C_I$, then the criterion for accepting H_0 will be

$$\frac{\int_\theta \pi(\theta)[1 - H_1(\theta)]\pi(\theta|\underset{\sim}{X})d\theta}{\int_\theta \pi(\theta)[1 - H_0(\theta)]\pi(\theta|\underset{\sim}{X})d\theta} > \frac{C_{II}}{C_I}$$

4 Examples

In the following, we present four examples. The first one is the same as Example 8.2 of Casals et al. [2], in which the hypotheses are ordinary. But in the others, we consider fuzzy hypotheses which are more realistic.

EXAMPLE 2 Consider a large population of insects, a proportion θ of which is infected with a virus. We want to test

$$\begin{cases} H_0 : \theta = \frac{1}{3} \\ H_1 : \theta = \frac{2}{3} \end{cases}$$

We take a random sample of three insects and examine each insect. Suppose we do not have a precise mechanism for an exact discrimination between the presence and the absence of virus, but rather they can inform us whether "with much certainty the insect present infection" or else "with much certainty the insect does not present infection".

Take up $X = \{0, 1\}$ for this Bernoulli type problem, where 0 (1) is denoted for absence (present) of virus. We can identify the information "with much certainty the insect presents (does not present) infection" with a fuzzy set $\underset{\sim}{X}_1$ ($\underset{\sim}{X}_2$) on X, for instance, in the following way:

$$\mu_{\underset{\sim}{X}_1}(x) = \begin{cases} 0.1 & x = 0 \\ 0.9 & x = 1 \end{cases}, \qquad \mu_{\underset{\sim}{X}_2}(x) = \begin{cases} 0.9 & x = 0 \\ 0.1 & x = 1 \end{cases}$$

Let us assume the prior $\pi = \{\frac{1}{2}, \frac{1}{2}\}$ on Θ. Now, we can decide to accept or reject H_0 based on the observations. For example if we observe $(\underset{\sim}{X}_1, \underset{\sim}{X}_2, \underset{\sim}{X}_1)$, then we have

$$\pi(\theta = \tfrac{1}{3}|\underset{\sim}{X}_1, \underset{\sim}{X}_2, \underset{\sim}{X}_1) = \frac{\pi(\theta=\frac{1}{3})\mathcal{P}(\underset{\sim}{X}_1,\underset{\sim}{X}_2,\underset{\sim}{X}_1|\theta=\frac{1}{3})}{\int_\theta \pi(\theta)\mathcal{P}(\underset{\sim}{X}_1,\underset{\sim}{X}_2,\underset{\sim}{X}_1|\theta)d\theta} = \frac{\frac{1}{2}.0.0194}{\frac{1}{2}.0.49035+\frac{1}{2}.0.0194} = 0.28.$$

So $\pi(\theta = \frac{2}{3}|\underset{\sim}{X}_1, \underset{\sim}{X}_2, \underset{\sim}{X}_1) = 0.72$. Therefore we accept H_1 versus H_0 on the basis of Definition 5. Degree of acceptance is 0.72, and the posterior odds ratio is 0.39.

EXAMPLE 3 In Example 2, suppose we want to test

$$\begin{cases} H_0 : \theta \textit{ is small} \\ H_1 : \theta \textit{ is large} \end{cases}$$

where $H_0(\theta) = 1 - \theta$ and $H_1(\theta) = \theta$. Suppose that $\pi(\theta) = 1, 0 \leq \theta \leq 1$. If we observe $(\underset{\sim}{X}_1, \underset{\sim}{X}_2, \underset{\sim}{X}_1)$, then, based on Definition 2, we have

$$\begin{aligned} \mathcal{P}(\underset{\sim}{X}_1, \underset{\sim}{X}_2, \underset{\sim}{X}_1|\theta) &= 0.081\theta^3 + 0.099\theta^2(1-\theta) + 0.163\theta(1-\theta)^2 + 0.009(1-\theta)^3 \\ &= 0.136\theta^3 - 0.2\theta^2 + 0136\theta + 0.009 \end{aligned}$$

Thus

$$\int_\theta \pi(\theta)\mathcal{P}(\underset{\sim}{X}_1, \underset{\sim}{X}_2, \underset{\sim}{X}_1|\theta)H_0(\theta)d\theta = \int_0^1 \mathcal{P}(\underset{\sim}{X}_1, \underset{\sim}{X}_2, \underset{\sim}{X}_1|\theta)(1-\theta)d\theta = 0.017,$$

and similarly

$$\int_\theta \pi(\theta)\mathcal{P}(\underset{\sim}{X}_1, \underset{\sim}{X}_2, \underset{\sim}{X}_1|\theta)H_1(\theta)d\theta = 0.027.$$

Therefore we accept H_1 versus H_0 with degree of acceptance 0.61. The posterior odds ratio is 0.63.

EXAMPLE 4 In Example 3, suppose that θ has a Beta density of the form $b(3, 1) = 3(1-\theta)^2$, and we want to test

$$\begin{cases} H_0 : & \theta \textit{ is small} \\ H_1 : & \theta \textit{ is approximately } \frac{1}{2} \end{cases}$$

where $H_0(\theta) = 1 - \theta$, $0 \leq \theta \leq 1$, and

$$H_1(\theta) = \begin{cases} 2\theta & 0 \leq \theta < \frac{1}{2} \\ 2 - 2\theta & \frac{1}{2} \leq \theta < 1 \end{cases}$$

In this case we have

$$\int_\theta \pi(\theta)\mathcal{P}(\underset{\sim}{X}_1, \underset{\sim}{X}_2, \underset{\sim}{X}_1|\theta)H_0(\theta)d\theta = 0.0197,$$

$$\int_\theta \pi(\theta)\mathcal{P}(\underset{\sim}{X}_1, \underset{\sim}{X}_2, \underset{\sim}{X}_1|\theta)H_1(\theta)d\theta = 0.0069.$$

Therefore, we accept H_0 versus H_1 with degree of acceptance 0.74. The posterior odds ratio is 2.86.

EXAMPLE 5 In Example 3, suppose that the rejection H_0 while it is true is more expensive than falsely rejecting H_1 while it is true. So we may take C_I greater than C_{II}, for example $C_I = 2$ and $C_{II} = 1$. On the other hand
$\int_\theta \pi(\theta)[1 - H_0(\theta)]\pi(\underset{\sim}{X}|\theta)d\theta = 0.027$, $\int_\theta \pi(\theta)[1 - H_1(\theta)]\pi(\underset{\sim}{X}|\theta)d\theta = 0.017$
Since $\frac{0.017}{0.027} = 0.63 > \frac{C_{II}}{C_I} = \frac{1}{2}$, we accept H_0.

5 Concluding Remarks

As we mentioned in Section 1, Casals [3] has worked on the problem of fuzzy hypotheses testing when the available information is fuzzy. The approach in present article is slightly similar to Casals's one, but in a different framework and with more advantages. Now, we are going to illustrate some differences and advantages of our approach w.r. to Casals's one.

1. We defined the concepts of "fuzzy sample space", and "fuzzy random sample", but Casals uses the concepts of "fuzzy information system", "sample fuzzy information", ... Our approach with using the mentioned concepts is more common and more suitable in statistical literators.
2. We introduced the concept of *degree of acceptance*, by which we can compare two hypotheses more carefully. In addition we defined the *posterior odds ratio* for having a frequentist interpretation of the results.
3. We allow $H_0(\theta)$ and $H_1(\theta)$ being any two fuzzy hypotheses, but Casals emphasizes to take $H_1(\theta) = 1 - H_0(\theta)$. Suppose that, we want to generalize testing the ordinary hypotheses $H_0 : \theta = \theta_0$ versus $H_1 : \theta = \theta_1$, to the case of fuzzy hypotheses $H_0 : \theta \simeq \theta_0$ versus $H_1 : \theta \simeq \theta_1$; then $H_1(\theta)$ will not be the complement of $H_0(\theta)$.
4. We take a loss function which depends on θ, through $H_i(\theta)$'s, $a(\theta)$, and $b(\theta)$. This makes loss function more realistic. Casals considers a loss function with two constants a and b, but combines $H_i(\theta)$'s with $\pi(\theta)$ to obtain $\pi_0 = \int_\theta \pi(\theta)H_0(\theta)d\theta$ and $\pi_1 = \int_\theta \pi(\theta)H_1(\theta)d\theta$ as the new prior density for θ (note that taking $H_1(\theta) = 1 - H_0(\theta)$, we have $\pi_0 + \pi_1 = 1$). So, $H_i(\theta)$'s have been considered in both approaches but in different ways.
5. Finally, if we take $a(\theta) = a$ and $b(\theta) = b$, and also $H_1(\theta) = 1 - H_0(\theta)$; then we will get the same result. In other words, the criterion in Theorem 4.1 of [3] and Corollary 1 of this article, leads to the same test.

References

1. Berger J. O.:Statistical Decision Theory and Bayesian Analysis. Springer-Verlag (1993).
2. Casals M. R., Gil M. R. and Gil P.:On the use of Zadeh's probabilistic definition for testing statistical hypotheses from fuzzy information. Fuzzy Sets and Systems **20** (1986) 175-190.
3. Casals M. R.:Bayesian testing of fuzzy parametric hypotheses from fuzzy information. RAIRO, Operations Research **27** (1993) 189-199.
4. Kruse R. and Meyer K. D.: Statistics with Vague Data. Reidal Publ. Netherlands (1987).
5. Taheri S. M. and Behboodian J.:Neyman-Pearson Lemma for fuzzy hypotheses testing. Metrika **49** (1999) 3-17.
6. Taheri S. M. and Behboodian J.:A Bayesian approach to fuzzy hypotheses testing. Fuzzy Sets and Systems, (2001) To appeare.
7. Zadeh L. A.:Probability measures of fuzzy events. J. Math. Anal. Appl. **23** (1968) 421-427.

Author Index

Lecture Notes in Artificial Intelligence (LNAI)

Vol. 2100: R. Küsters, Non-Standard Inferences in Description Logics. X, 250 pages. 2001.

Vol. 2101: S. Quaglini, P. Barahona, S. Andreassen (Eds.), Artificial Intelligence in Medicine. Proceedings, 2001. XIV, 469 pages. 2001.

Vol. 2103: M. Hannebauer, J. Wendler, E. Pagello (Eds.), Balancing Reactivity and Social Deliberation in Multi-Agent Systems. VIII, 237 pages. 2001.

Vol. 2109: M. Bauer, P.J. Gmytrasiewicz, J. Vassileva (Eds.), User Modeling 2001. Proceedings, 2001. XIII, 318 pages. 2001.

Vol. 2111: D. Helmbold, B. Williamson (Eds.), Computational Learning Theory. Proceedings, 2001. IX, 631 pages. 2001.

Vol. 2112: R. Kowalczyk, S.W. Loke, N.E. Reed, G. Williams (Eds.), Advances in Artificial Intelligence. Proceedings, 2000. IX, 363 pages. 2001.

Vol. 2116: V. Akman, P. Bouquet, R. Thomason, R.A. Young (Eds.), Modeling and Using Context. Proceedings, 2001. XII, 472 pages. 2001.

Vol. 2117: M. Beynon, C.L. Nehaniv, K. Dautenhahn (Eds.), Cognitive Technology: Instruments of Mind. Proceedings, 2001. XV, 522 pages. 2001.

Vol. 2120: H.S. Delugach, G. Stumme (Eds.), Conceptual Structures: Broadening the Base. Proceedings, 2001. X, 377 pages. 2001.

Vol. 2123: P. Perner (Ed.), Machine Learning and Data Mining in Pattern Recognition. Proceedings, 2001. XI, 363 pages. 2001.

Vol. 2132: S.-T. Yuan, M. Yokoo (Eds.), Intelligent Agents: Specification, Modeling, and Application. Proceedings, 2001. X, 237 pages. 2001.

Vol. 2143: S. Benferhat, P. Besnard (Eds.), Symbolic and Quantitative Approaches to Reasoning with Uncertainty. Proceedings, 2001. XIV, 818 pages. 2001.

Vol. 2148: A. Nareyek (Ed.), Local Search for Planning and Scheduling. Proceedings, 2000. IX, 171 pages. 2001.

Vol. 2155: H. Bunt, R.-J. Beun (Eds.), Cooperative Multimodal Communication. Proceedings, 1998. VIII, 251 pages. 2001.

Vol. 2157: C. Rouveirol, M. Sebag (Eds.), Inductive Logic Programming. Proceedings, 2001. X, 261 pages. 2001.

Vol. 2159: J. Kelemen, P. Sosík (Eds.), Advances in Artificial Life. Proceedings, 2001. XIX, 724 pages. 2001.

Vol. 2160: R. Lu, S. Zhang, Automatic Generation of Computer Animation. XI, 380 pages. 2002.

Vol. 2166: V. Matoušek, P. Mautner, R. Mouček, K. Taušer (Eds.), Text, Speech and Dialogue. Proceedings, 2001. XIII, 542 pages. 2001.

Vol. 2167: L. De Raedt, P. Flach (Eds.), Machine Learning: ECML 2001. Proceedings, 2001. XVII, 618 pages. 2001.

Vol. 2168: L. De Raedt, A. Siebes (Eds.), Principles of Data Mining and Knowledge Discovery. Proceedings, 2001. XVII, 510 pages. 2001.

Vol. 2173: T. Eiter, W. Faber, M. Truszczynski (Eds.), Logic Programming and Nonmonotonic Reasoning. Proceedings, 2001. XI, 444 pages. 2001.

Vol. 2174: F. Baader, G. Brewka, T. Eiter (Eds.), KI 2001: Advances in Artificial Intelligence. Proceedings, 2001. XIII, 471 pages. 2001.

Vol. 2175: F. Esposito (Ed.), AI*IA 2001: Advances in Artificial Intelligence. Proceedings, 2001. XII, 396 pages. 2001.

Vol. 2182: M. Klusch, F. Zambonelli (Eds.), Cooperative Information Agents V. Proceedings, 2001. XII, 288 pages. 2001.

Vol. 2190: A. de Antonio, R. Aylett, D. Ballin (Eds.), Intelligent Virtual Agents. Proceedings, 2001. VIII, 245 pages. 2001.

Vol. 2198: N. Zhong, Y. Yao, J. Liu, S. Ohsuga (Eds.), Web Intelligence: Research and Development. Proceedings, 2001. XVI, 615 pages. 2001.

Vol. 2203: A. Omicini, P. Petta, R. Tolksdorf (Eds.), Engineering Societies in the Agents World II. Proceedings, 2001. XI, 195 pages. 2001.

Vol. 2225: N. Abe, R. Khardon, T. Zeugmann (Eds.), Algorithmic Learning Theory. Proceedings, 2001. XI, 379 pages. 2001.

Vol. 2226: K.P. Jantke, A. Shinohara (Eds.), Discovery Science. Proceedings, 2001. XII, 494 pages. 2001.

Vol. 2246: R. Falcone, M. Singh, Y.-H. Tan (Eds.), Trust in Cyber-societies. VIII, 195 pages. 2001.

Vol. 2250: R. Nieuwenhuis, A. Voronkov (Eds.), Logic for Programming, Artificial Intelligence, and Reasoning. Proceedings, 2001. XV, 738 pages. 2001.

Vol. 2253: T. Terano, T. Nishida, A. Namatame, S. Tsumoto, Y. Ohsawa, T. Washio (Eds.), New Frontiers in Artificial Intelligence. Proceedings, 2001. XXVII, 553 pages. 2001.

Vol. 2256: M. Stumptner, D. Corbett, M. Brooks (Eds.), AI 2001: Advances in Artificial Intelligence. Proceedings, 2001. XII, 666 pages. 2001.

Vol. 2258: P. Brazdil, A. Jorge (Eds.), Progress in Artificial Intelligence. Proceedings, 2001. XII, 418 pages. 2001.

Vol. 2275: N.R. Pal, M. Sugeno (Eds.), Advances in Soft Computing – AFSS 2002. Proceedings, 2002. XVI, 536 pages. 2002.

Lecture Notes in Computer Science

Vol. 2232: L. Fiege, G. Mühl, U. Wilhelm (Eds.), Electronic Commerce. Proceedings, 2001. X, 233 pages. 2001.

Vol. 2233: J. Crowcroft, M. Hofmann (Eds.), Networked Group Communication. Proceedings, 2001. X, 205 pages. 2001.

Vol. 2234: L. Pacholski, P. Ružička (Eds.), SOFSEM 2001: Theory and Practice of Informatics. Proceedings, 2001. XI, 347 pages. 2001.

Vol. 2235: C.S. Calude, G. Păun, G. Rozenberg, A. Salomaa (Eds.), Multiset Processing. VIII, 359 pages. 2001.

Vol. 2236: K. Drira, A. Martelli, T. Villemur (Eds.), Cooperative Environments for Distributed Systems Engineering. IX, 281 pages. 2001.

Vol. 2237: P. Codognet (Ed.), Logic Programming. Proceedings, 2001. XI, 365 pages. 2001.

Vol. 2239: T. Walsh (Ed.), Principles and Practice of Constraint Programming – CP 2001. Proceedings, 2001. XIV, 788 pages. 2001.

Vol. 2240: G.P. Picco (Ed.), Mobile Agents. Proceedings, 2001. XIII, 277 pages. 2001.

Vol. 2241: M. Jünger, D. Naddef (Eds.), Computational Combinatorial Optimization. IX, 305 pages. 2001.

Vol. 2242: C.A. Lee (Ed.), Grid Computing – GRID 2001. Proceedings, 2001. XII, 185 pages. 2001.

Vol. 2243: G. Bertrand, A. Imiya, R. Klette (Eds.), Digital and Image Geometry. VII, 455 pages. 2001.

Vol. 2244: D. Bjørner, M. Broy, A.V. Zamulin (Eds.), Perspectives of System Informatics. Proceedings, 2001. XIII, 548 pages. 2001.

Vol. 2245: R. Hariharan, M. Mukund, V. Vinay (Eds.), FST TCS 2001: Foundations of Software Technology and Theoretical Computer Science. Proceedings, 2001. XI, 347 pages. 2001.

Vol. 2246: R. Falcone, M. Singh, Y.-H. Tan (Eds.), Trust in Cyber-societies. VIII, 195 pages. 2001. (Subseries LNAI).

Vol. 2247: C. P. Rangan, C. Ding (Eds.), Progress in Cryptology – INDOCRYPT 2001. Proceedings, 2001. XIII, 351 pages. 2001.

Vol. 2248: C. Boyd (Ed.), Advances in Cryptology – ASIACRYPT 2001. Proceedings, 2001. XI, 603 pages. 2001.

Vol. 2249: K. Nagi, Transactional Agents. XVI, 205 pages. 2001.

Vol. 2250: R. Nieuwenhuis, A. Voronkov (Eds.), Logic for Programming, Artificial Intelligence, and Reasoning. Proceedings, 2001. XV, 738 pages. 2001. (Subseries LNAI).

Vol. 2251: Y.Y. Tang, V. Wickerhauser, P.C. Yuen, C.Li (Eds.), Wavelet Analysis and Its Applications. Proceedings, 2001. XIII, 450 pages. 2001.

Vol. 2252: J. Liu, P.C. Yuen, C. Li, J. Ng, T. Ishida (Eds.), Active Media Technology. Proceedings, 2001. XII, 402 pages. 2001.

Vol. 2253: T. Terano, T. Nishida, A. Namatame, S. Tsumoto, Y. Ohsawa, T. Washio (Eds.), New Frontiers in Artificial Intelligence. Proceedings, 2001. XXVII, 553 pages. 2001. (Subseries LNAI).

Vol. 2254: M.R. Little, L. Nigay (Eds.), Engineering for Human-Computer Interaction. Proceedings, 2001. XI, 359 pages. 2001.

Vol. 2255: J. Dean, A. Gravel (Eds.), COTS-Based Software Systems. Proceedings, 2002. XIV, 257 pages. 2002.

Vol. 2256: M. Stumptner, D. Corbett, M. Brooks (Eds.), AI 2001: Advances in Artificial Intelligence. Proceedings, 2001. XII, 666 pages. 2001. (Subseries LNAI).

Vol. 2257: S. Krishnamurthi, C.R. Ramakrishnan (Eds.), Practical Aspects of Declarative Languages. Proceedings, 2002. VIII, 351 pages. 2002.

Vol. 2258: P. Brazdil, A. Jorge (Eds.), Progress in Artificial Intelligence. Proceedings, 2001. XII, 418 pages. 2001. (Subseries LNAI).

Vol. 2259: S. Vaudenay, A.M. Youssef (Eds.), Selected Areas in Cryptography. Proceedings, 2001. XI, 359 pages. 2001.

Vol. 2260: B. Honary (Ed.), Cryptography and Coding. Proceedings, 2001. IX, 416 pages. 2001.

Vol. 2262: P. Müller, Modular Specification and Verification of Object-Oriented Programs. XIV, 292 pages. 2002.

Vol. 2264: K. Steinhöfel (Ed.), Stochastic Algorithms: Foundations and Applications. Proceedings, 2001. VIII, 203 pages. 2001.

Vol. 2267: M. Cerioli, G. Reggio (Eds.), Recent Trends in Algebraic Development Techniques. Proceedings, 2001. X, 345 pages. 2002.

Vol. 2272: D. Bert, J.P. Bowen, M.C. Henson, K. Robinson (Eds.), ZB 2002: Formal Specification and Development in Z and B. Proceedings, 2002. XII, 535 pages. 2002.

Vol. 2273: A.R. Coden, E.W. Brown, S. Srinivasan (Eds.), Information Retrieval Techniques for Speech Applications. XI, 109 pages. 2002.

Vol. 2275: N.R. Pal, M. Sugeno (Eds.), Advances in Soft Computing – AFSS 2002. Proceedings, 2002. XVI, 536 pages. 2002. (Subseries LNAI).